Flow-Induced Vibrations

Classifications and Lessons from Practical Experiences

Flow-Induced Vibrations

Classifications and Lessons from Practical Experiences

Second Edition

Editors

Shigehiko Kaneko

Tomomichi Nakamura

Fumio Inada

Minoru Kato

Kunihiko Ishihara

Takashi Nishihara

Njuki W. Mureithi

Translating editor

Mikael A. Langthjem

ELSEVIER

AMSTERDAM • BOSTON • HEIDELBERG • LONDON
NEW YORK • OXFORD • PARIS • SAN DIEGO
SAN FRANCISCO • SINGAPORE • SYDNEY • TOKYO
Academic Press is an imprint of Elsevier

Academic Press is an imprint of Elsevier
32 Jamestown Road, London NW1 7BY, UK
225 Wyman Street, Waltham, MA 02451, USA
525 B Street, Suite 1800, San Diego, CA 92101-4495, USA

Second Edition 2014

British Library Cataloguing-in-Publication Data
A catalogue record for this book is available from the British Library

Library of Congress Cataloging-in-Publication Data
A catalog record for this book is available from the Library of Congress

ISBN: 978-0-08-101318-2

For information on all Academic Press publications
visit our website at elsevierdirect.com

Typeset by MPS Limited, Chennai, India
www.adi-mps.com

14 15 16 17 18   10 9 8 7 6 5 4 3 2 1

Contents

Preface

The soundness of the energy plant system and its peripheral equipment is attracting strong interest not only from people in industry engaged in nuclear power generation, thermal power generation, hydroelectric power generation and chemical plant operation etc., but also from ordinary citizens. In addition, abnormal vibrations and noise that arise in aircraft and automobiles, for example, are also important problems related to reliability.

One of the key factors inhibiting the soundness of the energy plant system and reliability of equipment is flow-induced vibration (FIV), which arises via flow and structural system coupling or flow acoustics and structural system coupling. Research on FIV, ranging from basic research in laboratories to practical research at the prototypical industrial scale in various research institutes around the world, has been widely carried out.

However, the number of incidents due to flow-induced vibration or noise generation leading to structural failure does not show any signs of significant decline. Indeed, there is an increase in the number of incidents, which might lead to loss of confidence from society as well as economic loss. Problems in nuclear power stations are a case in point.

The reason why such phenomena arise is because a network for the purpose of transmission of knowledge and information on flow-induced vibration between researchers, designers, builders and operation managers has not been constituted.

Once a vibration problem arises, it must be solved in a short time. To achieve this, investigation of similar past cases is energetically carried out. However, the number of well-documented previous cases with useful information is very limited due to the veil of business secrecy.

It is against this background that the researchers who gathered at the Japan Society of Mechanical Engineers FIV workshops undertook the task of extracting the most useful information from the literature reviewed at the series of workshops over the past 27 years.

The information has been put together in the form of a database in which fundamental knowledge on flow-induced vibration is explained in detail and the information vital for the designer is conveniently compiled and consolidated.

The first edition issued in 2003, which comprises the core part of the database, was composed of six chapters, which are:

1. Introduction, 2. Vibration induced by cross-flow, 3. Vibration induced by external axial flow, 4. Vibrations induced by internal fluid flow, 5.Vibration induced by pressure waves in piping, 6. Acoustic vibration and noise caused by heat (now titled: Heating-related oscillations and noise). In the second edition, two more chapters, i.e., 7. Vibrations in rotary machines and, 8. Vibrations in

fluid–structure interaction systems, are added to complete the systematization of flow-induced vibration analysis, evaluation and research.

We hope this volume will be useful to engineers and students both for the systematic evaluation and understanding of flow-induced vibration and in contributing to the creation of new research topics based on knowledge acquired in the field.

Representatives of the technical section on FIV,
Japan Society of Mechanical Engineers

Shigehiko Kaneko
(The University of Tokyo)

Tomomichi Nakamura
(Osaka Sangyo University)

Fumio Inada
(Central Research Institute of Electric Power Industry)

Minoru Kato
(Kobelco Research Institute, Inc.)

Kunihiko Ishihara
(Tokushima Bunri University)

Takashi Nishihara
(Central Research Institute of Electric Power Industry)

Njuki W. Murethi
(Polytechnique Montreal)

Mikael A. Langthjem
(Yamagata University)

List of Contributors

Takeshi Fujikawa
Ashiya University

Tsuyoshi Hagiwara
Toshiba Corporation

Itsuro Hayashi
Chiyoda Advanced Solutions Corporation

Kazuo Hirota
Mitsubishi Heavy Industries, Ltd.

Tohru Iijima
Muroran Technology Institute

Fumio Inada
Central Research Institute of Electric Power Industry

Kunihiko Ishihara
Tokushima Bunri University

Shigehiko Kaneko
University of Tokyo

Minoru Kato
Kobelco Research Institute, Inc.

Tatsuhiko Kiuchi
Toyo Engineering Corporation

Mikael A. Langthjem
Yamagata University

Hiroyuki Matsuda
Chiyoda Advanced Solutions Corporation

Shigeki Morii
MHI Soltech Corporation

Ryo Morita
Central Research Institute of Electric Power Industry

Hisayuki Motoi
IHI Corporation

Njuki W. Mureithi
Ecole Polytechnique Montreal

Hiroshi Nagakura
Mitsubishi Heavy Industries, Ltd.

Tomomichi Nakamura
Osaka Sangyo University

Akira Nemoto
Toshiba Corporation

Eiichi Nishida
Syonan Technical Institute

Takashi Nishihara
Central Research Institute of Electric Power Industry

Toru Okada
Kobe Stell, Ltd.

Noboru Saito
Toshiba Corporation

Masashi Sano
Shizuoka Institute of Science and Technology

Yoshihiko Urata
Shizuoka Institute of Science and Technology

Masahiro Watanabe
Aoyama Gakuin University

Kazuaki Yabe
Toyo Engineering Corporation

Kazuyuki Yamaguchi
Hitachi Corporation

Akira Yasuo
Central Research Institute of Electric Power Industry

Kimitoshi Yoneda
Central Research Institute of Electric Power Industry

Koichi Yonezawa
Osaka University

CHAPTER

Introduction

1

1.1 General overview

Vibration and noise problems due to fluid flow occur in many industrial plants. This obstructs smooth plant operation and in serious cases can lead to significant maintenance and repair costs and costly losses in productivity. These flow-related vibration phenomena are generally known as 'flow-induced vibrations' (FIV). The term 'flow-induced vibration and noise' (FIVN) is used when flow-induced noise is present.

It is fairly evident that the fluid force acting on an obstacle in flow will vary due to the flow unsteadiness and that the varying force, in turn, may cause vibration of the obstacle. In the case of piping connected to reciprocating fluid machines, for example, it is well-known that the oscillating (fluctuating) flow in the piping generates excitation forces causing piping vibration.

However, even for steady flow conditions, vibration problems may be caused by vortex shedding behind obstacles or by other phenomena. The drag direction vibration of the thermocouple well in the fast-breeder reactor at Monju in Japan is an example of a vibration problem caused by the symmetric vortex shedding behind the well. This kind of self-excited FIV occurs even in steady flow, making it much more difficult to determine the underlying mechanism and thus, one of the most difficult problems to deal with at the design stage or during troubleshooting. Flow-induced acoustic noise is also an important problem in industry.

1.1.1 History of FIV research

Two conferences (Naudascher [1]; Naudascher and Rockwell [2]) on fluid-related vibration were held in 1972 and 1979, on the initiative of Prof. Naudascher of the University of Karlsruhe, Germany. At the 1979 conference, many practical problems related to flow-induced vibration and noise in a wide variety of industrial fields, such as mechanical systems, civil engineering, aircrafts, ships, and nuclear power plants, were presented. The conference included many interesting results that are still pertinent today.

In 1977, Dr. Robert Blevins wrote the first book [3] in the field. The term flow-induced vibration became popular after he used it for the book's title. In the book and probably for the first time, FIV phenomena were classified based on the

Flow-Induced Vibrations.

two basic flow types: steady flow induced and unsteady flow related. Blevins went on to publish a handbook [4] focused on the frequencies and the eigenmodes of structural systems and fluid systems related to FIV. A second handbook (Blevins [5]), aimed at designers, gave systematic information for pipe flows, open water channels, separated flow, flow resistance, shear flow, etc. These handbooks contain valuable information on FIV-related problems and the underlying phenomena. The second edition of the book *Flow-Induced Vibration* (Blevins [6]) was published in 1990.

On the other side of the Atlantic, Prof. Naudascher in Germany wrote a book [7] on the hydraulic forces acting on dams and gates mainly from the civil engineering and hydraulics point of view. He and Prof. Rockwell co-authored a textbook [8] for designers where the vibration classifications are based on excitation mechanisms. The text is useful for the design of mechanical systems.

Several books focusing on specific phenomena have been written. These include a book [9] on cylindrical structures by Dr. S.S. Chen; a book [10] on the fluid–structure interaction (FSI) of pressure vessels by Dr. Morand and Dr. Ohayon; and two volumes [11] by Prof. Païdoussis dealing extensively with the subject of axial FIV. Païdoussis and Li [12] have written an important paper discussing pipes conveying fluids and the fundamental mechanisms behind the coupled fluid–structure oscillations. In the paper, the pipe conveying fluid problem was introduced as a model problem or paradigm useful for the understanding of the excitation mechanisms underlying FIV. Most recently, Païdoussis, Price and De Langre have published a book [13] dedicated to cross-flow-induced instabilities.

In Japan, a chapter [14] containing a detailed introduction to FIV was published in 1976. However, this formed only a small part of a large handbook on the vibration of fluid machines (including topics such as piping vibration and surging in reciprocating compressors). In 1980, the Japan Society of Mechanical Engineers (JSME) committee on FIV, under the leadership of Prof. Tajima, reviewed the state of the art in FIV, producing a report [15] which included many examples of experiences with flow-induced phenomena in Japanese industry. In 1989, another report [16] focusing on piping system pressure pulsations caused by reciprocating compressors was published under the leadership of Prof. Hayama. Other important papers have also been published on the excitation mechanisms underlying FIV (Iwatsubo [17]) or introducing examples from industrial experience (Fujita [18]).

On the specific topic of FIV in nuclear power plants, the Yayoi Research Seminar has been held since 1990 at the nuclear test facility of the University of Tokyo to introduce research studies and disseminate new information (Madarame [19]). The leader, Prof. Madarame, indicated at the first meeting that a FIV design support system such as depicted in Fig. 1.1 was required. The system incorporates analytical tools and a database on FIV. Such systems are already used in other fields. His idea later appeared in the form of published guidelines of the JSME. Examples include the guidelines [20] motivated by the steam generator tube failure in the Mihama nuclear power plant of the Kansai Electric Power Corporation

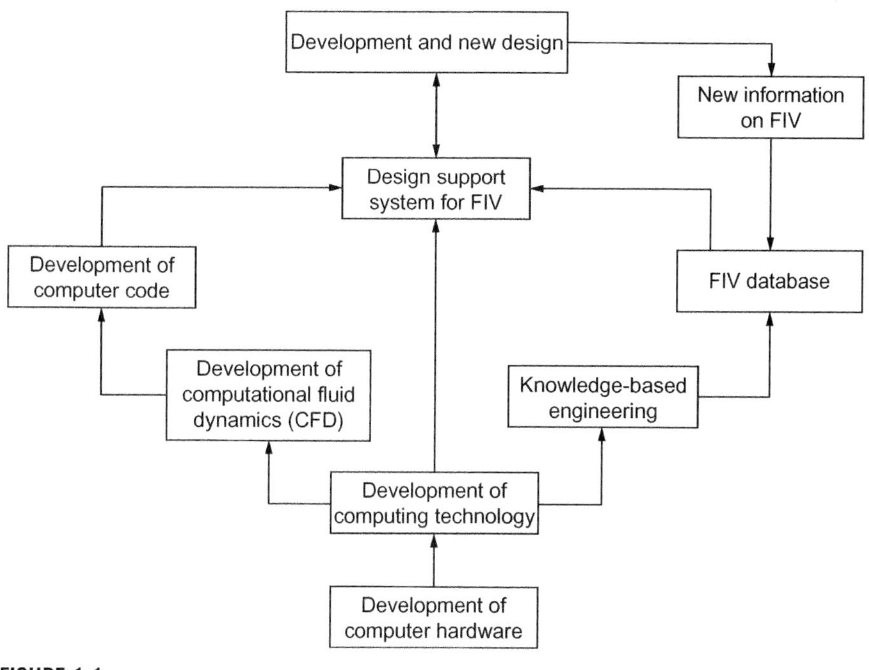

FIGURE 1.1

Design support system for FIV.

or the guidelines [21] developed following the thermocouple well failure in the Monju fast-breeder reactor.

With regard to flow-induced noise, an international conference was held in 1979 (Müller [22]), and a book on the subject was written by Blake [23] in 1986. In Japan, a book on *Examples of Noise & Vibration Resistance Systems* was published in 1990 by the Japanese Noise Control Academy [24]. The book includes many examples of flow-induced noise, making it very useful for designers. A paper (Maruta [25]) introducing the subject of FIV and including a review of recent research papers on noise was also published.

A number of important FIV meetings are held around the world. Within the American Society of Mechanical Engineers (ASME), the International Symposium on Fluid–Structure Interactions, Flow-Sound Interaction and Flow-Induced Vibration & Noise (FSI^2 & FIV+N) is held every 4 years. A FIV symposium is also held every second year at the ASME Pressure Vessel and Piping Conference. The latter is mainly geared toward the investigation of FIV problems in industrial pressure vessels & piping. In Europe, the International Conference on Flow-Induced Vibrations is also held every 4 years. This conference is related to the 1973 Keswick conference in England. The conference has since expanded to many fields. In Japan, a FIV session is held yearly during every JSME conference.

1.1.2 Origin of this book

The FIV seminar, which is held by the JSME, began in 1984 under the leadership of Profs. Hara (Science University of Tokyo) and Iwatsubo[1] (Kobe University at the time). The aim of the seminar is to collect and analyze worldwide information on FIV in the field of mechanical engineering, and to communicate this information to participants and Japanese researchers. More than 100 young researchers have attended this seminar. Its first-phase activities ended in February 1999. From April 1999, Prof. Kaneko took up the leadership of the second phase, to study and come up with the best method to use the information accumulated over more than 10 years of activities. The most important objective was to develop collaborations for future research and development using the technical information.

The main activity of the first phase was to gather information from around the world and present it to Japanese engineers and researchers. Reports on FIV problems and research papers from outside Japan were organized and translated into Japanese language summaries, which currently number more than 400. These organized files are considered to be of high quality because almost all Japanese specialists and researchers in FIV were involved. However, the information has, to date, not been compiled into a comprehensive document. This book is therefore intended to be an adequate reference documenting the present worldwide activities on FIV, but useful not only as an information source, but also as a resource for the creation of new knowledge.

FIV can be roughly divided into five fields. Similarly, the FIV seminar consists of five working groups who contributed to this book:

1. Vibration induced by cross-flow (Leader: Tomomichi Nakamura).
2. Vibration induced by parallel flow (Leader: Fumio Inada).
3. Vibration of piping conveying fluid, pressure fluctuation, and thermal excitation (Leader: Minoru Kato).
4. Vibrations in rotating structures in flow (Leader: Kunihiko Ishihara).
5. Vibrations in fluid–structure interaction systems (Leader: Takashi Nishihara).

The first edition of this book presented, as the first step, the results of work on the first three fields (above). In this second edition, material on these fields has been updated or revised and two new chapters have been added covering the final two subjects. Throughout the book, an effort has been made to include examples from actual practical experience to the extent possible. The description of a practical problem is followed by presentation of the evaluation method, the vibration mechanisms, and finally, practical hints for vibration prevention.

As outlined above, depending on the FIV problem, different evaluation methods are required. The different types of FIV problems are presented and explained in detail beginning in Chapter 2. It is recommended, however, to start by reading

[1]Currently at Nagahama corporation.

Sections 1.2 and 1.3 later in this chapter, where information on modeling and general basic mathematical methods that can be used, are first presented.

1.2 Modeling approaches

1.2.1 The importance of modeling

Figure 1.2 shows a semi circular pillar. The pillar is only lightly supported in the vertical and horizontal directions. How will the pillar respond when subjected to (wind) flow from left to right, as depicted in the figure?

A designer or technician with sufficient engineering knowledge would likely answer that the pillar will vibrate. We guess that the majority of engineers would answer that the pillar would vibrate in the direction transverse to the wind direction. Most, when asked to explain the cause of the vibration, are likely to suggest that vortex shedding behind the pillar is the source of the additional energy responsible for the vibrations.

For vortex-induced vibration (VIV), the typical analysis and resolution procedure is as depicted on the left in Fig. 1.3. This means proceeding as follows:

1. Determine the frequency of the vortex shedding responsible for the vibration excitation.
2. Measure the resonant frequency of the pillar.
3. Separate the structural frequency from the vortex shedding frequency as a countermeasure.

Changing the structural resonance frequency by changing the support conditions is more easily achievable than changing the vortex shedding frequency.

For the present problem though, the vortex excitation countermeasure presented in Fig. 1.3 would not work at all. It turns out that the pillar vibrates at a frequency far above the vortex shedding frequency. Thus, countermeasures based

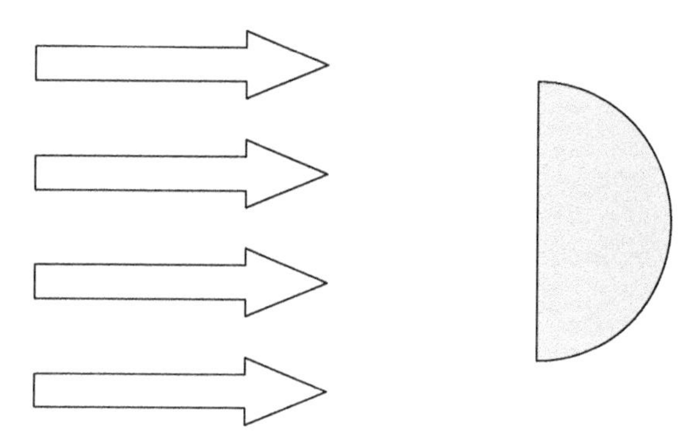

FIGURE 1.2

How will this half cylinder respond to wind flow?

on the premise of VIV would be inappropriate. This is an example of a diagnosis and solution failure. The failure is caused by mistakes in the assumed vibration excitation mechanism or in the assumed flow-structure interaction model.

Figure 1.4 depicts schematically the mechanism underlying the vibrations. When the pillar moves upward (Fig. 1.4(a)), the relative flow is oriented diagonally downward. The lift force therefore amplifies the motion. Since the pillar is elastically supported, the direction of motion reverses in the next half-cycle (Fig. 1.4(b)). The relative flow direction is now diagonally upward. Thus, a downward lift force acts on the semi-circular pillar. As a result, the pillar vibrates transverse to the upstream flow. The vibration itself is amplified and the vibration amplitude grows in time. This type of FIV is known as galloping. Vortex formation in the semicircular cylinder wake is irrelevant to the vibration phenomenon. It is clear then that any measurement on vortex shedding would be of little use.

Thus, other than analysis or experiments, modeling should also involve ascertaining the underlying mechanism in order to correctly determine the physical quantities involved.

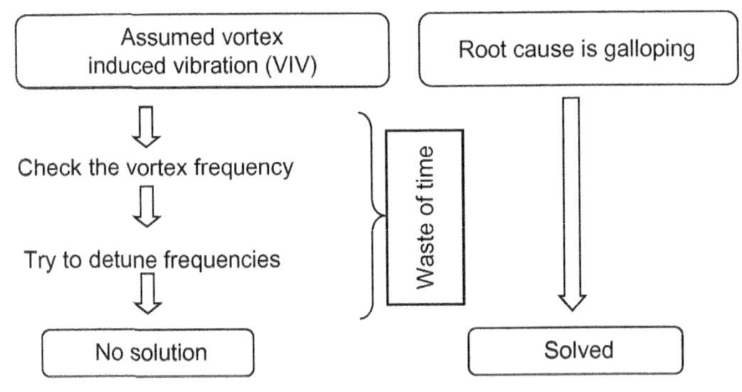

FIGURE 1.3

Route to solution.

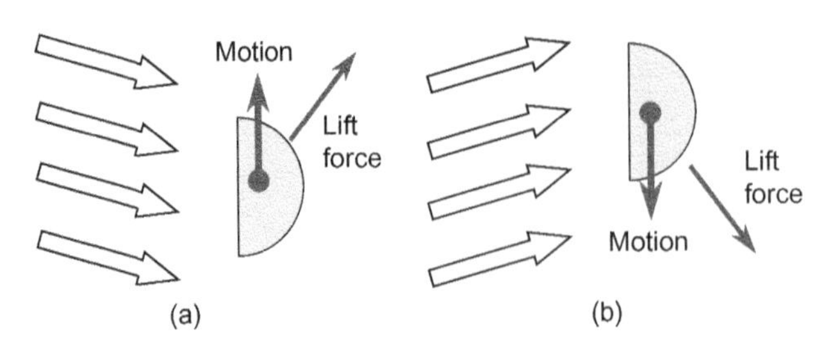

FIGURE 1.4

Mechanism of oscillation.

As will be demonstrated in the next section, many phenomena responsible for FIV have already been identified and classified. Many of the associated mechanisms are also well understood. For this reason, many problems usually encountered in FIV fall within one or more of the existing classifications. When difficulty is encountered solving a problem, it is likely that an incorrect modeling approach has been taken.

Modeling is the most important tool to investigate the causes of observed FIV phenomena, solve problems, confirm designs, and point to the shortest route to a solution.

1.2.2 Classification of FIV and modeling

Figure 1.5 classifies FIV according to the type of flow involved. Here, the types of FIV mechanisms that have already been confirmed are classified based on the

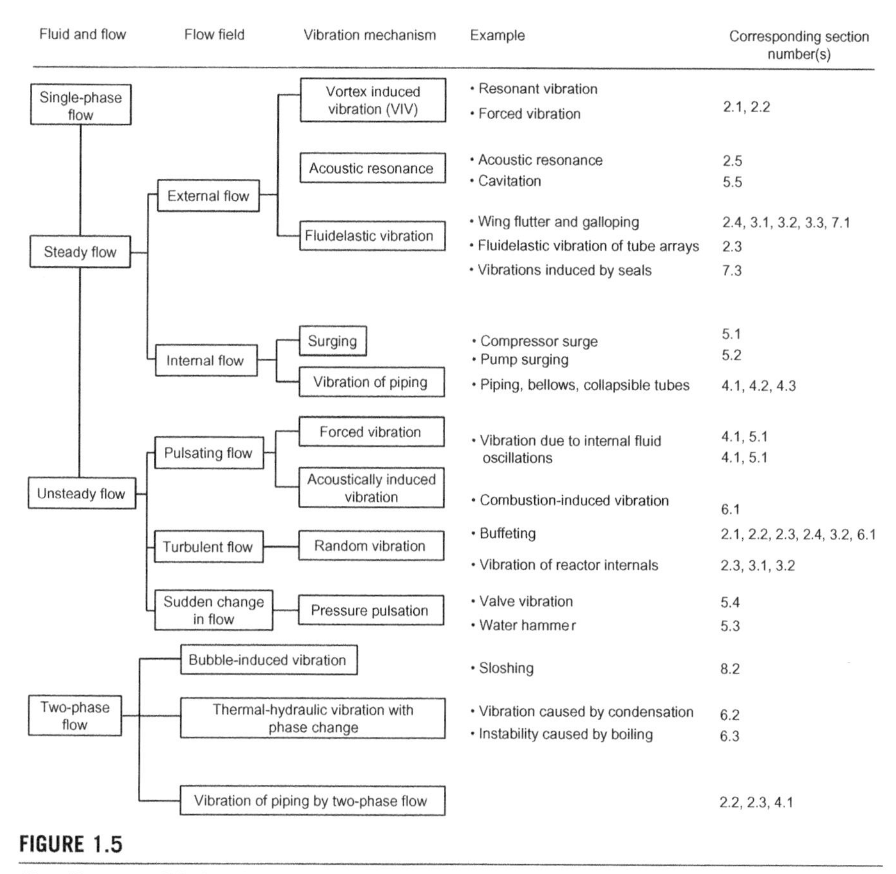

FIGURE 1.5

Classification of FIV and the corresponding sections.

following flow types: steady flow, unsteady flow, and two-phase flow. For vibration in steady flow, the mutual interaction between fluid and structure leading to increasingly large vibration amplitudes is the most commonly observed scenario. In unsteady flow, turbulence forces are the dominant source of structural vibration excitation. Since two-phase flow usually consists of a mixture of two fluids of different densities flowing together, the time variation of the flow momentum acts as a source of excitation for the structure. Furthermore, the variation of the momentum and pressure also contributes to the structural vibration.

Consider the example classified by: steady flow → external flow → VIV caused by steady flow. As we saw earlier (Fig. 1.4), there is often a tendency to associate flow-induced excitation with vortex shedding.[2] One must be careful to keep in mind that vortex shedding is but one of several mechanisms that may be responsible for the observed vibration phenomenon.

Figure 1.6 shows examples of FIV mechanisms as well as the corresponding models. The vibration mechanism depicted in Fig. 1.6(a), is governed by turbulent random pressure fluctuations. Flow instability underlies the (VIV) mechanism of Fig. 1.6(b). The FIV mechanism associated with the structure displacement

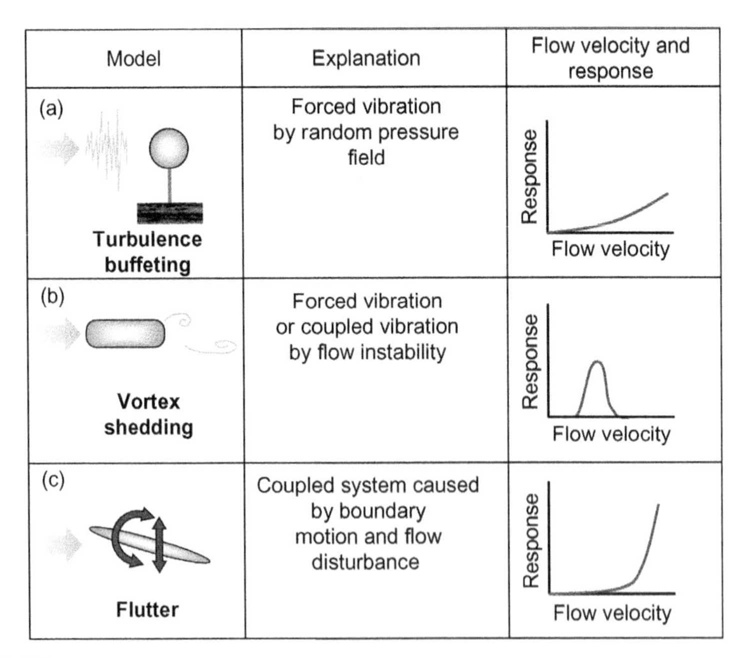

FIGURE 1.6

Examples of models and mechanisms.

[2]The classic example is the Tacoma-Narrows bridge failure in Washington State, USA.

induced boundary condition and temporal flow variations is shown in Fig. 1.6(c). In reality, these mechanisms will often interact resulting in complex flow and vibration phenomena [26].

For example, Fig. 1.7 shows a feed-water heater. The fluid components in the heater are steam (the heat source), heated water, and two-phase, steam-water flow mixture traversing the U-bend tube bank. Several factors must therefore be considered even for such a single piece of equipment.

Figure 1.8 shows the archetypical response curve for the cross-flow-induced vibration of a tube bundle. The vibration response is the sum of (i) turbulence excitation proportional to flow velocity (Fig. 1.8(b)), (ii) VIV due to resonance between the vortex shedding and structural frequencies (Fig. 1.8(c)), and (iii) fluidelastic instability beyond a critical flow velocity (Fig. 1.8(d)) which leads to an abrupt and large increase in the vibration amplitude. Thus, there is a possibility that several excitation mechanisms contribute to excitation of the same structure. In this case, it may not be easy to specify the root cause of the vibrations.

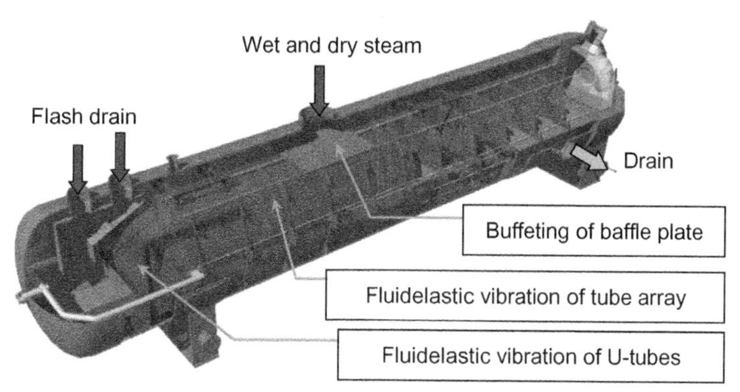

FIGURE 1.7

Vibration problems in design of feed water heater.

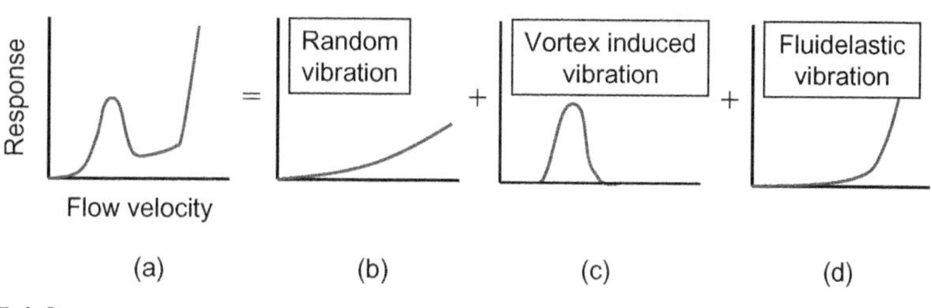

FIGURE 1.8

Vibration of tube array caused by cross-flow.

1.2.3 Modeling procedure

Engineering may be defined as the application of currently available scientific knowledge to approximately analyze complex physical phenomena, and using the results to develop useful machines, structures etc. Mathematical modeling know-how of physical phenomena is therefore at the core of engineering.

It is widely believed that modeling techniques and ability are acquired similarly to how experience is gained in a traditional (tradesman) apprenticeship. In reality though, due to the hierarchical nature of the analysis of physical phenomena, there is a danger of arriving at incorrect conclusions if mistakes are made at the initial modeling stage. Consultation with experienced engineers is therefore recommended as the appropriate approach.

The modeling procedure when dealing with a concrete physical FIV problem consists of the following steps:

1. Identification of the underlying physical phenomenon or mechanism.
2. Choice of evaluation approach.

There are two different approaches with regard to identification of the physical mechanism. The approaches are based on two opposing philosophies.

In the first, the problem at hand is simplified as much as possible and conclusions are drawn based on a basic analysis of the simplified problem. We shall refer to this approach as the 'simplified treatment.' In the second approach, the exact opposite is done. The problem is modeled as accurately as possible. The resulting model will normally be much more complex. We shall refer to this procedure, and the associated interpretation of results, as the 'detailed treatment.'

Which approach one should take depends on how important or critical the problem is. Thus, before modeling, there is the need to determine the importance of the problem. The decision regarding the choice of approach may be made following a flowchart such as Fig. 1.9. In what follows, assuming this decision has been made, we introduce a problem analysis approach depending on which of these two different positions is valid.

1.2.3.1 Simplified treatment

In order to simplify the problem, the essence of the associated physical phenomenon must be understood to the extent possible. Lack of understanding can lead to incorrect results, possibly contrary to the facts. Clearly this could lead to eventual problems in new designs while problems in existing designs would remain unresolved.

Simplifying the problem means, in general, extracting only the essential complexity of the fluid and the structure. Note, however, that without a good understanding of the physical phenomenon, any mathematical modeling is bound to lead to incorrect results and, hence, is doomed to fail.

In order to correctly perform a simplified assessment, it is vitally important to base the analysis on existing knowledge of the phenomenon under investigation.

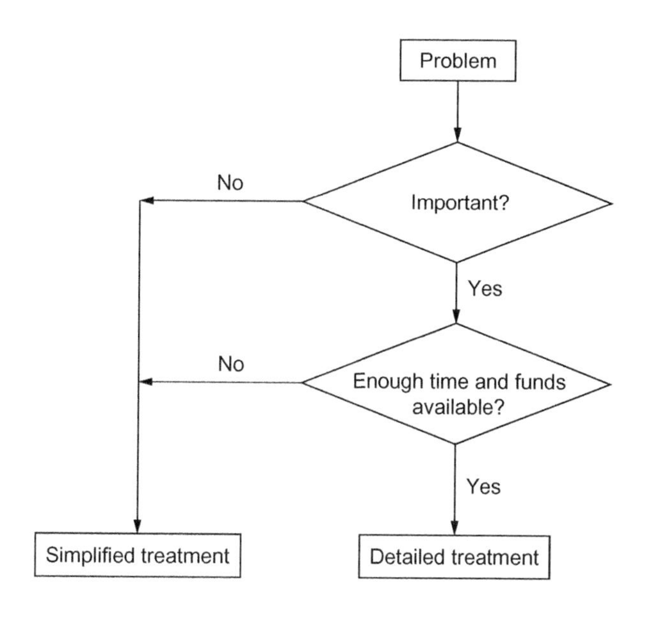

FIGURE 1.9

Decision based on importance.

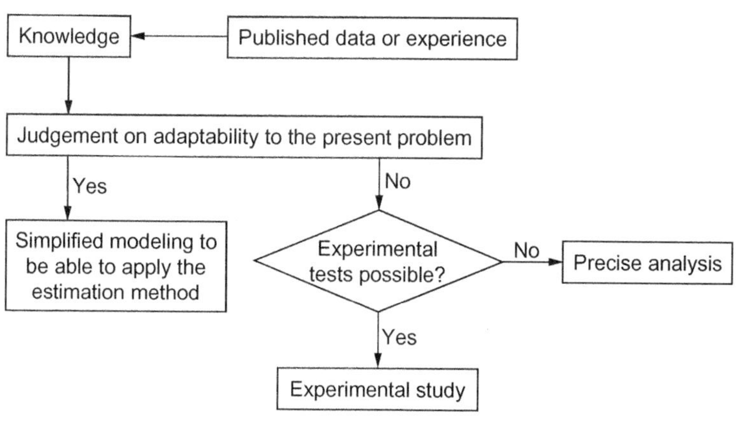

FIGURE 1.10

Flowchart of simplified treatment.

Furthermore, it is important to distinguish those aspects of the problem that can be analyzed theoretically or numerically from those that cannot.

In concrete terms, one must decide whether the decision procedure shown in Fig. 1.10 is applicable. If the answer is negative, then either a more detailed analysis or experimental investigation is called for.

1.2.3.2 Detailed treatment

This procedure is required either when the problem is very important or when it cannot be evaluated from previous knowledge. In both cases, this detailed treatment can be conducted by two methods: analytical approach or experimental approach. The former approach is adopted in general but the latter should be considered when the analytical method is unreliable.

1.2.4 Analytical approach

Two approaches to FIV and fluid-structure interaction (FSI) problems exist. As detailed in Section 1.3, theoretically, the problem can be modeled via a set of differential equations. Analysis taking into consideration the appropriate boundary conditions could then be carried out. The second approach is that shown in Table 1.1, where any one of a number of multipurpose finite element modeling (FEM) codes may be employed. For the non-expert, the latter approach might be advisable.

On the other hand, multipurpose programs do have their own drawbacks. Despite their versatility, the codes cannot handle every imaginable problem. Since each code tends to be specialized in a specific field, the user needs to be well informed of the code's capabilities and especially its limitations. Note also that the technical support associated with these large codes is usually considerable due to frequent software updates. It is clearly important to stay current and to develop the necessary in-house expertise.

When modeling FSI, employing a computer code capable of simultaneously computing the fluid flow field and the structural reponse may at first seem the natural choice. In general, however, pure fluid modeling codes are better adapted to flow field computations. This is also the case for finite element codes developed for structural analysis. In the majority of cases, most codes treat the fluid and the structure separately, as shown in Fig. 1.11.

It is important to understand the computational fluid dynamics and FEM theory underlying the multipurpose codes. In particular, computational fluid dynamics turbulence modeling ranges from the k-ε model to large eddy simulation (LES). The more accurate direct numerical simulation (DNS) requires no turbulence model. For coupled fluid–structure computations, some codes employ

Table 1.1 Examples of numerical codes

Name	Fluid	Structure	Coupling	Developer
ANSYS/CFX ANSYS/Mechanical	FEM	FEM	File	ANSYS
ANSYS/Fluent ANSYS/Mechanical	FVM	FEM	File	ANSYS
STAR-CCM+ ABAQUS	FVM	FEM	File	CD-adapco
FrontFlow Blue/Red FrontSTR	FVM	FEM	File	AdvanceSoft

moving boundary elements while others do not, thus leading to difficulty when dealing with moving geometries. Careful consideration must therefore be given to the choice of numerical code.

Besides the considerations above, one must keep in mind that time domain simulations based on a computed flow field will be subjective, their accuracy depending on the accuracy of the computed flow field.

In the case of time history analysis, it is often impractical to perform parametric studies to investigate the influence of different parameters. Since general conclusions cannot be derived from limited analysis, the numerical computations should be viewed as desktop numerical experiments rather than definitive physical results.

For a reliable analysis today, one should use only the steady flow computational results for the flow field, while the structural vibration analysis should be based on existing experimental evaluation techniques that have been developed for specific cases of FIV, as shown in the example of Fig. 1.12 [21].

1.2.5 Experimental approach

The experimental approach is the only remaining option when the underlying phenomenon cannot be identified via numerical analysis or when such analysis is not considered reliable. In principle, experimental testing should be carried out on the prototype component itself; very often though this is not possible. The remaining option is to build a test model to recreate the observed phenomenon.

1.2.5.1 Test facilities

The first important decision is with regard to the test facilities. As long as a large-scale project is not intended, efforts should be made to use existing test facilities rather than building new ones from scratch. The search for appropriate equipment can be extended from the organization to institutions (engineering

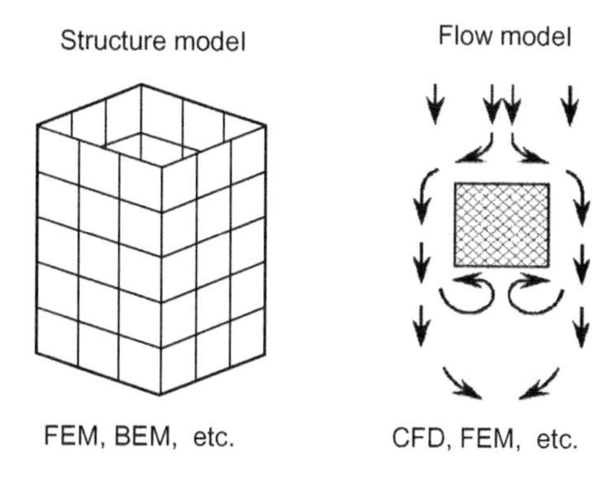

FIGURE 1.11

Separate and distinct modelling of structure and flow.

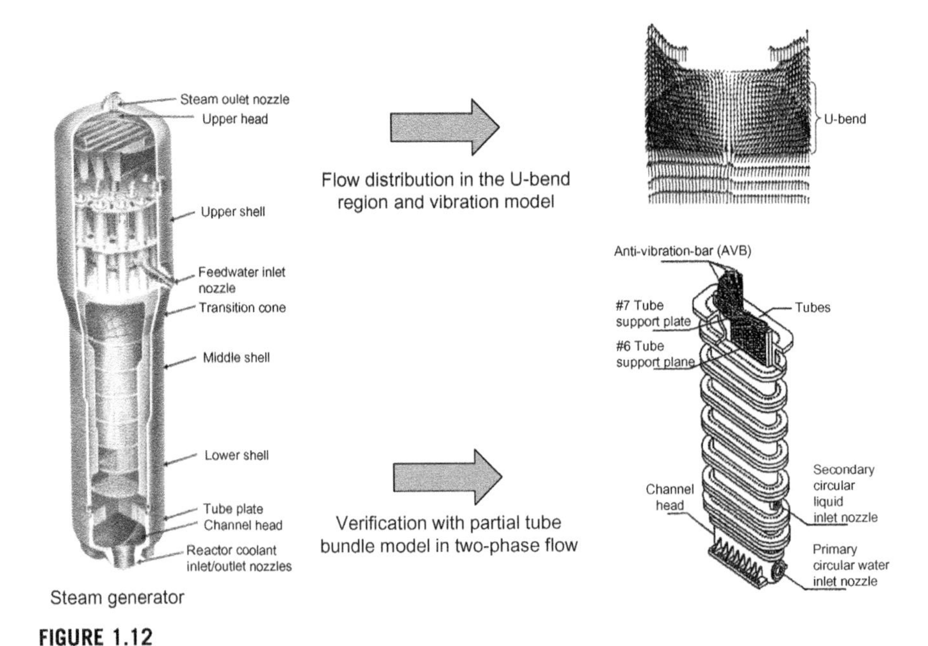

FIGURE 1.12

Example of flow analysis and vibration model.

schools are often equipped with basic test apparatus) or public research organizations if need be. The possibility of joint research can also be considered.

1.2.5.2 Similarity laws

It is rarely possible to recreate the prototypical operating conditions in a laboratory setting. It often happens that physical quantities such as temperature, pressure, and even the fluid itself are different for the test conditions. The test conditions should therefore be set to match dimensionless quantities such as Reynolds number, Froude number, Strouhal number, etc. in order to reproduce the observed phenomenon as accurately as possible.

The foregoing is the well-known similarity law-based approach [27]. Note that not only the fluid similarity law should be considered; structural similarity must also be modeled.

1.2.5.2.1 Structural model

Once the test facility has been identified, a test model matching the facility must be designed. The following points should be carefully considered at this stage:

1. With regard to the flow field–structure interaction problem, the flow field should approximate the prototype flow field as closely as possible. This means that the form of the test model should match the actual structure as closely as possible. It is usually not necessary to model the complete

structure; only the structural components pertinent to the phenomenon under investigation need be modeled.

2. When thermal effects are not of importance, only three independent physical parameters control FIV phenomena. These are characteristic length (L), time (T), and force (F) or mass (M). All other physical quantities are functions of these three parameters. Thus, once the scales or ranges of these three parameters are chosen (each one independently of the other two), the corresponding values of the other physical quantities are automatically determined by the underlying similarity laws. The existence of constraints associated with the similarity laws means that it may not be possible to match every dimensionless physical quantity of interest. Decisions must be made regarding the relative importance of the different dimensionless quantities that need to be matched between prototype and experiment. One could attempt to 'forcibly' match the quantities or, alternatively, quantities deemed to be of less importance can be ignored.

For the latter modeling procedure, the Buckingham-Pi theorem is generally used.

1.2.5.2.2 Fluid model

In reality, it is rarely possible to achieve perfect similarity for the physical quantities (flow speed, pressure, etc.) that characterize the flow field, while at the same time accurately matching the important structural quantities such as vibration frequencies, stresses, etc. The goal should therefore be to create a model that satisfies only the most important quantities based on engineering judgment.

In general, the important quantities referred to above are the normal dimensionless parameters that govern the physical phenomenon. Table 1.2 lists typical dimensionless quantities encountered in FSI problems.

As an example, Fig. 1.13 shows the problem of the Kármán wake flow behind a stationary circular cylinder. The relation between the associated physical quantities is demonstrated in the figure. The frequency f is a function of the steady flow velocity V, the cylinder diameter D, the fluid density ρ, and fluid viscosity μ. Based on these five physical quantities, a dimensional analysis yields two dimensionless groups corresponding to the Strouhal number (S) and Reynolds number (Re). Furthermore, combining the latter two relations, it can be shown that the Strouhal number is a function of the Reynolds number.

1.3 Fundamental mechanisms of FIV

In FIV, which couples fluid mechanics and vibration engineering, explaining the phenomena is of primary importance. For this reason, first a model to express the phenomenon is proposed. Next the model is expressed as a set of differential equations and finally solved taking into consideration the appropriate boundary

Table 1.2 Non-dimensional parameters and their physical meaning

Non-dimensional parameter and corresponding person	Definition	Physical meaning	Corresponding phenomena
Euler number Leonard Euler (1707–1783):* (Switzerland) Mathematician & physicist*	$Eu=\frac{p}{\rho \cdot V^2}$	$\frac{\text{Pressure}}{\text{Inertia force}}$	Cavitation in fluid machines
Reynolds number: Osborne Reynolds (1842–1912)* (England) Physicist	$Re=\frac{V \cdot D}{\nu}$	$\frac{\text{Inertia force}}{\text{Viscous force}}$	Fluid motion
Froude number: William Froude (1810–1879)* (England) Naval engineer	$Fr=\frac{V^2}{g \cdot D}$	$\frac{\text{Inertia force}}{\text{Gravity force}}$	Motion under gravity
Strouhal number: Vincenz Strouhal (1850–1925)* (Czech Republic) Physicist	$St=\frac{f_w \cdot D}{V}$	$\frac{\text{Local flow velocity}}{\text{Average flow velocity}}$	Periodic vortex shedding (Kármán vortex)
Scruton number (mass-damping parameter): Cristopher Scruton (1911–1990)* (England) Physicist	$\frac{m\delta}{\rho D^2}$	$\frac{\text{Mass of structure}}{\text{Mass of fluid}} \cdot \textit{Damping}$	Instability

Person (years), country of origin, and title.

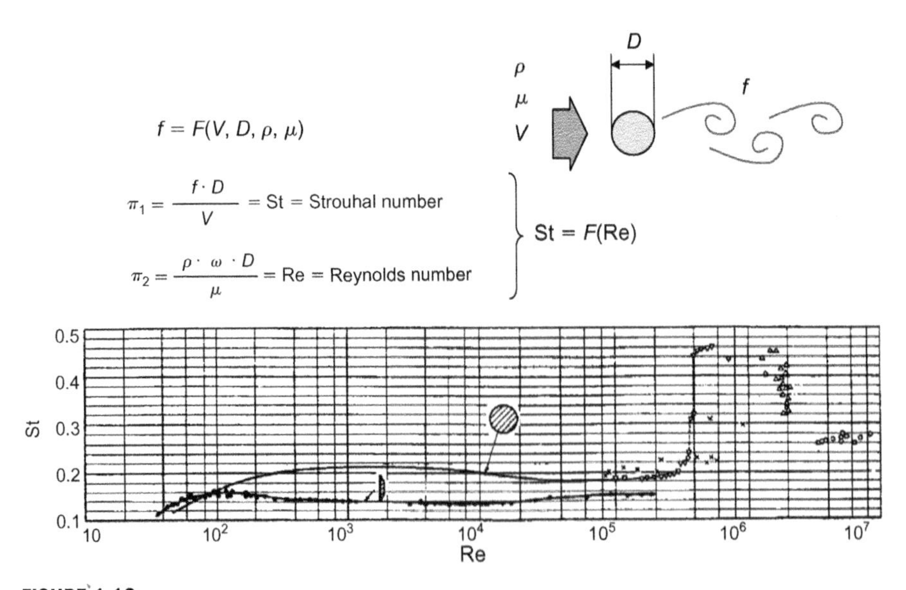

FIGURE 1.13

Dimensionless vortex shedding frequency dependence on Reynolds number.

conditions. In the modeling procedure, the equation of continuity, momentum equations (fluid flow governing equations), structural equations and the corresponding boundary conditions are described in a proper coordinate system. The equations are then expressed in non-dimensional form to obtain the ratios between the forces in the equations. Based on these ratios, a proper approximation is introduced to express the phenomenon. An analysis based on the approximate equations is then carried out.

Linear stability analysis may be performed employing one of several techniques including Fourier analysis, Laplace transforms, and traditional modal analysis. On the other hand, non-linear analysis is used to obtain the vibration response. The representative methods for the analysis of weak non-linear systems are the averaging method, the perturbation method, and the multiple scales method, as well as other analytical methods. The finite element method, the boundary element method, and the finite volume method are the standard numerical analysis approaches. From the linear analysis, the natural frequency, the vibration mode, the amplitude growth rate, the frequency response spectra, the transient response, etc. are obtained. The non-linear analysis yields the stability boundary, the post-instability limit cycle amplitude, and the time–history response.

In the following section, in order to understand the fundamental mechanisms of FIV, the basic mathematical treatment of the most important problem of self-excited oscillations is first presented.

1.3.1 Self-induced oscillation mechanisms

A system that oscillates under the influence of its own energy source due to its internal physical mechanisms is said to undergo self-excited vibration. Examples of energy sources are:

- Steady fluid flow.
- A system rotating at constant speed.
- Mechanical system forced by external load.

In the sections that follow, cases of a one-degree-of-freedom system, a two-degrees-of-freedom system, and a multi-degrees-of-freedom system are discussed.

1.3.1.1 One-degree-of-freedom system

In the case of the one-degree-of-freedom system, self-induced vibration occurs when the total damping of the combined fluid–structural system vanishes and then becomes negative. The vibration amplitude increases exponentially in time in the linear case. For this system, the governing equation of motion is:

$$m\ddot{x} + c\dot{x} + kx = f(x, \dot{x}, \ddot{x}) \tag{1.1}$$

Here, the right-hand term is the excitation force while the three left-hand terms correspond to the inertia force, the structural damping force, and the restoring force, respectively. The forces are linearly proportional to the acceleration, the velocity, and the displacement, respectively.

If $f(x, \dot{x}, \ddot{x}) = -c_0\dot{x}$ is assumed, the excitation force has a linear relation to the velocity. In this case, Eq. (1.1) can be re-written in the form:

$$m\ddot{x} + (c + c_0)\dot{x} + kx = 0 \tag{1.2}$$

If $(c + c_0) < 0$, the vibration amplitude increases with time. The condition for self-excited vibration may thus be recast in terms of c_0, as follows. A negative value of c_0 leads to self-excited vibrations when $|c_0| > c$. This case is referred to as negative-damping-induced self-excited vibration.

The foregoing phenomenon can be understood from an energy viewpoint. In Eq. (1.1), the energy variation is integrated for one period T as follows:

$$\int_0^T m\ddot{x}\dot{x}\,\mathrm{d}t + \int_0^T c\dot{x}^2\,\mathrm{d}t + \int_0^T kx\dot{x}\,\mathrm{d}t = \int_0^T f\dot{x}\,\mathrm{d}t \tag{1.3}$$

Here, the damped period T is given by:

$$T = \frac{2\pi}{\omega} = \frac{2\pi}{\sqrt{\frac{k}{m}}\cdot\sqrt{1-\left(\frac{c}{2\sqrt{mk}}\right)^2}} \tag{1.4}$$

Re-arranging Eq. (1.3) yields,

$$\int_0^T \left\{\frac{\mathrm{d}}{\mathrm{d}t}\left(\frac{1}{2}m\dot{x}^2 + \frac{1}{2}kx^2\right)\right\}\mathrm{d}t = \int_0^T (f\dot{x} - c\dot{x}^2)\mathrm{d}t \tag{1.5}$$

The left-hand term is the integral of the total energy variation during one period. Then, if the right-hand term is positive, the total energy increases during the cycle. In the case where $f = -c_0\dot{x}$ the right-hand term of Eq. (1.5) is:

$$\text{Right-hand term} = -(c_0 + c)\int_0^T \dot{x}^2\,\mathrm{d}t \tag{1.6}$$

Since the integral value is either zero or positive, the total energy increases when the coefficient of the integral is positive. When the condition $(c_0 + c) < 0$ is satisfied, self-excited vibrations occur. The latter condition physically means that the net damping of the fluid–structure system is negative.

1.3.1.2 Two-degrees-of-freedom system

In the case of the two-degrees-of-freedom system [28], coupled-mode self-excited vibration can occur in addition to the single-degree-of-freedom type instability described above. The equations of motion of the two-degrees-of-freedom system may be expressed in the following matrix form:

$$[M]\{\ddot{\mathbf{x}}\} + [C]\{\dot{\mathbf{x}}\} + [K]\{\mathbf{x}\} = \{F\} \tag{1.7}$$

where:

$$[M] = \begin{bmatrix} m_{11} & m_{12} \\ m_{21} & m_{22} \end{bmatrix}, [C] = \begin{bmatrix} c_{11} & c_{12} \\ c_{21} & c_{22} \end{bmatrix}, [K] = \begin{bmatrix} k_{11} & k_{12} \\ k_{21} & k_{22} \end{bmatrix},$$

and:

$$\{\mathbf{x}\} = \begin{bmatrix} x_1 & x_2 \end{bmatrix}^T, \{\mathbf{F}\} = \begin{bmatrix} F_1 & F_2 \end{bmatrix}^T$$

($m_{12} = m_{21}$; i.e. symmetrical mass matrix)

In equation (1.7), the excitation force is of the form $\{F(x, \dot{x}, \ddot{x})\}$ as in Eq. (1.1). The force term in the right-hand side of the above equation can be moved to the left-hand side, and incorporated into the matrices $[M]$, $[C]$, and $[K]$.

The excitation energy in this two-degrees-of-freedom system increases when the initial two natural frequencies of the system are close and the phase delay of the vibration is 90 degrees relative to the excitation force while the energy is mutually transferred between the two-degrees-of-freedom.

When off-diagonal cross-terms appear only in the elastic force (or stiffness) matrix, the system is said to have displacement or elastic coupling.

$$\begin{bmatrix} m_{11} & 0 \\ 0 & m_{22} \end{bmatrix} \begin{Bmatrix} \ddot{x}_1 \\ \ddot{x}_2 \end{Bmatrix} + \begin{bmatrix} c_{11} & 0 \\ 0 & c_{22} \end{bmatrix} \begin{Bmatrix} \dot{x}_1 \\ \dot{x}_2 \end{Bmatrix} + \begin{bmatrix} k_{11} & k_{12} \\ k_{21} & k_{22} \end{bmatrix} \begin{Bmatrix} x_1 \\ x_2 \end{Bmatrix} = 0 \quad (1.8)$$

Referring to Fig. 1.14, the force transferred from system 1 to system 2 is $k_{21}x_1$. As the equation of motion of system 2 is:

$$m_{22}\ddot{x}_2 + c_{22}\dot{x}_2 + k_{22}x_2 = -k_{21}x_1 \quad (1.9)$$

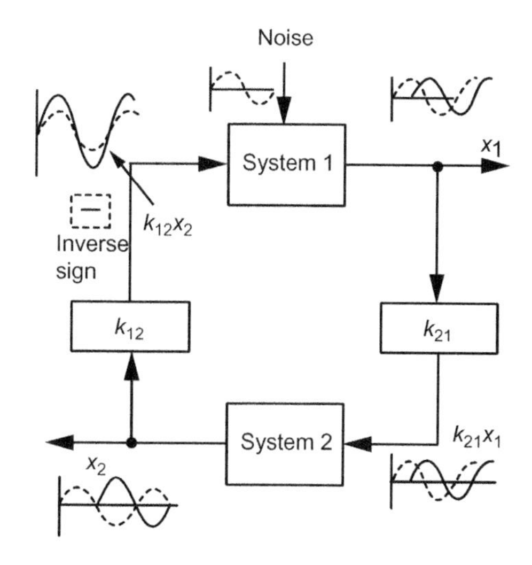

FIGURE 1.14

Instability mechanism of elastic force coupled system.

the input force is $-k_{21}x_1$. If this system is close to resonance, x_2 leads x_1 by 90 degrees in phase.

Recall the equation of motion for the one-degree-of-freedom system excited by an external force. The displacement response lags 90 degrees behind the excitation force at resonance. Since $-k_{21}x_1$ is considered to be the excitation force, x_2 lags 90 degrees behind the excitation force. This is equivalent to x_2 leading x_1 in phase by 90 degrees.

The equation of motion of system 1 is:

$$m_{11}\ddot{x}_1 + c_{11}\dot{x}_1 + k_{11}x_1 = -k_{12}x_2 \tag{1.10}$$

Based on a similar argument to the above, system 1 (x_1) now leads system 2 (x_2) by 90 degrees at resonance. When k_{12} and k_{21} have opposite signs, the phase angle becomes zero and the amplitude increases after each cycle around the system. This means that the two-degrees-of-freedom system can become unstable due to purely elastic coupling.

The same result can be obtained from energy considerations for the two-degrees-of-freedom system. Assuming a periodic solution of Eq. (1.8), the energy integrals satisfy:

$$\int_0^T \{\dot{\mathbf{x}}\}[M]\{\ddot{\mathbf{x}}\}dt + \int_0^T \{\dot{\mathbf{x}}\}[C]\{\dot{\mathbf{x}}\}dt + \int_0^T \{\dot{\mathbf{x}}\}[K]\{\mathbf{x}\}dt = 0 \tag{1.11}$$

Here, the first term corresponds to the kinetic energy, the second to the dissipated energy, and the third to the potential energy.

The two matrices $[M]$ and $[C]$ are assumed to be diagonal. The matrix $[K]$, on the other hand, is generally not. It can, however, be expressed as the sum of a symmetric matrix and a skew-symmetric matrix. Thus,

$$[K] = \begin{bmatrix} k_{11} & k_{12} \\ k_{21} & k_{22} \end{bmatrix} = \begin{bmatrix} k_{11} & k_0 \\ k_0 & k_{22} \end{bmatrix} + \begin{bmatrix} 0 & \Delta k \\ -\Delta k & 0 \end{bmatrix} = [K_0] + [\Delta K] \tag{1.12}$$

where:

$$k_0 = \frac{1}{2}(k_{12} + k_{21}),\ \ \Delta k = \frac{1}{2}(k_{12} - k_{21}) \tag{1.13}$$

From the fact that the symmetric stiffness matrix components do not contribute to the energy integral, the total energy change is obtained only from the skew-symmetric matrix components in the restoring force term.

$$E = \int_0^T \{\dot{\mathbf{x}}^T\}[\Delta K]\{\dot{\mathbf{x}}\}dt = \int_0^T \Delta k(\dot{x}_1 x_2 - x_1\dot{x}_2)dt \tag{1.14}$$

When the displacements are expressed in the form:

$$\begin{Bmatrix} x_1 \\ x_2 \end{Bmatrix} = \begin{Bmatrix} u_1 \sin(\omega t + \phi_1) \\ u_2 \sin(\omega t + \phi_2) \end{Bmatrix} \tag{1.15}$$

the total energy change per cycle becomes:

$$E = -(k_{12} - k_{21})u_1 u_2 \sin(\phi_1 - \phi_2) \tag{1.16}$$

This equation indicates that the signs of $k_{12} - k_{21}$ and $\phi_1 - \phi_2$ determine the sign of the energy E, which, in turn, means that the stability condition is determined by these terms.

For example, the case of $k_{12} > 0$, $k_{21} < 0$ under the condition of $\phi_2 = \phi_1 + \pi/2$ gives positive total energy change, which means negative energy dissipation or energy accumulation. This therefore leads to self-excited vibration of the two-degrees-of-freedom system.

It can therefore be concluded that self-excited vibration occurs when the non-diagonal components have opposite signs in the case of the elastic force coupling. Similarly, the possibility of self-excited vibration also exists in the case of damping force cross-coupling as well as inertia (mass) term cross-coupling. These possibilities are discussed in the next section where the more general multi-degrees-of-freedom systems are considered.

1.3.1.3 Multi-degrees-of-freedom system

For the multi-degrees-of-freedom system [29], variables representing the motion of the structure or of the fluid are defined as a vector $\{x\} = [x_1, x_2, \ldots, x_n \ldots]$. With the inertia, damping, and rigidity terms represented by the matrices $[M_s]$, $[C_s]$, $[K_s]$ and the fluid force acting on the structure by $\{F\}$, the equation of motion becomes:

$$[M_s]\{\ddot{x}\} + [C_s]\{\dot{x}\} + [K_s]\{x\} = \{F\} \tag{1.17}$$

The fluid force can be approximated as a linear function of the structural motion as follows:

$$\{F\} = -[M_f]\{\ddot{x}\} - [C_f]\{\dot{x}\} - [K_f]\{x\} + \{G\} \tag{1.18}$$

$[M_f]$, $[C_f]$, and $[K_f]$ are the added mass, fluid damping, and fluid rigidity matrix, respectively. $\{G\}$ is an external (time-dependent) force vector. Figure 1.15 shows a block diagram representation of Eqs. (1.17) and (1.18). The first three terms on the right-hand side of Eq. (1.18) express the feedback fluid force.

Introducing the following definitions:

$$\left.\begin{aligned} [M] &\equiv [M_s] + [M_f] = ([M_1] + [M_2]) \\ [C] &\equiv [C_s] + [C_f] = ([C_1] + [C_2]) \\ [K] &\equiv [K_s] + [K_f] = ([K_1] + [K_2]) \end{aligned}\right\} \tag{1.19}$$

where matrices with the subscript 1 correspond to the symmetric components while matrices with subscript 2 are the skew-symmetric components, Eq. (1.17) after substitution of Eq. (1.18) becomes:

$$[M_1]\{\ddot{x}\} + [C_2]\{\dot{x}\} + [K_1]\{x\} = -[M_2]\{\ddot{x}\} - [C_1]\{\dot{x}\} - [K_2]\{x\} + \{G\} \tag{1.20}$$

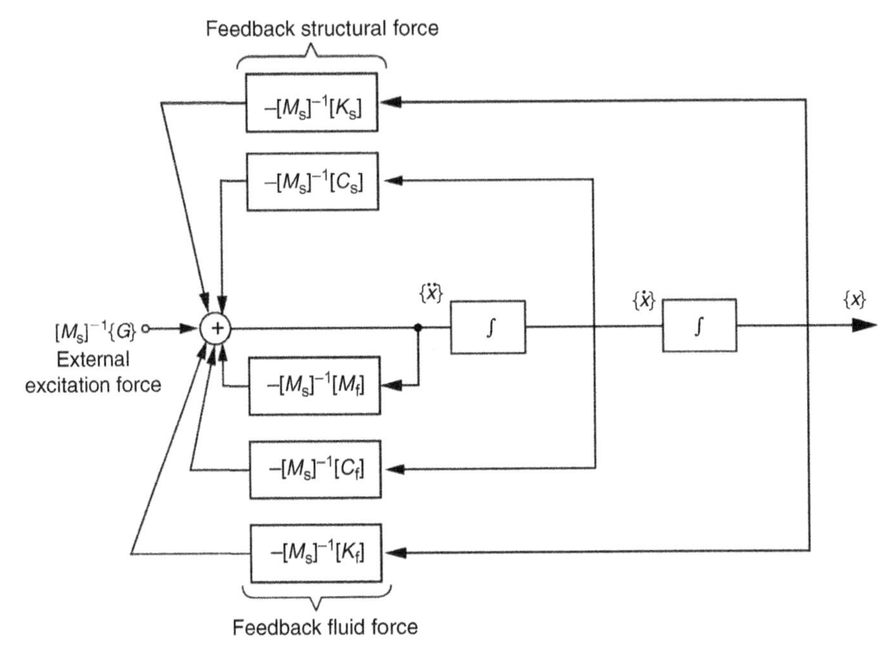

FIGURE 1.15

Feedback forces.

Multiplying Eq. (1.20) by $\{\dot{x}\}^T$, the energy change in one cycle is:

$$\left.\begin{aligned} &\int_0^T \{\dot{x}\}^T[M_1]\{\ddot{x}\}\mathrm{d}t + \int_0^T \{\dot{x}\}^T[C_2]\{\dot{x}\}\mathrm{d}t + \int_0^T \{\dot{x}\}^T[K_1]\{x\}\mathrm{d}t \\ = -&\int_0^T \{\dot{x}\}^T[M_2]\{\ddot{x}\}\mathrm{d}t - \int_0^T \{\dot{x}\}^T[C_1]\{\dot{x}\}\mathrm{d}t - \int_0^T \{\dot{x}\}^T[K_2]\{x\}\mathrm{d}t + \int_0^T \{\dot{x}\}^T\{G\}\mathrm{d}t \end{aligned}\right\} \tag{1.21}$$

The first and the third terms on the left-hand side above express the increase of the kinetic energy and potential energies. The sum of these terms corresponds to the net increase in the system energy E. The second term on the left-hand side is zero since $[C_2]$ is a skew-symmetric matrix. The right-hand side term of Eq. (1.21) equals the sum of the increase of the following energies:

$$E_\mathrm{M} = -\int_0^T \{\dot{x}\}^T[M_2]\{\ddot{x}\}\mathrm{d}t : \text{Work by } [M_2] \text{ in } [M] \tag{1.22}$$

$$E_\mathrm{C} = -\int_0^T \{\dot{x}\}^T[C_1]\{\dot{x}\}\mathrm{d}t : \text{Work by } [C_1] \text{ in } [C] \tag{1.23}$$

$$E_\mathrm{K} = -\int_0^T \{\dot{x}\}^T[K_2]\{x\}\mathrm{d}t : \text{Work by } [K_2] \text{ in } [K] \tag{1.24}$$

$$E_G = -\int_0^T \{\dot{x}\}^T\{G\}dt : \text{Work by the external force} \tag{1.25}$$

This means that the energy increase E:

$$E = E_M + E_C + E_K + E_G \tag{1.26}$$

must be positive for fluid-force-induced oscillations to occur.

When the structure oscillates as a one-degree-of-freedom system, the increase in the system energy E is:

$$E = E_C + E_G \tag{1.27}$$

When no external forces act on the structure, in the case of the negative fluid damping, $C_1 < 0$,

$$E = E_C > 0 \tag{1.28}$$

and self-excited vibration occurs. This case is called excitation by negative damping feedback.

On the other hand, when the system has two or more degrees-of-freedom, there is the case where E_K contributes to the increase of the kinetic energy E rather than the other components. The case where the matrix $[K_2]$ derives from fluid forces is called excitation by fluidelastic feedback.

Finally, vibration where $[E_G]$ mostly contributes to the increase of the system energy E corresponds to basic forced vibration.

1.3.2 Forced vibration and added mass and damping

1.3.2.1 Forced vibration system [16]

Eq. (1.18) expresses only the work done by the external force if the first three feedback fluid force terms are ignored. In this case, the equation of motion, Eq. (1.17), can be solved as a forced vibration system because the forcing is independent of the structural motion.

However, even in the presence of the first three terms of Eq. (1.18), the system may still be viewed as a forced vibration system, when the sum of the energies by Eqs. (1.22)–(1.24), $E_M + E_C + E_K$, is negative and hence no self-excited vibration occurs.

The forced vibration system appears as shown below in the case of turbulent flow, or when the mutual interaction between the structure and the fluid motion is ignored in the case of vortex shedding.

$$([M_s] + [M_f])\{\ddot{x}\} + ([C_s] + [C_f])\{\dot{x}\} + [K_s]\{x\} = \{F\} \tag{1.29}$$

Both responses of the one-degree-of-freedom system and the multi-degrees-of-freedom system can be obtained by solving Eq. (1.29) analytically or numerically.

In the case of continuous structures such as beams in flow, the excitation force is usually given by the summation of local fluid forces coupled with the correlation along the beam axis. After expanding the equation of motion with the undamped structural vibration modes, the continuous system can be decoupled into independent equations of motion, based on the orthogonality of the modes when the correlation effects between modes are negligible. The equation of motion for the j-th mode is:

$$\ddot{q}_j + 2h_j\omega_j\dot{q}_j + \omega_j^2 q = \frac{1}{m_j}\sum_{i=1} {}_j\phi_i F_i' \tag{1.30}$$

Here $\{q_j\}$ is the generalized time function (or generalized coordinate) of the j-th mode $(x_i = \sum_j {}_j\phi_i q_i)$, ω_j is the natural circular frequency of the j-th mode $(=2\pi f_j)$, and h_j the corresponding damping ratio. The modal mass is $m_j = \sum_{i=1} m_{ij}\phi_i^2$, where ${}_j\phi_i$ is the value of the j-th mode at position i and m_i and F_i' express the distributed mass and fluid forces.

In many practical cases, where the interaction with the surrounding structures can be assumed to be negligible, the diagonal terms of the added mass and the added damping matrices $[M_f]$ and $[C_f]$ are dominant, thus leading to Eq. (1.30).

When the right-hand term of Eq. (1.30) is due to excitation forces that are random (both in time and in space), the equation can be solved using statistical theory. When mode coupling terms are ignored, the following integral solution for the square average of the displacement response of the structure $\overline{X}_d^2$ is obtained:

$$\overline{X}_d^2(x) = \sum_{j=1} {}_j\phi^2(x)\int_0^\infty L_j(\omega)\left|H_j(\omega)\right|^2 d\omega \tag{1.31}$$

In Eq. (1.31) $H_j(\omega)$ is the transfer function of the j-th mode:

$$H_j(\omega) = \left[\left(1 - \frac{\omega^2}{\omega_j^2}\right) + 2ih_j\frac{\omega}{\omega_j}\right]^{-1} \tag{1.32}$$

and the correlation term $L_j(\omega)$, which is defined as the joint acceptance of the j-th mode, is given by:

$$L_j(\omega) = \frac{1}{m_j^2\omega_j^4}\iint_L R(x, x', \omega)_j\phi(x)_j\phi(x')dx\,dx' \tag{1.33}$$

where $R(x, x', \omega)$ is the correlation function of the excitation force:

$$R(x, x', \omega) = \frac{1}{\pi}S_F(\omega)\tilde{R}_{xx'}^2(\omega)\cos\theta_{xx'}(\omega) \tag{1.34}$$

which is frequency and space dependent. In the equation above, $S_F(\omega)$ is the power spectral density of the excitation force while $\tilde{R}_{xx'}(\omega)$ and $\theta xx'(\omega)$ are the correlation function and phase difference between the positions x and x'.

The joint acceptance $L_j(\omega)$ in Eq. (1.33) expresses the correlation of the fluid force along the beam axis. Empirical values of this quantity have, for example, been determined experimentally both for an isolated cylinder in cross-flow and for tubes and tube arrays in cross-flow.

Finally, Eq. (1.30) can be solved by methods similar to those used when the excitation force is not random; an example being the method used for case of periodic vortex shedding induced excitation.

1.3.2.2 Added mass

In many cases of flow-induced forced vibration, the response estimation can be divided into independent equations of motion for each mode (see Eq. (1.30)). However, the added mass and the added damping greatly affect the characteristics of the structures and therefore must be considered. In this section, the determination of the added mass for the case of a circular cylinder in still fluid is outlined.

The added mass [9] is basically obtained using potential flow theory for perfect fluids. When a circular cylinder is fully immersed in still fluid, the motion of the cylinder is coupled to that of the surrounding fluid. The continuity equation of the two-dimensional flow, assuming a perfect fluid, is:

$$\nabla^2 \phi(r, \theta, t) = 0 \tag{1.35}$$

where ϕ is the velocity potential and ∇^2 is the Laplacian operator.

The flow velocity and the associated pressure field are related to the velocity potential by:

$$\vec{U} = \nabla \phi \tag{1.36}$$

$$p = -\rho \frac{\partial \phi}{\partial t} \tag{1.37}$$

On the boundary between the circular cylinder and the fluid, the velocity of the fluid is equal to that of the cylinder. Thus,

$$u_{\mathrm{r}} = \frac{\partial u}{\partial t} \cos\theta \; (r = R) \tag{1.38}$$

while at infinity:

$$u_{\mathrm{r}} = 0 \quad (r = \infty) \tag{1.39}$$

The following forms for the velocity and velocity potential, respectively, are assumed:

$$u_{\mathrm{r}} = a e^{i\omega t} \tag{1.40}$$

$$\phi(r, \theta, t) = F_{\mathrm{r}}(r) F_{\theta}(\theta) e^{i\omega t} \tag{1.41}$$

Substituting Eq. (1.41) into Eq. (1.35), and employing separation of variables, the following ordinary differential equations are derived:

$$\frac{1}{r}\left\{\frac{\mathrm{d}}{\mathrm{d}r}\left(r\frac{\mathrm{d}F_{\mathrm{r}}}{\mathrm{d}r}\right)\right\} - \frac{1}{r^2}F_{\mathrm{r}} = 0 \tag{1.42}$$

$$\frac{d^2F_\theta}{d\theta^2} + F_\theta = 0 \tag{1.43}$$

Applying the boundary conditions (1.38) and (1.39) to the general solutions of Eqs. (1.42) and (1.43) yields the solution:

$$\phi(r,\theta,t) = -\frac{i\omega R^2 a}{r}\cos\theta e^{i\omega t} \tag{1.44}$$

Since the fluid force, g, acting on the cylinder in its direction of motion is;

$$g = -\int_0^{2\pi} P(r,\theta,t)|_{r=R} R\cos\theta \, d\theta \tag{1.45}$$

this force can be re-written in the form:

$$g = -M_f\frac{d^2u}{dt^2}, \quad M_f = \pi\rho R^2 \tag{1.46}$$

'M_f' in this equation is the cylinder 'added mass' per unit length. The added mass is also called the virtual mass. Practically, the added mass is expressed as the mass of fluid having the same volume as that of the immersed cylinder multiplied by the added mass coefficient C_m. For the circular cylinder in unconfined flow, C_m is unity. Note, however, that this is generally not the case for other configurations.

The added mass decreases the natural frequency of the structure. If the fluid is perfect, there is no fluid damping and only the change of the natural frequency characterizes the kinetic effect of the fluid on the structure.

1.3.2.3 Fluid damping

The fluid damping can be divided into two categories: damping in still fluid and damping induced by fluid flow. The latter damping depends on many mechanisms. Their detailed explanation is therefore given in later sections. Here, only the damping in still fluid is explained.

Damping in still fluid can be divided into two components: damping caused by fluid viscosity and energy dissipation by sound.

Fluid viscous damping may be obtained by the following method. The drag force by the surrounding fluid F_D, when the structure moves with velocity $\dot{x}$, is given by the following equation based on the characteristic length D and the non-dimensional drag coefficient C_D:

$$F_D = \frac{1}{2}C_D\rho_f|\dot{x}|\dot{x}D \tag{1.47}$$

Assuming sinusoidal motion with the circular frequency ω, and taking only the first term of the Fourier expansion:

$$|\dot{x}|\dot{x} \approx \frac{8}{3\pi} X_{\mathrm{d}}{}^2 \omega^2 \cos(\omega t) = -\frac{8}{3\pi} X_{\mathrm{d}} \omega \dot{x} \tag{1.48}$$

the damping ratio h_{f} $(=C_{\mathrm{f}}/2m\omega_n)$ in still fluid becomes:

$$h_f = \frac{2}{3\pi}\frac{\rho_{\mathrm{f}} D^2}{m}\frac{X_{\mathrm{d}}}{D}\frac{\omega}{\omega_n} C_{\mathrm{D}} \tag{1.49}$$

where ω_n is the undamped natural frequency.

This damping can therefore be estimated when the non-dimensional drag coefficient C_{D} is known. However, C_{D} does not take a simple form in FSI systems such as a cylinder in a narrow path or tubes in a tube array. Additional detailed information on the damping in still fluid is therefore presented later in Sections 3.3.2–3.3.4, 4.1.1, and 8.2.2.

The damping by energy dissipation through sound radiation cannot be ignored for light structures such as panels. The following estimation for this type of damping is proposed:

$$h_{\mathrm{f}} = \frac{1}{4\pi}\frac{\rho_{\mathrm{f}} a^2 b}{M}\frac{\lambda}{a}\Theta \tag{1.50}$$

$$\Theta = \begin{cases} \left(\frac{\pi}{2}\right)^2 (a^2+b^2)/\lambda, & for\ a/\lambda \ll 0.2 \\ 1, & for\ a/\lambda \gg 0.2 \end{cases} \tag{1.51}$$

Here, the panel having dimensions a by b and mass M radiates sound waves at the frequency ω_n having wavelength $\lambda = 2\pi c/\omega_n$ where c is the sound speed.

References

[1] E. Naudascher, Flow-Induced Structural Vibrations, Springer-Verlag, Berlin, 1972.

[2] E. Naudascher, D. Rockwell, Practical Experiences with Flow-Induced Vibrations, Springer-Verlag, Berlin, 1979.

[3] R.D. Blevins, Flow-Induced Vibration, Van Nostrand Reinhold, New York, 1977.

[4] R.D. Blevins, Formulas for Natural Frequency and Mode Shape, Van Nostrand Reinhold, New York, 1979.

[5] R.D. Blevins, Applied Fluid Dynamics Handbook, Van Nostrand Reinhold, New York, 1984.

[6] R.D. Blevins, Flow-Induced Vibration, second ed., Krieger, Malabar, 1990.

[7] E. Naudascher, Hydrodynamic Forces, A.A. Balkema, Rotterdam, 1991.

[8] E. Naudascher, D. Rockwell, Flow-Induced Vibrations, An Engineering Guide, A.A. Balkema, Rotterdam, 1993.

[9] S.S. Chen, Flow-Induced Vibration of Circular Cylindrical Structures, Springer-Verlag, Berlin, 1987.

[10] H.J.P. Morand, R. Ohayon, Fluid Structure Interaction, Wiley, New Jersey, 1995.
[11] M.P. Païdoussis, Fluid–Structure Interactions, Vol. 1, 2, Academic Press, London, 1998, 2004.
[12] M.P. Païdoussis, G.X. Li, Pipes conveying fluid: a model dynamical problem, J. Fluid. Struct. 7 (1993) 137–204.
[13] M.P. Païdoussis, S.J. Price, E. de Langre, Fluid-Structure Interactions: Cross-Flow-Induced Instabilities, Cambridge University Press, New York, 2011.
[14] S. Hayama, Chapter 23 Vibration Engineering Handbook (in Japanese), Yokendo Publication, Tokyo, 1976.
[15] JSME, Flow-Induced Vibration in Mechanical Engineering – Its Status and the Prevention, P-SC10 Report of Flow-Induced Vibration Organization Group, 1980.
[16] JSME, Pressure Pulsation in Piping Induced by Reciprocating Compressors (in Japanese), P-SC105 Report of Organization Group on Pressure Pulsation in Piping and Reciprocating Compressors, 1989.
[17] T. Iwatsubo, Flow-Induced Vibration (in Japanese), Proceedings of 63rd Annual Meeting of Kansai Branch in JSME, 884-2, 1988, p. 125.
[18] K. Fujita, Topic on Flow-Induced Vibration in Industry (in Japanese), Proceedings of the Annual Meeting of Kansai Branch in JSME, 884-2, 1988, p. 133.
[19] H. Madarame, Recent Flow-Induced Vibration in Nuclear Power Industry (in Japanese), Report of Organization Group on Nuclear Power Industry, Nuclear Engineering Facility in University of Tokyo, 1990, pp. 1–20.
[20] JSME, Guideline for Fluid-Elastic Vibration Evaluation of U-bend Tubes in Steam Generators (in Japanese), JSME S016, 2002.
[21] JSME, Guideline for Evaluation of Flow-Induced Vibration of a Cylindrical Structure in a Pipe, JSME S012, 1998.
[22] E.A. Müller, Mechanics of Sound Generation in Flows, Springer-Verlag, Berlin, 1979.
[23] W.K. Blake, Mechanics of Flow-Induced Sound and Vibration, Academic Press, Orlando, 1986.
[24] Examples of Prevention of Noise & Vibration, Japan Noise Control Society, 1990.
[25] Y. Maruta, Seminar on Flow-Induced Noise, Ebara Tech. Report, No. 181, 1998; No. 182, 1999; No. 183, 1999; No. 184, 1999; No. 185, 1999.
[26] E. Naudascher, D. Rockwell, Flow-Induced Vibrations, An Engineering Guide, Hydraulic Design Considerations, IAHRAIRH (Hydraulic Structures Design Manual 7).
[27] I. Emori, Theory and Application of Model Test (in Japanese), Gihodo Pub.
[28] T. Iwatsubo, Flow-Induced Vibration (in Japanese), Proceedings of #63 Annual Meeting of Kansai Branch of JSME, No. 884-2, 1988, p. 125.
[29] F. Hara, Dynamics Handbook (in Japanese), Asakura Shoten, 1993. p. 608.

CHAPTER

Vibration Induced by Cross-Flow

2

Cross-flow-induced vibration is the most important problem in various fields, and is known to have caused many failures in various industrial components. A classification based on the correspondence between targeted industrial products and basic structures discussed in this chapter is shown in Table 2.1. Examples are taken mainly from plant engineering, civil engineering, and marine engineering. Examples from aeronautical engineering are not considered.

The classification of Table 2.1 is certainly not absolute and the engineer should consider where his or her product lies within the classification. Depending on the product, two or more classes may be applicable.

The flow classification considered in this chapter is presented in Table 2.2. Fluids are usually classified as either 'gas' or 'liquid'. When dimensionless numbers, such as the Reynolds number, are equal in the two cases, there is no basic difference and both fluid flows may simply be treated as single-phase flow. On the other hand, when gas and liquid exist as a mixture due to evaporation and condensation of steam for instance, a treatment different from single-phase flow is required. This falls into the area of gas–liquid two-phase flow analysis.

Single-phase flow may be divided into steady flow and unsteady flow. The former may be either uniform or shear flow. While a similar classification should exist for two-phase flows, much less research has been done in comparison to single-phase flows. The detailed classification for single-phase flows is therefore more of an indicator of this difference.

2.1 Single circular cylinder

2.1.1 Structures under evaluation

This section describes the flow-induced vibration (FIV) problem in the case of a single structure having a circular cross-section. Examples in this category include components exposed to steady flow, such as thermo wells in pipes, and components exposed to oscillating flow or waves, such as marine pile structures. Multi-span slender structures such as cables or riser pipes also fall in this category.

Table 2.1 Classification of products considered in this chapter

Products			Classification	Sections
Structures in industrial plants	Plant components	Steam generators Boilers Heat exchangers Chillers Fuel assemblies	Multi-circular cylinders	2.3, 2.5
		Cross pipes	Two circular cylinders	2.2
	Structures in pipe	Thermo wells Sampling nozzles	Single circular cylinder	2.1
		Tandem Thermo wells	Two circular cylinders	2.2
		Guide vanes	Rectangular cylinder	2.4
Civil engineering structures	Structures exposed to wind	Stacks, towers	Single circular cylinder	2.1
		Tandem cables	Two circular cylinders	2.2
		Bundle of stacks	Multi-circular cylinders	
		Buildings Bridges	Rectangular cylinder	2.4
		Bridge cables Power cables	Single circular cylinder	2.1
			Rectangular cylinder	
	Structures exposed to tidal waves or river flow	Bridge support structures Marine cables Riser pipes Piles	Single circular cylinder	2.1
			Rectangular cylinder	2.4

Table 2.2 Classification of flows considered in this chapter

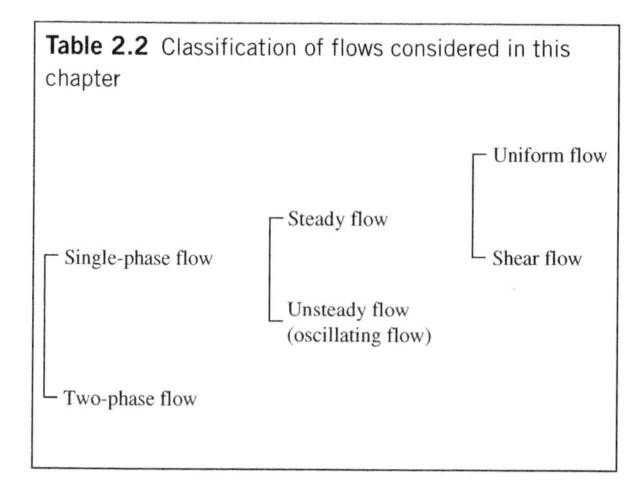

2.1.2 Vibration mechanisms and historical review

2.1.2.1 Vibration mechanisms

The vibration phenomena may be divided into two categories: (i) beam mode bending vibration of the circular cylindrical structures, and (ii) ovalling vibration of cylindrical shells. The former is classified into vibration induced by steady flow, for example the problem of cylindrical structures contained in pipes, and vibration caused by unsteady (oscillating) flow, as in the case of marine structures. The vibration of circular cylindrical structures in steady flow is further subdivided into four classes, as shown in Table 2.3. The mechanisms underlying each class of vibration are discussed next.

2.1.2.1.1 Bending vibration of a circular cylindrical structure in steady flow

Forced vibration by Kármán vortex shedding. When flow past a bluff body generates a vortex street in the wake region (this being the well-known Kármán vortex street), periodic shedding of these vortices from the surface of the body induces periodic pressure variations on the structure. Vibrations in two directions are possible, transverse and parallel to the flow. In the transverse direction, the excitation force has a dominant frequency called the Kármán vortex shedding frequency. In the drag direction the dominant frequency is at twice the Kármán vortex shedding frequency. The vortex shedding frequency f_W is expressed in dimensionless form by the Strouhal Number, St. The latter is defined as:

$$\mathrm{St} = \frac{f_W D}{V} \tag{2.1}$$

where D is a characteristic length (diameter) and V the flow velocity.

Table 2.3 Classification of vibration types in Section 2.1

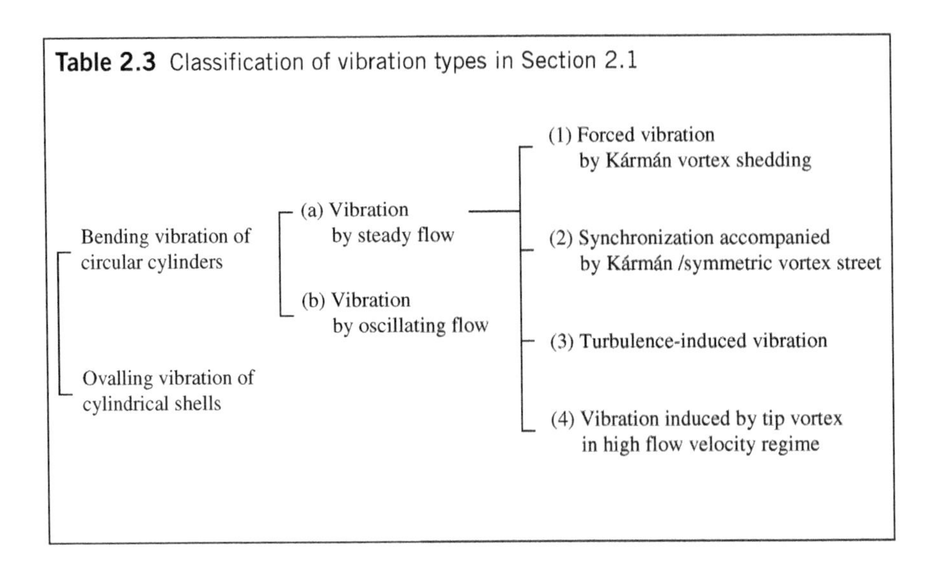

Synchronization accompanied by alternating Kármán or symmetric vortex street. When the natural frequency of the structure is close to the vortex shedding frequency, the latter frequency synchronizes with the natural frequency of the structure. This phenomenon is termed lock-in. Such synchronization can occur in both the transverse and parallel directions to the flow. Moreover, for certain conditions, in-line synchronization in the flow direction accompanied by symmetric vortex shedding may occur at lower flow velocities compared to transverse direction lock-in. Engineers should note that the critical velocity for this symmetric type of synchronization is lower than that for the Kármán vortex shedding synchronization. Note, however, that in-line synchronization may be suppressed in many cases when the fluid has low density (e.g. gas flows). The latter condition will come up when vibration suppression criteria are discussed later. Figure 2.1 depicts the vortex-induced vibration response amplitude as a function of upstream flow velocity.

Turbulence-induced vibration. A cylinder is inevitably excited by vortex-induced forces outside the synchronization region. Not only does this periodic component at a dominant frequency in the excitation force exist, but so do components over a wide frequency band. If the structural natural frequency is well separated from this dominant (shedding) frequency, the structure will instead be excited by the wide-band components closest to the natural frequency. The resulting response is termed turbulence-induced vibration. The latter includes the case where upstream turbulence may be high enough to induce cylinder vibration.

Vibration induced by tip-vortices in high flow velocity regime. Large-amplitude vibrations (comparable to synchronization-induced vibrations) may

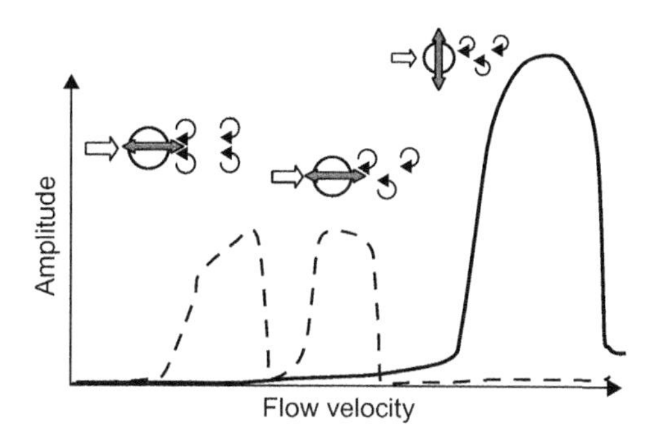

FIGURE 2.1

Vortex-induced synchronization.

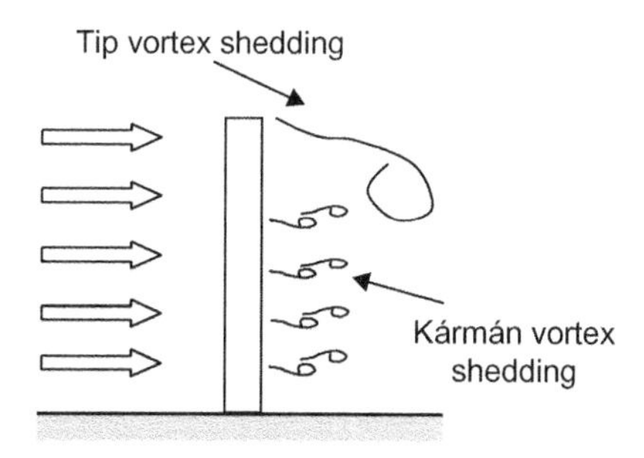

FIGURE 2.2

Tip-vortex shedding.

arise at high flow velocities, well beyond the Kármán shedding lock-in regime. This phenomenon is caused by vortices generated at the cylinder extremity (see Fig. 2.2). These tip-vortices are shed at a frequency roughly one-third the Kármán shedding frequency. The critical flow velocity for lock-in is therefore approximately three times the critical velocity for Kármán shedding.

2.1.2.1.2 Vibration of a circular cylinder in oscillating flow

When a circular cylinder is subjected to oscillating flow as shown in Fig. 2.3, the flow velocity oscillations result in variations of the drag and fluid inertia forces. Moreover, since vortex shedding from the circular cylinder is also present, excitation transverse to the flow direction also occurs. The result is superposed (in-flow + transverse) bending vibration of the circular cylinder induced by the oscillating flow. A feature of the resulting vibration is that the in-line vibration component induced by the flow oscillations is much more dominant.

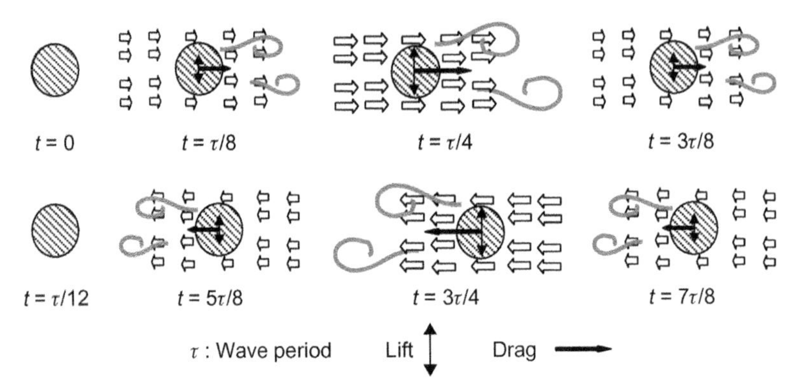

FIGURE 2.3

Cylinder motion and vortex shedding in oscillating flow.

Flow oscillations are normally characterized by a reduced velocity κ defined in Eq. (2.2) below, where V_m is the amplitude of the oscillating flow velocity, and f_{osc} the corresponding frequency of oscillation.

$$\kappa = \frac{V_m}{f_{osc}D} \tag{2.2}$$

The dimensionless quantity (κ) formed in Eq. (2.2) is known as the Keulegan–Carpenter number.

2.1.2.1.3 Ovalling vibrations of cylindrical shells in steady flow

Unlike the foregoing two types of vibrations, ovalling vibrations are accompanied by cross-sectional changes of the cylindrical shell. Although it was initially thought that vortex shedding was a key factor, it is now known that the aeroelastic effects due to interaction between cross-sectional modification and flow play a key role.

2.1.2.2 Historical background

2.1.2.2.1 Bending vibrations of a circular cylinder in steady flow

Kármán vortex shedding induced vibration. The Kármán shedding frequency has been an important subject of study due to its direct relation to the excitation force frequency. In particular, Von Kármán demonstrated theoretically, via a stability analysis of two vortex rows, that the Strouhal number is nearly constant. It is known, however, that the Strouhal number varies with Reynolds number, surface roughness, turbulence intensity, etc. Much research has been dedicated to the empirical determination of the dependence of the Strouhal number on these parameters.

Synchronization with Kármán/symmetric vortex shedding. Among vortex-induced vibration (VIV) mechanisms, lock-in in the direction transverse to the flow has been known for many years. Many studies have therefore been done.

Examples include studies on the hysteresis effect (of the vibration response amplitude) as a function of flow velocity, semi-empirical modeling of vortex-induced resonance, and studies on the influence of different parameters.

Although many researchers (e.g. Bishop and Hassan [1]) have reported on the hysteresis effect, it was Williamson and Roshko [2] who, while investigating the vortex shedding modes for a cylinder undergoing forced oscillations, concluded or discovered that the cause of hysteresis is the transition between vortex modes. Brika and Laneville [3] supported this conclusion by directly observing the mode transition and the corresponding hysteresis.

Van der Pol type wake oscillator models [4,5] and extensions of the Morison model [6] that 'mimic' the fluid force on a cylinder in oscillating flow are a popular approach to the theoretical modeling of lock-in in the lift direction. In addition to these models, there is also the model by Lowdon et al. [7] based on the time delay between the vibration of the structure and flow reconfiguration. Although most of these models agree qualitatively with experimental measurements, quantities such as the critical velocity for lock-in are not necessarily correctly predicted. However, since many models constitute semi-empirically 'tuned' formulae, the peak vibration magnitudes usually agree with experiments. This makes them at least useful for peak magnitude evaluation.

The effects of shear flow, wall effects, mass ratio, etc. have also been studied. For industrial structures the influence of shear flow is important. Griffin [8] pointed out in his review paper [8] that vortex shedding from a cylinder in shear flow consists of cell structures much like in the case of a tapered cylinder [9] because the vortex shedding corresponds to the local velocity and local diameter of the cylinder. In the lock-in region, cell structures tend to disappear (due to amalgamation) and vortices are shed at a single frequency for large enough vibration amplitudes.

For long-span structures, such as offshore risers, the influence of velocity distribution, as well as the active vibration mode shape in each span, plays an important role. Lock-in criteria [10,11] or evaluation formulae [12] for the response have therefore been proposed, taking these factors into account.

With regard to wall effects, Tsahalis [13] has reported that the critical velocity increases and that the resonant amplitude decreases when the distance from an adjacent wall becomes significantly small. King and Jones [14] have also reported that periodic vortex shedding diminishes when the distance from an adjacent wall falls below half the diameter of the cylinder.

Khalak and Williamson [15] have reported that the cylinder mass ratio and damping have independent effects for very small mass-damping parameter values.

Standards or codes on vortex-induced vibration, including symmetric vortex shedding excitation, include Code Sec. III, Div. 1, Appendix N, 1300 Series by the American Society of Mechanical Engineers (ASME) [16] and the Guideline for Evaluation of Flow-Induced Vibration of a Cylindrical Structure in a Pipe by the Japan Society of Mechanical Engineers (JSME) [17]. The latter was created in response to the thermo well failure accident in the Monju fast-breeder reactor. The latest knowledge is reflected in this code.

Turbulence-induced vibration. Many researchers have measured the power spectral density (PSD) of the excitation force in the broad frequency range outside the periodic vortex shedding resonance band. In particular, Mulcahy summarized the measured fluctuating fluid force coefficient results by many researchers as a function of Reynolds number [18]. Unlike vortex-induced vibration, the response to turbulence excitation is usually estimated from calculations based on empirical functions derived from measured PSDs. However, many studies [19–21] have recently appeared in the field of numerical analysis (computational fluid dynamics [CFD]) reflecting recent advances in computer technology.

Vibration induced by tip-vortices in high flow velocity regime. Fox and Apelt [22] have reported that the tip-vortex shedding frequency is about one-third the natural shedding frequency along the cylinder span. The tip region spans roughly two diameters at the cylinder extremity. Kitagawa et al. [23] reported that vibration induced by tip-vortex shedding occurs at far higher flow velocities relative to the range in which lock-in by Kármán vortex shedding occurs. Although it is known that the resonance region is determined by reduced velocity calculations similar to the type of vibration discussed in the section on synchronization with Kármán/symmetric vortex shedding above, this knowledge is not reflected in either the ASME or the JSME design guidelines.

2.1.2.2.2 Vibration of a circular cylinder in oscillating flow

Studies on the vibration of circular cylinders in oscillating flow began with research on structures exposed to ocean waves in marine engineering. In the 1950s, Morison et al. [24] proposed a modeling method where the force on the structure in oscillating flow is expressed as the sum of inertia and drag forces. Most researchers still use this method. Besides Morison's approach, Longoria et al. [25] proposed a method where the wave-motion-induced force is represented in the frequency domain in the form of a PSD. This method yields a good approximation of the excitation force in the low-frequency range. Vibration control methods based on the Morison model have also been reported [26,27].

There are also numerous CFD-based vortex-induced vibration studies [28–30]. The latter are usually for comparatively low reduced velocities. In Zhang's study [28], the predicted added mass and drag coefficients are in good agreement with experimental data (Table 2.4).

2.1.2.2.3 Ovalling vibrations of cylindrical shells in steady flow

For ovalling vibration of a cylindrical shell in cross-flow, aeroelastic flutter theory is generally considered valid although there is a dispute between two opinions (one espousing a vortex-induced vibration theory [29–33] and the other, an aeroelastic flutter theory [34–38]), shown in Table 2.5. These discussions are explained in full detail in the review paper by Païdoussis et al. [39].

Table 2.4 Studies on vortex-induced vibration

Year	Studies (researchers)
1970	Van der Pol type lock-in model using fluctuating lift coefficient and displacement of cylinder as variables (Hartlen and Currie)
1972	Observation of cell structures in wake behind tapered cylinder (Vickery and Clark)
1974	Van der Pol type lock-in model using displacements of cylinder and fluid as variables (Iwan and Blevins)
1980	Vortex-induced vibration of a cylinder near a wall (King)
1982	Extension of Morison models to lock-in (Sarpkaya)
1984	Wall effect on lock-in (Tsahalis)
1985	Review paper on vortex-induced vibration in shear flow (Griffin)
1986	Modified lock-in model for vibration in shear flow (Wang)
1988	Observation of vortex mode transition in lock-in (Williamson and Roshko)
1991	Time-delay model of fluid force at lock-in; model of unsteady boundary layer (Lowdon)
1993	Lock-in criteria for marine structures (Vandiver)
	Vortex mode transition and hysteresis (Brika and Laneville)
1995	Revision of ASME Code Sec. III, Div. 1, Appendix N, 1300 Series (ASME)
	Response estimation method for vortex-induced vibration of multi-span circular structure (Bokaian)
1996	Lock-in for low mass ratio and low damping ratio (Khalak and Williamson)
1998	Publication of Guideline for Evaluation of Flow-Induced Vibration of a Cylindrical Structure in a Pipe (JSME)

Table 2.5 Studies on ovalling vibration of a cylindrical shell

Year	Studies (researchers)
1956	Observation of ovalling vibration at half the frequency of vortex shedding frequency (Dockstader)
1974	Observation of ovalling vibration with frequency ratio 1:6 (a vortex-induced vibration theory is asserted) (Johns)
1979	Observation of vibration with non-integral frequency ratio (Païdoussis)
	Discovery of non-suppression effect of a splitter plate for ovalling vibration (Païdoussis)
	Discussion on splitter plate in journals (Johns)
1982	Refutal of the vortex-induced vibration theory by detailed measurement of vortex shedding (Païdoussis)
	Proposal of an aeroelastic flutter theory (Païdoussis)
1985	Measurement of aeroelastic damping (backing an aeroelastic flutter theory) (Katsura)

2.1.3 Evaluation methods

2.1.3.1 *Bending vibrations of a circular cylinder in steady flow*

Although a flowchart of the evaluation procedure for bending vibrations of a circular cylinder is given in Fig. 2.4, the engineer need not perform all evaluations according to this chart. For example, in the JSME guidelines, the evaluation of forced vibration by Kármán vortex shedding is not included because of the mechanism's low contribution to the vibration amplitude. The resonant amplitude is not evaluated because the design criteria avoid the resonance region. In a similar manner, the designer must make judgment calls on parameters that must be considered in design and their appropriate values. This determination is directly linked to the problem at hand.

2.1.3.1.1 Vibration induced by single-phase flow

Forced vibration induced by Kármán vortex shedding. The fluid excitation force on a cylinder subjected to cross-flow is periodic, and its frequency f_W is expressed by the Eq. (2.3):

$$f_W = \mathrm{St}\frac{V}{D} \tag{2.3}$$

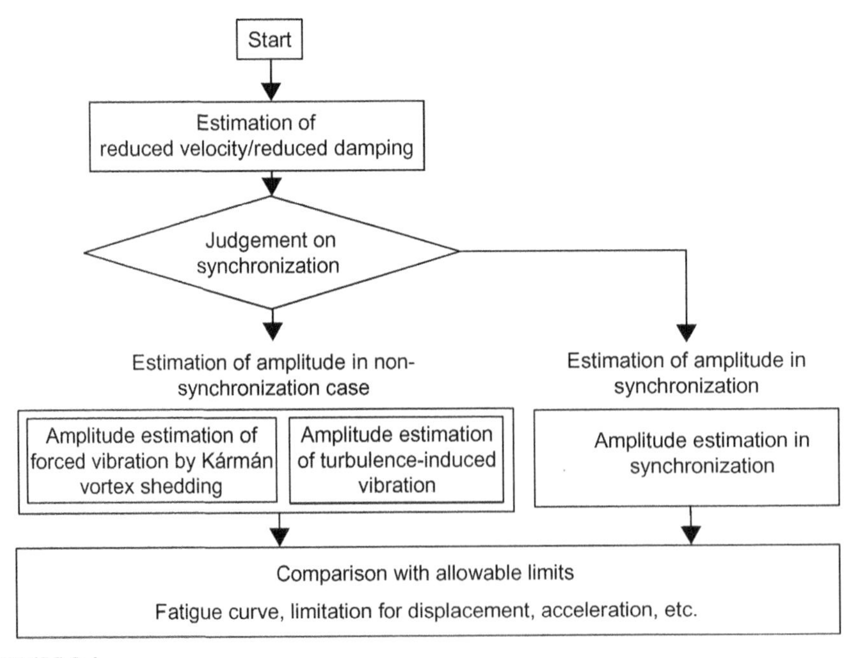

FIGURE 2.4

Evaluation approach for vibration of a circular cylinder in cross-flow.

Equation (2.3) gives the frequency of the excitation force in the lift direction. In the drag direction, the excitation frequency is twice that in the lift direction. Unless these frequencies approach the cylinder natural frequency, no significant vibration occurs. Conversely, when either forcing frequency approaches the cylinder frequency, synchronous vibrations can usually be expected.

Synchronization with Kármán/symmetric vortex shedding. The condition for avoidance or suppression of synchronization for a single circular cylinder is expressed in terms of reduced velocity V_r and reduced damping C_n defined in Eqs. (2.4) and (2.5) where f_0 is the fundamental natural frequency of the circular cylinder. The avoidance or suppression region for synchronization is shown in Fig. 2.5.

$$V_r = \frac{V}{f_0 D} \tag{2.4}$$

$$C_n = 2m^*\delta = \frac{2m\delta}{\rho D^2} = \frac{4\pi m\zeta}{\rho D^2} \tag{2.5}$$

The threshold reduced damping for in-line synchronization in Fig. 2.5 is based on the study and guidelines of JSME [17]. The supporting data are presented in Fig. 2.6. The parameter bounds for vibration avoidance in the JSME guideline are presented together with the equivalent ASME guideline bounds in Table 2.6. The JSME bounds correspond to Fig. 2.5.

Next, the vibration amplitude evaluation method during synchronization (lock-in) is described. Here, the model by Iwan and Blevins [5] is presented as an example of a typical modeling method. In the model, the system is expressed by

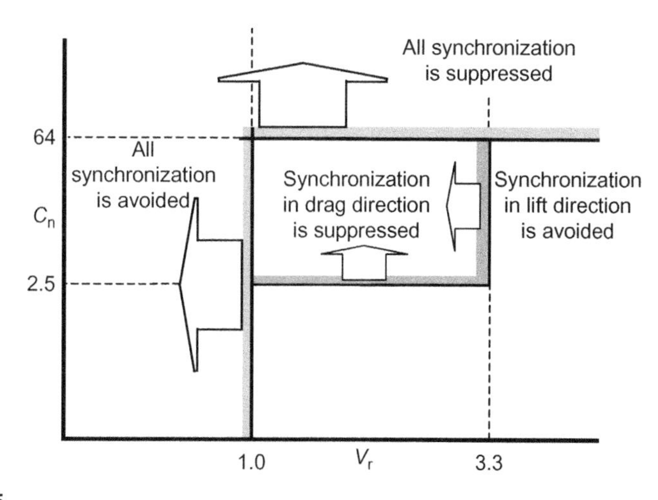

FIGURE 2.5

Range of avoidance and suppression of synchronization.

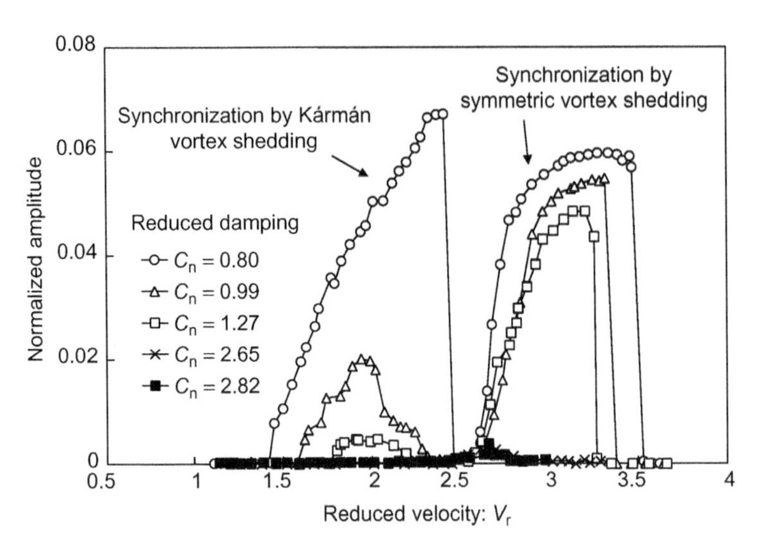

FIGURE 2.6

Suppression of in-line synchronization [17].

Table 2.6 Conditions for avoidance or suppression of synchronization

	JSME guidelines	ASME guidelines	
	Conditions to be satisfied		
1	$V_r < 1$	$V_r < 1$	Avoidance
2	$C_n > 64$	$C_n > 64$	Suppression
3	$V_r < 3.3$ and $C_n > 2.5$	$V_r < 3.3$ and $C_n > 1.2$	Avoidance of lift oscillation Suppression of drag oscillation
4		$f_0/f_w < 0.7$ or $f_0/f_w > 1.3$	Avoidance of only lift oscillations

the following set of simultaneous equations for the cylinder displacement y and the hidden fluid parameter z corresponding to the center of mass of the fluid:

$$\alpha_0 \rho D^2 \ddot{z} - \left\{ \alpha_1 \rho D V - \alpha_2 \rho \frac{D}{V} \dot{z}^2 \right\} \dot{z} + \omega_W{}^2 z = -F \tag{2.6}$$

$$m\ddot{y} + c\dot{y} + ky = F \tag{2.7}$$

$$F = \alpha_3 \rho D^2 (\ddot{z} - \ddot{y}) + \alpha_4 \rho D V (\dot{z} - \dot{y}) \tag{2.8}$$

Here, α_0, α_1, α_2, α_3, α_4 are empirical constants. Using such a model, the resonant vibration amplitude can be estimated. In addition to the semi-empirical formula by Iwan and Blevins [5], other formulae exist such as those by Griffin [40] and Sarpkaya [41]. These formulae are presented in Table 2.7.

Table 2.7 Semi-empirical formulae for resonant amplitude

	Semi-empirical formulae	
Griffin	$\frac{y}{d_0}=\frac{1.29\gamma}{[1+0.43(2\pi St^2C_n)]^{3.35}}$	Gives conservative amplitude in low C_n region
Blevins	$\frac{y}{d_0}=\frac{0.07\gamma}{(C_n+1.9)St^2}\left[0.3+\frac{0.72}{(C_n+1.9)St}\right]^{1/2}$	
Sarpkaya	$\frac{y}{d_0}=\frac{0.32}{[0.06+(2\pi St^2C_n)^2]^{1/2}}$	Mode factor not included

Turbulence-induced vibration. The response to turbulence excitation is treated as follows in the JSME S 012 guideline [17]. Random vibration theory is used for response estimation. The spatial dependence of the excitation force is taken into consideration by introducing a turbulence fluctuation fluid force coefficient that incorporates the influence of the spatial correlation length. The resulting root-mean-square (rms) value of the turbulence-induced vibration response is given by:

$$w_{mean}^2=\frac{\beta_0 G(f_0)}{64\pi^3 m^2 f_0^3 \zeta}\phi_0^2(z) \tag{2.9a}$$

where,

$$\beta_0=\frac{\int_{Le}\phi_0(z)dz}{\int_L\phi_0^2(z)dz} \tag{2.9b}$$

is the participation factor of the fundamental mode $\phi_0(z)$. The PSD per unit length of the turbulence excitation force $G(f)$ is expressed in terms of the normalized PSD Φ and the turbulence excitation force coefficient C' as shown in Eqs. (2.10)–(2.12).

$$G(f)=\left(C'\frac{1}{2}\rho V^2 D\right)^2\Phi(f_r)\frac{D}{V} \tag{2.10}$$

$$\Phi(f_r)=4/(1+4\pi^2 f_r^2)\quad (V_r\leq 3.3: f_r\geq 0.3) \tag{2.11}$$

$$\Phi(f_r)=7.979\times 10^{-3}/f_r^4\quad (3.3<V_r<5.0: 0.2<f_r<0.3) \tag{2.12}$$

In the equations above, f_r is the reduced frequency defined as $f_r=fD/V$.

$\Phi(f_r)$ is a function enveloping most of the existing experimental data. A value of C' approximately 0.13 is adopted based on Fung's data. In the case where turbulence is generated upstream, or in the regime beyond the critical Reynolds number, PSD measurements in model experiments are required.

Vibration induced by tip-vortices in the high flow velocity regime. Research on this type of vibration (e.g. [23]) is relatively new and has yet to find its way into the ASME or JSME design guidelines. Tip-vortex-induced vibration occurs starting near the reduced velocity $V_r = 10$. The resulting vibration amplitude attains a maximum value at $V_r = 15$. Thus, it is a phenomenon manifested in the region beyond about 3 times the critical velocity for classic lock-in in the lift direction due to Kármán vortex shedding. Usually a cylinder undergoes Kármán shedding lock-in before the onset of tip-vortex-induced vibration as the flow velocity is increased. The tip-vortex-induced vibration depends on tip shape. According to Kitagawa [23], vibration can be suppressed by attaching, at the cylinder extremity, a circular plate of diameter $D_p > 1.6D$, where D is the circular cylinder diameter.

2.1.3.1.2 Vibration induced by two-phase flow

Although Kármán vortex shedding has not been clearly observed in most two-phase flow regimes, Hara and Ohtani [42] reported that Kármán vortices are formed for void fractions below 15%, much like in single-phase flow. The critical void fraction for Kármán vortex shedding depends on the ratio of the average bubble diameter to the cylinder diameter, and increases with decreasing bubble diameter. In addition, Hara [43] measured the fluctuating lift and drag forces for various void fraction values. He reported that these forces increase dramatically beyond a certain void fraction. This feature is seen in the frequency characteristics of the flow, which takes on a random character for high void fractions.

As for synchronization in two-phase flow, lock-in in the lift direction tends to be suppressed when compared with the same phenomenon in single-phase flow. On the other hand, at high void fractions in-line vibration amplitudes are more significant in comparison to single-phase flow in-line excitation (see Fig. 2.7) [44]. Besides Hara and Ogawa's data, little else has been reported on vibration in two-phase flow. For this reason, an established method for the analysis of

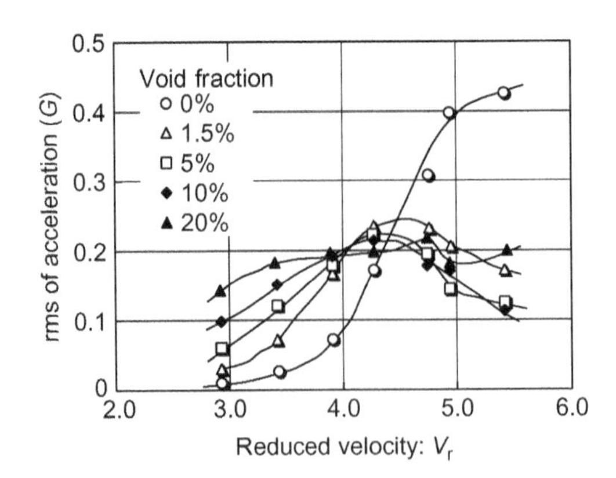

FIGURE 2.7

Lock-in in two-phase flow [44].

two-phase flow-induced vibration of a solitary circular cylinder does not currently exist.

2.1.3.2 Vibration of a circular cylinder in oscillating flow

Morison's model [24] is well known as an estimation method for the fluid force acting on a circular cylinder in oscillating flow. In this model, the fluid inertia force F_I is expressed as the sum of buoyancy and added mass components as follows:

$$F_I = \rho A \dot{V} + C_m \rho A(\dot{V} - \ddot{x}) \tag{2.13}$$

The fluctuating drag force F_D derived from flow velocity oscillation is expressed as:

$$F_D = \frac{1}{2}\rho |V - \dot{x}| \cdot (V - \dot{x}) D C_D \tag{2.14}$$

The cylinder governing equation of motion therefore becomes:

$$(m + \rho A C_m)\ddot{x} + 2m\zeta\omega_i \dot{x} + kx = \rho A C_I \dot{V} + \frac{1}{2}\rho |V - \dot{x}|(V - \dot{x}) D C_D \tag{2.15}$$

Here, the inertia coefficient is defined as $C_I = 1 + C_m$.

Because this equation is non-linear, simplified solution methods employing linearization have been proposed. Using the simplified methods, the estimated amplitudes tend to be larger than those obtained by solving the non-linear equation.

2.1.3.3 Ovalling vibrations of cylindrical shells in steady flow

The ovalling vibration of a cylindrical shell is considered to be caused by aeroelastic flutter. However, in the *Recommendations for Loads on Buildings* published (in Japanese) by the Architectural Institute of Japan [45], designers are required to ensure that the design flow velocity falls below the following two critical velocities: The first is the critical flow velocity for vortex-induced vibration, U_r, as shown in Eq. (2.16). The second is the aeroelastic flutter critical velocity, U_{thr}, by Uematsu et al. [46] given in Eq. (2.17).

$$U_r = \frac{f_i D}{j\mathrm{St}} \tag{2.16}$$

In Eq. (2.16) above, f_i is natural frequency of the i-th ovalling mode and j is a positive integer ($j = 2$ when $i = 3$).

$$U_{thr} = 1.03\pi \frac{f_i D}{i} \tag{2.17}$$

Eq. (2.17) is based on experimental results. A theoretical method [36] for estimating the critical velocity based on aeroelastic flutter theory also exists. Unfortunately, this model cannot accurately estimate the critical velocity due to the influence of pressure fluctuations in the wake region [47]. Modified models [48] have also been proposed.

2.1.4 Examples of component failures due to vortex-induced vibration

Since lock-in in the lift direction is well-known among designers, there have been few failures attributable to this mechanism in the recent past. Although a slit or spiral strake [49,50] on the surface of the cylinder is often mentioned as a countermeasure for vortex-induced vibration, these cannot completely eliminate vortex-induced vibration. In fact, these countermeasures only shift the resonant velocity regime (see Fig. 2.8).

Few well-documented examples of failures due to in-line shedding synchronization exist [51,52]. Engineers are therefore likely to overlook this type of vibration at the design stage. The first report of this type of failure was that of a marine pile supporting oil pipelines during construction of the oil terminal for the North Sea oil field at Immingham. In this case, the pile vibrated in the tidal wave direction in the low flow velocity region (see Fig. 2.9) [53].

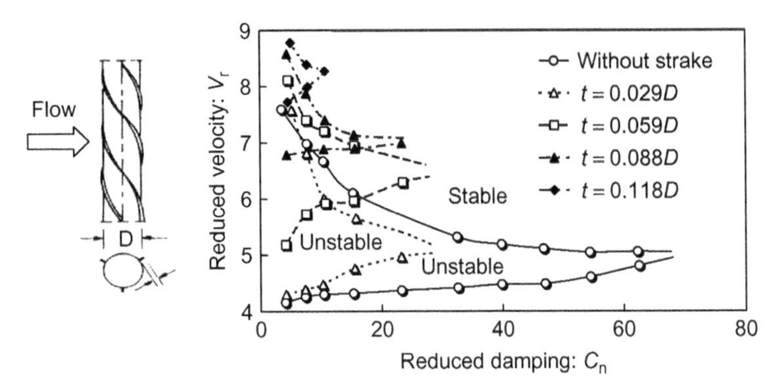

FIGURE 2.8

Suppression of synchronization by spiral strake [50].

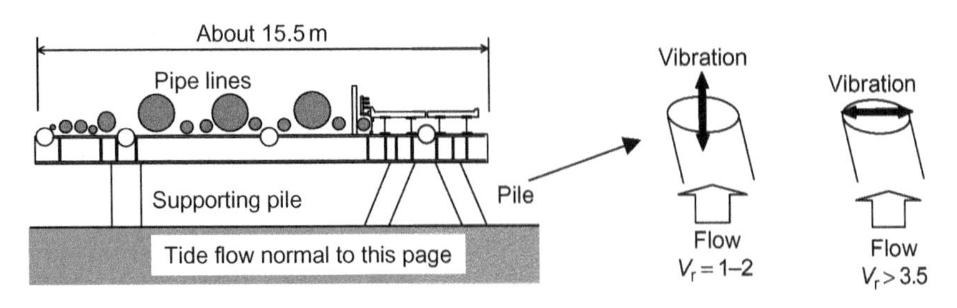

FIGURE 2.9

In-line vibration of marine pile at Immingham [53].

In addition to the phenomena classified and explained above, other problems include rain-induced vibration as well as the coupling of shell and beam mode vibrations. Rain-induced vibrations occurred on the Meiko West Bridge [54] (Japan) as shown in Fig. 2.10. Vibration occurred because rainwater flowing down the inclined cables changed the effective cross-section of the cables to a non-circular shape, thus generating significant excitation lift forces. Vibrations involving coupling between shell and beam modes occurred in a thermo well in a fast breeder experimental reactor in the USA, as shown in Fig. 2.11 [55].

From these examples it is clear that, in addition to the simple application of the design guidelines, one must consider and make judgments regarding the

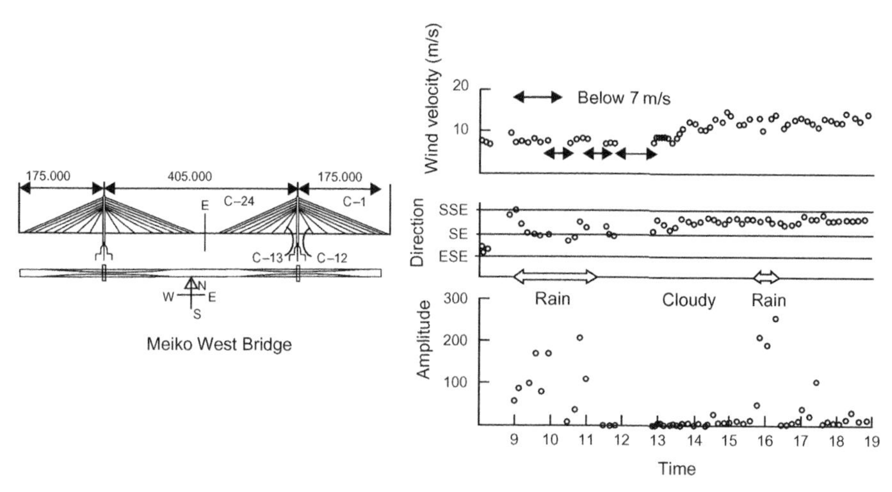

FIGURE 2.10

Rain-induced vibration in Japan [54].

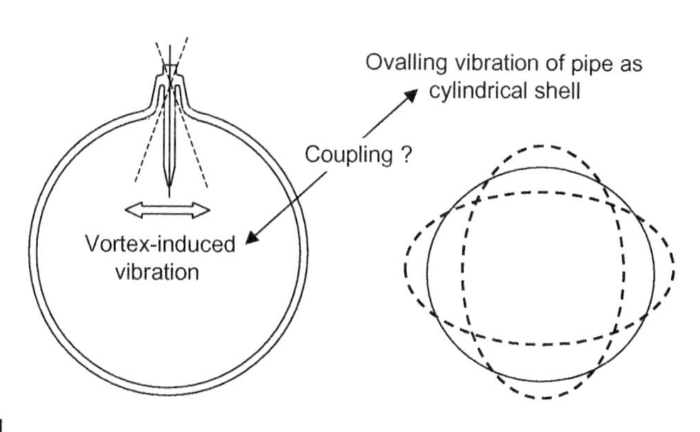

FIGURE 2.11

Example of coupling of a pipe and a thermo well [53].

environment in which a circular cylindrical structure is installed to ensure vibration safety.

2.2 Two circular cylinders in cross-flow

2.2.1 Outline of structures of interest

2.2.1.1 Examples

Vibration problems are frequently encountered for structures consisting of multiple cylinders such as electric power lines, flow sensor tubing, etc. The resulting vibrations depend strongly on cylinder configuration (relative to flow), pitch spacing, cylinder diameters, and flow conditions.

2.2.1.2 Classification based on flow type

In many engineering applications, these structures will be subjected to gas or liquid flow. The flow type is therefore determined by the Reynolds number of the single-phase flow. In power plants on the other hand, liquid and gas coexist, due to evaporation or condensation, as multi-phase flows.

2.2.1.3 Classification based on spatial configuration

When the two cylinders are separated by a large enough distance, each cylinder can be treated as an isolated body from an FIV viewpoint. For closely spaced cylinders, the resulting dynamic behavior is strongly dependent on the cylinder pair configuration relative to the flow. The important configurations are shown in Fig. 2.12. As explained in the previous section on the single cylinder case, two types of vibration excitation can occur in steady flow: in-line vibrations due to

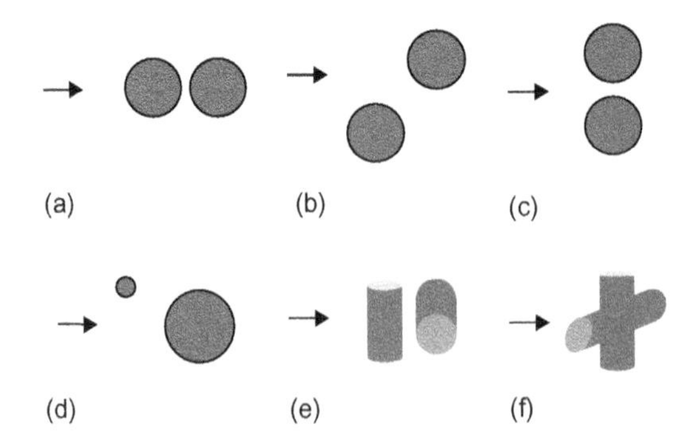

FIGURE 2.12

Possible configurations of cylinder pairs in cross-flow: (a) two cylinders in tandem, (b) staggered arrangement of two cylinders, (c) two cylinders in parallel, (d) two staggered cylinders with different diameters, (e) two cylinders in criss-crossed configuration, and (f) intersecting cylinders.

symmetrical vortex shedding and transverse vibrations in the lift direction due to Kármán vortex shedding.

2.2.2 Historical background

2.2.2.1 Excitation phenomena

2.2.2.1.1 Vibration of cylinder pairs subjected to steady cross-flow

As in the single cylinder case, the vibration mechanisms for cylinder pairs are classified into three types.

Forced vibrations induced by Kármán vortex shedding. Kármán vortex shedding occurs due to the flow separation around the structures. The resulting flow resembles that observed for single cylinders when the cylinder separation is large enough. The alternate shedding of vorticity behind the structures leads to periodic pressure oscillations in both the transverse and flow directions. The oscillation frequency in the transverse direction matches the vortex shedding frequency. However, in the flow direction pressure oscillations occur at double the vortex shedding frequency. When the two cylinders are in close proximity, the locations of the flow separation points on the cylinder surface, which affect the stage at which a vortex is shed, differ from the single cylinder case depending on cylinder configuration, pitch-to-diameter ratio (P/D), and tube natural frequencies. In this case, there is strong coupling between the cylinders which generally leads to vibrations in both the lift and drag directions.

Self-excited vibrations (lock-in with Kármán and symmetric vortex shedding). When the cylinder natural frequency is close to the vortex shedding frequency, synchronization occurs, leading to lock-in and consequently large amplitude vibrations. Lock-in occurs in both the cross-flow and in-flow directions, similar to the forced vibration case above. Compared to the single cylinder case, larger amplitude self-excited vibrations in the flow direction may occur at lower flow velocities.

Turbulence-induced vibrations. The cylinders will be subjected not only to the periodic vortex shedding induced forces but also to broadband turbulence. The latter excitation becomes relatively more important when the periodic shedding frequency is far from the structural frequency. Both turbulence propagated from upstream and turbulence generated in the cylinder wakes may play an important role.

2.2.2.1.2 Oscillatory-flow-induced vibration

Oscillatory flow, such as may occur at a coastline, can lead to forced vibrations. A periodic excitation force is generated in the drag direction due to the resulting time-dependent inertia and drag forces. Furthermore, transverse vibrations can be induced by shear layer separation and vortex shedding similar to the steady flow case. For oscillatory flow, it has been observed that the excitation force in the drag direction is significantly larger than that in the lift direction. The cylinder drag and inertia forces vary strongly for small cylinder spacings, $P/D < 3$.

2.2.2.2 Research background

2.2.2.2.1 Steady-flow-induced cylinder vibration

Vortex excited vibrations. Vibration excitation by Kármán shedding, for a cylinder pair, is fundamentally similar to that for single isolated cylinders. The reader is therefore referred to Section 2.1.3.1 dealing with the case of isolated cylinders.

Self-excited vibration. For the design of single cylindrical structures subjected to cross-flow the ASME Code Sec. III, Div. 1, Appendix N, 1300 Series may be used. However, when a multiple-cylinder design is based on the single cylinder design code above, the risk of vibration caused by lock-in in the stream-wise direction due to symmetrical vortex shedding or transverse direction excitation by Kármán shedding will remain.

The vortex flow structure for multiple cylinders differs from the single cylinder case and depends on cylinder diameters and configuration relative to the flow. The vibration modes for cylinder pairs are categorized and shown below.

Two cylinders with equal diameters. Zdravkovich et al. [56–58] reported a detailed classification of the flow structure for cylinder pairs as shown in Table 2.8 and Fig. 2.13 and highlighted differences with the single isolated cylinder case.

The flow structure for cylinder pairs may be roughly categorized into the following three regions:

1. Proximity interference region (proximity).
2. Wake interference region (wake).
3. Proximity and wake interference region (proximity and wake).

The resulting self-excited vibration is determined by the following flow, structure, and geometrical parameters: *L/D* (stream-wise cylinder separation), *T/D* (transverse cylinder separation), ω (dimensionless frequency), the Scruton number Sc ($=2M\delta/\rho D^2H$), DF (tube vibration direction and degrees-of-freedom), Re (Reynolds number), SR (surface roughness), and TI (turbulence intensity). (Note: "?" in Table 2.8 indicates that no clear flow structure can be identified for the corresponding parameters.)

Cylinder pair with unequal diameters. For large cylinder diameter difference and low Reynolds numbers, positioning the smaller diameter cylinder in the wake of the larger cylinder has a vibration control (suppression) effect. Vibration of the smaller cylinder effectively modifies the larger cylinder's wake, leading to stabilization of the latter cylinder. Numerical analysis results on the effective region or position within the wake where the smaller cylinder stabilizes the larger one have been presented by Strykowski et al. [59]. Figure 2.14 shows the region within which control can be achieved depending on the diameter ratio *D/d*. The smaller cylinder can stabilize the oncoming shear wake flow by diffusing or destabilizing vortex formation. The formation of vortices behind the larger cylinder is coupled to the wake interference region near the smaller cylinder. The effectiveness of the control mechanisms described above is limited to Reynolds

Table 2.8 Classification of interference flow-induced oscillations

Category	Cylinder configuration	Streamwise spacing L/D	Transverse spacing T/D	Flow regimes for stationary cylinders		Type of excitation			
						Low V_r		Medium V_r	High V_r
Proximity Interference	Side by side (parallel)	0	1.0 1.2	Single vortex street		?	?		Jet switching
		0	1.1 2.2	Biased jet			;	;	
		0	2.0 4.0	Coupled vortex streets		?	?		
	Staggered	,3.8	$,I_B$			?	?		
Wake and Proximity Interference	Tandem	1.1 1.5	0	Synchronized reattachment				max	Gap flow switching
		1.5 3.8	0	Quasi-steady reattachment			max	max	
	Staggered	1.1 3.2	0.2 0.3	Gap flow		?	?	max	
Wake Interference	Tandem	.3.8	0	Coupled vortex streets		?		max	None
		.8	0	Disturbed vortex street			?	max	
		.3.2	$,C_L$max	Displaced wake			?	max	Wake displacement
	Staggered	.8	$,I_B$	Outside wake boundary		No Interference			Wake galloping

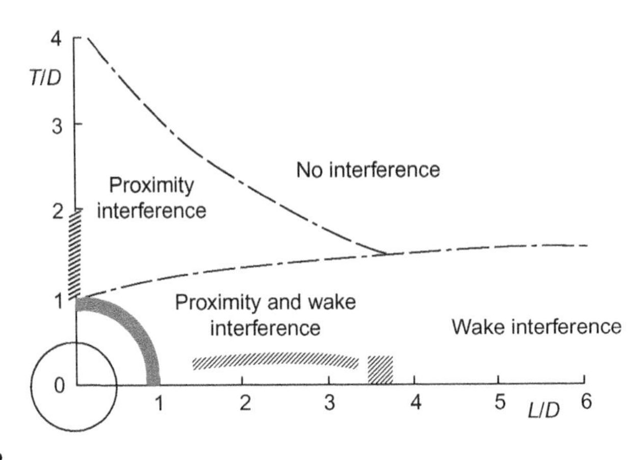

FIGURE 2.13

Cylinder interaction regimes based on in-flow and transverse spacings.

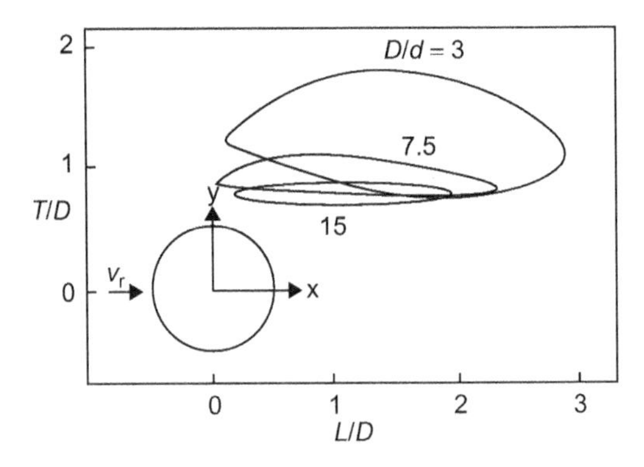

FIGURE 2.14

Region within which control can be achieved for different diameter ratios D/d.

numbers up to approximately Re = 80. Above this Reynolds number, the vortex wake depends on the position of the control cylinder. Effective control is possible for stream-wise spacings in the range $L/D = 1 \sim 3$. When L/D is larger than 4, the larger cylinder's vibration is no longer controllable.

The flow of rainwater along inclined cylindrical structures, such as power transmission lines or suspension bridge cables, changes the effective cross-section seen by the approaching wind flow. The result is what is termed rain-induced vibrations. As the effective cross-section of a cable is changed by the flowing rainwater, the resulting fluid force leads to large-amplitude transverse vibrations of the downstream cable. A report by Yoneta et al. [60] is a representative

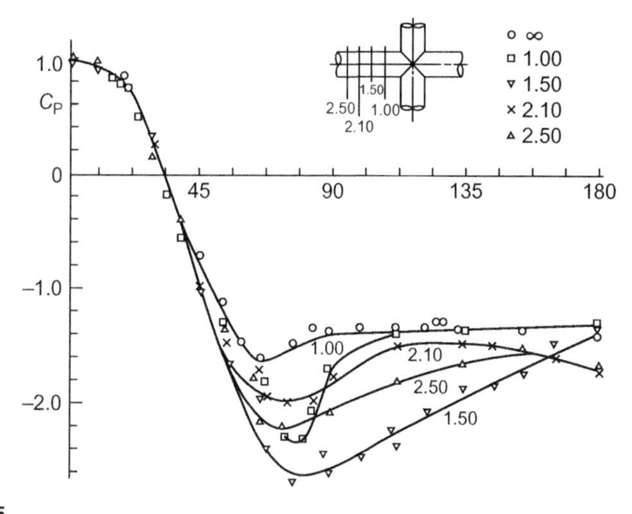

FIGURE 2.15

Pressure coefficient C_p variation and dependence on Z/D.

example. Other examples of rain-induced vibrations have been reported by Nakamura et al. [61]. In the report, coupled-mode flutter involving translation and torsional motions is shown to occur for bundled overhead cables or power transmission lines. For instance, in 1965, power transmission lines in Misaki (Japan) suffered rain-induced vibrations for wind speeds ranging from 6 to 18 m/s. The cable spacing was in the range $P/D = 10 \sim 20$.

Criss-crossed circular cylinders. For criss-crossed cylinders (i.e. cylinders oriented with axes at right angles) Bae et al. [62] reported that the flow separation points have a diagonal orientation relative to the cylinders. Tomita et al. [63] showed that noise by Kármán vortex shedding from a single cylinder in steady cross-flow is stabilized by a cylinder located downstream at right angles. When the upstream cylinder is supported elastically, the vibration can also be controlled or stabilized by the same mechanism. Note, however, that for high flow velocities it is possible for much larger vibrations to occur compared to the case without the downstream cylinder.

Intersecting circular cylinders. There have been relatively few studies reporting on vibrations of intersecting cylinders. The work of Zdravkovich in 1984 is noteworthy. The author presented experimental work on the variation of the drag coefficient and vortex shedding with distance-to-diameter ratio Z/D (Z = distance from the center of cylinder intersection, D = diameter). The angular position of the flow separation point and the pressure coefficient C_p showed significant dependence on Z/D as shown in Fig. 2.15. For $Z/D = 2.5$ or less, three-dimensional effects were found to become important. The separated shear layers move inward and become entwined, leading to highly three-dimensional flow.

2.2.2.2.2 Oscillatory flow

Nagai et al. [64] reported on the mutual interference effects of marine structures consisting of cylinders subjected to oscillatory flow (e.g. ocean waves). Ookusu et al. [65], on the other hand, reported theoretical work on two cylinders of the same diameter oriented in tandem to the direction of wave propagation. Spring and Monkmeyer [66] reported similar work for two cylinders having different diameters subjected to oscillatory ocean wave flow.

2.2.2.2.3 Vibration of cylinder pairs in two-phase flow

In real machinery both single-phase and two-phase flows are encountered. Two-phase flows introduce phenomena different from those found in single-phase flows. Iijima and Hara [67–71] reported on the interaction between two tandem cylinders subjected to two-phase cross-flow by comparing the effects of the spacing P/D and bubble size. When the reduced velocity is below the critical velocity at which lock-in occurs in single-phase flow, air bubbles cause random excitation of the cylinders resulting in larger vibration amplitudes than in single-phase flows. At the same time it is observed that the accumulation of air bubbles in the cylinder wake leads to increased damping. This is an important point to consider when dealing with two-phase flows.

2.2.3 Evaluation methodology

2.2.3.1 Experimental evaluation

2.2.3.1.1 Vibration of cylinder pair in single-phase flow

Vortex excited vibration. The fluid force acting on the cylinders is periodic, like in the single cylinder case. The fluid force frequency is related to the flow and cylinder parameters by the following relation:

$$f_W = \mathrm{St}\frac{V}{D} \tag{2.18}$$

Self-excited vibration (with lock-in to Kármán and symmetric vortex shedding). The same parameters (V_r and C_n) as in the single cylinder case are used. See Table 2.8 for the classification of cylinder dynamics based on configuration and spacing. Zdravkovich et al. [57] have studied different configurations of elastically supported cylinder pairs subjected to transverse flow. They observed that the cylinders may be treated as isolated structures for the spacings $L/D > 7$ and $T/D > 4$. When both cylinders are supported elastically, inter-cylinder interaction must be considered when the spacing $L/D > 7$ and $T/D > 4$ [56–58]. Furthermore, it has been found that for the same spacing, but with the downstream cylinder fixed, there are cases where the vibration amplitude of the upstream cylinder differs from the cases when both cylinders are flexible. Figure 2.16 shows the case where the downstream cylinder is rigidly fixed while the upstream cylinder is elastically supported. For the spacing labeled A9, for example, the upstream cylinder is stable when the downstream cylinder is also elastically mounted; the downstream cylinder on the other hand vibrates. This suggests that the downstream cylinder does not necessarily affect its upstream counterpart [72]. Figure 2.16 gives a clear idea of the effect of a fixed cylinder located downstream of a flexible cylinder.

2.2.3.2 Theoretical modeling

Taking into account experimental results such as those presented above, some theoretical work has also been done.

2.2.3.2.1 Wake interference mathematical model

Zdravkovich et al. [56] suggested a mathematical model of wake interference in 1985. The fluid dynamic forces F_x and F_y are expressed in the mathematical form of Eq. 2.19. This form is obtained assuming quasi-static conditions and linearity.

$$F_X = \frac{1}{2}\rho V^2 DH\left(\frac{\partial C_D}{\partial X}X + \frac{\partial C_D}{\partial Y}Y\right) + \frac{\rho VDH}{2K}(C_L\dot{Y} - 2C_D\dot{X})$$

$$F_Y = \frac{1}{2}\rho V^2 DH\left(\frac{\partial C_L}{\partial X}X + \frac{\partial C_L}{\partial Y}Y\right) + \frac{\rho VDH}{2K}(C_D\dot{Y} - 2C_L\dot{X}) \tag{2.19}$$

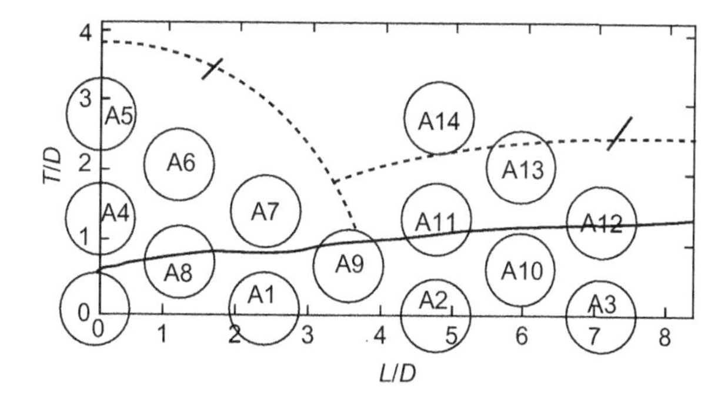

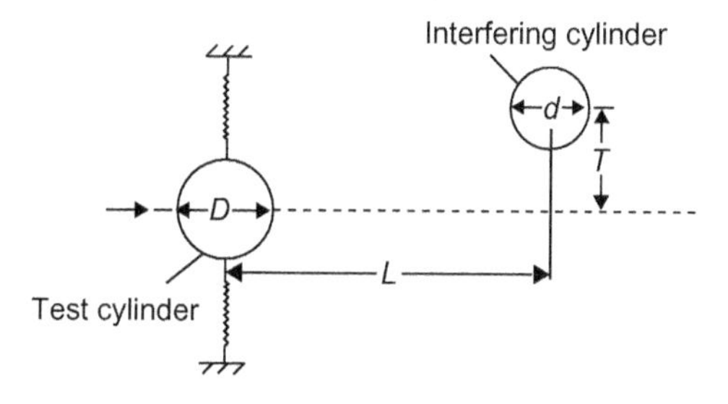

FIGURE 2.16

Experiments on upstream cylinder dynamics in the case where the downstream cylinder is fixed.

Introducing these forces on the right-hand side of the structural equation of motion, the resulting system can be analyzed for the cylinder dynamics.

2.2.3.2.2 Fluid–structure coupled analysis

Ichioka et al. [73] have presented numerical results on the coupled fluid–structure problem for isolated cylinders or cylinder arrays, where the two-dimensional Navier–Stokes equations for incompressible flows were discretized by a finite difference scheme. The resulting system of equations was solved simultaneously by an iterative method, taking into account moving boundaries in the fluid–structure interaction problem, using appropriate conformal coordinates.

2.2.3.2.3 Determination of instability boundary by unsteady fluid force models

Theoretical models based on inter-cylinder coupling unsteady fluid forces have been developed by Tanaka et al. [74] and Chen [75]. The unsteady fluid forces are measured experimentally. In the tests the rigid reference cylinder is oscillated periodically at a known frequency and small amplitude. The resulting force amplitude and phase on the neighboring cylinder are measured. The force magnitude and phase determine the net work done during a cycle of oscillation. Instability occurs when the net work done is positive and large enough to overcome damping. In the model by Chen, the drag force h_j and lift force g_j acting on a fixed cylinder are given, respectively, by:

$$\begin{aligned} g_j &= \frac{1}{2}\rho U^2 D C_{\mathrm{D}j} + \frac{1}{2}\rho U^2 D C'_{\mathrm{D}j}\sin(\Omega_{\mathrm{D}j}t + \varphi_{\mathrm{D}j}) + g'_j \\ h_j &= \frac{1}{2}\rho U^2 D C_{\mathrm{L}j} + \frac{1}{2}\rho U^2 D C'_{\mathrm{L}j}\sin(\Omega_{\mathrm{L}j}t + \varphi_{\mathrm{L}j}) + h'_j \quad j = 1,2 \end{aligned} \tag{2.20}$$

The first terms are the steady drag and lift forces while the second terms are the unsteady periodic fluid force components. The third terms represent the random turbulence-induced forces. When the cylinders are supported elastically, hence free to vibrate, the following inter-cylinder coupling forces must be added:

$$\begin{aligned} \tilde{g}_j &= \sum_{k=1}^{2}\left(\overline{\alpha}_{jk}\frac{\partial^2 u_k}{\partial t^2} + \overline{\sigma}_{jk}\frac{\partial^2 v_k}{\partial t^2} + \overline{\alpha}'_{jk}\frac{\partial u_k}{\partial t} + \overline{\sigma}'_{jk}\frac{\partial v_k}{\partial t} + \overline{\alpha}''_{jk}u_k + \overline{\sigma}''_{jk}v_k\right) \\ \tilde{h}_j &= \sum_{k=1}^{2}\left(\overline{\tau}_{jk}\frac{\partial^2 u_k}{\partial t^2} + \overline{\beta}_{jk}\frac{\partial^2 v_k}{\partial t^2} + \overline{\tau}'_{jk}\frac{\partial u_k}{\partial t} + \overline{\beta}'_{jk}\frac{\partial v_k}{\partial t} + \overline{\tau}''_{jk}u_k + \overline{\beta}''_{jk}v_k\right) \end{aligned} \tag{2.21}$$

The wake-induced flutter of a cylinder in the wake of a fixed upstream cylinder (Fig. 2.17) may be analyzed by the unsteady theory above. Assuming the downstream cylinder lift and drag coefficients C_L and C_D are functions of position

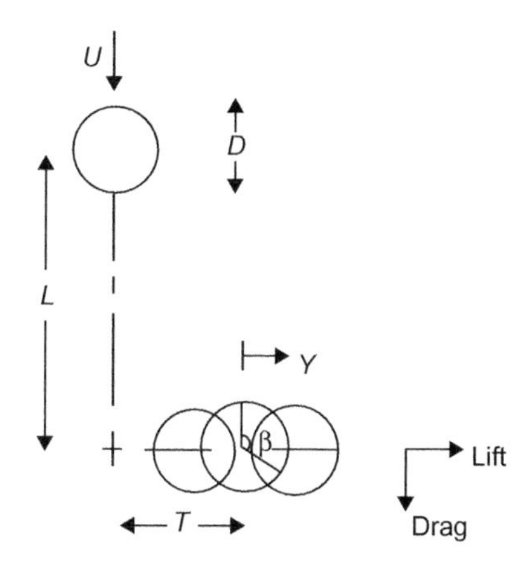

FIGURE 2.17

Cylinder configuration and coordinates.

only, the stability boundary for the downstream cylinder for small *P/D* spacings is given by:

$$\frac{U}{fD} = 3.54\left(\frac{P/D}{-(\partial C_{L2}/\partial\beta)\sin\theta}\right)^{0.5}\left(\frac{2\pi\zeta m}{\rho D^2}\right)^{0.5} \tag{2.22}$$

For the general problem, however, where cylinder dynamical behavior has been classified based on in-line and transverse spacing (Fig. 2.12) much work remains to be done.

2.2.3.2.4 Quasi-steady theory

Price et al. [76] have proposed the quasi-steady model in which the relative velocity vector between the moving cylinder and the flow is taken into consideration when determining the fluid forces acting on the cylinder. The underlying quasi-static assumption is valid for high reduced velocities but may lead to significant error at low reduced velocities. The force coefficients in the effective lift and drag directions are given by:

$$\begin{aligned} C_{\mathrm{x}} &= (V_r^2/V^2)\{C_{\mathrm{D}}\cos\alpha + C_{\mathrm{L}}\sin\alpha\} \\ C_{\mathrm{y}} &= (V_r^2/V^2)\{C_{\mathrm{L}}\cos\alpha - C_{\mathrm{D}}\sin\alpha\} \end{aligned} \tag{2.23}$$

where V_{r} is the magnitude of the flow velocity relative to the cylinder. Blevins [77] suggests that a reasonable limit value of the reduced velocity for design based on the quasi-static assumption is $V_{\mathrm{r}} > 10$. However, this latter value does not take into account the changes in the fluid force orientation during cylinder motion. This suggested limit is therefore currently not recommended by the authors.

Table 2.9 Evaluation and classification in steady flow

Two cylinders in parallel	Two cylinders in tandem
For $L/D = 1.5$, added mass is 20% higher than that for an isolated cylinder	For $2 < L/D < 3$ the added mass is slightly lower than isolated cylinder case
For $3 < L/D < 4$ proximity interference between the cylinders is negligible; each cylinder may be treated as an isolated body	For $3 < L/D < 4$ proximity interference between the cylinders can be ignored as a first approximation; however, added mass and drag forces, which increase by 20–50%, must be considered

Table 2.10 Evaluation and classification for sinusoidal flow

Equal-diameter cylinders	$P/D < 5$	$V_r < 1$	No self-excited vibration occurs
		$V_r < 3.3$	Lift-force lock-in does not occur
	Use mass-damping parameter C_n	$C_n > 64$	All self-excited vibration is eliminated
		$C_n > 2.5$	In-flow direction self-excited vibration is eliminated
Unequal-diameter cylinders	Diameter ratio > 5	Isolated cylinder approximation valid	Vibration of larger cylinder damped as a function of smaller cylinder position
Criss-crossed cylinders	Take $Z/D > 1$ (Z = cylinder separation)	Isolated cylinder approximation valid	See Section 2.1.3

Cylinder pair dynamics in steady and oscillating flows have been presented in the foregoing. The key results are summarized in Tables 2.9 and 2.10, respectively.

2.2.4 Examples of practical problems

1. The majority of cases reporting on vibrations due to steady flow have been for cable-stayed bridges and power transmission lines. These may, however, be treated as isolated structures in most cases and are therefore not presented here. Few problems caused by oscillating flow have been reported. However, there are cases where large added mass values have been found depending on cylinder diameter ratio (Table 2.10). In such cases, it is prudent to increase the structural strength by 50% relative to the isolated cylinder case. Vibrations have also been reported in the case of criss-crossed pipes. As a countermeasure, $Z/D > 1$ is recommended for design. In single-phase flow, special care should be taken when Z/D is of the order of $0.2 \sim 0.3$ since severe vibrations could potentially occur.

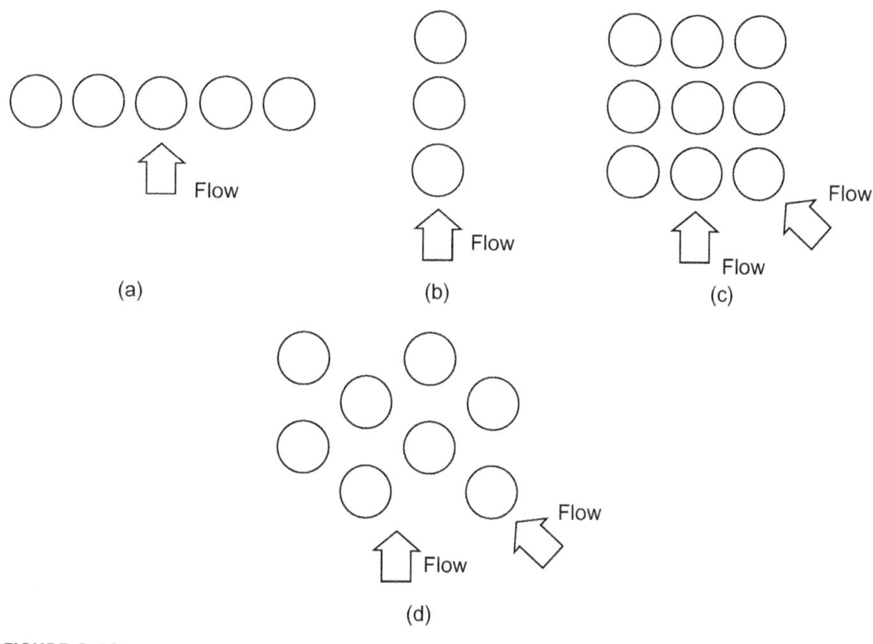

FIGURE 2.18

Tube array patterns: (a) tube row, (b) tube column, (c) square tube array, and (d) triangular tube array.

2. Vibrations in two-phase cross-flow and countermeasures: in bubbly flow, fluid damping continuously changes, making it difficult to adjust the reduced damping. Theoretical analysis has not yet progressed.

When Kármán vortex shedding-induced vibrations occur, large-amplitude motions result even away from the lock-in flow rate. Operating conditions (flow rates) should therefore be carefully monitored to stay far from this 'extended' lock-in regime.

2.3 Multiple circular cylinders

2.3.1 Outline of structures considered

FIV problems of structures that consist of multiple circular cylinders are described in this section. There are many cases encountered in practical products, but existing studies and evaluation methods are limited to a few cases. Only the four configurations of cylinder groups shown in Fig. 2.18 are treated in this book. When one is faced with a different pattern, it is recommended to adapt the most similar pattern. Although structures having complex forms, such as spiral

U-bends, are usually encountered in practice, only simple structures are considered in this section. The reason for this is the fact that nearly all experimental tests have been done on only simple straight structures. For complex structures the procedure presented in Section 1.2 is recommended.

2.3.2 Vibration evaluation history

2.3.2.1 Excitation mechanisms

For groups of cylinders subjected to cross-flow, three types of potential vibration mechanisms exist, as shown in Fig. 1.8 [78]:

1. *Vortex-induced vibration (due to Kármán vortex shedding)*: this phenomenon is the same as that encountered for single cylinders and discussed in Section 2.1.
2. *Self-induced oscillations (fluidelastic vibration)*: arrays which consist of multiple cylinders can oscillate with large amplitudes when they are exposed to high-velocity fluid flow. This phenomenon is not observed for a single cylinder. This is the most dangerous vibration mechanism and the root cause of many problems in industry.
3. *Random vibration*: vibration is considered random when it does not correspond to either of the two phenomena above. This therefore includes the excitation by vorticity generated by upstream structures. In the latter case, the random force excitation spectra will contain periodic components related to upstream vorticity shedding.

2.3.2.2 Historical background

In many heat exchangers, heat transfer tubes are designed to work in cross-flow. The FIV problem of circular cylindrical structures is strongly associated with FIV problems for heat exchanger tubes. Numerous problems, including tube rupture and tube surface wear, have been encountered for many years. Therefore, the design procedure includes some experience-based knowledge, such as keeping span lengths short, to avoid severe accidents [79]. However, this knowledge has been kept as closely guarded corporate know-how.

Vortex shedding, as explained in Section 2.1, not only occurs for single cylinders, but is also observed in cylinder arrays. In 1970, Y.N. Chen [80] proposed a map of the non-dimensional vortex flow periodicity frequency based on the cylinder array geometry pitch spacing. This map has become an important reference for designers.

In 1969, Connors, in his doctoral dissertation [81], introduced a phenomenon where a tube array could become a self-excited system capable of absorbing energy from the fluid flow. Connors then introduced a simple formula to evaluate the stability of the system by assuming the fluid excitation force to be a linear function of the tube displacement. His theory on this instability, later named fluidelastic vibration, was adopted and expanded in the 1970s and 1980s. Earlier on, this phenomenon had been analyzed and confirmed experimentally by Roberts [82] in 1966. However, it was Connors who introduced a simple formula for the instability which has been adopted in the design and development of nuclear power plants over the years (Table 2.11).

Other researchers who have worked on fluidelastic instability include Price and Païdoussis, Weaver and Lever, and Chen and his colleagues. Price and Païdoussis [83] developed a quasi-static model for the excitation force while Weaver and co-workers [84] modeled fluidelastic instability by relating it to the movement of the separation point on the tube surfaces. Chen [78] proposed a model expressing the fluid force as a first-order function of the tube displacement, velocity, and acceleration. Tanaka [85] presented the measured data for this function. This method has the weak point, that any tube array needs the measured excitation force data for the estimation; however, the model is considered theoretically sound.

Based on these studies, the first model by Connors is now known to correspond to one mechanism, called the displacement mechanism, among the two mechanisms underlying fluidelastic vibration. The second mechanism is called the velocity mechanism. This mechanism cannot be expressed by a simple formula such as that proposed by Connors.

Due to the lesser importance of the random excitation forces, there are only a few papers published on the subject. In 1973, Pettigrew and Gorman [86]

Table 2.11 History of research on FIV of tube arrays

Year	Works and studies (Lead Author)
1966	Proposal of jet switch mechanism (Roberts)
1968	Strouhal number map for tube arrays (Y.N. Chen)
1969	Introduction of fluidelastic instability (Connors)
1970	Guideline for tube wear in heat exchangers (Thorngren)
1972	Strouhal number map for tube arrays (Fitz-Hugh)
1973	Random force and two-phase flow estimation (Pettigrew)
1980	Measurement of fluidelastic forces (Tanaka and Takahara)
1981	Fluidelastic model related to displacement and velocity (S.S. Chen)
1982	Fluidelastic model due to movement of separation point (Lever and Weaver)
1983	Quasi-static fluidelastic force model (Price and Païdoussis)
1984	Non-existence of vortex shedding in two-phase flow (Hara)
1987	Vibration test in steam–water two-phase flow (Axisa)
1988	Correlation estimation of random force (Au-Yang)
1992	Introduction of two instability mechanisms in two-phase flow (Nakamura)
1994	Revision of ASME's guideline for FIV (Au-Yang)
1996	Measurement of fluidelastic force in steam–water two-phase flow (Mureithi et al.)
1997	Measurement of fluidelastic force in air–water two-phase flow (Inada)
2000	U-bend tube bundle model test using freon fluid (Takai et al.)
2001	JSME guideline for fluidelastic vibration in steam generators (Iwatsubo et al.)

presented estimated random force spectra, which showed dependence on the location of the tube-within-tube array. These data are incorporated in the ASME design guideline. Although the correlation evaluation of the excitation force along the tube axis is required for random vibration analysis, there had been no work reported on this subject until Au-Yang [87] published his study in 1999. To date, available data on the subject remain very limited.

Studies cited above pertain to vibration in single-phase flow. However, many tube arrays in heat exchangers are subjected to gas–liquid two-phase flow. Faced with this reality, Pettigrew and co-workers [88] started to study the two-phase FIV problem in 1973. They focused primarily on fluidelastic vibration and random vibration; vortex-induced vibration was not considered since two-phase flow in many heat exchangers does not produce stable periodic vortex shedding. Their approach was based on methods similar to those employed for single-phase flow. However, the idea that the vibration induced by two-phase flow could be treated with this method was original. When the void fraction of the two-phase flow is low enough, vortex shedding can be observed. However, in this case the flow can be approximately treated as single-phase, as shown by Hara and Ohtani [89].

Initial studies in the 1970s were based on air–water two-phase flow tests at atmospheric conditions. However, many heat exchangers operate at high temperature and high pressure conditions. Axisa et al. [90] have measured the two-phase flow damping in steam–water two-phase flow tests. Nakamura et al. [89] proposed an original estimation method for the two-phase flow instability velocity based on measured data by tests in steam–water two-phase flow. This theory proposes a unique estimation method for the fluidelastic instability boundary, and for the random excitation force caused by gas–liquid two-phase flow. Two possible instability boundaries for the absolute instability and intermittent instability are predicted.

In the 1990s, freon two-phase mixtures were used to simulate the steam–water flow, because they are easier to handle due to low temperature and pressure operating conditions. A large-scale vibration test [92] of the partial model of a prototypical steam generator has been performed using freon.

On the other hand, measurement of the fluidelastic forces, which has successfully been achieved in single-phase flow, has been attempted by several researchers, including Inada et al. [93] and Mureithi et al. [94]. However, the measurement of fluidelastic forces in two-phase flow remains a difficult challenge.

Many papers continue to be published on fluidelastic instability even in the twenty-first century. In 2012, a new problem was reported for a steam generator, where U-bend tubes vibrated in the in-plane direction, resulting in tube-tube impacting and significant fretting wear. The vibration has been attributed to fluidelastic instability. In this case, however, instability occurred (apparently purely) in the in-flow or drag direction. In view of the resulting tube damage, it is clear that in-flow instability can be as strong as the more classically encountered (primarily) transverse direction fluidelastic instability. Research on purely in-flow fluidelastic instability has only recently been reported [95–97]. The instability is yet to be widely considered in equipment design guidelines.

2.3.3 Estimation method

2.3.3.1 Vibration in single-phase flow

2.3.3.1.1 Steady flow

Vortex-induced vibration. When vortex shedding occurs in the flow, a periodic excitation force acts on the cylinder. The force frequency, f_W, can be expressed in the form of Eq. (2.1). The non-dimensional number, St, in this equation has been experimentally determined for tube arrays and is shown in Fig. 2.19 [98]. The Strouhal number is known to be a function of the tube pitch ratio.

For isolated cylinders, the Strouhal number, St, is known to be a function of the Reynolds number, Re. However, for tube arrays, no work has been reported on the variation of the Strouhal number with Reynolds number. Equation (2.1) corresponds to the case of lift direction (transverse) excitation. However, other possibilities, for example drag direction (in-flow) excitation at double the transverse frequency and symmetric vortex shedding with much larger Strouhal number, are often encountered for single cylinders. These other frequencies are expected to be considered in new designs. In addition, a Strouhal number map for finned-tube arrays has been proposed [99].

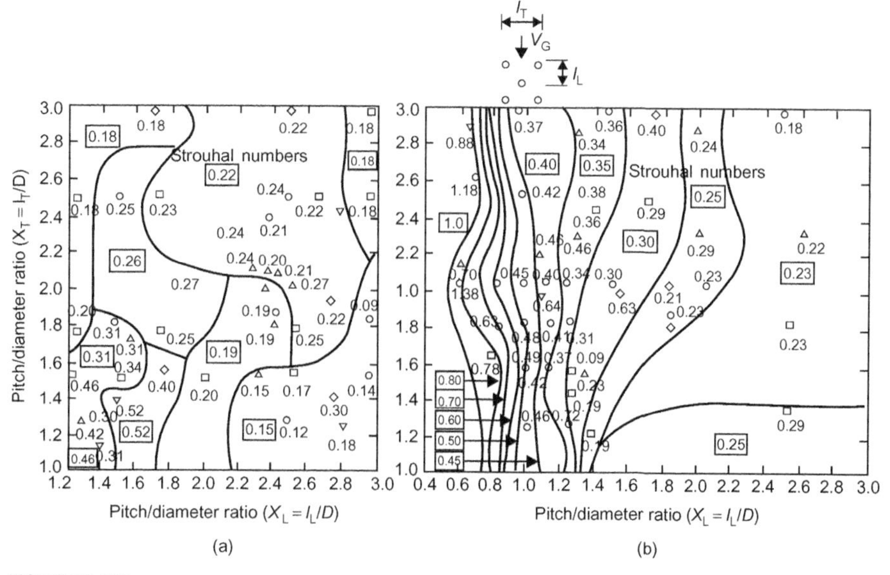

FIGURE 2.19

Strouhal numbers for tube arrays related to pitch/diameter ratio: (a) in-line array, and (b) staggered array.

The lock-in phenomenon can also be observed in tube arrays, much like for the single cylinder, where the vortex shedding frequency, f_W, synchronizes to the cylinder's frequency, f_S. However, only limited data for tube arrays exist. To avoid lock-in the following empirical equation is suggested for design usage in the ASME standard [100]:

$$0.7\ (or\ 0.8) \leq \lambda \leq 1.3\ (or\ 1.2) \quad \lambda = f_W/f_S\text{: Frequency-ratio} \tag{2.24}$$

The tube response, in the lock-in-region, is estimated in a manner similar to that for an isolated cylinder. The excitation fluid force, F, is assumed to be sinusoidal and is defined using measured non-dimensional force coefficients, c_L, c_D, as follows:

$$F = \begin{cases} \frac{1}{2}\rho c_L V_G^2 DL \quad \sin(2\pi f_W t) \\ \frac{1}{2}\rho c_D V_G^2 DL \quad \sin(2\pi f_W t) \end{cases} \tag{2.25}$$

Besides the data published by Chen [78], limited information on the force coefficients c_L and c_D exists.

When the excitation force is assumed to be independent of cylinder motion, the response of the tube can be obtained by direct solution of the equation of motion. The analysis is generally quite complex, thus requiring numerical computation. The simple case of a simply supported tube is estimated with the following equation:

$$X_d(x) = H(j)X_s(x) \tag{2.26}$$

where $X_d(x)$ is the root-mean-square-response at the position x, $X_s(x)$ is the deformation when the excitation force statically acts on the tube, and $H(\lambda)$ is the transfer function of the tube given by:

$$\left(H(\lambda) = \left[\frac{1}{(1-\lambda^2)^2 + (2\zeta\lambda)^2} \right]^{1/2} \right)$$

Fluidelastic vibration (self-excited vibration). The self-excited vibration of tube arrays due to cross-flow was named 'fluid-elastic vibration' by Connors [81]. He proposed a simple criterion for the occurrence of this instability by introducing a non-dimensional factor K as shown in the following Eq. (2.27) and Fig. 2.20:

$$\frac{V_c}{f_S D} = K \left[\frac{m\delta}{\rho D^2} \right]^{1/2} \tag{2.27}$$

Here, V_c is the critical flow velocity. The gap flow velocity is usually used.

Tube array	a	b	δ_s
Tube row	1.35 (*T*/*D*-0.375)	0.06	$0.05 < \delta_s < 0.3$
	2.30 (*T*/*D*-0.375)	0.5	$0.3 < \delta_s < 4.0$
	6.00 (*T*/*D*-0.375)	0.5	$4.0 < \delta_s < 300$
Square array	2.1	0.15	$0.03 < \delta_s < 0.7$
	2.35	0.5	$0.7 < \delta_s < 300$
Rotated square array	3.54 (*T*/*D*-0.5)	0.5	$0.1 < \delta_s < 300$
Triangular array	3.58 (*T*/*D*-0.9)	0.1	$0.1 < \delta_s < 2$
	6.53 (*T*/*D*-0.9)	0.5	$2 < \delta_s < 300$
Rotated triangular array	2.8	0.17	$0.01 < \delta_s < 1$
	2.8	0.5	$1 < \delta_s < 300$

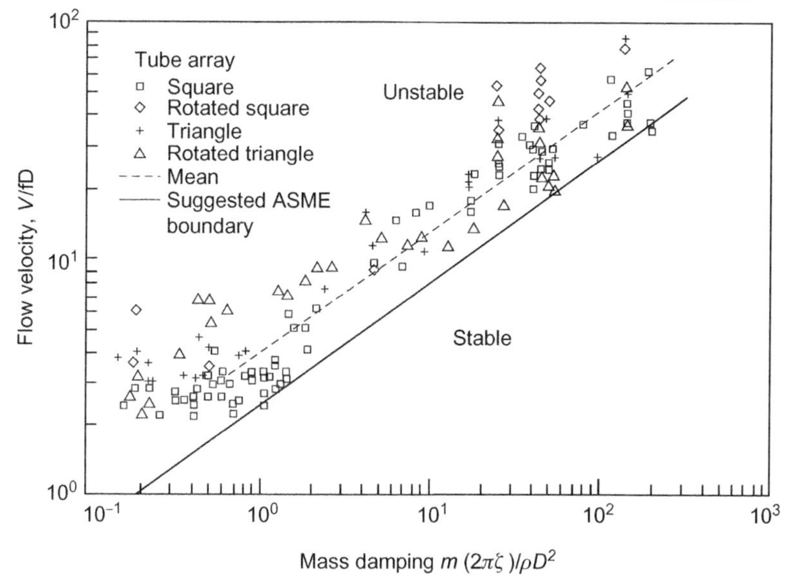

FIGURE 2.20

Example of stability boundary for fluidelastic instability.

There are two major estimation methods for fluidelastic instability:

1. Method based on measured fluid force data:
   The fluid forces are measured experimentally [85] or determined by numerical simulations [73] using CFD. Assuming the fluid forces to be functions of cylinder displacement, velocity, and acceleration, the measured forces are estimated in terms of the coefficients of Eq. (2.21). This method is considered the most accurate.

2. Simplified estimation method for designers:

Chen [78] proposed the criterion below for the occurrence of fluidelastic instability based on many experimental data. Since he used the lowest flow velocity among the available data, this formula gives a conservative value for the critical velocity.

$$\frac{V_c}{f_S D} = a\left[\frac{m_V \delta_V}{\rho D^2}\right]^b \tag{2.28}$$

The constants a and b are given in the table in Fig. 2.20, where T is the array pitch, and δ_s the non-dimensional mass-damping parameter, given by:

$$\delta_s = \frac{m_V \delta_V}{\rho D^2}$$

The subscript "v" indicates the value in vacuum. There is ongoing debate on which values should be used for the criterion, values in vacuum or those in practical flow conditions. Since the critical factor K depends on the definition of these parameters, care must be exercised in their definition. One should especially be careful about the fact that the natural frequency of tubes in liquid has many coupled values due to the fluid–structure interaction effect.

In addition, the tube damping ratio depends on support conditions. Table 2.12 shows examples; the values used will usually be average values.

Random vibration. The random excitation fluid force originates from the turbulent flow both due to the upstream flow turbulence and due to turbulence generated in the tube array, which includes vortex-type flow. The upstream turbulence depends on the situation of the tube array. To estimate a general trend of the effect of the turbulent flow, the turbulence intensity, TI, is defined as the ratio of the fluctuating flow velocity components to the average flow velocity. Figure 2.21 shows measured data [78] for the non-dimensional frequency spectra of the fluid force coefficients c'_D, c'_L (where c'_D, c'_L correspond to the drag force

Table 2.12 Natural frequency and average damping ratio of circular structures

Boundary condition	Approximate fundamental natural frequency	Moderate damping ratio ζ_v in vacuum
Welded at both ends	$f_v = \frac{(4.73)^2}{2\pi L^2}\left(\frac{EI}{m_v}\right)^{1/2}$	0.05–0.2%
With gap at both ends	$f_v = \frac{\pi}{2L^2}\left(\frac{EI}{m_v}\right)^{1/2}$	1.0–1.5%
Welded at one end and with gap at the other	$f_v = \frac{(3.93)^2}{2\pi L^2}\left(\frac{EI}{m_v}\right)^{1/2}$	Between above values

and the lift force, respectively) as functions of Reynolds number. These coefficients are defined similarly to the force coefficients in Eq. (2.25).

When the random fluid force F' is estimated, the response of the tubes can be calculated using computational schemes based on random vibration theory. However, a simple equation similar to Eq. (2.6) can be derived by making some assumptions. The fluid excitation force is assumed uniform along the tube axis, the tube oscillates in one fundamental vibration mode, and the fluctuating fluid force has a constant value near this fundamental frequency. The root-mean-square-response $\overline{X}_d$ at the midpoint of a simply supported tube is then given by the following equation:

$$\overline{X}_d^2(x = L/2) = \frac{S_F(\omega_1)}{4\pi^5 f_1^3 m^2 h_1} \tag{2.29}$$

The maximum response is estimated to be the value $\overline{X}_d$ multiplied by a factor between 3 and 5.

2.3.3.1.2 Non-uniform flow

Vortex shedding. In the case of non-uniform flow, we similarly estimate the frequency of the periodic vortex shedding by Eq. (2.1). However, the frequency changes, depending on axial location, since the flow velocity is not uniform. When the flow velocity greatly varies with axial location, no resonance may be observed even though there may be a coincidence of the tube natural frequency with the shedding frequency over a narrow area on the tube. Designers must treat this problem according to the following method, when there is no prior experience.

Consider the case where the flow velocity changes in steps with spanwise location. The flow region can be divided into several, say k, intervals. As the

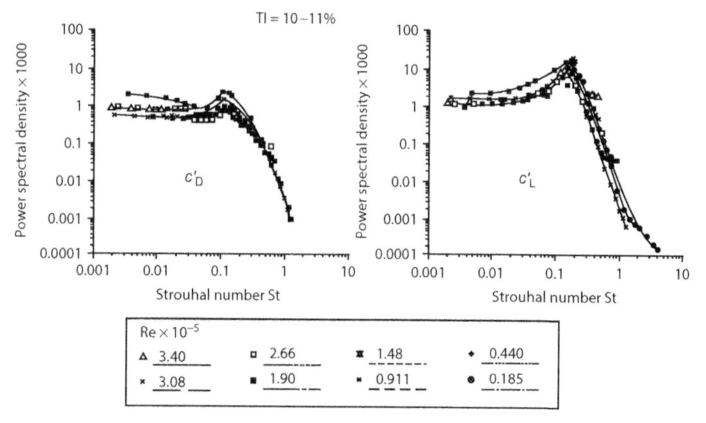

FIGURE 2.21

Example of measured random force acting on array of circular tubes.

equation of motion due to the vortex shedding is assumed linear, the response of the tube can be calculated as the sum of the maximum response components, ${}_kX_d(x)$, due to the excitation forces, F_k, at each interval k.

$$X_d(x) = \sum_{k=1}^{k} {}_kX_d(x) \tag{2.30}$$

The reason for using the sum of maximum responses comes from the fact that the excitation frequencies due to vortex shedding differ from interval to interval. Considering the simultaneous occurrence of maximum vibration response over each interval therefore leads to a conservative estimate for design safety.

Fluidelastic vibration. Eq. (2.27) is valid for uniform flow. For non-uniform flow, Connors [101] proposed the following Eq. (2.31) for the stability ratio SR based on an energy balance analysis, assuming the excitation force to be a linear function of the tube displacement. Although this assumption is not valid in some cases, Eq. (2.31) is commonly used.

$$S_F(x,\omega) = \left(\frac{1}{2}\rho c' V_G^2(x)\right)^2 \overline{S}_F(\omega)$$

$$V_e = \left[\frac{\int_0^L \frac{\rho(x)}{\rho_{\text{average}}} V_G^2(x)\phi^2(x)\,dx}{\int_0^L \frac{m(x)}{m_{\text{average}}}\phi^2(x)\,dx}\right]^{1/2} \tag{2.31}$$

V_e is called the effective flow velocity, and V_c is given by Eq. (2.27). Even if the average mass of the tube, m_{average}, and the average fluid density, ρ_{average}, are used in Eq. (2.27), these average values cancel out in Eq. (2.31). When the value of SR defined in Eq. (2.31) is smaller than 1, the system is stable; when it is larger than 1, the system will be unstable. The latter condition should be prevented by employing appropriate countermeasures (see Section 2.6).

Random vibration. Eq. (2.29) is not mathematically valid in the case of non-uniform flow because the power spectral density (PSD) of the excitation force, $S_F(\omega)(=2G(\omega))$ depends on the axial location along the tube. To avoid this problem, the following two approximate methods are applied:

1. The frequency composition of the excitation force is considered to be constant even if the flow velocity changes with location. Thus, only the magnitude of the excitation force is considered in the response calculation using the following equation:

$$S_F(x,\omega) = \left(\frac{1}{2}\rho c' V_G^2(x)\right)^2 \overline{S}_F(\omega) \tag{2.32}$$

   Here, $\overline{S}_F(\omega)$ is the non-dimensional PSD of the excitation force.
2. Dividing the span into intervals, $k = 1, 2\ldots$, along the cylinder within each of which the PSD of the excitation force is considered uniform, the average

square of the response, ${}_k\overline{X}_d^2$, in each interval k, is calculated. Finally, the total response is obtained by summation, thus:

$$\overline{X}_d^2(x) = \sum_{k=1} {}_k\overline{X}_d^2(x) \tag{2.33}$$

2.3.3.2 Vibration induced by two-phase flow

The vibration phenomenon in two-phase flow depends on the so-called flow regime or flow pattern. Detailed explanation on the flow regime will not be given here, since the two-phase flow regimes are well known in the field of thermal-hydraulics. Although it is not an easy task to determine the flow regimes, Grant [102] and Baker [103] have proposed flow regime maps for cross-flow in tube arrays. The void fraction (the volume fraction of gas in the mixture) is one of the quantities used to identify the flow regime. Air–water two-phase flow shows the following flow regimes as the void fraction is progressively increased from low values:

- Bubbly flow (mainly liquid flow mixed with bubbles).
- Intermittent flow (strongly unsteady flow, slug flow, froth flow, or annular flow).
- Mist flow (mainly gas flow mixed with liquid mist).

2.3.3.2.1 Vortex shedding

Hara and Ohtani [89] reported that periodic vortex shedding is observed in the low void fraction regime, below 15%, as in the case of single-phase flow. However, no periodic vortex shedding has been observed at higher void fractions. Although there are no data in the extremely high void fraction range, the possibility of periodic vortex shedding should be considered in the mist flow regime.

Vortex shedding can therefore be neglected in the void fraction range of 15–95%, but its effect must be estimated otherwise, even in two-phase flow conditions. The estimation method is similar to that for single-phase flow, but use is made of the following homogeneous density and flow velocity for two-phase flow:

$$\begin{cases} \rho = \rho_g\alpha + \rho_l(1-\alpha) \\ V_G = j_g + j_l \end{cases} \tag{2.34}$$

Here, j_g and j_l are the superficial gas and liquid flow velocities, respectively, which are defined by the flow velocities assuming each fluid flows as a single-phase flow through the total flow path.

Damping in two-phase flow is much greater than that in single-phase flow, as shown in Fig. 2.22 [86]. In the figure, the volumetric fraction β is used as the reference parameter instead of the void fraction α. The designer should use this higher damping value for two-phase flow.

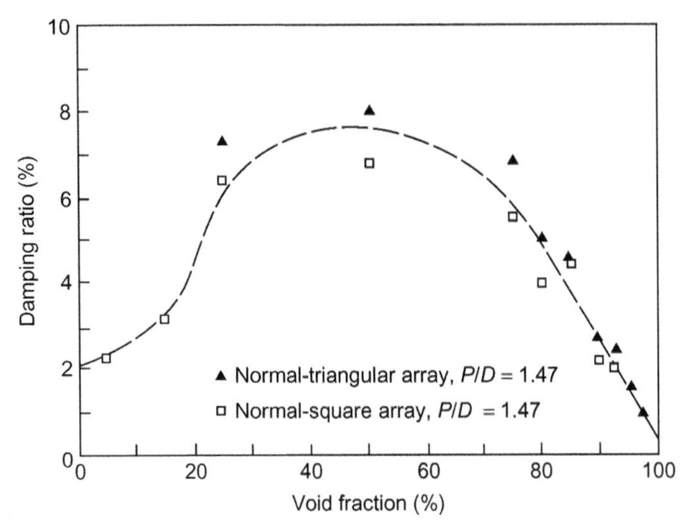

FIGURE 2.22

Example of measured damping ratio for tubes in two-phase flow.

2.3.3.2.2 Fluidelastic vibration

Fluidelastic instability can occur even in two-phase flow. The instability does not depend on the flow pattern. However, it is not easy to judge the occurrence of instability in the intermittent flow regime due to the severe random excitation, as explained later.

Generally, the Connors equation (2.27) is used by experimentally estimating the critical factor K for two-phase flow. Pettigrew et al. [104] recommend the factor $K = 3.0$ for all tube arrays.

For the estimation of the two-phase flow-induced instability, values used for the tube frequency, damping ratio, and mass are those corresponding to two-phase flow conditions, instead of those in vacuum. The damping ratio is estimated to be the sum of the two-phase flow damping ζ_{TP} (shown in Fig. 2.22) and the structural damping of the tube in vacuum ζ_V. The frequency of the tube is that in two-phase flow, the mass per unit length of the tube m includes the inertia effect of the surrounding two-phase fluid.

In addition, a design criterion using the critical factor $K = 7.3$ in Connors' equation for U-bend tubes is proposed by the JSME [105], based on the experimental data of a large-scale test.

2.3.3.2.3 Random vibration

The most important difference between two-phase flow and single-phase flow is the random buffeting force in the two-phase intermittent flow regime. When not in the intermittent flow regime (e.g. in bubbly flow or mist flow), the vibration estimation is similar to that for single-phase flow except for the fluid density.

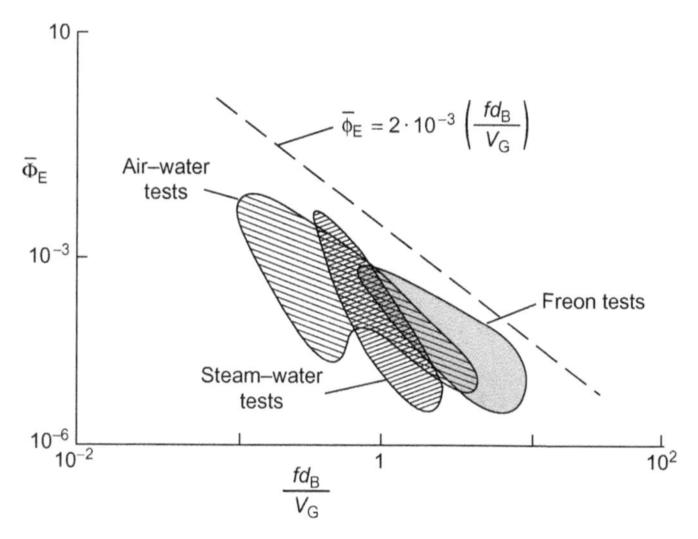

FIGURE 2.23

Example of proposed random force for liquid–gas two-phase flow [106].

A smaller response may be obtained due to the larger damping ratio in two-phase flow. However, the large excitation force must be considered in the intermittent flow regime.

There are two excitation forces proposed by researchers:

General power spectral density (PSD). A non-dimensional PSD [106] of the two-phase flow excitation force is proposed as the following equation. This equation was determined empirically considering numerous experimental data (Fig. 2.23). It is therefore valid for the majority of cases.

$$\begin{cases} \overline{\Phi}_E(f_r) = 10 f_r^{-0.5} \quad (\text{for } 0.06 \geq f_r) \\ \overline{\Phi}_E(f_r) = 2 \cdot 10^{-3} f_r^{-3.5} \quad (\text{for } 0.06 \leq f_r) \end{cases} \tag{2.35}$$

Here, f_r is the non-dimensional frequency, defined as fd_B/V_G, based on the bubble diameter d_B.

PSD in the intermittent flow regime. For the intermittent flow regime [91] only, where the excitation force is large, the equation below is proposed, based on the practical flow velocity in this regime. From the experimental fact that the excitation force has a linear relation with the liquid slug velocity V_S, the PSD of the excitation force due to the liquid slugs is modeled as shot noise.

$$G_F(\omega) = \frac{1}{2} S_F(\omega) = \frac{4\nu}{\omega^2 + \nu^2} \tag{2.36}$$

Here, ν is the measured damping value of the impulse response. Depending on the flow regime (2.35) or (2.36) should be substituted into Eq. (2.32).

Table 2.13 Examples of Tube Failures within Tube Banks

		Example	Excitation mechanism	Solution
1.		Vibration and impact of tube in U-bend tube bundle to shell	Fluidelastic vibration $\frac{U}{fD} = 1.9 - 5.2$ $\frac{m\delta}{\rho D^2} = 0.35$	Remove 4 outer tubes, insert wire, set impingement baffle, and increase flow path between shell and bundle
2. Gas		Tube wear after 35 h, without support	Mainly by vortex shedding, but possibility of acoustic resonance in shell	Change frequency by inserting bars between tubes
3. Water		Tube rapture in three units at the same position	Possibly vortex shedding, not fluidelastic vibration	Plug the raptured tube and restrict flow volume
4. Water	Damaged Fuel Rods; Flow Path; Centre Baffle Plate; Fuel Rods; Corner Baffle Plate	Rapture of a few fuel rods at a corner	Initially random vibration from leakage flow at the corner suspected, but eventually determined to be fluidelastic vibration	Decrease gap in plate

2.3.4 Examples of component failures

Although few examples have been published despite the many problems encountered in industry, a large amount of data on this problem exists, especially for heat exchangers. Table 2.13 shows some examples [107] that can serve as a reference for designers.

2.4 Bodies of rectangular and other cross-section shapes

In this section, we discuss FIV caused by flow around structures having rectangular and rectangular-like cross-sections commonly encountered in mechanical engineering. In addition, we briefly touch on the FIV of wing-shaped cross-section guide vanes and airplane wings, as well as bridges and towers in civil engineering. Many FIV studies in aerospace and civil engineering are conducted because safety is a primary priority in these engineering fields. References in the fields (for example: [108–111]) are invaluable in order to avoid FIV issues for bodies having the type of cross-sections cited above.

2.4.1 General description of cross-section shapes

At first, several examples of products treated in this section are given for a better understanding of the objective of the study. Next, the flow conditions and cross-section shapes are presented in order to identify the different FIV mechanisms.

2.4.1.1 Products focused on

In this section, we focus on mechanical engineering structures subjected to flow in power plants and civil engineering structures subjected to wind, river flow, and tides. Furthermore, we limit ourselves to components having simple rectangular geometry or structures common in mechanical engineering.

2.4.1.2 Classification based on fluid flow

Single-phase flow is ubiquitous. Whether the fluid is gas or liquid, the flow behaves similarly if all non-dimensional numbers, such as Reynolds number, are nearly equal.

On the other hand, gas–liquid two-phase flow in power plants has different characteristics from a single-phase flow and must be treated in a special way. There are few studies on two-phase FIV for bodies having the type of cross-sections discussed in this section. The effect of two-phase flow is therefore not discussed here.

Broadly, two types of single-phase flow conditions may be considered. The first condition is steady flow. The second is unsteady flow where the flow varies (oscillates) sinusoidally in the flow direction.

There are only a few studies on FIV in oscillating flows. Such flows may not be important in actual applications for the bodies treated here; hence, they are not considered.

It may also be necessary to consider the influence of sheared flows. Significant research has been done on such flows in marine engineering. These flows are, however, of less importance in mechanical engineering. FIV specifically related to sheared flows are therefore not discussed.

2.4.1.3 Classification by vibration mode of structure

In classifying the vibration of rectangular-cross-section structures, two types of vibration are focused on (Fig. 2.24). The first type involves the vibration modes restricted 'within' the cross-section plane. In the second type, vibration is not restricted to the cross-section plane. Vibration modes normal to the cross-section (e.g. airplane wing beam modes) are excited. The former vibration, within the rectangular-cross-section plane, includes lateral vibration, rotational vibration, and combined lateral–rotational vibration.

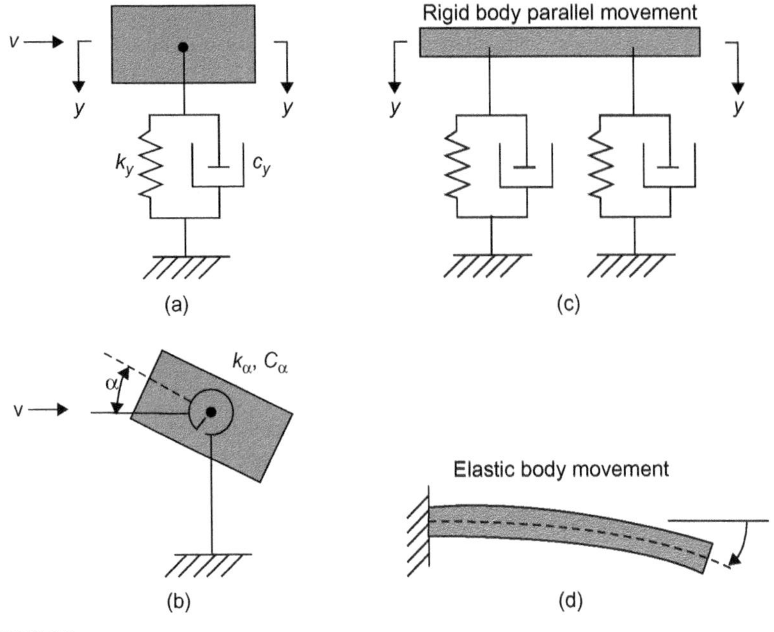

FIGURE 2.24

Vibration modes for various types of structures: (a) parallel vibration, (b) rotational vibration, (c) in plane vibration, and (d) out of plane vibration.

2.4.1.3.1 Two-dimensional vibration within the cross-section plane

When the rotational stiffness is comparatively large, as is often the case for mechanical components and civil engineering structures such as chimneys and buildings, the rotational vibration caused by fluid dynamic rotational moments may be small and therefore negligible. In this case, only vibrations transverse to the flow and in the in-flow direction are of importance.

On the other hand, for civil engineering structures such as bridges, or aircraft wings in aeronautical engineering, the mass per unit length is typically not large. In this case, the rotational stiffness is not significantly larger than the lateral stiffness. The rotational moment induced by fluid flow is then no longer negligible. In addition to lateral vibrations, rotational and coupled lateral–rotation vibrations can also occur.

2.4.1.3.2 Axial vibration

In some cases, it is necessary to conduct multi-degrees-of-freedom axial vibration analysis, considering the effects of multi-span or long-span structure mode shapes. In addition, for some structures, the cross-section varies in the axial direction. Surface properties (e.g. roughness) may also vary axially. In this case, multi-dimensional vibration analysis using, for example, finite element methods, transfer matrix methods, etc. must be conducted. Besides the lateral–rotational vibration discussed above, shell-type vibration problems of rectangular-cross-section bodies can arise. These special vibrations are not discussed here.

2.4.1.4 Objectives

In this section, we shall mainly discuss the vibration phenomena of isolated rectangular-cross-section bodies subjected to steady cross-flow. Such vibrations are commonly encountered in industry. Typical applications of vibration phenomena of two rectangular-cross-section-bodies [112] in tandem and vortex-induced vibration problems of a single row of rectangular bodies in cross-flow with or without downstream splitter plates [113,114] exist. However, these two special cases will not be discussed.

2.4.2 FIV of rectangular-cross-section structures and historical review

2.4.2.1 FIV phenomena

An isolated rectangular-cross-section body may undergo three types of vibrations: vortex-induced vibrations, turbulence-induced vibrations, and galloping.

2.4.2.1.1 Vortex-induced vibration

Around a rectangular-cross-section body, vortices are shed at the corners and flow downstream. The behavior of the vortices is similar to that observed in the case of a circular cylinder. The frequency of Kármán vortex shedding behind a rectangular-cross-section body is affected by the body's aspect ratio and the flow attack angle.

When the natural frequency of the body is close to the vortex shedding frequency, lock-in of the shedding frequency to the structural frequency may occur. Lock-in is possible transverse to the flow direction as well as in the flow direction.

Careful examination is needed for transverse vibration since the vibration can be large enough to cause structural failure. Various kinds of vibration can be observed depending on the aspect ratio of the cross-section, the angle of attack, and the vibration direction (parallel/normal to the flow [112]). There is a lack of comprehensive and systematic knowledge on the effects of large parameter changes, and variations in configuration and flow conditions.

2.4.2.1.2 Turbulence-induced vibration

Similar to a circular cylinder, a rectangular-cross-section structure vibrates mainly due to turbulence when the structure's natural frequency is well separated from the dominant vortex-shedding frequencies. Turbulence-induced vibration amplitudes are generally larger than amplitudes for a circular cylinder. In addition, the vibration varies with the strength of upstream flow turbulence.

2.4.2.1.3 Galloping

Suppose that an elastically supported rectangular-cross-sectional body subjected to cross-flow has an attack angle to the cross-flow. The body, subjected to a static lift force, drag force, and torsional moment in cross-flow, will normally vibrate only slightly in the lift, drag, and/or torsional directions. However, under certain conditions, the vibratory motion itself induces negative dynamic damping leading to dynamic instability. The resulting vibration is also known as galloping. Galloping can arise in the lift and drag directions as well as in torsion. Furthermore, coupled-mode galloping involving a combination of modes is also possible. We adopt the terms 'flutter' and 'galloping' to suit the engineering field. Galloping is used mainly in civil engineering where the focus is primarily on the single-degree-of-freedom self-excited vibrations of bluff bodies subjected to wind or ocean tides. Whereas, flutter is used mainly in the aerospace engineering field, where bending-torsional coupled self-excited vibrations of a streamlined wing subjected to airflow are mainly focused on. Galloping occurs usually far above the critical flow velocity for self-excited vibration by Kármán vortex shedding. If the Scruton number, Sc, is small, the flow velocity range for Kármán vortex vibration and that for galloping vibration may coincide. In this case, the phenomenon becomes complicated and evaluation becomes difficult [115].

2.4.2.2 Historical background

2.4.2.2.1 Vortex-induced vibration

The Strouhal number is defined by:

$$\mathrm{St} = \frac{f_{\mathrm{W}} d}{V} \tag{2.37}$$

f_{W} being the vortex-shedding frequency, V the approach flow velocity, and d the projected width normal to the flow direction.

The Strouhal number, St, for a rectangular body is affected by the aspect ratio, attack angle, as well as the radius of curvature of the corners. It is also affected by flow conditions (Reynolds number) and turbulence intensity in the mainstream. Strouhal numbers for rectangular bodies, obtained experimentally by many researchers, seem to fit within the range $0.1 < \mathrm{St} < 0.2$ [112], with the exception of bodies rounded off at the corners. The Strouhal number, St, for the latter falls within the range $0.2 < \mathrm{St} < 0.3$.

Few data exist on Strouhal numbers for high Reynolds number flows commonly encountered in actual industrial processes. Clearly much more work is needed for these flows.

Many researchers have attempted to theoretically predict the maximum vibration amplitude for circular cylinders at lock-in. There are, however, few investigations on rectangular-cross-section bodies. According to the study by Naudascher [112], a rectangular-cross-section body can vibrate in various vibration modes. The vibration depends on: (i) cross-section shape (aspect ratio, attack angle, and the corner radius of curvature), (ii) characteristics of the body (mass, damping, and elastic characteristics in the flow direction, direction orthogonal to the flow, and rotational direction), and (iii) flow characteristics such as Reynolds number and turbulence intensity. Thus far, a systematic database which is needed for evaluating vortex-induced vibration, has not been established. Therefore, if one wishes to evaluate the vibration of a rectangular-cross-section body, a search through past studies on the vibration behavior of similar bodies must be done to determine the expected vibration characteristics. It is especially difficult to predict vibration characteristics in high Reynolds number flow regimes; little is known about Strouhal number characteristics here.

2.4.2.2.2 Turbulence-induced vibration

To analyze the turbulence-induced vibration of bluff bodies, the PSD of the turbulence-induced forces acting on the body is measured experimentally and expressed in normalized form. The response of the body is computed using the normalized PSD.

Recently, in the architectural and civil engineering fields, experimental PSD data for flexible structures (tall buildings, etc.) which vibrate fairly easily when subjected to wind turbulence, have been actively accumulated. PSD data for rectangular-cross-section bodies, however, depend on many parameters, such as the cross-section shape (aspect ratio, attack angle, and the radius of curvature of corners of the body), and flow characteristics (Reynolds number, turbulence intensity). This is much like the case of a circular cylinder. The existing database is still quite limited, especially for high Reynolds number flows. To evaluate the turbulence-induced vibration of a rectangular body, past studies must be searched to find PSDs as closely relevant as possible to the body under study.

With the continuing increase of computational power, CFD methods for the determination of turbulence-induced force PSDs for rectangular-cross-section bodies are under development. The methods numerically evaluate PSDs for very complicated cross-section bodies without experimental input. Practical application may be expected in the near future.

2.4.2.2.3 Galloping

When ice builds up on power transmission lines (thus changing the effective cross-sectional geometry) large-amplitude vibrations can arise. The phenomenon is named 'galloping', due to the similarity of the power line vibration to a galloping horse. Galloping, which was likely the cause of the Tacoma Narrows Bridge collapse in 1940, was explained theoretically by Den Hartog [116]. Thereafter, researchers such as Novak [117,118], Nakamura [111,113,114], and Naudascher and Rockwell [107,119] investigated the galloping of bodies of various cross-section shapes for a wide range of flow conditions. This work was done in the four-decade period from the 1950s to the 1980s. Regions of self-excited vibrations such as galloping can be evaluated to examine the negative damping condition using the lift and drag coefficients obtained experimentally or theoretically. The vibration amplitude during galloping can also be evaluated analytically/

quantitatively, with reasonable accuracy, using non-linear differential equation analysis methods based on the assumption of slowly varying amplitude and phase. It is extremely important to accurately determine the flow condition for the onset of bending-torsion coupled-mode flutter for airplane wings. The critical speeds for flutter and prevention of airplane flutter were investigated actively from the 1950s to the 1980s. Evaluation methods are now well established for many types of airfoil.

2.4.3 Evaluation methods

The ASME Code Sec. III, Div. 1, Appendix N, 1300 Series are the first standards of evaluation methods concerning vortex- and turbulence-induced vibration of circular cylinders in steady cross-flow. The same evaluation methods can be applied to the evaluation of vortex- and turbulence-induced vibrations of a rectangular-cross-section body. However, the standard contains only data (e.g. Strouhal numbers) for the evaluation of circular cylinder vibration. One may therefore either perform approximate analysis based on circular cylinder data or conduct experiments to obtain data for non-circular cross-section bodies.

There is no detailed description of galloping in the ASME Code. Refs. [24,77,119] may serve as useful sources of information.

2.4.3.1 Vortex-induced vibration

The frequency f_W at which vortices induce a periodic force on a rectangular body is given by:

$$f_W = \mathrm{St}\frac{V}{d} \tag{2.38}$$

where V is the upstream flow velocity, and d a characteristic length dimension (e.g. cross-section width transverse to the flow).

The Strouhal number, St, which is an important non-dimensional coefficient, is obtained experimentally. Generally, large-amplitude vortex-induced vibrations of a rectangular-cross-section body will not be observed and the integrity of the structure will be maintained, unless the frequency of vortex shedding is locked into a frequency of the body. If the frequency of vortex shedding coincides with a structural natural frequency, the lock-in phenomenon discussed above can occur, resulting in resonant vibration. Evaluation methods for self-excited resonant vibrations are discussed below.

The expression (2.38) estimates the excitation frequency of the lift direction vibration. Excitation forces at twice the lift direction frequency can act dynamically in the flow direction. The vibration amplitude of vortex-induced self-excited vibration in the flow direction is generally small compared to its transverse direction counterpart [120,121]. However, the vibration in the flow direction should be taken into account in design.

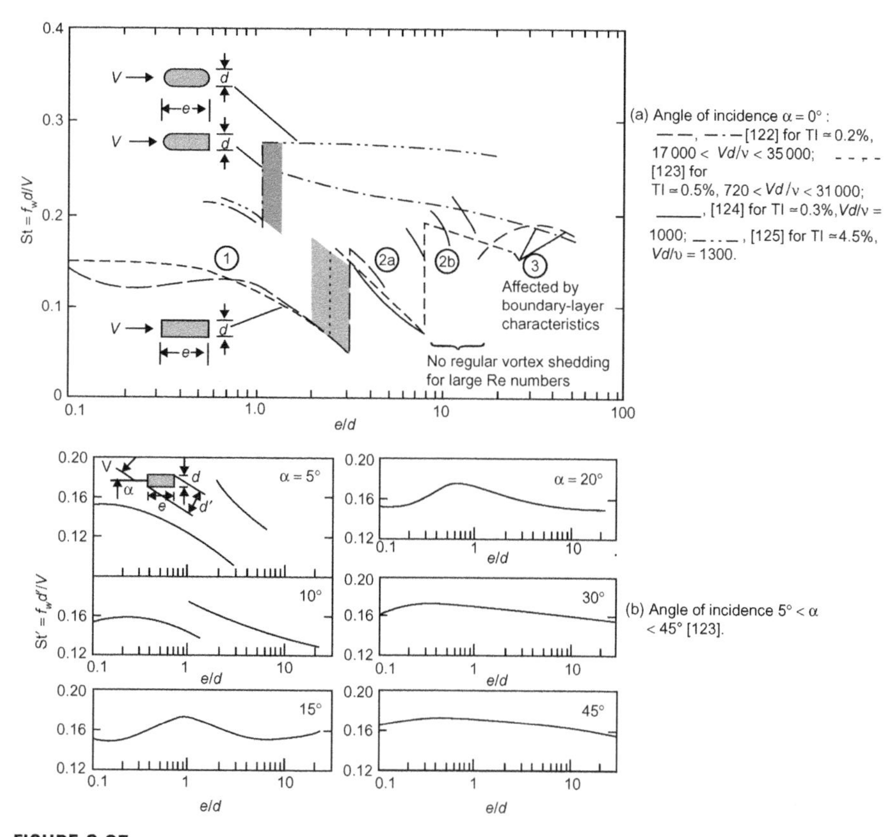

FIGURE 2.25

Strouhal numbers for rectangular-cross-section bodies for various aspect ratios, attack angles, and rounded corners [112].

2.4.3.1.1 Strouhal number for zero attack angle of rectangular-cross-section body

The Strouhal number varies with the shape of the rectangular cross-section (i.e. aspect ratio e/d) and the curvature radius of the cross-section corners. The Strouhal number also varies with flow characteristics (i.e. Reynolds number [Re], and Turbulence Intensity [TI]) (Fig. 2.25(a)). The numerical value of the Strouhal number is approximately 0.05–0.2.

2.4.3.1.2 Strouhal number for a rectangular-cross-section body inclined to the flow

Examples of the dependence of the Strouhal number on the angle of attack, α, are shown in Fig. 2.25(b) and Fig. 2.26. Note that d' in both figures is the projected width instead of the real width, d, of the body.

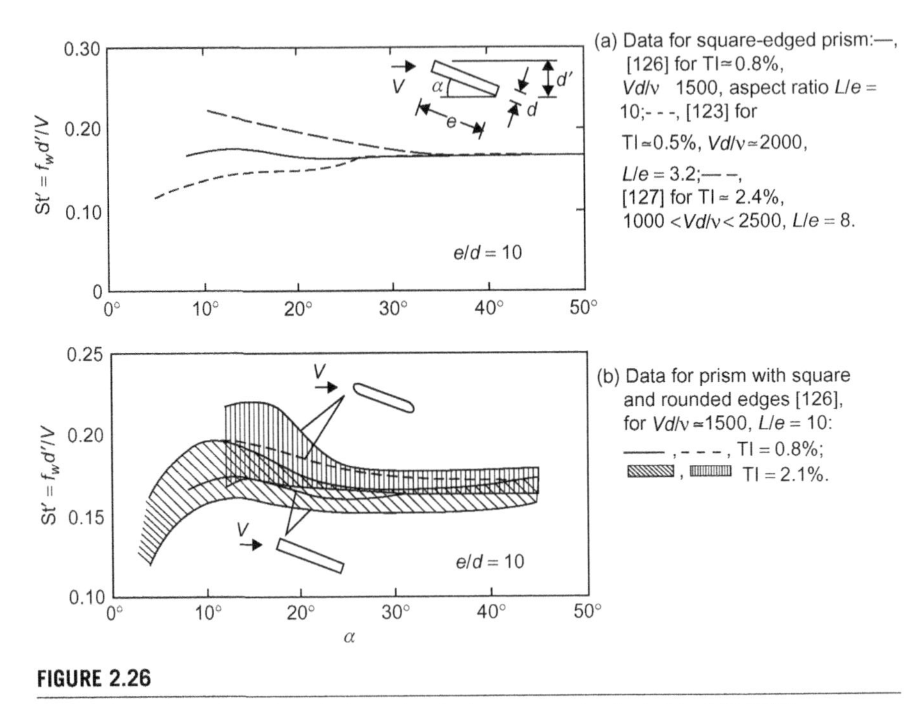

FIGURE 2.26

Effect of attack angle on Strouhal number for large aspect ratio ($e/d = 10$) [112].

2.4.3.1.3 Flow speed range of vortex-induced vibrations for a rectangular-cross-section body

When the non-dimensional flow speed, V_r, is increased from a low value, the four different vibration mechanisms in Table 2.14 are possible depending on the angle of attack, α, and the aspect ratio, e/d (see also Figs. 2.27 and 2.28).

2.4.3.1.4 Vibration amplitude of vortex-induced vibrations for a rectangular-cross-section body

The amplitude of vortex-induced vibrations for a rectangular-cross-section body may be predicted using the evaluation methods developed for a circular cross-section body. However, there seem to be only a few studies, including work by Naudascher and Wang [112]. The latter discuss the effect of aspect ratio, attack angle, corner radius of curvature, and vibration direction in the flow on vibration amplitudes measured experimentally.

2.4.3.2 Turbulence-induced vibration

The turbulence-induced vibration response of a rectangular-cross-section body can also be evaluated with the method presented in the Japanese JSME S 012 guideline. The method has already been detailed in Section 2.1.3 of this chapter.

Table 2.14 Types of vortex-induced vibration of a rectangular cross-section body in cross-flow [112]

Aspect ratio: *e/d*	Attack angle: α	Kinds of vortex shedding	Possible vibration
$0 < e/d < 2$	–	LEVS: Leading-edge vortex shedding (separation bubble at leading edge of the body [Fig. 2.27])	• Kármán vortex-induced vibration in the transverse direction • Galloping caused by negative damping with the body movement
$2 < e/d < 16$	–	ILEV: Impinging leading-edge vortex shedding (vortices detached from leading edge attach again on the body before passing trailing-edge [Fig. 2.27])	Kármán vortex-induced vibration in the transverse flow direction as well as the rotational direction
$16 < e/d$	–	TEVS: Trailing-edge vortex shedding (vortices detach from the trailing edge of the body [Fig. 2.27])	The vibration characteristics are similar to those of a flat board, but the amplitude might not be large
$10 < e/d$	$10° < \alpha$	AEVS: Alternate-edge vortex- shedding (vortex shedding in turn from leading/trailing edges is observed [Fig. 2.28])	When both attack angle and aspect ratio are large, vortex shedding in the condition of IEVS/TEVS causes vibrations in the flow direction or in the chord direction

However, for the reader's convenience, the essence of the method, applied to rectangular-cross-section bodies, is outlined below.

The turbulence-induced vibration response is obtained based on random vibration theory. The root-mean-square value, w_{mean}, of the turbulence-induced vibration amplitude is evaluated using the following expression:

$$w^2_{\text{mean}} = \frac{\beta_0 G_F(f_0)}{64\pi^3 m^2 f_0^3 \zeta} \phi_0^2(z) \tag{2.39}$$

where,

$$\beta_0 = \frac{\int_{Le} \phi_0(z) dz}{\int_L \phi_0^2(z) dz}$$

m is the mass per unit length and ζ is the damping ratio of the structure.

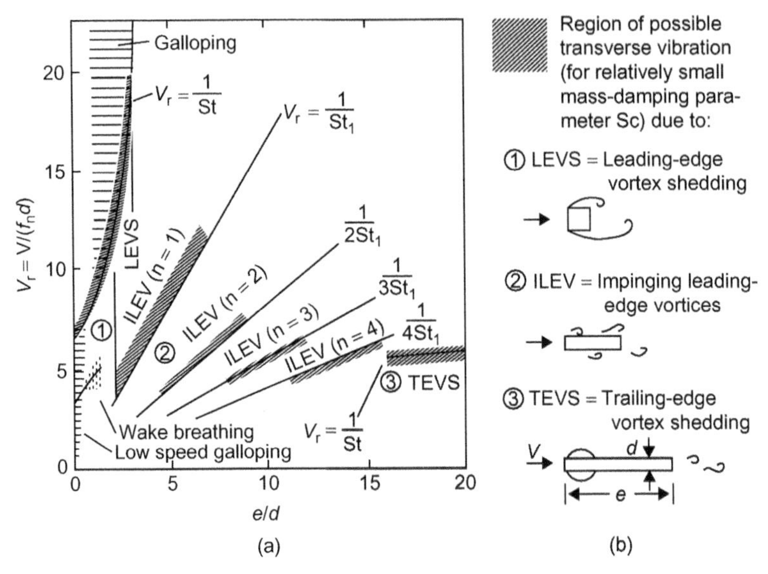

FIGURE 2.27

Effect of aspect ratio on possible vortex-induced vibration modes with zero attack angle [112]: (a) ranges of possible FIV for lightly damped rectangular prisms in low-turbulence cross-flow ($\alpha = 0$), cross-hatched and dotted regions represent prisms with transverse and streamwise degrees-of-freedom, respectively, and (b) effects of aspect ratio on the mode of vortex formation. St_1 can be approximated by $St_1 \simeq 0.6(d/e)$ [127].

After introducing the conversion coefficient C_0, the peak value of the turbulence-induced vibration is evaluated using Eq. (2.39):

$$w_R(z) = C_0\sqrt{\frac{\beta_0 G_F(f_0)}{64\pi^3 m^2 f_0^3 \zeta}}\phi_0(z) \tag{2.40}$$

where C_0 is approximately 3–5.

Using the standardized PSD, Φ, and unsteady fluid force coefficient, C', the PSD $G_F(f)$ of the random excitation force per unit length is given by:

$$\begin{aligned} G_F(f) &= (C'\tfrac{1}{2}\rho V^2 d)^{2\prime}(f) \\ &= (C'\tfrac{1}{2}\rho V^2 d)^{2\prime}(\bar{f})\frac{d}{V} \end{aligned} \tag{2.41}$$

where ρ is the fluid density and $\bar{f} = \frac{fd}{V}$ is a non-dimensional frequency.

General expressions for the standardized PSD $\Phi(f)$ in Eq. (2.41) are generally unknown for rectangular-cross-section structures. It is thus necessary to estimate $\Phi(f)$ based on results of similar past experimental studies, or to conduct model tests to determine $\Phi(f)$.

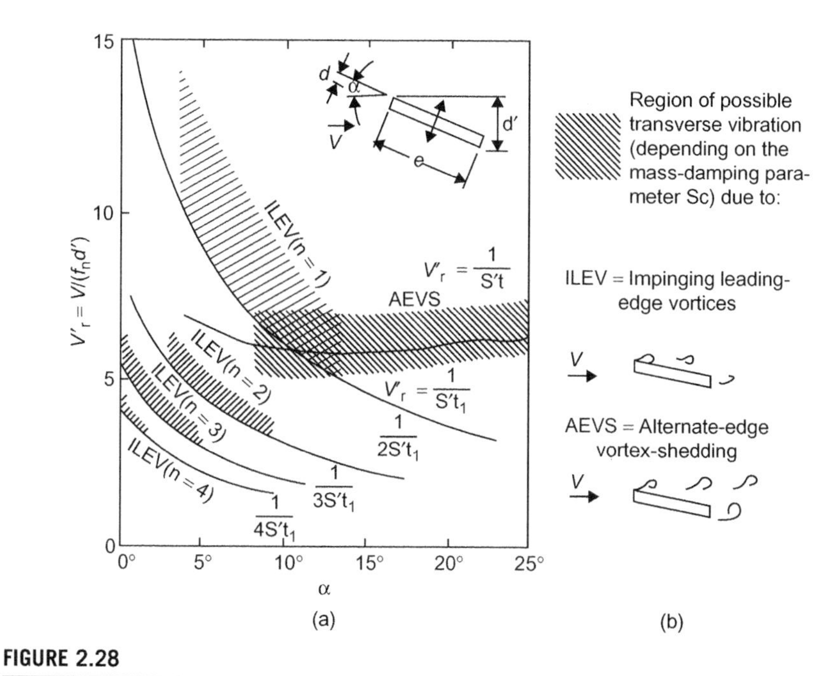

FIGURE 2.28

Effects of attack angle on possible vortex-induced vibration modes for large aspect ratio (e/d = 10) [112]: (a) effects of attack angle on possible transverse vibration of rectangular cross-section body in cross-flow (lightly damped, low-turbulence flow, zero attack angle), and (b) effects of attack angle on the mode of vortex formation.

2.4.3.3 Galloping

Consider the galloping phenomenon in the lift (transverse) direction. The transverse motion of the body at the velocity $\dot{y}$ relative to an upstream velocity V induces a relative angle of attack given by:

$$\alpha \approx \frac{\dot{y}}{V} \tag{2.42}$$

The resultant fluid force in the lift direction, for small α, is then:

$$F_y = -\frac{1}{2}\rho V^2 dC_L\bigg|_{d=0} - \frac{1}{2}\rho V^2 d\left(\frac{\partial C_L}{\partial \alpha} + C_D\right)\bigg|_{\alpha=0} \alpha + 0(\alpha^2) \tag{2.43}$$

In the Eq. (2.43), $\partial C_y/\partial\alpha = -\partial C_L/\partial\alpha - C_D$ is an important coefficient of the fluid force, where C_y is the vertical force coefficient [77]. A positive value of the vertical force coefficient derivative ($\partial C_y/\partial\alpha > 0$) indicates that the fluid force acts in the direction of motion of the body. The fluid force thus amplifies the body's vibration in the lift direction if $\partial C_L/\partial\alpha$ is negative and the absolute value of $\partial C_L/\partial\alpha$ is larger than the drag coefficient (i.e. $\partial C_L/\partial\alpha + C_D < 0$). Self-excited vibrations finally occur when the negative fluid damping force overcomes the internal damping of the structure. The critical velocity is given by:

$$V = -\frac{4m\zeta\omega}{\rho d\left(\frac{\partial C_L}{\partial \alpha} + C_D\right)|_{\alpha=0}} \tag{2.44}$$

Table 2.15 [77] shows values of $\partial C_y/\partial\alpha = -\partial C_L/\partial\alpha - C_D$ for various cross-section shapes. These data may be useful to the reader wishing to determine the critical velocity for galloping.

The vibration amplitude of galloping in the lift direction can be evaluated assuming that the amplitude and phase of the non-linear vibrations vary slowly. The equation of motion in the lift direction is:

$$m\ddot{y} + 2m\zeta_y\omega_y\dot{y} + k_y y = \frac{1}{2}\rho V^2 d C_y \tag{2.45}$$

Expanding the force coefficient C_y in a Taylor series with respect to the angle of attack α defined in Eq. (2.42) yields:

$$\begin{aligned} C_y(\alpha) &= a_0 + a_1\alpha + a_2\alpha^2 + a_3\alpha^3 + \cdots \\ &= a_0 + a_1\left(\frac{\dot{y}}{V}\right) + a_2\left(\frac{\dot{y}}{V}\right)^2 + a_3\left(\frac{\dot{y}}{V}\right)^3 + \cdots \end{aligned} \tag{2.46}$$

Several examples of the Taylor approximation are presented in Table 2.16. An approximate solution of the limit cycle amplitude of the non-linear equation (2.45) is obtained based on Eq. (2.46) with the assumption that both the amplitude and phase are slowly varying. Novak [117] derived the following simple expression for the vibration amplitude by assuming C_y could be approximated by a summation of the first four terms in Eq. (2.46):

$$A^* = \left[4(1 - V^* a_1)\frac{V^*}{3a_3}\right] \tag{2.47}$$

In the equation above, A^* and V^* are the non-dimensional amplitude and non-dimensional flow speed respectively, defined by:

$$A^* = \frac{A_y}{d}\frac{\rho d^2}{4m\zeta_y} \tag{2.48}$$

$$V^* = \frac{V}{f_y d}\frac{\rho d^2}{4m(2\pi\zeta_y)} \tag{2.49}$$

While an approximate evaluation is possible using Eq. (2.47) for a square-cross-section body, a more accurate solution is obtained when a seventh-order polynomial for C_y is used in Eq. (2.46) (see Fig. 2.29).

Table 2.15 Values of partial derivative of C_y with respect to angle of attack for several cross-section bodies in steady flow [129,130–133]

Section	$\partial C_y/\partial\alpha$[a]		
	Smooth flow	**Turbulent flow**[b]	**Reynolds number**
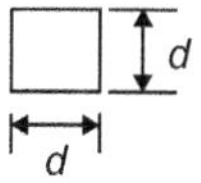 d; d	3.0	3.5	10^5
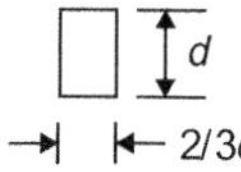 d; 2/3d	0	0.7	10^5
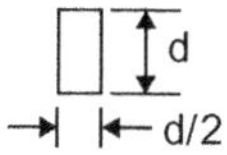 d; d/2	−0.5	0.2	10^5
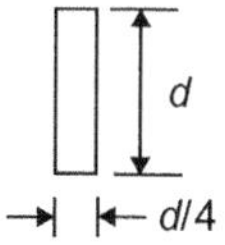 d; d/4	−0.15	0	10^5
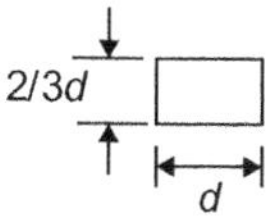 2/3d; d	1.3	1.2	66 000
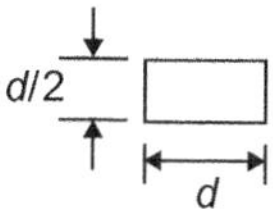 d/2; d	2.8	−2.0	33 000
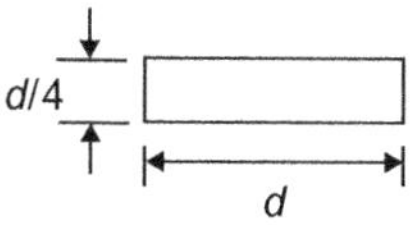 d/4; d	−10	–	2000–20 000
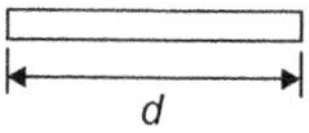 d	−6.3	−6.3	>10^3
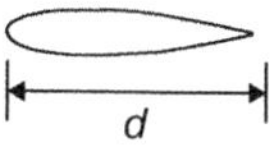 d	−6.3	−6.3	>10^3
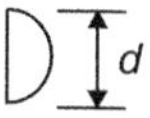 d	−0.1	0	66 000
	−0.5	2.9	51 000
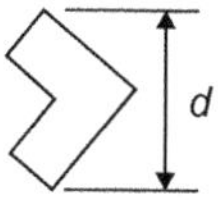 d	0.66	–	75 000

[a] α is in radians; flow is from left to right. $\partial C_y/\partial\alpha = -\partial C_L/\partial\alpha - C_D$. C_y based on the dimension d. $\partial C_y/\partial\alpha < 0$ for stability.
[b] Approximately 10% turbulence.

Table 2.16 Polynomial approximation expressions of C_y for several cross-section bodies [117,118]

Coefficients[a]	3/2 Rectangle (d, 2/3d)		2/3 Rectangle (3/2d, d)		D Section (d)		Square (d)
	Smooth flow	Turbulent flow	Smooth flow	Turbulent flow	Smooth flow	Turbulent flow	Smooth flow
a_1	0	0.74285	1.9142	1.833	−0.097431	0	2.69
a_2	−3.2736 + 1[b]	−0.24874	3.4789 +1	5.2396	4.2554	−0.74824	0
a_3	7.4467 + 2	1.7482 + 1	−1.7097 + 2	−1.4518 + 2	−2.8835 + 1	5.4705	−1.684 + 2
a_4	−55834 + 3	−3.6060 + 2	−2.2074 + 1	3.1206 + 2	6.1072 + 1	−6.3595	0.
a_5	1.4559 + 4	2.7099 + 3	0	0	−4.8006 + 1	2.6844	6.27 + 3
a_6	8.1990 + 3	−6.4052 + 3	0	0	1.2462 + 1	−0.3903	0
a_7	−5.7367 + 4	−1.1454 + 4	0	0	0	0	−5.99 + 4
a_8	−1.2038 + 5	6.5022 + 4	0	0	0	0	0
a_9	3.37363 + 5	−6.6937 + 4	0	0	0	0	0
a_{10}	2.0118 + 5	0	0	0	0	0	0
a_{11}	−6.7549 + 5	0	0	0	0	0	0

[a] *C_y based on* d.
[b] *$Re = 5 \times 10^4$. Flow left to right. The notation +2, etc. denotes powers of 10: for example, 3.445 + 2 = 344.5.*

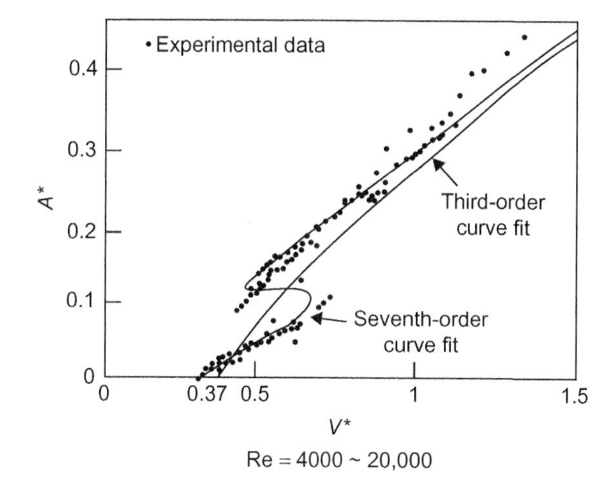

FIGURE 2.29

Galloping vibration amplitude in the transverse degree-of-freedom of a square body [134].

Rotational mode vibration or coupled lift-torsional vibrations can also occur for a rectangular-cross-section structure. The critical speed for torsional vibration, skipping the detailed derivation, is given by the following equation [77]:

$$\frac{V}{f_\alpha d} = -\frac{\left(\frac{4J_\alpha(2\pi\zeta_\alpha)}{\rho d^3 R}\right)}{\left(\frac{\partial C_M}{\partial \alpha}\right)} \tag{2.50}$$

where $f_\alpha = (1/2\pi)(k_\alpha/J_\alpha)^{1/2}$ is the natural frequency in torsion, ζ_α, the corresponding damping ratio, and R the characteristic radius [77]. Examples of $\partial C_M/\partial\alpha$ in Eq. (2.50) are given in Table 2.17. The values of $(\partial C_M/\partial\alpha)$ for airfoils, obtained by other researchers are also presented in Section 3.2.1.

The lift/torsion coupled-mode-unstable vibration can occur for bridges or aircraft wings as these structures are fairly flexible in both torsion and flexure. Evaluation of the coupled-mode instability is thus essential for safety in civil and aircraft engineering. Coupled-mode vibrations of bridges and aircrafts have been investigated extensively in Ref. [77].

2.4.4 Examples of structural failures and suggestions for countermeasures

In the mechanical engineering field, vortex-induced vibration and galloping vibration for a square-cross-section body or a rectangular-cross-section body of aspect ratio *e/d* of nearly one can be evaluated fairly quantitatively. However, it is only recently that researchers have begun to classify complicated vortex shedding

Table 2.17 Values of partial derivative of C_M with respect to angle of attack for several bodies of different cross-sections, in steady flow [129]

Section	$\partial C_M/\partial \alpha$[a]	Reynolds number
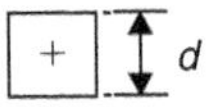 Square, side d	−0.18	10^4–10^5
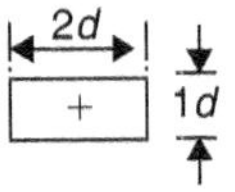 Rectangle $2d \times 1d$	−0.64	5×10^3–5×10^5
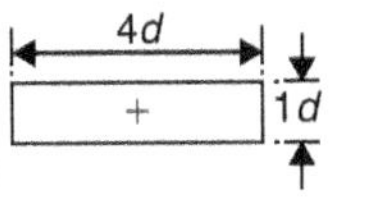 Rectangle $4d \times 1d$	−18	2×10^3–2×10^4
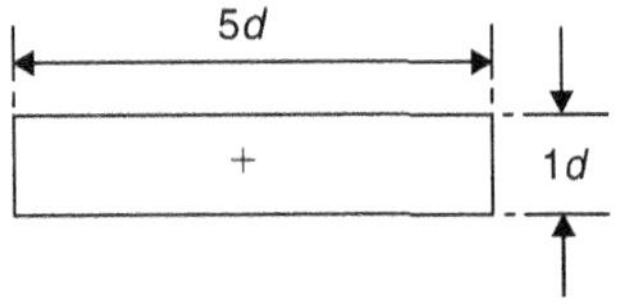 Rectangle $5d \times 1d$	−26	2×10^3–2×10^4
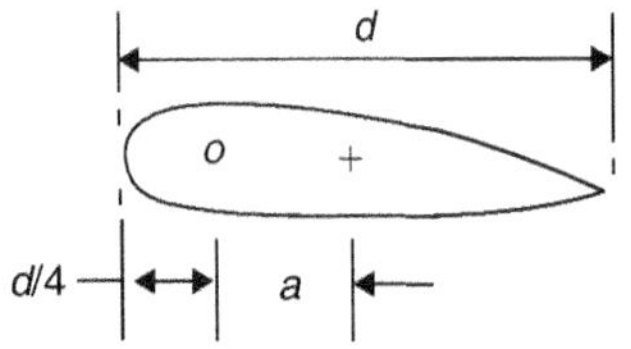 Airfoil, chord d; $d/4$; a	$2\pi a/d$[b]	$>10^3$

[a] α is in radians; flow is left to right.
[b] For angles up to about 8 degrees.

behavior of rectangular bodies of fairly large aspect ratios ($e/d = 2-16$) [112]. Future studies on the vibration of large-aspect-ratio bodies will contribute to decrease vibration problems in this area.

In industrial plants, structures having cross-shaped sections can sometimes be subjected to parallel/cross-flow. The vibration characteristics of such bodies have recently been investigated [132]. There are many examples of vibration problems for guide vanes and elbow splitters (Fig. 2.30). FIV treated in this section should be considered at the design stage of such structures. Tall steel structures, buildings/bridges, in civil and architectural engineering fields, which are large but relatively flexible can be destroyed if large-amplitude self-excited vibration (Kármán vortex vibration, galloping, etc.) occurs. The Tacoma Narrows Bridge in the USA is a classic example. Recently, self-excited vibration evaluation of flexible buildings/bridges has been extensively conducted at the design stage using scale model tests and computer numerical simulations. As a result, such vibration has caused very few problems in recent years.

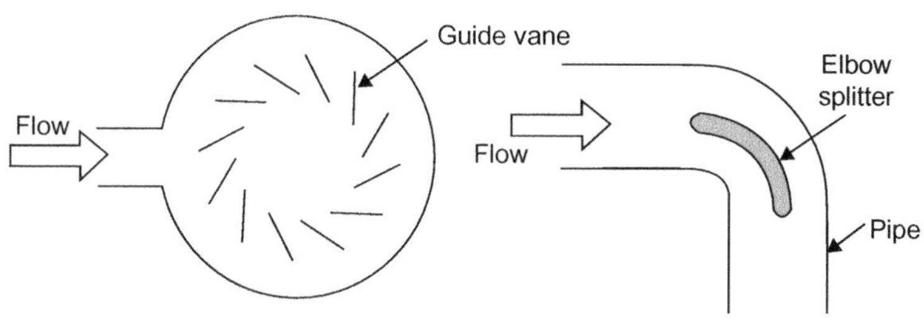

FIGURE 2.30

Structures of guide vane and elbow splitter.

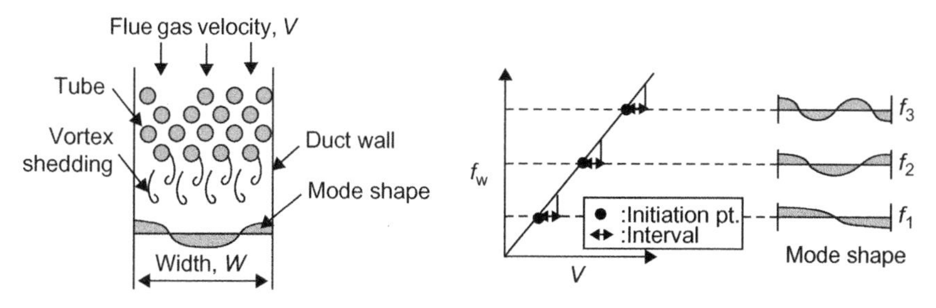

FIGURE 2.31

Overview of acoustic resonance in tube bundle.

2.5 Acoustic resonance in tube bundles

2.5.1 Relevant industrial products and a brief description of the phenomenon

2.5.1.1 Relevant industrial products

Acoustic resonance may occur in heat exchangers such as gas heaters or boilers which contain tube bundles. This phenomenon is related to three factors: tube bundle geometry, gas flow normal to tube axes, and the acoustic space limited by duct or vessel walls. High-level noise generated by resonance has concerned heat exchanger designers for a long time. The phenomenon is still under investigation. The goal is to predict the occurrence of resonance at the design stage. Countermeasures to suppress acoustic resonance are also under investigation.

2.5.1.2 Mechanism underlying resonance

When the gas flow rate in the duct (or vessel) surpasses a critical level, generation of high-level noise by acoustic resonance may occur. In the extreme case, this phenomenon may make it impossible to operate the plant or may cause structural damage. Figure 2.31 shows the mechanism leading to acoustic resonance. As the gas flow velocity increases, the Kármán vortex shedding frequency f_w increases, eventually reaching the natural frequency of an acoustic mode within the duct. The resulting resonance can potentially generate high levels of noise. The acoustic natural frequency f_i in the duct is estimated by:

$$f_i = \frac{ic}{2W} \quad i = 1, 2, 3, \ldots \tag{2.51}$$

where c, W, and i are sound speed, duct width, and order of acoustic mode, respectively. The vortex shedding frequency f_W is estimated by:

$$f_W = \mathrm{St}\frac{V_g}{D} \tag{2.52}$$

where V_g, D, and St are gap velocity (flow velocity between tubes), outer diameter of the tube, and Strouhal number, respectively. The Strouhal number is dependent on the configuration of the tube bundle. As the flow velocity increases, resonances at increasingly higher modes occur in sequence. Hereafter, flow velocity refers to the gap flow velocity, unless otherwise noted.

The acoustic resonance phenomenon is classified as a self-excited oscillation with the following feedback mechanism. Vortices shed from the tube bundle excite an acoustic mode. The resulting regular acoustic pressure fluctuations caused by this mode in turn promote and strengthen the synchronization of vortex shedding which in turn increases the net excitation energy of the acoustic mode. The acoustic resonance occurs when the following two conditions are satisfied:

1. *Frequency condition*: coincidence of vortex shedding frequency with the natural frequency of any acoustic mode in the duct.
2. *Energy condition*: energy supplied by vortex shedding to an acoustic mode exceeds the energy dissipation of this mode in the acoustic field.

2.5.1.3 Classification of resonance phenomena

As shown in Fig. 2.32, acoustic resonances are classified into two types depending on the acoustic mode shape; Fig. 2.32(a) shows the form of the first type of excitation where the acoustic resonant modes are dominant in the direction

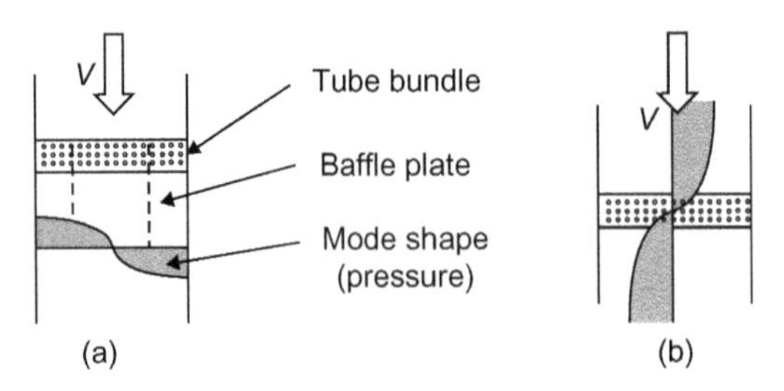

FIGURE 2.32

Classification of acoustic resonance by mode shape: (a) transverse mode, and (b) longitudinal mode.

perpendicular both to the gas flow and the tube axes. The second type is shown in Fig. 2.32(b). In this case, the acoustic modes parallel to the gas flow direction are dominant. Hereafter, the first type is referred to as the transverse mode resonance while the second type is referred to as the longitudinal mode resonance.

The transverse mode resonance is more commonly encountered than its longitudinal counterpart. It is further classified into two types depending on the location of resonance in the tube bundle; the type where the main part of the resonance exists within the tube bundle and the other where the resonance occurs mainly in the cavity between upstream and downstream tube bundles. To eliminate these types of acoustic resonances, baffle plates are introduced. Subdivision of the acoustic space effectively increases the resonant frequencies of the acoustic modes separating them from the vortex shedding frequency. For the second type of resonance, which occurs in the cavity, baffle plates must be located within the cavity, or within the cavity and tube bundle itself. Therefore, when countermeasures are required, it is recommended to measure the acoustic pressure distribution in the duct.

For the longitudinal mode resonance, the foregoing baffle plate method is generally not considered an effective countermeasure. Countermeasures will be presented later in Section 2.5.2.2. Resonance with both types of acoustic modes involves lock-in. Once resonance occurs, the vortex shedding frequency is captured by the acoustic natural frequency over a range of gas flow velocities. For reference see Section 2.1.2; a similar phenomenon was discussed where lock-in occurred between the vortex shedding frequencies and the structural natural frequencies.

2.5.2 Historical background

Key research issues include prediction of resonance and estimation of the critical gas flow velocity for a given tube bundle configuration and acoustic space geometry in ducts. Many studies have attempted to clarify the relation between the occurrence of resonance and the two foregoing frequency and energy conditions.

Since the early 1950s there have been many research articles on resonance phenomena in heat exchangers [135]. The work of Chen [99] in 1968 is widely considered a breakthrough in the field. Chen was the first to identify the relationship between resonant noise and Kármán vortex shedding. To study this relationship, hot-wire measurements coupled with excitation force measurements, using strain gauges mounted on a tube were conducted. The tests revealed that the vortex shedding was synchronized with the acoustic resonance modes. Furthermore, the lock-in effect was identified where vortex shedding was captured by the acoustic resonant modes. Chen proposed a unique baffle plate configuration where plates were placed in a staggered configuration or with unequal spacing between the adjacent plates so as to stagger the acoustic natural frequencies of the subdivided spaces. The vortex shedding frequencies for bare- and finned-tube bundles were measured in various pitch ratio configurations, resulting in a map depicting Strouhal number distribution with tube array spacing. Based on these studies, Chen and Young proposed parameters which can be used to predict potential resonances at the design stage [136]. Based on these achievements, Chen's studies are regarded as the pioneering work in this field. Details will be presented in Section 2.5.3.2.

In what follows, a summary of the research on resonances excited by transverse and longitudinal modes is presented. Resonance prediction methods at the design stage, countermeasures for existing resonant cases, and mechanism of resonance generation with a focus on feedback phenomena will be discussed. Details may be found in the review papers by Weaver and Fitzpatrick [137], Païdoussis [138], Blevins [139], and Eisinger and Sullivan [140].

2.5.2.1 Review of research on transverse mode resonance

2.5.2.1.1 Resonance prediction methods

At the design stage, the most fundamental method to predict the likelihood of acoustic resonance consists of checking the proximity of the computed acoustic natural frequency to the estimated vortex shedding frequency. As shown by Eq. (2.52), the vortex shedding frequency can be easily estimated based on Strouhal number, which is dependent on tube bundle configuration. Strouhal numbers for tube bundles consisting of bare tubes have been studied by Chen [99] and Weaver and Fitzpatrick [141]. The Strouhal number map of Fitz-Hugh [98] shown in Fig. 2.19 is most commonly used. For finned tubes, the results by Chen [99], Mair et al. [142], Okui et al. [143] and, Hamakawa et al. [144,145] are available. This issue will be discussed in detail in Section 2.5.3.2.

An alternative method to predict acoustic resonances based on an energy balance concept has also been developed. The balance between the energy supplied by vortex shedding to an acoustic mode and energy dissipated in the acoustic field is considered. It is, however, very difficult to estimate, theoretically, these two kinds of energy. Consequently, empirical design rules have been proposed based on accumulated experimental or in-situ plant data on the critical flow velocity for a variety of tube bundle configurations. Representative works have been done by Grotz and Arnold [146], Chen [99], Fitzpatrick [147], Ziada et al. [148], Blevins and Bressler [149], and Eisinger [150]. Comparison and evaluation studies on the validity of these empirical rules have been done by Eisinger and Sullivan [140], and Blevins and Bressler [149]. In the next section two representative methods proposed by Chen [99] and Eisinger et al. [150,151] will be discussed in some detail.

These empirical studies yielded another important result: high-density tube bundles are less likely to suffer acoustic resonance. Figure 2.33 [149] is a representative result which shows the resonance region as a function of array spacing. In-line bundle data are shown in Fig. 2.33(a) and staggered bundle data in Fig. 2.33(b). In both maps, stable regions exist near the origin, showing high-density tube bundles are less likely to undergo acoustic resonance. This trend

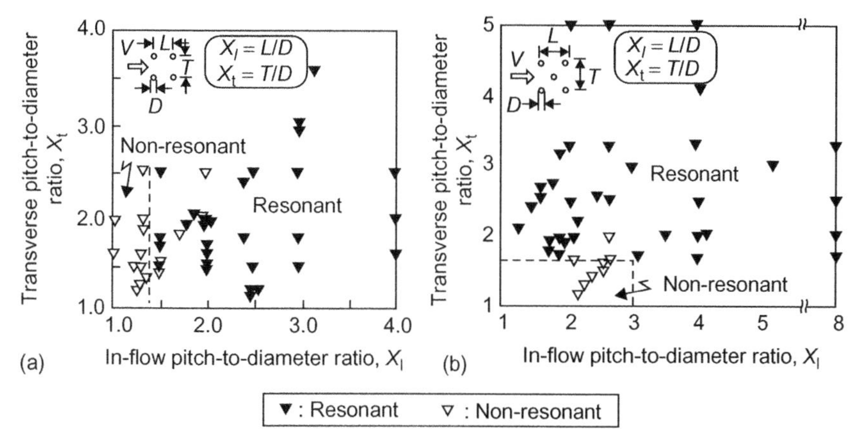

FIGURE 2.33

Resonance map for tube bundles, based on pitch-to-diameter ratio [149]: (a) in-line, and (b) staggered array.

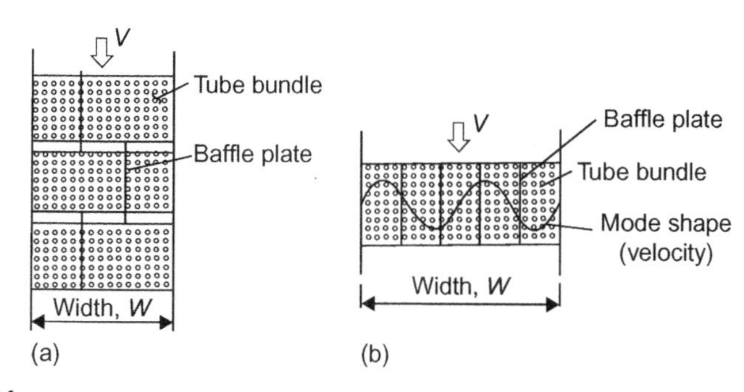

FIGURE 2.34

Examples of baffle placement for countermeasures against transverse acoustic modes [153]: (a) irregular pitch, and (b) regular pitch.

might be explained by the fact that the vortices shed from tubes need a certain minimal inter-tube spacing for formation and amplification.

Practical experience-based methods exist and can be used to estimate the maximum sound pressure level during resonance [152]. Details of these methods will not be discussed here.

2.5.2.1.2 Baffle plate design method

The following method is usually used to suppress transverse mode resonances. In principle, as shown by Fig. 2.34(b), baffle plates should be installed near the pressure node of the resonant modes of interest. Therefore, in a plant suffering resonance, experienced field engineers will simply walk along the duct listening to

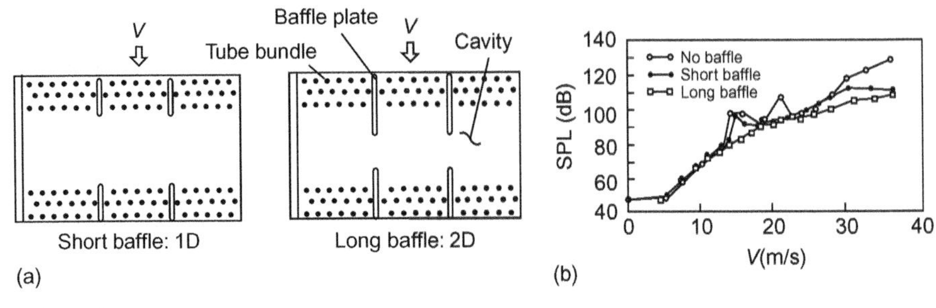

FIGURE 2.35

Resonance suppression effect of cavity baffle [154]: (a) baffle structure, and (b) sound pressure level.

the sound level and could point out suitable positions for baffle plate location. Figure 2.34 shows typical baffle plate configurations. Figure 2.34(b) shows an example of a common regular pitch placement, while Fig. 2.34(a) shows a typical irregular case [99]. Note, however, that Eisinger [153] does not consider the configuration of Fig. 2.34(a) to be effective.

In cases where resonance has occurred mainly in the cavity region between the upstream and downstream tube bundles, many methods of baffle plate placement have been attempted. In one case, plates were extended from the tube bundle to the cavity. In another, plates were installed only within the cavity because of the difficulty of insertion into the tube bundle. Examples will be presented in Section 2.5.4. Some research work has also been done on the problem. Blevins and Bressler did not find significant effectiveness associated with plates installed in the tube bundle upstream or downstream of the cavity [149]. Meanwhile, as shown in Fig. 2.35, Nemoto and Yamada [154,155] conducted experiments using finned-tube bundles and insisted that extension of baffle plates into the upstream and downstream tube bundles by twice the longitudinal tube pitch had a resonance suppression effect.

There are some articles suggesting the effectiveness of baffle plates for longitudinal acoustic modes [156]. However, other studies [157] showed negative results, that is, baffle plate insertion promoted the longitudinal acoustic modes. It is therefore considered that, as a countermeasure for longitudinal modes, the baffle plate method is ineffective.

2.5.2.1.3 Studies on the feedback mechanism

It is well-known that acoustic resonance in tube bundles is induced by the vortex/acoustics interaction. Many researchers, including Chen [99] and Blevins [158], have studied this phenomenon. Figure 2.36 shows the basic mechanism. In the stable state, the acoustic pressure is low and has random characteristics in time and space. In this state, the vortex shedding frequency is unique but is not synchronized in space as shown in Fig. 2.36(a). On the other hand, under the

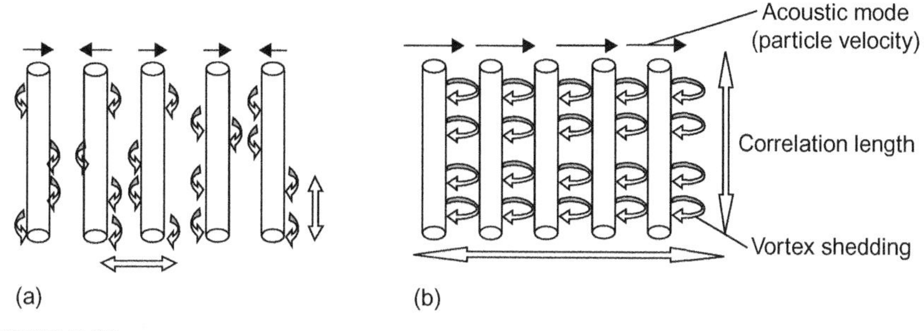

FIGURE 2.36

View showing vortex–acoustic interaction: (a) stable, and (b) resonance.

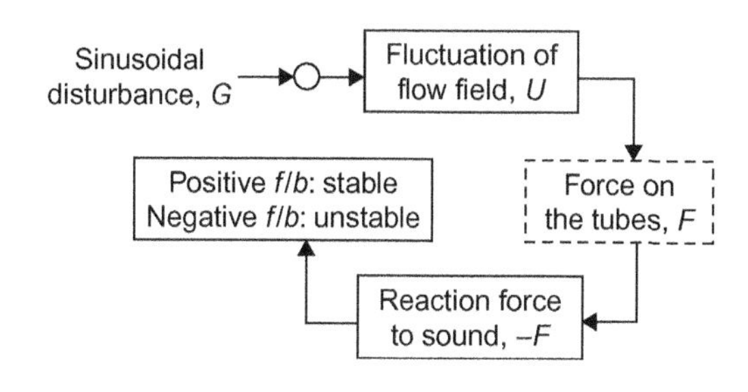

FIGURE 2.37

Feed-back mechanism between flow and acoustic field.

resonance conditions, the acoustic mode generates high-level synchronized pressure fluctuations, thus affecting the vortex shedding mechanism in two ways: first, the vortex strength is increased; second, the correlation length of vortex shedding (region of vortex shedding synchronization) is expanded in three dimensions. Synchronization of vortex shedding occurs between adjacent tubes or any two points along the tube axis.

The study of this phenomenon is fundamentally based on the idea shown in Fig. 2.37. The acoustic excitation force is regarded as a reaction fluid force acting on the tube. Therefore, introducing a sinusoidal fluctuation to the flow field and evaluating the phase relation between the flow fluctuation and the fluid force on the tube, it is possible to evaluate the stability of this system. A number of publications are based on this idea. Refs. [158–162] adopt the experimental approach while Ref. [162] is based on numerical simulations. Tanaka et al. have also derived a stability estimation formula [163]. A representative experimental apparatus is shown in Fig. 2.38 [160,161]. The apparatus is setup in a water channel. A sinusoidal fluctuation (U) is introduced to the uniform flow (V). The force acting on the tube is measured and, based on the phase relationship between the force and fluctuating flow, stability evaluation is achieved.

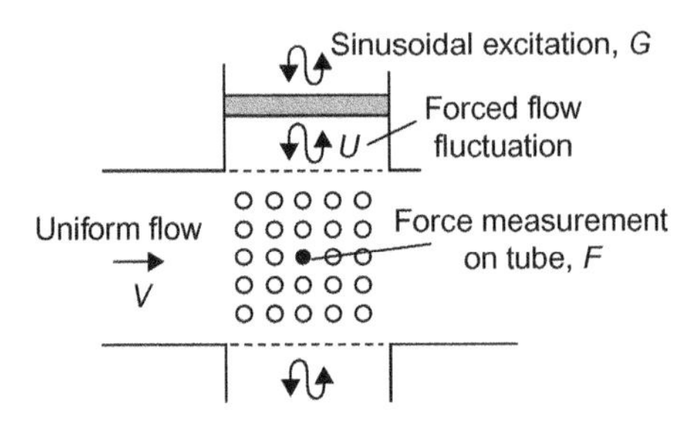

FIGURE 2.38

Experimental setup for stability evaluation by forced water-flow fluctuation [158].

In order to apply this method to the anti-resonance design of heat exchangers, there are many issues to be resolved. These include correlation length effect; data accumulation for various types of tube configurations; evaluation of acoustic energy dissipation due to radiation at duct inlet and exit; and absorption on duct walls.

2.5.2.2 Review of research on longitudinal mode resonance

This type of resonance had not been considered as important as its transverse counterpart. Recently though, it has attracted much attention because the common countermeasure, baffle plate installation, does not work well. Furthermore, the conditions under which the longitudinal mode resonance may occur are not yet fully understood. The resonance is therefore regarded as a serious issue for heat exchanger designers.

In the early stages of the study of this phenomenon, Hayama and Watanabe [164] discovered longitudinal acoustic modes using a fundamental one-dimensional duct and showed that the pressure waves of this mode have a node (correspondingly an antinode in the particle velocity) at the tube bundle location. This shows that the tube bundle was acting as a dipole acoustic excitation source. Meanwhile, Katayama et al. used a scale model of an actual gas heater industrial component which has single [157] or multiple [165] tube bundles and studied practical countermeasures using baffle plates. They identified the resonant mode shapes and showed that the resonance frequencies of longitudinal modes are almost double those of transverse modes. On the other hand, Nishida et al. [166] used plant boiler scale models. Their results on resonance phenomena are shown in Figs. 2.39 and 2.40. In Fig. 2.39(a), a series of resonances first occur along the line almost double the Strouhal number from the Fitz-Hugh map (Fig. 2.19). Near a velocity of 20 m/s, the resonance frequency suddenly shifts to a point on the line which agrees with the map. At the same time the mode shapes also change from the longitudinal type to the transverse type. Figure 2.40 shows representative mode shapes for both types of resonance. Although we have named these mode shapes longitudinal and transverse, these results indicate that the two types of mode shapes are actually coupled to each other. It is presumed that the vortex shedding pattern is different for longitudinal and transverse modes. Using a test

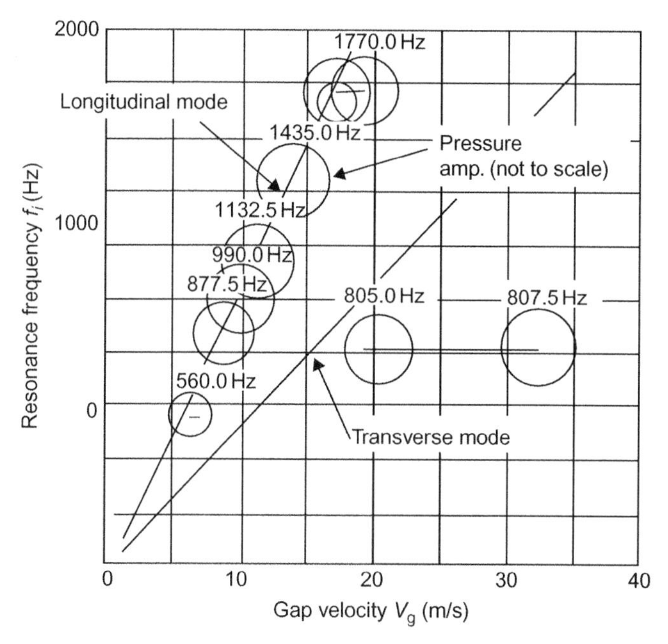

FIGURE 2.39

Resonance occurrence in boiler scale model apparatus [166].

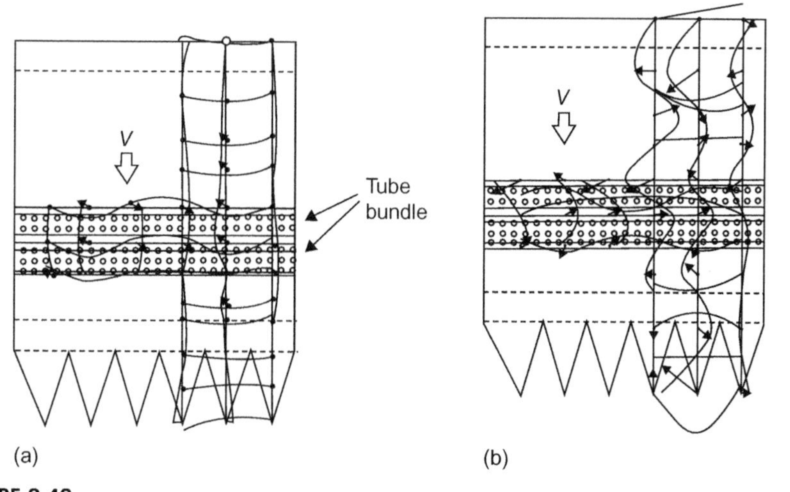

FIGURE 2.40

Mode shapes in boiler scale model apparatus: (a) transverse mode – 805 Hz, and (b) longitudinal mode – 990 Hz.

apparatus of the same structure, Hamakawa et al. measured the vortex shedding and showed that the first series of resonances accompanies symmetric vortex shedding and the second series, the usual Kármán vortex shedding [167]. As for symmetric vortex shedding, refer to Fig. 2.1 in section 2.1.2.1.

The study by Ikebe and Funakawa proposes interesting countermeasures [168]. The researchers used a one-dimensional duct with multiple tube bundles. Their results are shown in Fig. 2.41. Attaching a bell-mouth at the duct exit was found to increase the resonance flow velocity from 5 to 9 m/s. To explain this effect it was presumed that the bell-mouth, by decreasing radiation impedance, increased the effective damping inside the duct. An alternative countermeasure is the installation of fins in front of the tubes. This has the same effect as increasing the resonance flow velocity.

Nevertheless, for the longitudinal mode, the prediction of the possibility of resonance and a common method of countermeasure remain under investigation. Katayama et al. emphasized the fact that adding baffle plates makes mode shapes more complex and there are some cases in which this countermeasure might increase the possibility of longitudinal mode resonance [157]. They also pointed out an important fact with regard to tube configuration: staggered tube bundles are more resistant to resonance than in-line bundles [165].

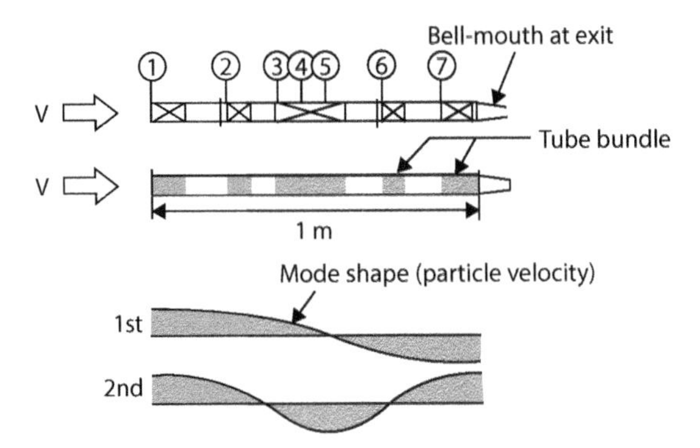

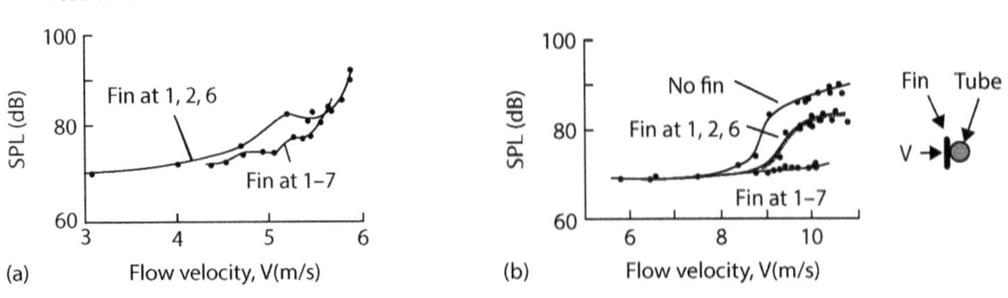

FIGURE 2.41

Typical experimental results of longitudinal mode suppression: (a) without bell-mouth, and (b) with bell-mouth.

2.5.3 Resonance prediction method at the design stage

Hereafter, we discuss mainly the transverse modes unless otherwise noted and use the gap velocity as the reference flow velocity.

2.5.3.1 Prediction based on the first condition: coincidence of vortex shedding frequency with natural frequencies of acoustic modes

The most fundamental method to suppress resonance is to avoid the occurrence of the first condition specified above. In order for designers to apply this method, they must know the following frequencies: the vortex shedding frequency and the natural frequencies of the acoustic modes in the duct.

The vortex shedding frequency of bare-tube bundles can be estimated easily by the method presented in the previous section and Eq. (2.52). Figure 2.42 shows two types of finned tubes: the first one has conventional solid fins, while the second has serrated fins where a spiral fin is split up and consists of many short sections twisted so as to increase heat transfer efficiency. Hamakawa et al. proposed a method for predicting the vortex shedding frequencies in a bundle having either type of finned tubes [144,145]. This method is based on the idea that the blockage effect by fins can be expressed by an increase of the bare-tube diameter. The new equivalent diameter, D^*, is estimated by:

$$D^* = D + (D_f - D)\frac{nt}{25.4} \tag{2.53}$$

where D, D_f, n, and t are, respectively, the core tube diameter, the outside diameter of the finned tube, the number of fins per inch along the tube axis and the fin

FIGURE 2.42

Finned tubes: (a) serrated fin, and (b) solid fin.

thickness. For serrated fins, the increased flow blockage effect due to twisting must be included in the estimation of t based on the increase in the fin's projected area in the flow direction. Using the equivalent diameter and Fitz-Hugh's Strouhal number map in Fig. 2.19, estimation of vortex shedding frequencies for a bundle with serrated finned tubes can be achieved. Figure 2.43 shows the validity of this method. Two kinds of Strouhal numbers are compared: one is based on the proposed method and is shown by dashed lines and the other (shown by symbols) is based on direct measurement of vortex shedding frequencies by a hot-wire sensor. The results agree very well. It is known that tube bundles having both types of finned tubes suffer acoustic resonance. A comparative study of resonance in the two types of finned-tube bundles has been done [169].

For the evaluation of acoustic natural frequencies of a duct, Eq. (2.51) can be used in the case of a one-dimensional acoustic mode oriented width-wise (normal to the flow direction) in the duct. When the duct has a non-uniform temperature distribution, complex geometry, or indefinite boundary conditions at the exit, analysis by a FE model is required. Additionally, in order to increase the accuracy of the model, it is important to consider the decrease in sound speed within the tube bundle. This can be done using Parker's equation [170]:

$$c' = c/(1+\sigma)^{0.5} \tag{2.54}$$

where c' and c are sound speed with and without the tube bundle in the duct, and σ is the solidity ratio, which is given by the volume fraction occupied by the tubes.

As to the degree of proximity between the vortex shedding frequency and the acoustic natural frequency, Blevins and Bressler [171] proposed that if the i-th mode natural frequency f_i satisfies the equation below, resonance is likely to occur:

$$\mathrm{St}(1-\alpha)V_g/D < f_i < \mathrm{St}(1+\beta)V_g/D \tag{2.55}$$

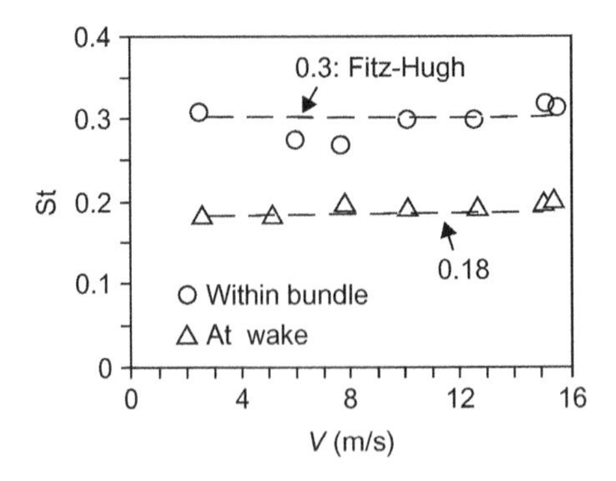

FIGURE 2.43

Experimental validation of estimated Strouhal number using equivalent diameter and Fitz-Hugh Strouhal number.

Here, St, V_g, and D are the Strouhal number, gap velocity, and tube diameter, respectively. The values of α and β depend on tube configuration in the bundle. Average values for the staggered and in-line configurations are 0.19 and 0.29, respectively. This result implies the possibility of resonance at flow conditions where the vortex shedding frequency is lower than the corresponding acoustic natural frequency. This equation also indicates the lock-in region where the vortex shedding frequency is captured by the acoustic natural frequency. When this equation is applied to the longitudinal mode resonance, one must take note of the fact that the Strouhal number is almost double the vortex shedding frequency.

2.5.3.2 Prediction based on the second condition: balance between energy supplied by vortex shedding and the energy dissipated in the acoustic field

The first condition is the most fundamental for design against resonance in heat exchangers. But there are many cases where this condition demands impractical countermeasures. For example, it may require too many baffle plates, making the duct structure more complex, which is not well suited for maintenance work. Sometimes engineers may need to decrease the number of baffle plates due to cost. In these cases the second condition, based on energy balance, can be an effective tool. However, as mentioned before, the accurate estimation of the two kinds of energy is almost impossible. Consequently, almost every method proposed for heat exchanger design is based on empirical knowledge or experimental data. Two representative resonance prediction methods are presented here: Chen's method and Eisinger's method. Chen's [99] method is considered relatively reliable based on experimental proof [140] and used by many engineers. Eisinger's method [150,151] is regarded as an improved version of Chen's method.

2.5.3.2.1 Y.N. Chen's method

Chen [99] proposed a parameter of resonance likelihood, Ψ (hereafter referred to as Chen's parameter), which depends on the structural and flow conditions:

$$\Psi = \frac{\mathrm{Re}}{\mathrm{St}}\left(1 - \frac{1}{X_l}\right)^2 \frac{1}{X_t} \tag{2.56}$$

where Re, St, X_l, X_t are the Reynolds number, Strouhal number, longitudinal tube pitch, and transverse tube pitch, respectively. This parameter has a critical value which, when exceeded, indicates that the probability of resonance is high. This method implies that an increase of Reynolds number or longitudinal tube pitch promotes the possibility of resonance occurrence. This trend shows good correspondence with experimental data such as those in Fig. 2.33. Initially, Chen set the critical value at 600. Later he increased this value to 2000 based on subsequent studies on boiler plant tube bundles (super heater and reheater) in which the flow speed had a substantially non-uniform distribution [136]. This fact implies that when engineers apply this method, they should use their own critical values depending on research on their own plant or product data.

2.5.3.2.2 Eisinger's method

The engineer may find it inconvenient using Chen's method because it is difficult to take the effect of baffle plates into account. Eisinger [150,151] solved this problem and his method is considered an improved version of Chen's method. He asserts that his method is effective for both transverse and longitudinal mode resonances. Here, we restrict its application to the transverse mode resonances. This method is applicable for staggered and in-line tube configurations.

The idea of the method is as follows. The energy loss for uniform flow by a tube bundle per unit time and per unit volume is evaluated as the product of flow velocity and pressure loss in the bundle. This energy loss is considered to be fed into the acoustic field. Based on this idea, Eisinger proposed the following parameter (hereafter referred to as Eisinger's parameter):

$$M\Delta P = (V_g/c')\cdot\Delta P \tag{2.57}$$

where M is the Mach number given by the ratio of the gap velocity V_g to the sound speed in the bundle, c'. ΔP is the pressure loss in the bundle. Based on the accumulation of plant data and experimental data, Eisinger arrived at the following procedure: evaluate Eisinger's parameter at the flow velocity which corresponds to resonance for the i-th acoustic mode (determine resonance velocity using Eq. (2.60)), and if this value exceeds the higher of the values from Eqs. (2.58) and (2.59), resonance in the i-th mode is likely to occur.

$$(M\Delta P)p,i = 0.07\times 10^{0.4375[(D/W\mathrm{St})/0.0172-1+i]} \tag{2.58}$$

$$(M\Delta P)v,i = 0.035c'(D/W\mathrm{St})\mathrm{Re}\ i/\psi \tag{2.59}$$

In the mathematical derivation, which is not detailed here, $(M\Delta P)p,i$, and $(M\Delta P)v,i$ are referred to as the critical value based on acoustic pressure and acoustic particle velocity, respectively. In Eqs. (2.58) and (2.59) D, W, St, i, c', and Re are the tube diameter, duct width, Strouhal number, mode number, sound velocity within the tube bundle (2.54), and Reynolds number, respectively. Ψ is Chen's parameter given in Eq. (2.56) and includes transverse and longitudinal tube pitch. Parameter definition for a staggered configuration is shown in Fig. 2.44(a).

A flowchart showing the procedure for application of Eisinger's method is presented in Fig. 2.45. First, the heat exchanger design parameters are set as indicated in double frames. For the pressure loss evaluation, reference should be made to the articles of Grimison [172], Jacob [173], and Blevins [174]. The next step is to select the resonance modes which have a probability of excitation based on the condition of frequency coincidence between the vortex shedding frequency and the acoustic mode natural frequencies. In principle, any acoustic mode below the highest one (identified by i_{max}) which corresponds to the maximum flow velocity of the plant under consideration must be taken into account. The orders

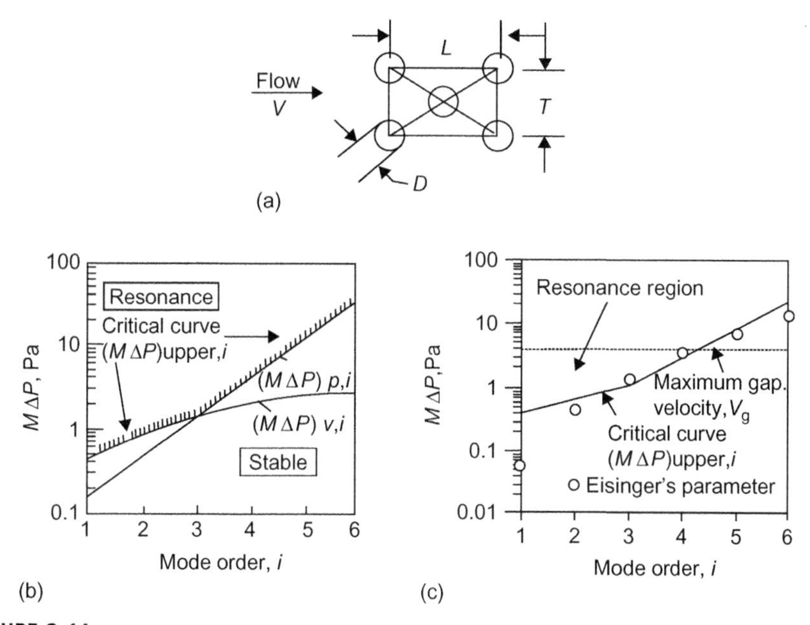

FIGURE 2.44

Prediction of resonance based on Eisinger's method [151]: (a) parameter definition for staggered array, (b) setting of critical region, and (c) example of application.

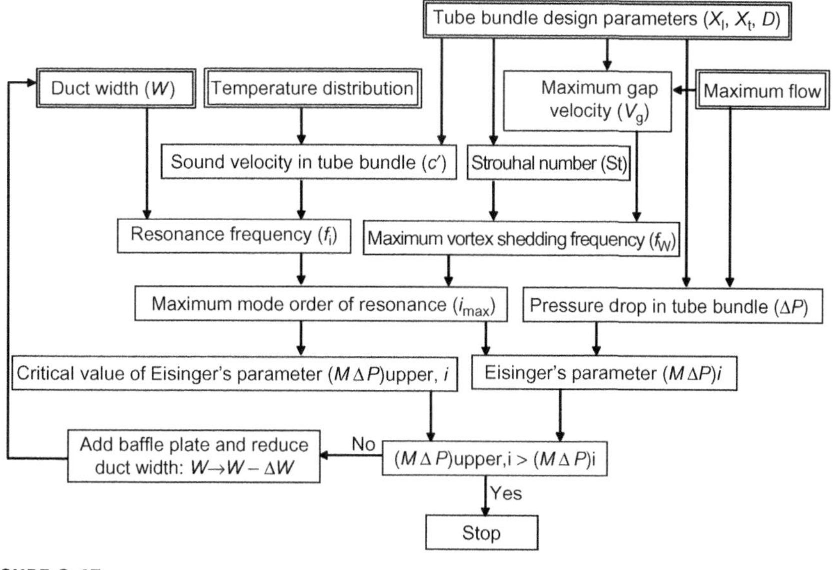

FIGURE 2.45

Flowchart of Eisinger's resonance suppression design.

of these modes are simply determined from the condition of coincidence between the acoustic resonance frequencies of Eq. (2.51) and the vortex shedding frequency in Eq. (2.52), as follows:

$$f_i = ic'/(2W) = \mathrm{St}\ V_g/D \qquad (2.60)$$

When the value of Eisinger's parameter for this i-th mode $(M\Delta P)i$, exceeds the critical value shown below and also shown in Fig. 2.44(b),

$$(M\Delta P)\mathrm{upper}, i = \max[(M\Delta P)p, i,\ (M\Delta P)v, i] \qquad (2.61)$$

baffle plate installation will be required to suppress resonance in this mode. Fig. 2.44(c) is an example showing the application of this method. The area bounded by the limit line evaluated by Eq. (2.61) and the horizontal line which shows Eisinger's parameter at maximum flow (referred to as maximum gap velocity) shows the region of possible resonance. Here, the third and fourth modes are within this region. In order to suppress the fourth mode, baffle plates should be installed as shown in Fig. 2.34(b).

The same kind of attention should be paid here as in the application of Chen's method. This method is also based on accumulation of experimental or in-situ plant data, so, when engineers apply the method, they should use their own critical values depending on a search of their own plant or product data.

2.5.4 Examples of acoustic resonance problems and hints for anti-resonance design

In order to suppress acoustic resonance, engineers must counteract the first condition related to coincidence of the two kinds of frequencies or the second condition related to energy balance.

This is done as follows:

1. *Frequency coincidence condition*: increase resonance frequency by installation of baffle plates, or reduce maximum flow velocity in operation.
2. *Energy balance condition*: increase acoustic damping by installation of absorbent material. Alternatively, set dummy tubes upstream of tube bundles to perturb flow field and to disrupt the synchronization of vortex shedding. This alternative is included in the second condition because the dummy tubes work to cut off the feedback loop between the acoustic modes and vortex shedding [157,174,175].

Hereafter, representative cases of troubleshooting based on these ideas are presented.

2.5.4.1 Waste heat recovery boiler

The structure of the boiler is shown in Fig. 2.46(a) [176]. The boiler suffered a strong acoustic resonance as shown in Fig. 2.46(b). In-situ measurement data indicated that resonance of the first transverse mode (resonance frequency 68 Hz)

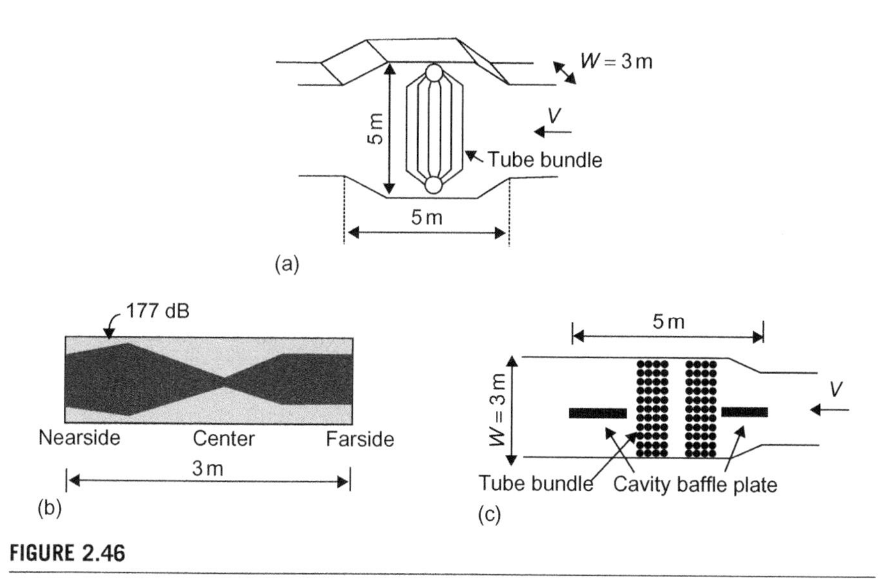

FIGURE 2.46

Example of countermeasure by cavity baffle [176]: (a) side view of boiler, (b) standing wave data, and (c) countermeasure (top view).

was dominant. Using the duct width (approximately 3 m) and gas temperature (approximately 200°C), the natural frequency of this mode was estimated by Eq. (2.51) to be 70 Hz. On the other hand, the vortex shedding frequency was estimated by Eq. (2.52) with gap velocity (6.5 m/s) and tube diameter (31.8 mm) to be 63 Hz. Because the two frequencies were close, it was concluded that Kármán vortex shedding excited the acoustic mode. Baffle plates were installed in the cavity zone as shown in Fig. 2.46(c). This countermeasure worked effectively to resolve the problem.

2.5.4.2 Shell and tube type heat exchanger

The structure of the air cooler is shown in Fig. 2.47(a) [177]. As shown in Fig. 2.47(b), acoustic pressure measurements inside the shell indicated that the in-plane resonance mode across the shell cross-section was dominant and had a resonance frequency of 120 Hz. The plane of the mode is normal to both the tube axes and flow direction. The calculated vortex shedding frequency at the operating flow rate coincided with this resonance frequency. Based on these facts, the cause of this problem was deduced to be a transverse mode resonance. Baffle plates were installed in the vertical direction as shown in Fig. 2.47(b), which solved the problem.

A similar of problem was reported by Eisinger et al. [178]. In this case, a standing wave was detected in the axial direction of the shell structure. The countermeasure was to add filler (obstacles) in the open area of the tube support sheets

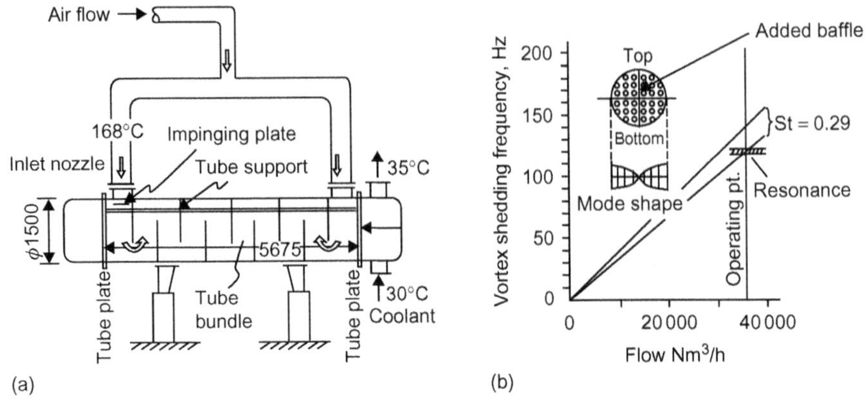

FIGURE 2.47

Example of countermeasure in shell and tube type heat exchanger [177]: (a) structure of air cooler, and (b) relation between resonance frequency and flow.

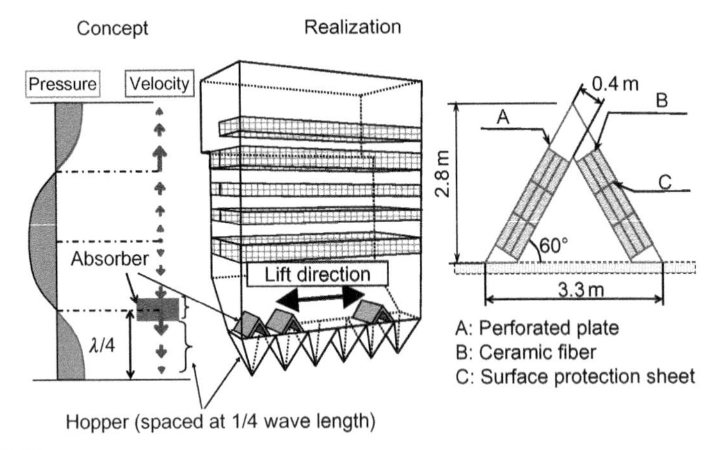

FIGURE 2.48

Example of countermeasure with acoustic absorbers for coal-fired boiler [166].

to increase acoustic damping. As a result, the noise level reduced to 30% of the original level but it led to a remarkably high pressure loss of the flow.

2.5.4.3 Coal-fired boiler

The boiler shown in Fig. 2.48 [166] suffered a resonance in the transverse and longitudinal modes. A scale model experiment discovered these modes. Figure 2.39 shows the relation between the flow velocity and resonance

Table 2.18 Hints on prevention of FIV

Phenomenon	Prevention	Note	Example
Vortex shedding	Non-resonant type:		
	1. Change frequency (usually by shortening of the span length L)	1. Should have higher frequency than vortex shedding frequency	
	2. Change flow velocity (decrease the flow velocity or change the distribution along tube span)		
	3. Add slit or spiral projection on the surface (for tripping wires should be careful, because these only shift the frequency)	3. Has limited effect. When rounding off corners for rectangular cylinders, be careful because this can increase FIV	3. As for case 3, a spacer with a damping device has been set to keep the space between wires in electric power transmission cables more than 4 times the diameter
	4. The most popular prevention for acoustic resonance is installation of baffle plates (see Section 2.5)	4. Adding dummy tubes upstream of the tube array is another preventive measure	
	To decrease response:		
	5. Add damping to decrease the response amplitude (by controlling wake using heating or sound pressure)	5. Some methods [181–182] have been proposed, but they are not very practical. For rectangular cross-section structures, the critical flow velocity for vortex-induced vibration is minimally influenced by increased structural damping. The lock-in region is shown in Fig. 2.49	5. Active mass dampers have been installed on bridge cables. On power cables, eccentric mass dampers or torque restriction type galloping dampers are used. However, there are some problems, such as increased mass or aging of spacers. Friction dampers usually have little effect. V-shaped cables have been used to control the flow of rainwater [183]
Self-induced vibration	1. Increase natural frequency of structure	1. Shorten the span length L	
	2. Decrease flow velocity to be below the critical flow velocity (when the flow is not uniform,		

(Continued)

Table 2.18 (Continued)

Phenomenon	Prevention	Note	Example
	an alternative may be to redirect the high velocity flow toward support points, leading to low velocity for the middle span region)		
	3. Increase damping to increase the critical flow velocity	3. Some active control methods exist but these may not be practical	
	4. Add a protrusion axially on the surface of structure to prevent the fluidelastic instability by fixing the separation point	4. An example [184] shows the case of a cylindrical shell, where a small pipe is attached in the wake region	
	5. Based on the critical flow velocity conditions (2.44), (2.50) for bending and twisting, the critical flow velocity for galloping can be estimated knowing the mass m and rigidity k as follows: $$V_{cr} \propto m^{1/2} k^{1/2} \quad (2.59)$$ This equation is valid even in the case of coupled bending and twisting [75]. Prevention methods include increasing the equivalent mass m and equivalent rigidity k in order to raise the critical flow velocity. By adding some attachment on the rectangular cylinder, the negative gradient of lift force can be reduced. Note also that if the modification of section properties increases the drag coefficient, this will introduce a damping effect		5. Table 2.19 shows the relation between the relative positions of the gravity center, elastic axis, and the total fluid force from the upstream edge of the wing considered to design for the prevention of a/c wing bending-twisting coupled flutter and static divergence. The wing is stable when the elastic axis is downstream of the gravity center, and is upstream of the total fluid force

Random vibration	While these vibrations cannot be completely eliminated, it is possible to decrease the vibration level:	
	1. Increase natural frequency of structures	1. Generally shortening the span length is effective
	2. Decrease flow velocity. Redistributing the flow such that high-flow velocity is closer to the support points and low velocity closer to the middle of the spans is especially effective	2. An alternative method involves setting an impingement baffle to restrict the flow
Acoustic resonance	Two methods are available: (i) changing the vortex shedding, or (ii) changing the acoustic field	
	To modify the acoustic field, the following approaches exist:	
	1. Insert plates into the duct to change the acoustic natural frequency	1. Sometimes higher modes appear
	2. Install acoustic absorbers in the duct to increase the field damping	2. Includes perforated plate
	3. Change the flow condition (temperature, etc.) to change the frequency of the acoustic field	3. Can be achieved by changing operating conditions

Table 2.19 The conditions of occurrence of wing flutter/divergence, depending on the relative positions of the gravity center (*G*), the elastic axis (*E*), and the total fluid force on the wings (*F*): along the fluid flow direction (Case of small attack angle [77])

<table>
<tr><td>Case of
[Leading edge]→(F)→(E)→(G)→[Trailing edge]
(Note: '→' indicates flow direction)

Then
• Flutter: Bad
• Divergence: Bad</td><td>Case of
[Leading edge]→(F)→(G)→(E)→[Trailing edge]
Or
[Leading edge]→(G)→(F)→(E)→[Trailing edge]
Then
• No flutter: Good
• Divergence: Bad</td></tr>
<tr><td>Case of
[Leading edge]→(E)→(G)→(F)→[Trailing edge]
Or
[Leading edge]→(E)→(F)→(G)→[Trailing edge]
Then
• Flutter: Bad
• No divergence: Good</td><td>Case of
[Leading edge]→(G)→(E)→(F)→[Trailing edge]

Then
• No flutter: Good
• No divergence: Good</td></tr>
</table>

frequencies. The corresponding mode shapes are shown in Fig. 2.40, which indicates the excitation of both types of acoustic modes. In order to suppress both modes simultaneously, acoustic absorbers were installed in the coal ash storage space (hopper). The most desirable location of the absorbers is the antinode of the acoustic mode (expressed by the sound particle velocity). By making active use of the hopper space, this was achieved. In addition, absorbers were formed in a triangular shape which enabled the ash to slide off into the hopper and also provided a wide absorption surface area for both the translational and longitudinal modes.

The number of absorbers and their placement were decided as follows. First, the resonance oscillation strength was identified using data acquired by in-situ measurements. Second, the increase of the acoustic modal damping ratio by installation of absorbers was estimated using FE complex eigenvalue analysis. A parametric study on the number and placement of absorbers was then conducted. As a result of this study, absorbers were designed so as to achieve a higher modal damping increment than the resonance oscillation strength. Structure and location of absorbers are shown in Fig. 2.48. In the plant test, transverse mode resonance was completely suppressed while noise generated by the longitudinal mode decreased by 10 dB, which achieved the noise level below the allowable maximum.

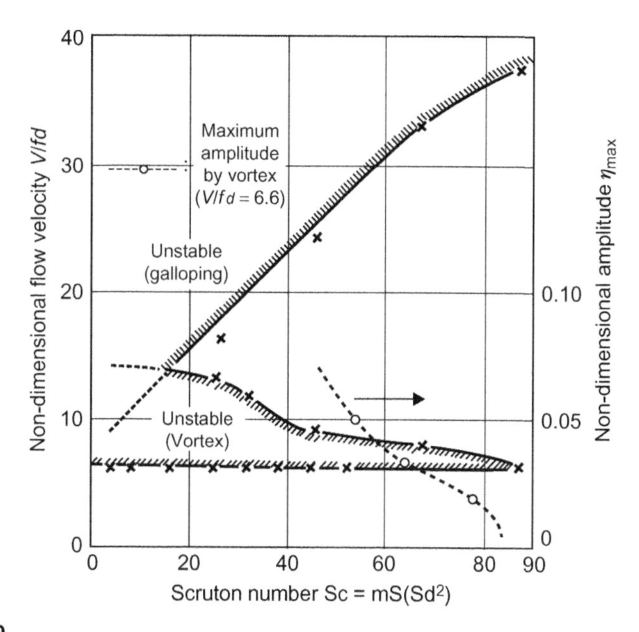

FIGURE 2.49

Region of occurrence of vortex shedding and maximum response of square cylinder as functions of Scruton number [115]. (Solid line: boundary of occurrence; broken line: maximum response.)

2.6 Prevention of FIV

There exists a variety of excitation forces associated with the phenomena encountered in FIV problems. The prevention methods can be divided into two categories: (i) reduction or elimination of the excitation forces, and (ii) modification of the structure susceptible to vibrations. These methods are summarized in Table 2.18 (Table 2.19 and Fig. 2.49 are cited in Table 2.18).

References

[1] R.E.D. Bishop, A.Y. Hassan, Proc. Roy. Soc. A. 277 (1964) 51–75.
[2] C.H. Williamson, A.J. Roshko, Fluid. Struct. 2 (4) (1988) 355–381.
[3] D. Brika, A.J. Laneville, Fluid. Mech. 250 (1993) 481–508.
[4] R.T. Hartlen, I.G. Currie, ASCE J. Eng. Mech. Div. 96 (1970).
[5] W.D. Iwan, R.D. Blevins, J. Appl. Mech. 40 (1974) 518–586.
[6] T. Sarpkaya, Proceedings of the International Conference on FIV in Fluids Engineering, 1982, pp. 131–139.
[7] A. Lowdon, et al. C416/041, IMechE, 1991, p. 283.
[8] O.M. Griffin, J. Fluid. Eng. 107 (1985) 298–306.

[9] B.J. Vickery, A.W. Clark, ASCE J. Struct. Div. 98 (ST1) (1972) 1.
[10] E. Wang, et al., Proceedings of OMAE/Tokyo, Vol. III, ASME, 1986, pp. 393–401
[11] J.K. Vandiver, J. Fluid. Struct. 7 (1993) 423–455.
[12] A. Bokaian, J. Sound Vib. 175 (5) (1994) 607–623.
[13] D.T. Tsahalis, J. Energ. Resour., 106, *ASME*, 1984, pp. 206–213.
[14] R. King, R. Jones, in: E. Naudascher, D. Rockwell (Eds.), Practical Experiences with Flow-Induced Vibrations, Springer-Verlag, Berlin, 1980.
[15] A. Khalak, C.H. Williamson, J. Fluid. Struct. 10 (1996) 455–472.
[16] ASME, Boiler and Pressure Vessel Code Section III, Division I Appendix N, Article N-1300, 1995, pp. 370–397.
[17] JSME Standard, Guideline for Evaluation of Flow-Induced Vibration of a Cylindrical Structure in a Pipe (in Japanese), JSME S 012, 1998.
[18] T.M. Mulchay, ASME FIV Symposium, Vol. 1, (1984) pp. 15–28.
[19] D.J. Newman, G. Karniadakis, in: Bearman (Ed.), Flow-Induced Vibration, Balkema, Rotterdam, 1995.
[20] A.R. Massih, K. Forsberg, FSI,AE,FIV + N Vol. I ASME, AD-vol. 53–1, 1997.
[21] P. Anagnostopoulos, et al., J. Fluid. Struct. 12 (3) (1998) 225–258.
[22] T.A. Fox, C.J. Apelt, J. Fluid. Struct. 7 (1993) 375–386.
[23] J. Kitagawa, et al., J. Fluid. Struct. 13 (1999) 499–518.
[24] J.R. Morison, et al., AIME Petrol. Trans. 189 (1950) 149–154.
[25] R.G. Longoria, et al., J. Offshore Mech. Arct. 115 (1993) 23–30.
[26] K. Yoshida, et al., J. Offshore Mech. Arct. 112 (1990) 14–20.
[27] M.S. Hall, O.M. Griffin, J. Fluid. Eng. 115 (1993) 283–291.
[28] J. Zhang, et al., J. Fluid. Struct. 7 (1993) 39–56.
[29] J.R. Meneghini, P.W. Bearman, J. Fluid. Struct. 9 (1995) 435–455.
[30] H.M. Badr, et al., J. Fluid. Mech. 303 (6) (1995) 215–232.
[31] E.A. Dockstader, et al., Trans. ASCE 121 (1956) 1088–1112.
[32] D.J. Johns, C.B. Sharma, In Flow-Induced Structural Vibrations, Naudascher (ed.), 1974, pp. 650–662.
[33] D.J. Johns, J. Sound Vib. 67 (1979) 432–435.
[34] M.P. Païdoussis, C. Helleur, J. Sound Vib. 63 (1979) 527–542.
[35] M.P. Païdoussis, J. Sound Vib. 103 (1985) 201–209.
[36] M.P. Païdoussis, J. Sound Vib. 83 (1982) 533–553.
[37] M.P. Païdoussis, D.T. M-Wong, J. Fluid. Mech. 115 (1982) 411–426.
[38] S. Katsura, J. Sound Vib. 100 (1985) 527–550.
[39] M.P. Païdoussis, et al., Int. Conf. FIV (1987) 377–392.
[40] O.M. Griffin, et al., Paper No. OTC-2319, Offshore Technology Conference, 1975.
[41] T. Sarpkaya, J., WWPC & Ocean Div. 104, ASCE, 1978, pp. 275–290.
[42] Hara, Ohtani, *J. Jpn. Soc. Mech. Eng.* (in Japanese) 48 (431) (1982 C) 962–971.
[43] Hara, *J. Jpn. Soc. Mech. Eng.* (in Japanese) 48 (433) (1982 C) 1371–1379.
[44] Hara, Ogawa, *J. Jpn. Soc. Mech. Eng.* (in Japanese) 49 (445) (1983 C) 1624–1629.
[45] Architectural Institute of Japan, Recommendations for Loads on Buildings (in Japanese), Maruzen, 2000, pp. 323–325.
[46] Y. Uematsu, et al., J. Fluid. Struct. 2 (1988).
[47] M.P. Païdoussis, J. Sound Vib. 83 (1982) 555–572.
[48] A. Mazouzi, et al., J. Fluid. Struct. 5 (1991) 605–626.

[49] H.Y. Wong, R.N. Cox, third ed., *Proceedings of Colloquium* on *Industrial Aerodynamics*, Vol. 2, Fachhochschule Aachen, Germany, 1978.

[50] C. Scruton, D.E.J. Walshe, NPL, UK, Aero Report No. 335, 1957.

[51] Oarai Engineering Center, Power reactor and nuclear fuel development corporation, Report Cause identification of sodium leakage accident in Monju; Vibration of thermo well by fluid force (in Japanese), 1997.

[52] R. King, et al., J. Sound Vib. 29 (2) (1973) 169.

[53] R.N. Sainsbury, D. King, Proc. Inst. Civ. Eng. 49 (1971).

[54] Hikami, *Wind Eng.* (in Japanese) 11 (27) (1986) 17–28.

[55] H. Halle, W.P. Lawrence, in: E. Naudascher, D. Rockwell (Eds.), Practical Experiences with Flow-Induced Vibrations, Springer-Verlag, Berlin, 1980.

[56] M.M. Zdravkovich, et al., J. Sound Vib. 101 (4) (1985) 511–521.

[57] M.M. Zdravkovich, et al., J. Fluid. Eng. 1 (107) (1985) 507–511.

[58] M.M. Zdravkovich, et al., Proceedings of ASME Winter Meeting, vol. 2, 1984, pp. 1–18.

[59] P.J. Strykowski, et al., J. Fluid. Mech. 71 (1990) 107.

[60] M. Yoneta, et al., Bridge Eng. (in Japanese) (1992) 54–62.

[61] T. Nakamura, et al., J. Wind Eng. (20) (1984) 129–140.

[62] K. Bae, et al., *J. JSME* (in Japanese) 55 (519) (1989) 3328–3332.

[63] Y. Tomita, et al., *J. JSME* (in Japanese) 51 (468) (1987) 2571–2580.

[64] S. Nagai, et al., *J. JCCE* (in Japanese) 24 (1997) 352–356.

[65] M. Ookusu, et al., *J. JASNAOE* (in Japanese) 131 (1972) 53–64.

[66] B.H. Spring, P.L. Monkmeyer, Proceedings of 14th International Conference on Costal Engineering, 1974, 1828–1847.

[67] F. Hara, et al., ASME FIVN 2 (1988) 63–78.

[68] T. Iijima, F. Hara, *JSME* (in Japanese) 54 (497) (1988) 80–86.

[69] T. Iijima, F. Hara, *JSME* (in Japanese) 56 (525) (1990) 1087–1093.

[70] T. Nojima, et al., *JSME* (in Japanese) 56 (527) (1990) 1665–1672.

[71] T. Iijima, et al., *JSME* (in Japanese) 57 (537) (1991) 1469–1476.

[72] B.H. Lakshmana, et al., JSV 122 (1988) 497–514.

[73] T. Ichioka, et al., Proc. Fluid Mech. JSME (1993) 370–372.

[74] H. Tanaka, et al., Proceeding of JSME (in Japanese), 1999, pp. 370–372.

[75] S.S. Chen, ASME J. PVP 108 (1986) 382–393.

[76] S.J. Price, et al., FIVN 1 (1998) 91–111.

[77] R.D. Blevins, Flow-Induced Vibration, second ed., Krieger Publishing Co., Malabar, FL, 1990.

[78] S.S. Chen, Flow-Induced Vibration of Circular Cylindrical Structures, Hemisphere Publishing Corporation, Washington, 1987.

[79] J.T. Thorngren, Hydrocarb. Process. (1970).

[80] Y.N. Chen, Trans. ASME Vol. 2 (1970) 134–146.

[81] H.J. Connors, An Experimental Investigation of the Flow-Induced Vibration of Tube Arrays in Cross Flow, Ph.D. Report, University of Pittsburg, 1969.

[82] B.W. Roberts, Mechanical Engineering Science, Monograph No. 4, 1966.

[83] S.J. Price, M.P. Païdoussis, J. Sound Vib. 97 (4) (1984) 615–640.

[84] J.H. Lever, D.S. Weaver, ASME J. Pressure Technol. 104 (1982) 147–158.

[85] H. Tanaka, JSME J. 46 (408) (1980) 1398–1407.

[86] M.J. Pettigrew, D.J. Gorman, Proceedings of International Symposium Vibration Problems in Industry, No. 424 Part I, 1973.
[87] M.K. Au-Yang, ASME PVP 389 (1999) 17–33.
[88] M.J. Pettigrew, et al., Nucl. Eng. Des. 48 (1978) 97–115.
[89] F. Hara, I. Ohtani, JSME J. 48 (431) (1984) 962–969.
[90] F. Axisa, et al., ASME PVP 133 (1988).
[91] T. Nakamura, et al., ASME J. Pressure Vessel Technol. 114 (1992) 472–485.
[92] M. Takai, et al., Proceedings of Eigth International Conference on Nuclear Engineering, ICONE-8090, 2000.
[93] F. Inada, et al., ASME AD 53 (2) (1997) 357–364.
[94] N.W. Mureithi, et al., ASME PVP 328 (1996) 111–121.
[95] N.W. Mureithi, et al., J. Fluid. Struct. 21 (2005) 75–87.
[96] R. Violette, N.W. Mureithi, M.J. Pettigrew, ASME J. Pressure Vessel Technol. 128 (2006) 148–159.
[97] G. Ricciardi, et al., ASME J. Pressure Vessel Technol. 133 (6) (2011) 061301.
[98] Fitz-Hugh, J.S., Proceedings of UKAEA/NPL International Symposium on Vibration Problem in Industry, Keswick, England, Paper 427, 1973, pp. 1–17.
[99] Y.N. Chen, Trans. ASME J. Eng. Ind. (1968) 134–146.
[100] ASME, ASME Boiler and Pressure Vessel Code – Section III Rules for Construction of Nuclear Power Plant Components Division 1 – Appendices, 1995, pp. 370–397.
[101] H.J. Connors, ASME J. Mech. Des. 100 (1978) 347–353.
[102] I.D.R. Grant, NEL Report No. 590, 1975, pp. 1–22.
[103] O. Baker, Oil Gas J. 53 (1954) 185.
[104] M.J. Pettigrew, et al., ASME J. PVP 111 (1989) 478–587.
[105] JSME, Guideline for Fluidelastic Vibration Evaluation of U-bend Tubes in Steam Generators, JSME S016, 2002.
[106] E. de Langre, B. Villard, Proceedings in EAHA Conference, 1994.
[107] E. Naudascher, D. Rockwell (Eds.), Practical Experiences with Flow-Induced Vibrations, Springer-Verlag, New York, 1980.
[108] Japanese Society of Steel Construction, *Wind-Resistant Engineering of Structures* (in Japanese), Tokyo Electric University Publication Station, 1997.
[109] Architectural Institute of Japan, *Building Load Standard/Guidance* (in Japanese), Maruzen, 1993.
[110] T. Balendra, Vibration of Building to Wind and Earthquake Loads, Springer-Verlag, New York, 1995.
[111] Y. Nakamura, et al., Reports of Research Institute for Applied Mechanics, No. 40 *The Aerodynamic Forces and Moment Characteristics of Rectangular and H-shaped Cylinders* (in Japanese), Kyushu University, 1982.
[112] E. Naudascher, Y. Wang, J. Fluid. Struct. 7 (1993) 341–373.
[113] Y. Nakamura, K. Hirata, T. Urabe, J. Fluid. Struct. 5 (1991) 521–549.
[114] Y. Nakamura, J. Fluid. Struct. 10 (1996) 147–158.
[115] C. Scruton, On the wind excited oscillations of stacks, towers, and masts, Proceedings of the International Conference of Wind Effects on Buildings and Structures, Her Majesty's Stationery Office, Teddington, 1963.
[116] J.P. Den Hartog, Mechanical Vibrations, fourth ed., McGraw-Hill, New York, 1984Reprinted by Dover, New York.

[117] M. Novak, ASCE J. Eng. Mech. Div. 96 (1969) 115–142.
[118] M. Novak, H. Tanaka, ASCE J. Eng. Mech. Div. 100 (1974) 27–47.
[119] E. Naudascher, D. Rockwell, Flow-Induced Vibrations – An Engineering Guide, A.A. Balkema, Rotterdam, 1994.
[120] A. Okajima, et al., *Trans. Jpn. Soc. Mech. Eng.* (in Japanese), Series B 65 (635) (1999) 2196–2222.
[121] A. Okajima, *Trans. Jpn. Soc. Mech. Eng.* (in Japanese), Series B 65 (635) (1999) 2190–2195.
[122] R. Parker, M.C. Welsh, The Effect of Sound on Flow Over Bluff Bodies, University of Wales, Swansea, Mechanical Engineering Report MR/87/81, 1981. (Also in Int. J. Heat Fluid Flow, Vol. 4, 1983, pp. 113–127.)
[123] C.W. Knisely, Report SFB 210/E/13. Sonderforschungs-bereich 210 Strouhal Number of Rectangular Cylinders at Incidence, Universitat Karlsruhe, Karlsruhe, Germany, 1985.
[124] Y. Nakamura, Y. Ohya, H. Tsuruya, J. Fluid. Mech. 222 (1991) 437–447.
[125] D.T. Nguyen, K. Naudasher, J. Hydraulic Eng. 117 (1991) 1056–1076.
[126] Y. Wang, Report SFB 210/E/74, Sonderforschungsbereich 210 Transverse and Plunging Vibrations of Trashrack Bars in Inclined Flow (in German), Universitat Karlsruhe, Karlsruhe, Germany, 1992.
[127] J. Novak, Acta Technica CSAV (1972) 372–386.
[128] N. Shiraishi, M. Matsumoto, On classification of vortex-induced oscillation and its application for bridge structures, J. Wind Eng. Ind. Aero Dyn. 14 (1983) 419–430.
[129] Y. Nakamura, T. Mizota, ASCE J. Eng. Mech. Div. 101 (2) (1975) 125–142.
[130] G.V. Parkinson, N.P.H. Brooks, J.D. Smith, J. Appl. Math. 28 (1961) 252–258.
[131] A.S. Richardson, et al., Proceedings of the First International Conference on Wind Effects on Buildings and Structures, vol. 2, 1965, pp. 612–686.
[132] J.E. Salter, Aero-elastic Instability of Structural Angle Section, Ph.D. Thesis, University of British Columbia, 1969.
[133] Y. Nakamura, Y. Tomonari, J. Sound Vib. 52 (1977) 233–241.
[134] G.V. Parkinson, J.D. Smith, Quart. J. Math. Appl. Math. 17 (1964) 225–239.
[135] R.C. Baird, Combustion 25 (10) (1954) 38–44.
[136] Y.N. Chen, W.C. Young, Trans. ASME (1974) 1072–1075.
[137] D.S. Weaver, J.H. Fitzpatrick, International Conference on Flow Induced Vibration, Paper A1. BHRA., 1987, pp. 1–17.
[138] M.P. Païdoussis, Nucl. Eng. Des. 74 (1) (1983) 31–60.
[139] R.D. Blevins, J. Sound Vib. 92 (4) (1984) 455–470.
[140] F.L. Eisinger, R.E. Sullivan, J. Pressure Vessel Technol. 166 (1994) 17–23.
[141] D.S. Weaver, J.A. Fitzpatrick, ASME J. Pressure Vessel Technol. 109 (1987) 219–223.
[142] W.A. Mair, P.D.F. Jones, R.K.W. Palmer, J. Sound Vib. 39 (3) (1975) 293–296.
[143] K. Okui, M. Iwabuchi, H. Oda, K. Shimada, *Proc. JSME VS Tech.'95* (in Japanese) 22 (1995) 132–135.
[144] H. Hamakawa, T. Fukano, M. Aragaki, E. Nishida, *JSME J.* (in Japanese) 65(B) (635) (1997) 18–25.
[145] H. Hamakawa, T. Fukano, E. Nishida, Y. Ikuta, T. Morooka, *JSME J.* (in Japanese) 66(B) (646) (2000) 1301–1308.

[146] B.J. Grotz, F.R. Arnold, Technical Report, No. 31, Mechanical Engineering Department, Stanford University, Stanford, CA, 1956.
[147] J.A. Fitzpatrick, J. Sound Vib. 99 (3) (1985) 425–435.
[148] S. Ziada, A. Oengoren, E.T. Buhlmann, International Symposium on Flow Induced Vibration and Noise, Winter Annual Meeting Chicago, ASME, vol. 3, 1988, pp. 245–254.
[149] R.D. Blevins, M.M. Bressler, Trans. ASME 109 (1987) 282–288.
[150] F.L. Eisinger, *Flow-Induced Vibration*, ASME PVP 298 (1995) 111–120.
[151] R.E. Sullivan, J.T. Francis, F.L. Eisinger, ASME PVP 363 (1998) 1–9.
[152] R.D. Blevins, M.M. Bressler, Symposium on Flow-Induced Vibration and Noise, ASME, 4, vol. 243, 1992, pp. 59–79.
[153] F.L. Eisinger, ASME J. Pressure Vessel Technol. 102 (1980) 138–145.
[154] A. Nemoto, M. Yamada, Symposium on Flow-Induced Vibration and Noise, ASME, PVP, vol. 243 (4) 1992, pp. 137–152.
[155] A. Nemoto, M. Yamada, *Flow-Induced Vibration*, ASME, PVP 273 (1994) 273–282.
[156] F.L. Eisinger, R.E. Sullivan, J. Fluid. Struct. 10 (1996) 99–107.
[157] K. Katayama, M. Tsuboi, T. Kawaoka, K. Ohta, Y. Sato, *JSME J.* (in Japanese) 65 (C) (640) (1999) 4633–4639.
[158] R.D. Blevins, J. Fluid. Mech. 161 (1985) 217–237.
[159] G. Yamanaka, T. Adachi, *J. Acoust. Soc. Jpn.* (in Japanese) 27 (5) (1971) 246–256.
[160] H. Tanaka, Y. Imayama, K. Koga, K. Katayama, *JSME J.* (in Japanese) 55(B) (509) (1989) 120–125.
[161] T. Sato, Y. Imayama, K. Katayama, *JSME J.* (in Japanese) 61(C) (585) (1995) 1763–1768.
[162] T. Sato, Y. Imayama, F. Nakajima, *JSME J.* (in Japanese) 61(C) (585) (1995) 1769–1775.
[163] H. Tanaka, K. Tanaka, F. Shimizu, T. Iijima, *JSME J.* (in Japanese) 64(C) (626) (1998) 3293–3298.
[164] S. Hayama, T. Watanabe, Proceedings of the 73rd JSME's Annual Meeting (in Japanese), V, 1995, pp. 155–156.
[165] K. Katayama, S. Morii, M. Tsuboi, T. Kawaoka, Y. Sato, Y. Imayama, *JSME J.* (in Japanese) 66(C) (641) (2001) 60–66.
[166] E. Nishida, M. Miki, N. Sadaoka, T. Fukano, H. Hamakawa, ASME IMECE (2002) 2002-33405, 17–22.
[167] H. Hamakawa, T. Fukano, E. Nishida, N. Fujimura, Proceedings of Dynamics and Design Conference 2005, No. 117, JSME, 2005 (on CD ROM).
[168] H. Ikebe, M. Funakawa, *Proc. JSME's DMCD* (in Japanese) A (930-42) (1993) 370–375.
[169] A. Nemoto, A. Takakuwa, M. Tsutsui, AD-Vol. 53-2 Fluid–Structure Interaction, Aeroelasticity, Flow-Induced Vibration and Noise, Vol. 11, ASME, 1997, pp. 311–320.
[170] R. Parker, J. Sound Vib. 57 (1978) 245–260.
[171] R.D. Blevins, Applied Fluid Dynamics Handbook, Van Nostrand Reinhold, New York, 1984.
[172] E.D. Grimison, Trans. ASME 59 (1937) 583–594.
[173] M. Jacob, Trans. ASME 60 (4) (1938) 384–386.

[174] B. Rolsma, B. Nagamatsu, Design Engineering Technical Conference, Hartford, Connecticut, 1981, pp. 20–23.
[175] R.D. Blevins, M.M. Bressler, J. Pressure Vessel Technol. 190 (1987) 275–281.
[176] P.J. Strykowski, Ph.D. Thesis, Engineering and Applied Science, Yale University, 1986.
[177] Proceedings of [v-BASE] Forum, JSME, [No. 920-55], 1997, pp. 22–23.
[178] Proceedings of [v-BASE] Forum, JSME, [No. 920-65], 1992, pp. 9–10.
[179] F.L. Eisinger, Symposium on Flow-Induced Vibration and Noise, ASME PVP, vol. 243, (4), 1992, pp. 27–43.
[180] P.A. Monkewitz, IUTAM Symposium, Gottingen, Bluff-Body Wakes Dynamics and Instabilities, 1992, pp. 227–240.
[181] J.E.F. Williams, B.C. Zhao, Flow-Induced Vibration and Noise, vol. 1, 1988, p. 51.
[182] A. Baz, M. Kim, ASME PVP 206 (1991) 75.
[183] K. Roussopoulos, J. Fluid. Mech. 248 (1993) 267–296.
[184] T. Miyazaki, Proceedings of Symposium on Wind Direction, 1988, pp. 145–150.
[185] M.P. Païdoussis, S.J. Price, S.Y. Ang, International Conference on Flow Induced Vibrations, 1987, pp. 377–392.

CHAPTER

Vibration Induced by External Axial Flow

3

In this chapter, straight tubes, tube bundles, flexible plates, and shells subjected to external parallel flow are presented. Leakage-flow-induced vibration is also discussed. In the case of straight tubes and tube bundles, random vibrations are predominant. Evaluation methods for vibration amplitudes in single-phase as well as two-phase flow are discussed in Section 3.1. The critical flow velocity for flutter and divergence is very high for external parallel flows. On the other hand, the structural stiffness is relatively low for flexible plates and shells. Consequently, the feedback force to the structural vibration can be relatively large for structures subjected to leakage-flow. In this case, divergence and flutter can easily occur at relatively low velocities. Flutter and divergence analysis is presented in Sections 3.2 and 3.3. Evaluation methods for these instabilities are also discussed.

3.1 Single cylinder/multiple cylinders

3.1.1 Summary of objectives

Nuclear fuel bundles (Fig. 3.1(a)) and steam generators (Fig. 3.1(b)) subjected to external parallel flow may undergo flutter and/or divergence instability at high flow velocities. Furthermore, even below these high instability velocities, vibration due to random pressure fluctuations may occur. However, compared to the cross-flow case presented in Chapter 2, the level of random vibration in axial flow is small. Also, the critical flow velocity for flutter is much higher than the instability velocity in cross-flow. Random vibration in external axial flow is caused by flow noise, consisting of both near-field and far-field components. The near-field components are boundary-layer turbulence pressure fluctuations. The sources of far-field flow noise include pulsations, vortex shedding over submerged objects, turbulence generated at pipe bends, cavitation, etc.

While boundary-layer pressure fluctuations may be evaluated via correlation of statistical characteristics, similar correlation analysis cannot be applied to far-field noise because the latter depends on the source of the excitation.

The evaluation method for random vibration due to near-field flow noise is presented in Section 3.1.2. Flutter and/or divergence instability analysis is presented in Section 3.1.3.

Flow-Induced Vibrations.

3.1.2 Random vibration due to flow turbulence

3.1.2.1 Historical background of evaluation methodologies

External parallel-flow-induced random vibration of tubes and tube bundles in two-phase flow has been investigated since the advent of nuclear reactors (in the 1960s).

Tests simulating boiling water reactor (BWR) fuel bundles subjected to steam-water flow were performed. Quinn [1] found the relationship between the natural frequencies of fuel rods and the maximum amplitude of vibration in the early 1960s. In the 1970s, Gorman [2], Pettigrew [3] and others studied vibration in air–water and steam–water two-phase flows. Chen [4] proposed a semi-theoretical method to determine tube random vibration in single-phase flow. The method employs random vibration theory and measured pressure fluctuation data.

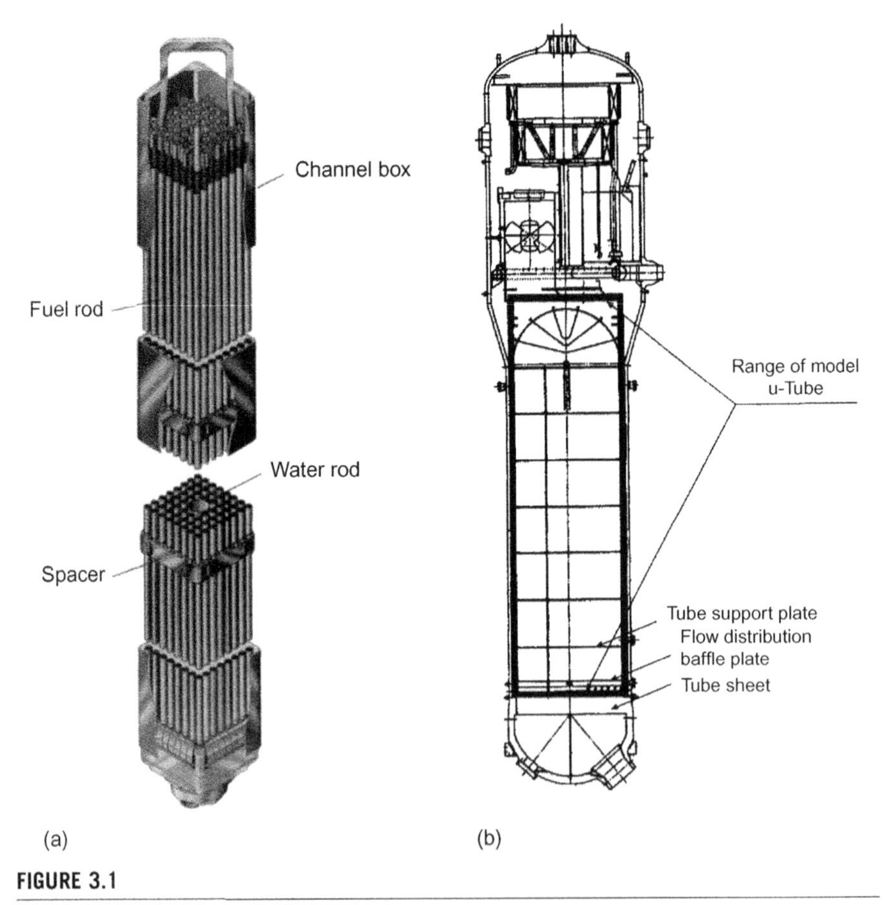

FIGURE 3.1

Schematic of (a) BWR fuel bundle and (b) steam generator.

In the 1980s, Païdoussis and Curling [5] developed a method for random vibration analysis of tube bundles in which inter-tube coupling was taken into consideration.

3.1.2.2 Evaluation of random vibration in single-phase flow

The evaluation of random vibration due to near-field flow noise is described in this section. The evaluation method for the vibration of a cylinder in axial single-phase flow is described quoting the papers of Chen and Wambsganss [6] and others cited in the work of Chen [4]. Random vibration theory is employed by Chen and co-workers. The spatial correlation of the fluid excitation in the axial and circumferential directions is taken into consideration. A phenomenological model of the cross-correlation function of the wall surface pressure field by Corcos [7], is usually used (see Fig. 3.2). Mathematically, this model is expressed as follows:

$$\Psi_{\mathrm{pp}}(\omega, z_1, z_2, \theta_1, \theta_2) = \Phi_{\mathrm{pp}}(\omega) A\left(\frac{\omega|z_2 - z_1|}{V_{\mathrm{c}}}\right) B\left(\frac{\omega D|\theta_2 - \theta_1|}{2V_{\mathrm{c}}}\right) \exp\left(\frac{i\omega|z_2 - z_1|}{V_{\mathrm{c}}}\right) \tag{3.1}$$

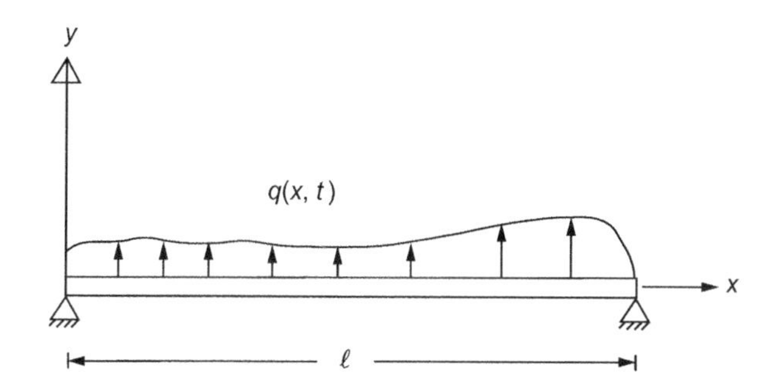

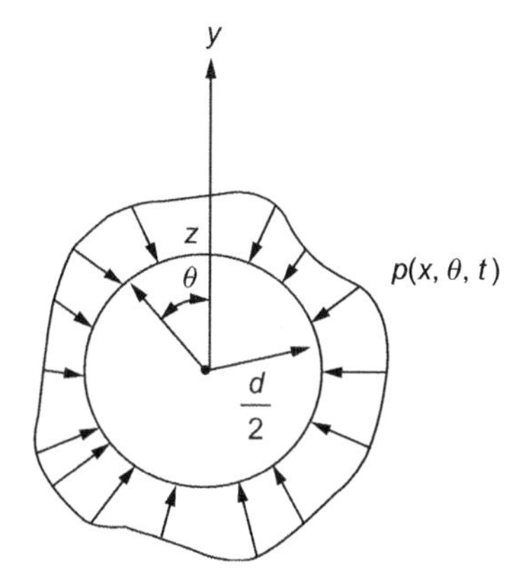

FIGURE 3.2

Circular cylinder subject to turbulent pressure fluctuations.

Here, Ψ_{pp} is the cross-spectral density of the wall pressure field; $\Phi_{pp}(\omega)$ is the wall pressure power spectral density at a point; A and B are spatial functions describing the axial and circumferential decay of the correlation, respectively; and V_c is the convection velocity. Chen and Wambsganss proposed the functional forms for A, B, and V_c in Eq. (3.1) and compared their results with experimental measurements. The wall pressure power spectral density from experimental measurements is shown in Fig. 3.3, and is given by:

$$\begin{aligned}\Phi_{pp}(f_r) &= \frac{\Phi_{pp}(f)}{\rho^2 V^4}\frac{V}{d_h} \\ &= 0.272\times 10^{-5}/f_r^{0.25} \quad f_r<5 \\ &\quad 22.75\times 10^{-5}/f_r^3 \quad f_r>5\end{aligned} \tag{3.2}$$

Here, d_h is the hydraulic diameter and f_r is the reduced frequency defined by $f_r = fd_h/V$, where $f = \omega/2\pi$. The decay functions A and B are given by:

$$\begin{aligned}A\left(\frac{\omega|z_2-z_1|}{V_c}\right) &= \exp\left(-0.1\left|\frac{\omega(z_2-z_1)}{V_c}\right|\right) \\ B\left(\frac{\omega D|\theta_2-\theta_1|}{2V_c}\right) &= \exp\left(-0.55\left|\frac{\omega D(\theta_2-\theta_1)}{2V_c}\right|\right)\end{aligned} \tag{3.3}$$

Comparisons with the experimental results of Willmarth and Wooldridge [8] and Clinch [9] are shown in Figs. 3.4 and 3.5.

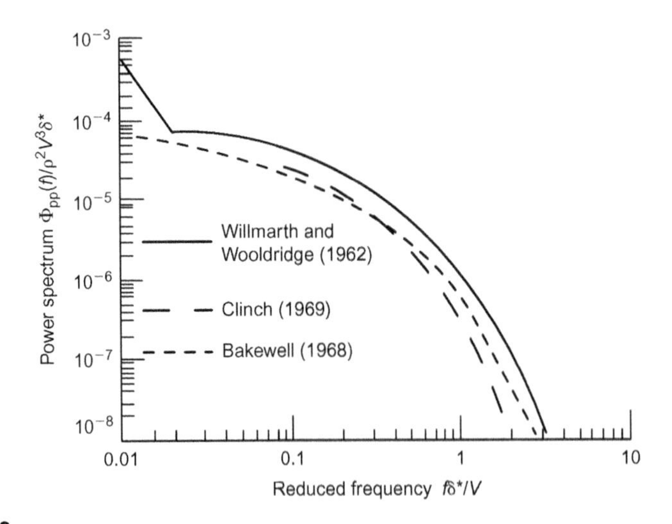

FIGURE 3.3

Turbulent wall pressure spectra [6].

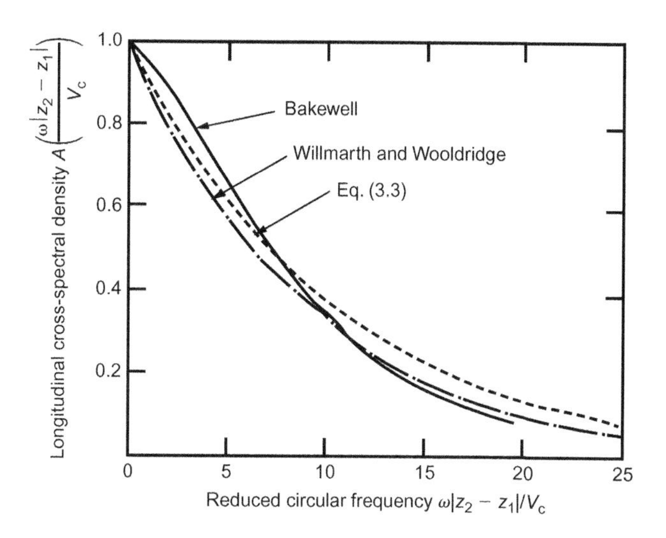

FIGURE 3.4

Magnitude of cross-spectral density of turbulent wall pressure (longitudinal) [6].

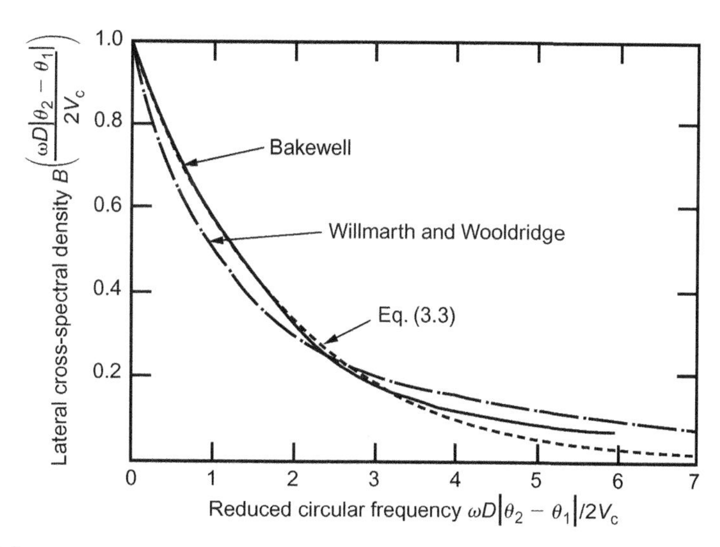

FIGURE 3.5

Magnitude of cross-spectral density of turbulent wall pressure (lateral) [8,9].

The average convection velocity of the boundary layer is given by:

$$\frac{V_c}{V} = 0.6 + 0.4 \exp\left(-2.2\frac{\omega\delta^*}{V}\right) \tag{3.4}$$

This empirical relation is compared with the experimental results of Bakewell [10] and Schloemer [11] in Fig. 3.6. In the equation, δ^* is the displacement thickness of the boundary layer, given by:

$$\delta^* = \frac{d_h}{2(n+1)} \text{ (for large pipe or flow channel)}$$

where,

$$n = 0.125m^3 - 0.181m^2 + 0.625m + 5.851$$

and,

$$m = \log_{10}(\text{Re}) - 3$$

The cross-spectral density of the wall pressure field of Eq. (3.1) is derived based on Eqs. (3.2), (3.3), and (3.4). The root-mean-square (rms) vibration amplitude and tube displacement power spectral density can be calculated using random vibration theory via this equation.

From the mathematical model and the pressure field characteristics, the following conclusions can be drawn:

1. The response of the first mode is dominant in comparison with the higher modes. In practice, a one-mode approximation is therefore sufficiently accurate.
2. Since the magnitude of the power spectral density of the pressure field is approximately proportional to the mean axial flow velocity cubed, the rms cylinder vibration amplitude is proportional to $V^{1.5} \sim V^{3.0}$.

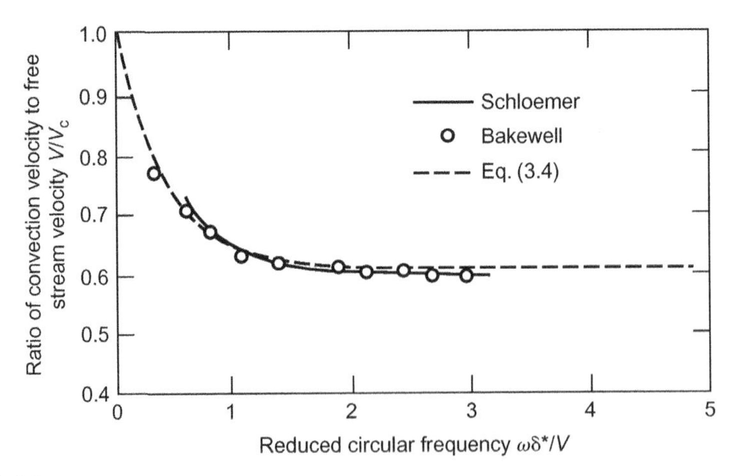

FIGURE 3.6

Dependence of convection velocity on dimensionless frequency [10,11].

The rms response can be expressed approximately, depending on the reduced natural frequency of the cylinder, as follows:

$$
\begin{aligned}
&y_{\text{rms}} \propto V^{1.5} \quad && f_{\text{r}} < 0.2 \\
&y_{\text{rms}} \propto V^{2.0} \quad && 0.2 < f_{\text{r}} < 3.5 \\
&y_{\text{rms}} \propto V^{3.0} \quad && 3.5 < f_{\text{r}}
\end{aligned}
\tag{3.5}
$$

An example of the comparison between random vibration theory prediction and experimental data is shown in Fig. 3.7, where the rms response at the mid-point of a clamped-clamped cylinder is compared with the predicted value [12]. The empirical model underestimates the experimental data at low flow velocity. This may be due to far-field flow noise, which is not considered in the model.

As the flow velocity increases, the pressure fluctuations in the turbulent boundary layer become dominant and the predicted values are in better agreement with the experiment values. However, the predicted values overestimate the experimental values in TEST 1-C in comparison with other conditions. The reason for this is that the pressure power spectral density of Eq. (3.2) used in the evaluation is based on measurements on a single cylindrical rod with a hydraulic diameter of 25.4 mm while a cylindrical rod with a hydraulic diameter of 12.7 mm is used in TEST 1-C.

As the hydraulic diameter decreases, the turbulence intensity also decreases, resulting in lower vibration amplitudes. It is clear that the hydraulic diameter is an important parameter in the prediction of the vibration amplitude.

In the case of multiple cylinders, the influence of the interaction with the adjacent cylinders is evaluated by Païdoussis and Curling [5]. The authors also make comparisons between theory and experiments.

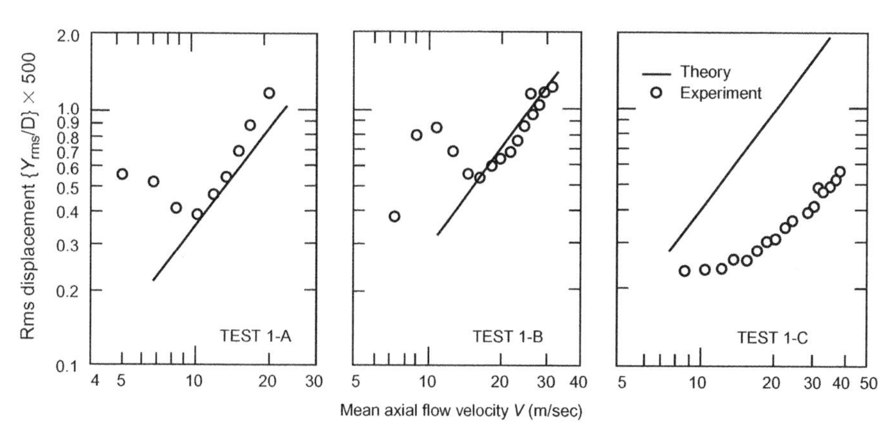

FIGURE 3.7

Mid-span rms displacement of fixed-fixed cylinders [12].

3.1.2.3 Random vibration evaluation in two-phase flow

The following experimental research has been done on the vibration of cylinders in axial two-phase flow.

Quinn [1] performed fundamental experiments on single and multiple pipes in steam-water under average thermal-hydraulic conditions of 7.0 MPa pressure and a temperature of 260°C. The two-phase flow quality (ratio of gas-phase mass flow rate to total mass flow rate) was kept below 0.15 to simulate BWR fuel rod vibration. A schematic of the BWR fuel bundle is shown in Fig. 3.1(a). The amplitude of fuel rod vibration was measured by strain gauges mounted on the inner surface of the instrumented rod, which contained lead pellets to simulate an actual fuel rod. This instrumented rod was used for both single-rod and multi-rod tests.

The maximum vibration amplitude and the dominant frequency of the rod were measured changing flow velocity as a parameter under single-phase flow conditions for single-rod tests. Test model parameters were: outer diameter of the rod, 10 mm; inner diameter of the flow channel, 32 mm; and distance between end supports, 1875 mm.

The experiments on the 6×6 multi-rod bundle were performed in single-phase and steam-water two-phase flow with a prototypical flow velocity of about 2.1 m/s, for three dominant frequencies, 3.2 Hz, 11.0 Hz, and 27.5 Hz. By varying the distance between the end supports the relationship between the maximum amplitude and dominant frequency was determined.

Gorman [2] performed fundamental experiments with a single rod in air–water two-phase flow to evaluate BWR fuel rod vibration. These tests were carried out in two-phase flow under the following conditions: 0.28 MPa average pressure, a quality range between 0.04 and 0.44, and constant mass flow rate. The test rod had an outer diameter of 19 mm and a span of 520 mm between supports. The vibration response was measured by strain gauges mounted on the inner surface of the rod similar to Quinn's experiments. The excitation force was determined indirectly by differential pressure measurements in between the opposite inner surfaces of the flow channel (inner diameter, 32 mm) rather than directly on the outer surfaces of the rod.

Pressure fluctuation correlations in the axial and circumferential directions were derived from the measurement results of the differential pressure. It was shown that the pressure fluctuations are more highly correlated in the circumferential direction for two-phase flow than for single-phase flow. The rms value of the differential pressure fluctuations attained a maximum value in the quality range 0.1–0.2. The rms vibration amplitude at the rod mid-span also attained a maximum value in the same quality range. The damping ratio of the rod was measured by tapping tests. It was found to be relatively independent of quality but was approximately four times the damping in single-phase flow.

Pettigrew and Gorman [3] performed fundamental experiments with a single heated rod having an outer diameter of 20 mm. The inner diameter of the flow channel was 29 mm, and the distance between the end supports was 1700 mm.

Tests were done in boiling water two-phase flow at an average pressure of 2.8–5.5 MPa, a maximum mass flow rate of 1760 kg/m^{2}s, and a maximum quality of 0.65. They selected these conditions in order to evaluate heat exchanger pipes used in CANDU-type nuclear reactors. The maximum rod vibration amplitude was found to be approximately proportional to the mass flux, but decreased with increasing average pressure. The quality at which the maximum amplitude occurs decreased as mass flux increased and/or average pressure decreased. The vibration amplitude versus quality graph was found to have two peaks near 0.1–0.25 and 0.4–0.5 quality for low-mass flux, suggesting an important effect of the two-phase flow regime on the vibration amplitude.

In a recent study, Pettigrew and Taylor [13] concluded that nucleate boiling on the surface of a heated rod had no effect on vibration. The authors performed experiments under BWR fuel operating conditions of 2.8 MPa to 9.0 MPa average pressure, up to 1000 kW/m^2 heat flux, up to 25% quality, and a mass flux up to 4600 kg/m^{2}s. The results of their steam-water experiment may be summarized as follows: the excitation force power spectral density is proportional to $V^{1.56} \sim V^{2.7}$ while the vibration amplitude is in proportion to $V^{0.78} \sim V^{1.35}$. Pettigrew and Taylor noted that the single-phase flow turbulence spectrum for which the vibration amplitude of the fuel rod is nearly proportional to the square of the flow velocity cannot be applied for two-phase flow where the maximum vibration amplitude is approximately proportional to mass flux.

In 2002, Saito et al. [14] performed tests on BWR fuel bundles. The fundamental experiment of the 2×2 bundle and actual scale experiments on 8×8 and 9×9 bundles were performed in this study under the following BWR fuel conditions: up to 7.2 MPa in average pressure, 270°C temperature, 25% quality, and up to 2110 kg/m^{2}s mass flux. The vibration of the fuel rod was measured with an instrumented rod in which biaxial accelerometers were installed at several axial positions. Moreover, pressure fluctuations on the adjacent inner surface of the flow channel wall were simultaneously measured. The experimental results showed that the fuel rod vibration was caused by boiling water/two-phase flow turbulence. The rms value of the wall pressure fluctuations was found to be proportional to the mass flux with the constant of proportionality being dependent on two-phase quality. The fuel rod rms vibration amplitude varied in proportion to the magnitude of the wall pressure fluctuations. This same vibration amplitude was found to decrease with increasing average pressure in the bundle.

Kawamura et al. [15] performed steam-water flow-induced vibration tests on full-scale steam-generator tubes and supports under prototypical PWR thermal-hydraulic conditions. Employing random vibration theory, they expressed their test results in the form of general excitation force spectra. The resulting computed rms vibration is given by:

$$y_{\mathrm{rms}}^2 = \sum_n \frac{\phi_n^2(z)}{64\pi^3 m_s^2 f_n^3 \zeta_s} \Phi_{\mathrm{e}}(f_n) \tag{3.6}$$

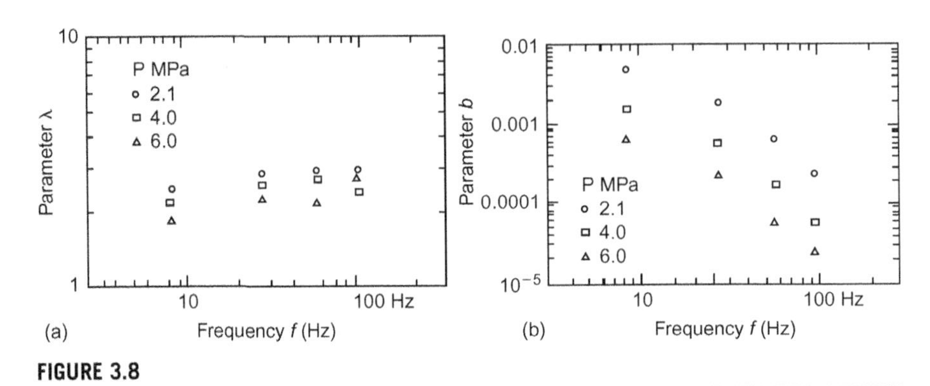

FIGURE 3.8

Parameters used for random vibration correlation [15]: (a) parameter λ, and (b) parameter *b*.

The excitation force spectrum Φ_e in Eq. (3.6) is determined from the experimentally measured model response. The local pressure power spectral density $\Phi_{pp}(f)$ associated with Φ_e (see Eq. [3.1]) was found to be expressible in the following empirical form:

$$\Phi_e(f)/L = b\alpha(1-\alpha)^{\lambda} \tag{3.7}$$

Here, b and λ depend on the pressure and the vibration frequency as shown in Fig. 3.8(a) and (b), respectively. From Fig. 3.8, it is seen that the vibration amplitude shows a decreasing trend with increasing frequency and increasing average pressure.

Although the analysis above is based on generalized excitation force spectra, it is necessary to obtain the spatial correlation in order to evaluate vibration amplitudes in cases which have different mode shapes from those measured by Kawamura et al. Gorman [2] and Inada et al. [16] have measured the spatial correlation in air–water flow. Inada et al. [16] suggested that the spatial correlation of the fluid force acting on the pipe per unit length has a form similar to that of Eq. (3.1), thus arriving at the following formula:

$$\Psi_{FF} = \frac{\overline{F}^2}{f_0} \exp(-\xi|z_2 - z_1|)\,\exp\left(-j\frac{2\pi f(z_2 - z_1)}{V_c}\right) \tag{3.8}$$

Here, $\overline{F}^2$ is the mean square of the liquid force acting at a local location in the frequency range below 20 Hz, while f_0 is the frequency bound (20 Hz in this case). The parameters ξ and V_c were obtained experimentally, as shown in Fig. 3.9.

Based on these results, it is deduced that the correlation length $1/\xi$ is much longer for axial flow than for cross-flow and that it decreases with increasing homogeneous void fraction. The convection velocity V_c was found to be close to the phase velocity of air (see Fig. 3.9).

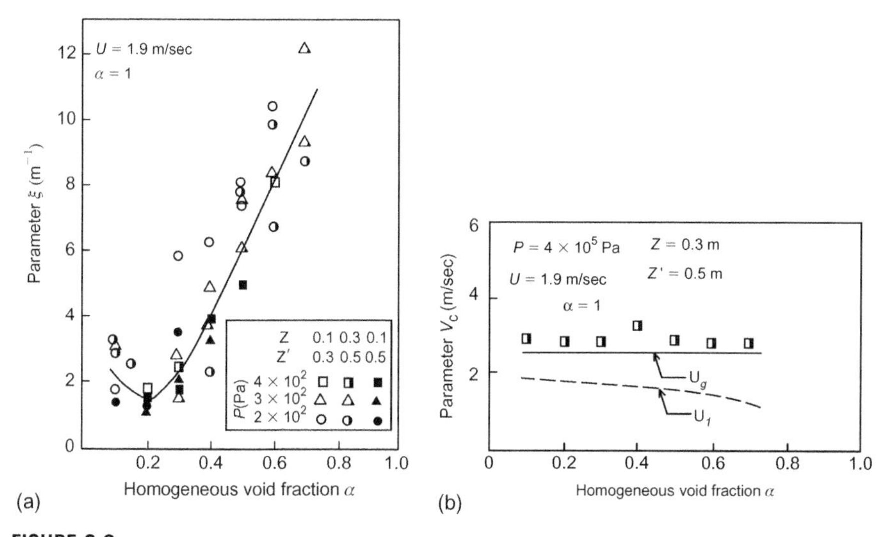

FIGURE 3.9

Parameters for spatial correlation in air–water flow condition [16]: (a) parameter ξ, and (b) parameter V_C. Permission is granted by the Council of the Institution of Mechanical Engineers.

It is very likely that random vibrations in air–water flow and steam–water flow have very different characteristics.

3.1.2.4 Semi-empirical formulae for vibration amplitude prediction

Several empirical correlations have been proposed to predict the random vibration response of a pipe due to axial fluid flow [17–21]. A summary of the correlations is shown in Table 3.1.

These correlation models have generally been developed using dimensional analysis and experimental results. The formulae are useful for predicting approximate pipe response. However, due to the significant scatter in the reported data, correlation factors have inevitably been incorporated into the models. Before employing these correlations, one must take into consideration the uncertainty inherent in the actual problem analyzed (see Fig. 3.10).

3.1.3 Flutter and divergence

3.1.3.1 Historical background

Studies on the stability of fluid conveying pipes were conducted in the 1950s and 1960s by Païdoussis and co-workers. This work will be described in Section 4.1.

With regard to instability due to external parallel flow, Hawthorne [22] was the first to demonstrate the occurrence of the divergence-type instability.

Table 3.1 Formulae for vibration amplitude estimation

Investigator	Correlation	Comments
Burgreen, Byrnes, and Benforado [17]	$\left(\frac{y_{pp}}{d_h}\right)^{1.3} = 0.83 \times 10^{-10}\kappa\left(\frac{\rho V^2 L^4}{EI}\right)\left(\frac{\rho V^2}{\mu\omega}\right)$ Dimensionless parameters $\rho V^2 L^4/EI$, $\rho V^2/\mu\omega$, Reynolds number y_{pp}: vibration amplitude (peak-peak) κ: end fixity factor ($\kappa = 5$) (for simply supported rods)	This equation was the first attempt to correlate test data and it is the simplest correlation. From a dimensional analysis, the vibration amplitude is found to be a function of three dimensionless parameters In comparison with test data, the discrepancies can be up to two orders of magnitude (Païdoussis, Pavlica and Marshall)
Reavis [18]	$y_{pp} = C_\alpha \eta_D\, \eta_h\, \eta_L \frac{D}{mf^{1.5}\zeta^{0.5}} V\rho\nu^{0.5}$ η_D, η_h, η_L: scale factors, respectively fD/V, fd_h/V, fL/V C_α: disparity factor m : mass of cylinder per unit length	The forcing function is derived from Bakewell's measurements for wall pressure fluctuations in pipe flow. When compared with test data, the theoretical prediction is found to underestimate measured maximum displacements by a factor of 3.5 to 4.0. A disparity ratio C_α is introduced in the correlation.
Païdoussis [19]	$\frac{y_{max}}{D} = \alpha_1^{-4}\left[\frac{v_r^{1.6} L_r^{1.8} Re^{0.25}}{1 + v_r^2}\right]\left[\left(\frac{d_h}{D}\right)^{0.4}\left(\frac{\beta^{2/3}}{1 + 4\beta}\right)(5 \times 10^{-4}\kappa)\right]$ α_1 : dimensionless first mode eigenvalue of the cylinder Re : Reynolds number based on the hydraulic diameter ($=Vd_h/\nu$) κ: effect factor of upstream disturbance and mechanically transmitted vibration	The vibration is supposed to be caused by axial steady uniform flow turbulence. Figure 3.10 shows the comparison between the calculation results and experimental data The overall agreement is reasonable. However, a large discrepancy exists due to the mechanically transmitted vibration at low flow velocities

	$\kappa = 1$: for low upstream disturbance and low mechanically transmitted vibration level $\kappa = 5$: for realistic industrial environments v_r : reduced flow velocity ($=(m_a/EI)^{0.5}VL$) m_a : added mass ($=(\pi D^2/4)\rho C_m$) C_m : added mass coefficient L_r : reduced length ($=L/D$) β: added mass ratio ($=m_a/(m+m_a)$)	
Chen and Wever [20]	$\frac{y}{d_h} = \left[1-\left(\frac{V}{V_{cr}}\right)^2\right]^{-1}\left(\frac{\kappa V}{V_{cr}}\right)^2$ κ: initial turbulence factor ranging from 0.5 for ideally quiet flow to 2.0 for practical flow conditions $V_{cr} = \left[\frac{(\pi^2/L^2)EI}{\frac{\pi}{4}C_f\rho LD+m_a}\right]^{0.5}$ C_f: surface drag coefficient	The vibration is supposed to be caused by parametric excitation. Introducing a sinusoidally varying velocity fluctuation into the equation, the critical buckling velocity V_{cr} is determined
Chen and Wambsganss [21]	$y_{rms}(z,V) = \frac{0.018\kappa D^{1.5}d_h^{1.5}V^2\phi(z)}{L^{0.5}f^{1.5}(m+m_a)\zeta^{0.5}}$ κ: constant derived from experiment $\phi(z)$: modal function of the cylinder	This equation is derived from a theoretical relationship for calculating the rms displacement of a cylinder in nominally axial flow

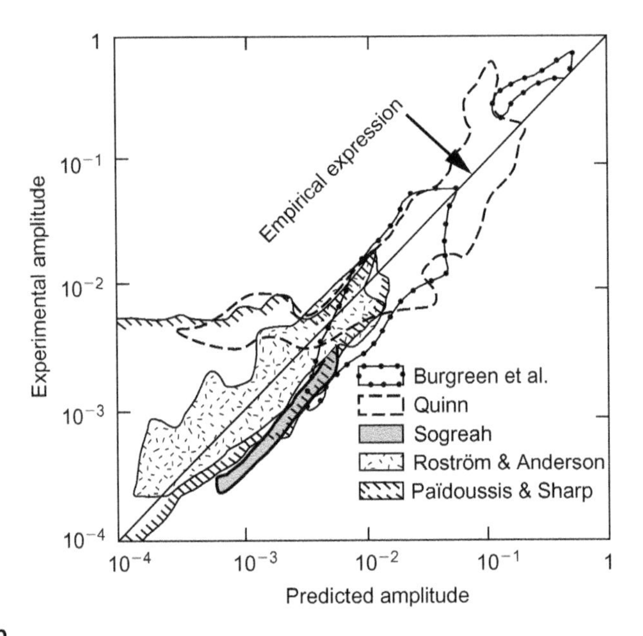

FIGURE 3.10

Relationship between measured and predicted amplitudes of vibration according to Païdoussis's empirical formula [4].

Païdoussis et al. [23–26] derived the basic equations for the first time, and investigated the critical velocity for flutter and divergence analytically. The analytical results were also verified experimentally. Until the 1980s [27], many studies were conducted in which more detailed analytical models were developed and the mechanisms of instability further investigated. The dependence of the fluid force on structural motion and the dynamic characteristics of the pipe in external parallel flow have similar basic characteristics to those found for fluid-conveying pipe systems.

3.1.3.2 Evaluation methods and countermeasures against instability

In general, divergence instability for supported tubes occurs when the flow velocity surpasses a critical value. When the flow velocity is increased further, various types of instabilities can occur depending on the flow field surrounding the tube, and the tube support conditions (see Table 3.2) [24,28].

Friction on the tube surface can introduce tension and stabilize the system when the downstream end is free. Lee [29] and Triantafyllou and Chryssostomidis [30] reported that the system became stable when either of the requirements shown in Table 3.3 is satisfied.

Table 3.2 Tube support conditions and instability phenomena

Support condition	Instability phenomena
Pinned-pinned tube	Coupled-mode flutter
Cantilevered tube	Single-degree-of-freedom flutter: critical velocity depends on the shape of the downstream end. The critical velocity with smooth tapered end is lower than that without it
Flexible tube bundle	Critical velocity decreases because of the tube-coupling through fluid

Table 3.3 Conditions for prevention of instability induced by external parallel flow

When either of the conditions below is satisfied, the system is stabilized
Tension force at the downstream end > ρAU^2 or $L/D > \pi/(2C_f)$

3.1.4 Examples of reported component-vibration problems and hints for countermeasures

Axial flow-induced random vibration in nuclear plant components has been reported. Examples include the vibration of in-core neutron flux monitoring tubes causing neutron flux fluctuation, fretting wear of BWR fuel cladding, CANDU reactor fuel rod vibration, and tube fretting wear at tube supports caused by vibration in heat exchangers and steam generators. Note, however, that there is no reported case of tube failure.

As mentioned above, the random vibration response due to near field turbulence noise in external axial flow varies in proportion to $V^{1.50} \sim V^{3.0}$ in single-phase flow, and to $V^{0.75} \sim V^{1.35}$ in two-phase flow. The excitation force is such that the amplitude decreases with increasing frequency. Research findings show that decreasing the flow velocity and increasing the natural frequency of the tube or tubes are effective measures in decreasing the amplitude of random vibration.

It is reported that both the divergence-type and flutter-type instabilities can be eliminated when tension develops in the pipe, and when the length of the pipe is increased. If the conditions given in Table 3.3 are satisfied, stabilization can be achieved. In addition, considering that the instability occurs above a certain non-dimensional critical velocity, increasing the natural frequency by thickening the tube walls or by reinforcement and decreasing the flow velocity are effective measures to control instability phenomena.

3.2 Vibration of elastic plates and shells

In this section, self-excited vibration and turbulence-induced vibration of elastic plates and shells subjected to parallel flow are described.

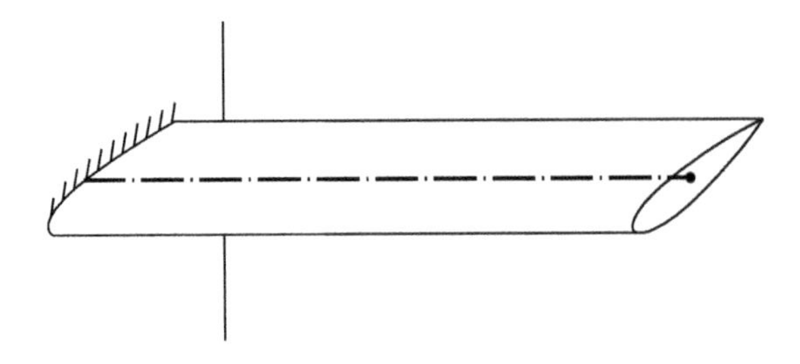

FIGURE 3.11

A two-dimensional airfoil.

FIGURE 3.12

Coupling between the bending and torsional motion.

3.2.1 Bending–torsion flutter

3.2.1.1 Historical background

Bending–torsion flutter is a dynamic instability which results from coupling between the bending and torsional modes of a flexible structure such as shown in Fig. 3.11. The instability occurs if the flow velocity exceeds a critical value, while the bending and torsional modes have a phase and amplitude combination that extracts energy from the flow [31], as shown in Fig. 3.12.

Theories on aerodynamic forces acting on two-dimensional airfoil sections were developed mostly in the mid-1940s. These theories are reviewed by Bisplinghoff et al. [32] and Fung [33]. Recently, intensive studies have been carried out on aeroelastic tailoring techniques [34], computations of unsteady aerodynamic loads [35], active flutter control [36], etc.

3.2.1.2 Evaluation method

Consider the spring-supported wing section in incompressible flow of velocity U, shown in Fig. 3.13. Let the bending and pitching displacements be resisted by springs at the elastic axis with spring constants K_h and K_α, respectively. The restoring force corresponding to a displacement h is $-K_h(1+jg_h)h$. The restoring moment against a is $-K_\alpha(1+jg_\alpha)\alpha$. The equations of motion for bending and torsional motions then, are given by:

$$\left.\begin{aligned} m\ddot{h} + S_\alpha\ddot{\alpha} + K_h(1+jg_h)h &= -L \\ S_\alpha\ddot{h} + I_\alpha\ddot{\alpha} + K_\alpha(1+jg_\alpha)\alpha &= M_\alpha \end{aligned}\right\} \tag{3.9}$$

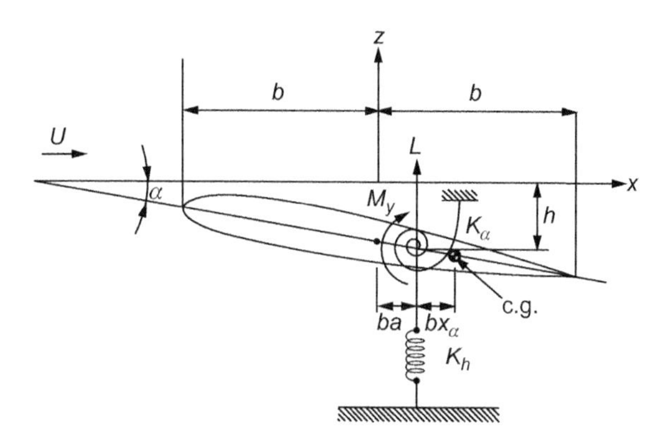

FIGURE 3.13

Spring-supported rigid wing section in two-dimensional flow.

where m, I_α, and $S_\alpha = mbx_\alpha$ are the mass, mass moment of inertia about elastic axis, and static mass moment about the elastic axis, respectively. L and M_a denote the externally applied force and moment. Assuming simple harmonic motion and letting $h = \overline{h}e^{j\omega t}, \alpha = \overline{\alpha}e^{j\omega t}, L = \overline{L}e^{j\omega t}$, and $M_\alpha = \overline{M}_\alpha e^{j\omega t}$, Eq. (3.9) can be written as:

$$\left.\begin{array}{l} -m\omega^2\overline{h} - S_\alpha\omega^2\overline{\alpha} + m\omega_h^2\overline{h}(1 + jg_h) = -\overline{L} \\ -S_\alpha\omega^2\overline{h} - I_\alpha\omega^2\overline{\alpha} + I_\alpha\omega_\alpha^2\overline{\alpha}(1 + jg_\alpha) = \overline{M}_\alpha \end{array}\right\} \tag{3.10}$$

where, $\omega_h = \sqrt{K_h/m}$ and $\omega_\alpha = \sqrt{K_\alpha/S_\alpha}$ are the bending and torsional natural circular frequencies, respectively. Aerodynamic lift and moment are given by:

$$\left.\begin{array}{l} \overline{L} = -\pi\rho b^3\omega^2\{A_{hh}(\overline{h}/b) + A_{h\alpha}\overline{\alpha}\} \\ \overline{M}_\alpha = \pi\rho b^4\omega^2\{A_{\alpha h}(\overline{h}/b) + A_{\alpha\alpha}\overline{\alpha}\} \end{array}\right\} \tag{3.11}$$

where, A_{hh}, $A_{h\alpha}$, $A_{\alpha h}$, and $A_{\alpha\alpha}$ are functions of the reduced frequency k; $\omega b/U$. Substituting Eq. (3.11) into Eq. (3.10), we obtain:

$$\begin{bmatrix} \mu\left\{1 - \left(\dfrac{\omega_h}{\omega}\right)^2(1 + jg_h)\right\} + A_{hh} & \mu x_\alpha + A_{h\alpha} \\ \mu x_\alpha + A_{\alpha h} & \mu r_\alpha^2\left\{1 - \left(\dfrac{\omega_\alpha}{\omega}\right)^2(1 + jg_\alpha)\right\} + A_{\alpha\alpha} \end{bmatrix} \begin{Bmatrix} \overline{h}/b \\ \overline{\alpha} \end{Bmatrix} = \begin{Bmatrix} 0 \\ 0 \end{Bmatrix} \tag{3.12}$$

where, $\mu = m/\pi\rho b^2$, $x_\alpha = S_\alpha/mb$, and $r_\alpha^2 = I_\alpha/mb^2$. Setting the determinant to zero leads to a characteristic equation. The flutter speed U_F and flutter frequency ω_F are obtained by solving the eigenvalue problem.

In Ref. [32], various graphs demonstrate how the dimensionless flutter speed U_F varies with frequency ratio ω_h/ω_α, mass ratio $\mu = m/\pi\rho b^2$, and dimensionless static unbalance $x_\alpha = S_\alpha/mb$. An example is shown in Fig. 3.14. The dimensionless flutter speed U_F is plotted against ω_h/ω_α, for $\mu = 20$. The flutter speed increases when $x_\alpha = 0$, and a marked dip can be seen near unit frequency ratio for $x_\alpha = 0.05$ and 0.1.

3.2.2 Panel flutter

3.2.2.1 Historical background

Panel flutter is a type of self-excited vibration of flexible plates with supported edges subjected to supersonic flow over one side of the plate, as shown in Fig. 3.15. It is well known that the early German V-2 rockets crashed because of panel flutter during World War II. Theoretical research on panel flutter has been carried out since the early 1950s. It is known that the thermal compressive stress induced in the skin due to aerodynamic heating, lowers the critical speed [33]. The stability of the skin plate subjected to water flow over one side has also been studied (see Fig. 3.16) [37].

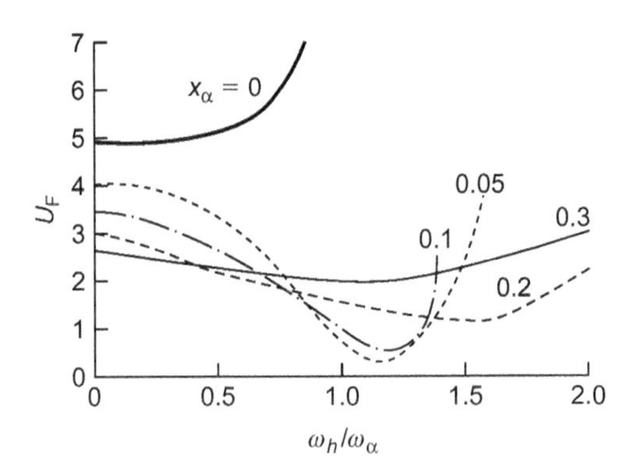

FIGURE 3.14

Dimensionless flutter speed versus frequency ratio.

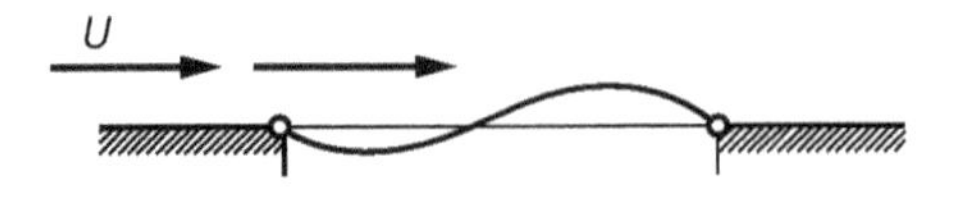

FIGURE 3.15

Panel flutter.

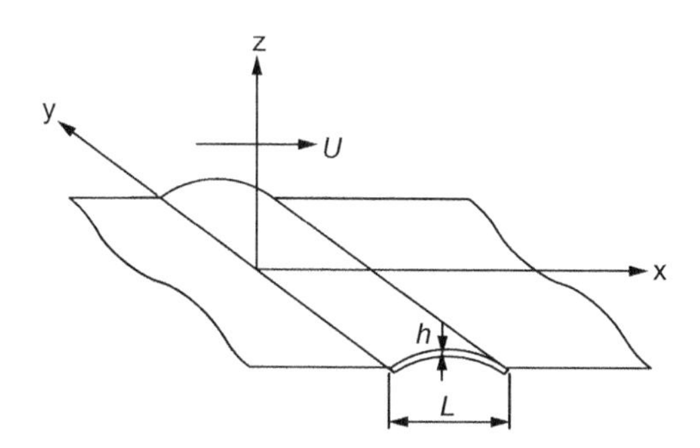

FIGURE 3.16

Simply supported plate of infinite width subjected to fluid flow over one side.

3.2.2.2 Evaluation method

Consider the simply supported elastic plate of length L, thickness h, and density ρ_m, subjected to inviscid incompressible flow of density ρ_0, and velocity U. The equation of motion of the plate is:

$$D\nabla^4 W + \rho_m h \frac{\partial^2 W}{\partial t^2} + \rho_m h \gamma \frac{\partial W}{\partial t} + P = 0 \tag{3.13}$$

where, $W(x, t)$ is the plate deformation. The boundary conditions are:

$$W(0) = W(L) = 0, \quad (\partial^2 W/\partial x^2)_{x=0} = (\partial^2 W/\partial x^2)_{x=L} = 0 \tag{3.14}$$

The pressure on the plate may be obtained from Bernoulli's equation for unsteady flow:

$$P = -\rho_0 \{(\partial\phi/\partial t) + U(\partial\phi/\partial x)\}_{z=0} \tag{3.15}$$

where ϕ is the velocity potential. The governing equation for the potential flow is Laplace's equation:

$$\nabla^2 \phi(x, z, t) = 0 \tag{3.16}$$

The impermeability condition at the surface is given by:

$$(\partial\phi/\partial z)_{z=0} = (\partial W/\partial t) + U(\partial W/\partial x), \quad 0 \le x \le L \tag{3.17}$$

Assuming that the plate deformation can be expressed as:

$$W(x, t) = \sum_{n=1}^{\infty} A_n \sin\frac{n\pi x}{L} \cdot e^{i\omega t} \tag{3.18}$$

and combining these equations leads to linear algebraic equations in the unknown constants A_n. Letting the determinant of the coefficients matrix vanish, the characteristic equation is obtained [37].

The resulting non-dimensional critical flow velocity is plotted against the mass ratio in Fig. 3.17. As the flow velocity is increased, divergence occurs first. Flutter then follows with further increase of the flow velocity. The critical velocity decreases with increasing mass ratio.

3.2.3 Shell flutter due to annular flow

A thin cylindrical shell in axial flow may lose stability by flutter [38]. The severe damage of the heat-shielding shell of a jet engine afterburner [39] is one example of a practical occurrence of such an instability (see Fig. 3.18). In the situation where a cylinder is contained in a narrow coaxial cylindrical duct and thus, subjected to annular flow, instability tends to occur at very low flow velocities. The mechanism of instability is, however, not yet fully understood.

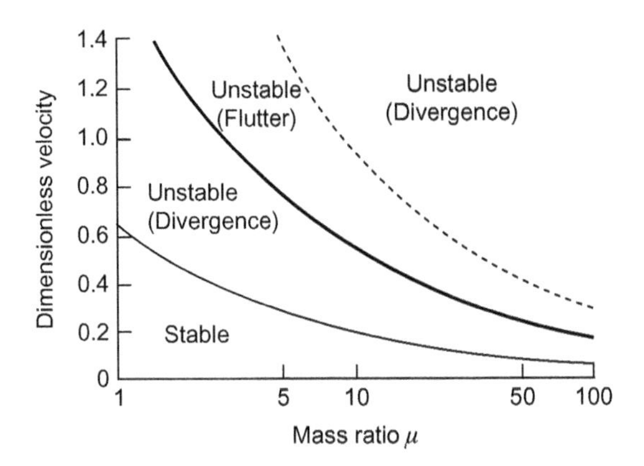

FIGURE 3.17

Dimensionless critical flow velocity versus mass ratio.

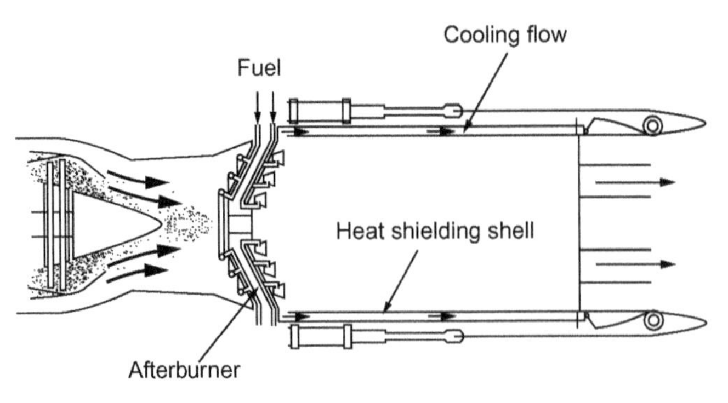

FIGURE 3.18

Jet engine afterburner.

3.2.4 Turbulence-induced vibration

3.2.4.1 Summary

There are cases where flexible structures subjected to parallel flow vibrate due to turbulence produced by the upstream wake. Buffeting of the aircraft tail is a typical example. In the area of mechanical engineering, structures such as heat-shielding liners or guide vanes in pipes and ducts may vibrate at their natural frequencies due to the broadband pressure disturbance created by elbows, branches, and orifices located upstream. Some normalized sound power spectra for elbows and branches, e.g. Fig. 3.19, are presented in Ref. [40]. Analytical methods for predicting the structural vibration response to random fluid forces are presented in Ref. [41].

3.2.4.2 Evaluation method

The equation of motion of a plate subjected to random pressure disturbance, P, can be written as:

$$m\frac{\partial^2 W}{\partial t^2} + L[W(x,y,z,t)] = P(x,y,z,t) \tag{3.19}$$

where $m(x, y, z)$ is the mass per unit area, and L is the linear operator which represents the relation between the force and the deformation of the plate. Let the deformation of the plate be expressed as the superposition of eigenfunctions as follows:

$$W(x,y,z,t) = \sum_{i=1}^{N} \tilde{w}_i(x,y,z)w_i(t) \tag{3.20}$$

and assume that the pressure perturbations can be expressed as:

$$P(x,y,z,t) = \tilde{p}(x,y,z)p(t) \tag{3.21}$$

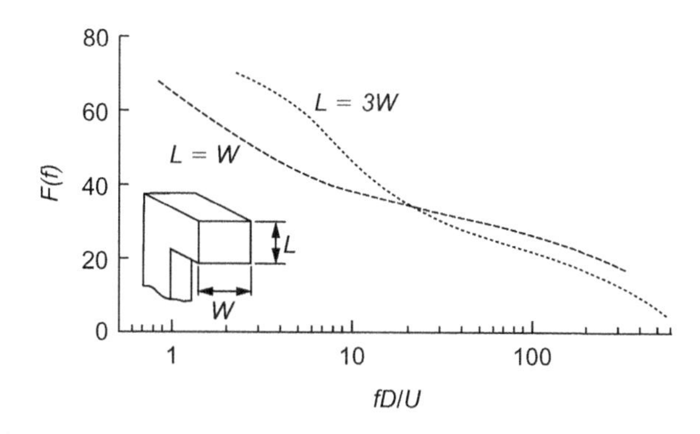

FIGURE 3.19

Sound spectrum for 90 degree elbow duct.

Substituting Eqs. (3.20), (3.21) into Eq. (3.19) and taking the orthogonality of eigenmodes into account, we obtain the following equation of motion:

$$\ddot{w}_i(t) + 2\zeta_i\omega_i\dot{w}_i(t) + \omega_i^2 w_i(t) = \frac{\int \tilde{p}(x,y,z)\tilde{w}_i(x,y,z)\mathrm{d}s}{\int m(x,y,z)\tilde{w}_i^2(x,y,z)\mathrm{d}s}\cdot p(t) \tag{3.22}$$

Consider the case where the modal natural frequencies are distinct and each mode can be treated as a single-degree-freedom-system. For simplicity, assuming that the coefficient of the right-hand side of Eq. (3.22) is equal to ω_i^2, the power spectral density of $w_i(t)$ can be written as:

$$S_{wi}(f) = \frac{S_\mathrm{p}(f)}{[1-(f/f_i)^2]^2 + (2\zeta_\mathrm{i} f/f_i)^2} \tag{3.23}$$

where $S_\mathrm{p}(f)$ is the power spectral density of $p(t)$. The mean square value of $w_i(t)$ can be obtained by integrating Eq. (3.23) from the lower limit to the upper limit of the pressure perturbation frequency. When the pressure perturbation power spectrum is uniform between the frequencies f_l and f_2, the rms value of $w_i(t)$ can be approximated as:

$$w_{i,\mathrm{rms}} = \sqrt{\int_{f_1}^{f_2} \frac{S_\mathrm{p}(f)}{[1-(f/f_i)^2]^2 + (2\zeta_\mathrm{i} f/f_i)^2}\mathrm{d}f} \approx \sqrt{\frac{\pi}{4\zeta_\mathrm{i}}\cdot f_i \cdot S_\mathrm{p}(f_i)} \tag{3.24}$$

It should be noted that the rms vibration response for random excitation is proportional to the inverse of the square root of the damping ratio.

3.2.5 Hints for countermeasures

3.2.5.1 Self-excited vibrations

Taking into account the fact that instability occurs if the reduced velocity U/f_nD exceeds the critical value, it is clear that increasing the structural rigidity and hence, the natural frequencies, stabilizes the system. Adding structural damping also increases the critical velocity.

3.2.5.2 Turbulence-induced vibrations

For periodic fluid forces, the structural vibration amplitude can be reduced significantly by moving the natural frequency away from the excitation frequencies. However, for the broadband fluid forces, it is difficult to move the natural frequency away from the excitation frequencies. Therefore, the structural amplitude reduction is only in proportion to the increase in the structural rigidity. Adding structural damping also reduces the vibration amplitude, as shown in Eq. (3.24). It is also obviously effective to displace or separate the structure itself from the source of turbulence, if possible.

3.3 Vibration induced by leakage flow

3.3.1 General description of the problem

Large-amplitude self-excited vibrations can occur in structures forming leakage-flow passages. One of the major features of leakage-flow-induced vibration is that the sensitivity of the pressure loss can be very high compared to the displacement of the structures forming the leakage-flow passage. This can induce the vibration of huge structures such as the core barrel of a PWR, feed-water spargers in BWRs (see Fig. 3.20 [42]), and so on.

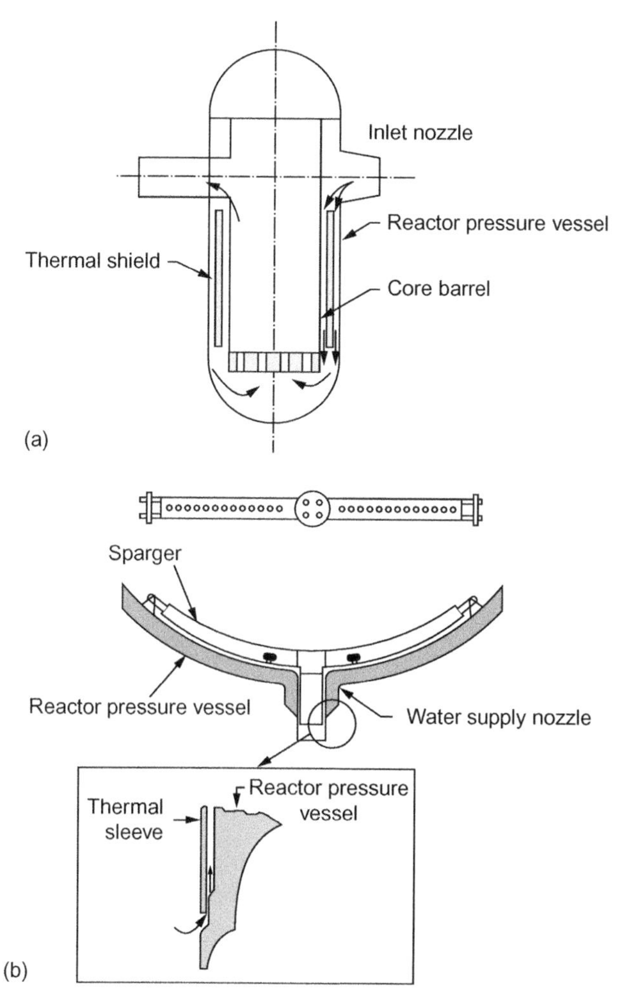

FIGURE 3.20

Examples of leakage-flow-induced vibration events [42]: (a) PWR core barrel, and (b) feed-water sparger.

3.3.1.1 *Single-degree-of-freedom translational systems*

In this section, we assume incompressibility of the gas or liquid, when the Mach number is less than 0.3. One-dimensional tapered as well as arbitrary-shaped passages are considered. The structure forming the leakage passage is assumed to vibrate as a single-degree-of-freedom translational system.

3.3.1.2 *Other cases*

We also consider other cases such as single-degree-of-freedom rotational systems, two-degrees-of-freedom translational and rotational systems, and continuous systems. The results are compared with the case of single-degree-of-freedom translational systems. At the end of this section, an explanation is given for the case of the annular gap, where neither the inner tube nor the outer tube rotates.

3.3.2 Evaluation method for single-degree-of-freedom translational system

In this section, a method is described to model the fluid dynamic forces acting on the wall of a tapered passage of area increment ratio $\alpha(=\{\overline{H}(L)-\overline{H}(0)\}/\overline{H}(0)$; over-bars $^{-}$ indicate a steady state condition). Using this method, the characteristics of the system are considered [43].

3.3.2.1 *Basic equations of leakage flow*

As shown in Fig. 3.21(a), the upper wall of the tapered passage is assumed to be vibrating in the normal direction relative to the lower wall with a circular frequency Ω. When Q is the flow rate per unit width of the passage, the continuity and momentum equations of the leakage flow can be written as:

$$\frac{\partial Q}{\partial Y}+\frac{\partial H}{\partial t}=0,\quad \frac{\partial Q}{\partial t}+\frac{\partial}{\partial Y}\left(\frac{Q^2}{H}\right)=-\frac{H}{\rho}\frac{\partial P}{\partial Y}-\frac{1}{4}\lambda\frac{Q^2}{H^2} \tag{3.25a}$$

The boundary conditions at the entrance and the exit of the passage are expressed as:

$$P(0)=P_{\mathrm{in}}-\xi_{\mathrm{in}}\frac{1}{2}\rho\frac{Q^2(0)}{H^2(0)},\quad P(L)=P_{\mathrm{ex}}+\xi_{\mathrm{ex}}\frac{1}{2}\rho\frac{Q^2(L)}{H^2(L)} \tag{3.25b}$$

where ξ_{in} and ξ_{ex} denote the entrance and the exit pressure loss factors, and λ denotes the friction factor of the passage. The values for pipes can be used for ξ_{in} and ξ_{ex} for qualitative evaluation; the loss factor $\xi_{\mathrm{in}}=1.5$ for entrance into a passage mounted flush with the wall, $\xi_{\mathrm{in}}=1.0$ for circular bellmouth entrance, and $\xi_{\mathrm{ex}}=0$ for outlet from a passage mounted flush with the wall. λ is the friction factor of the passage and is given as a function of the Reynolds number, Re; $\lambda=48/\mathrm{Re}$ (Re $<$ 1000) for laminar flow, and $\lambda=0.266\ \mathrm{Re}^{20.25}$ (Re $>$ 1000) for turbulent flow where $\mathrm{Re}=Q/\nu$.

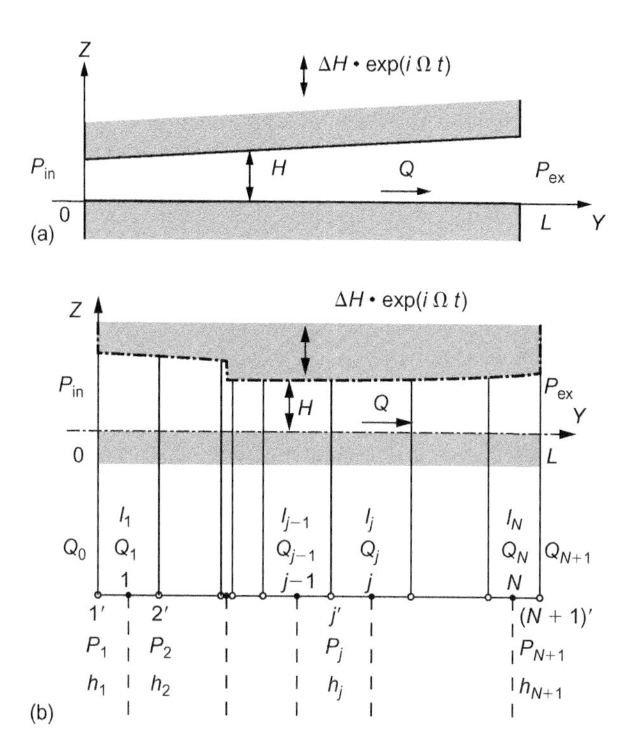

FIGURE 3.21

One-dimensional leakage passage under study and coordinate system: (a) tapered leakage-flow passage, and (b) arbitrary-shaped leakage-flow passage.

3.3.2.2 Relation between steady flow rate and pressure difference across the passage

The relationship between the steady flow rate and the steady pressure difference across the leakage-flow passage is expressed as:

$$\frac{\overline{Q}}{H_0} = \sqrt{\frac{P_{\text{in}} - P_{\text{ex}}}{\rho\{(\beta - \alpha)I_3(1) + \xi_{\text{in}}/2 + \xi_{\text{ex}}/\{2h^2(1)\}\}}} \tag{3.26}$$

where, H_0 is the steady inlet gap of the passage, and the effective pressure loss factor of the passage β, non-dimensional axial location y, and the function $I_n(y)$, ($n = 3$ in Eq. (3.26)), are defined as:

$$\beta = \overline{\lambda}L/(4H_0), \quad y = Y/L, \quad h = \overline{H}/H_0 = 1 + \alpha y$$

$$I_n(y) = \frac{1}{(n-1)\alpha}\left[1 - \frac{1}{h(y)^{n-1}}\right] \quad (n \neq 1) \quad \ln(h(y))/\alpha \quad (n = 1) \tag{3.27}$$

3.3.2.3 Unsteady fluid forces of a tapered passage

The basic equations are linearized with respect to the unsteady components. The unsteady pressure distribution is determined and integrated from the entrance to the outlet. The unsteady fluid dynamic force, ΔF, acting on the wall of unit width is obtained assuming that the wall vibrates with infinitesimally small amplitude, ΔH. Defining the non-dimensional frequency $\omega = \Omega H_0 L/\overline{Q}$, the dynamic force becomes:

$$\Delta F = -\frac{\rho \overline{Q}^2 L}{H_0^2}\left\{\left(i\omega m + c_1 + \frac{c_2}{1+i\omega T}\right)i\omega + \left(k_1 + \frac{k_2}{1+i\omega T}\right)\right\}\frac{\Delta H}{H_0} \tag{3.28}$$

The parameters m, c_1, c_2, k_1, k_2 are defined in Table 3.4. The first term in the braces in Eq. (3.28) corresponds to the squeeze-film effect, in which the fluid force can be generated by oscillation of the relatively large axial velocity induced by the wall oscillation of the leakage passage. The second term is the fluid force due to oscillations of the axial velocity induced by the variation of the flow resistance. Equation (3.28) includes a first-order delay term due to fluid inertia. When we estimate stability, $i\omega$ is replaced by the Laplace transform parameter s, and ΔF is substituted into the basic equation of the structure forming the leakage-flow passage. If the equation of the structure is expressed in terms of a second-order system, the characteristic equation is expressed by a third-order polynomial.

Equation (3.28) is rearranged as:

$$\Delta F = -\{M_a \Delta\ddot{H} + (C_w + C_e)\Delta\dot{H} + K_a \Delta H\} \tag{3.29}$$

where a dot denotes differentiation with respect to time. M_a, $(C_w + C_e)$, and K_a are added mass, added damping, and added stiffness coefficients defined as functions of ω:

$$M_a = \frac{\rho L^3}{H_0}\left(m - \frac{c_2 T}{1+\omega^2 T^2}\right), C_w = \frac{\rho \overline{Q} L^2}{H_0^2}\left(c_1 + \frac{c_2}{1+\omega^2 T^2}\right), C_e = -\frac{\rho \overline{Q} L^2}{H_0^2}\frac{k_2 T}{1+\omega^2 T^2},$$
$$K_a = \frac{\rho \overline{Q}^2 L}{H_0^3}\left(k_1 + \frac{k_2}{1+\omega^2 T^2}\right) \tag{3.30}$$

C_w, the damping component due to squeeze-film effects, is generally positive. C_e, the damping coefficient due to the flow resistance variations, can be negative in the case of divergent passages. The negative damping can generate self-excited vibrations. In the case of divergent passages, the added stiffness coefficient can be negative, which can induce divergence-type (buckling-type) instability. Figure 3.22 shows the ranges where $C_w + C_e$ and K_a become positive or negative on the (α, β) plane. It is recommended that the area increment ratio α be set to zero or negative, the latter corresponding to the leakage-flow passage being changed to a convergent passage to control the self-excited vibration.

Table 3.4 Definition of parameters in Eq. (3.28)

$$m = J_1 N_1 / I_1 - N_1^*$$

$$c_1 = 2(\beta\gamma - \alpha)\left(\frac{N_1 J_3}{I_1} + \frac{N_3 J_1}{I_1} - \frac{J_1 N_1 I_3}{I_1^2} - N_3^*\right) + 2\left(\frac{I_2 N_1}{I_1} - N_2\right) + \xi_{in}\frac{J_1}{I_1}\left(1 - \frac{N_1}{I_1}\right) + \frac{\xi_{ex}}{h^2(1)}\frac{N_1}{I_1}\left(1 - \frac{J_1}{I_1}\right)$$

$$c_2 = (J_1 N_1 / I_1^2)(B_1 B_3 / r)$$

$$k_1 = 3(\beta - \alpha)\left(\frac{N_1 I_4}{I_1} - N_4\right) - \xi_{in}\left(1 - \frac{N_1}{I_1}\right) + \frac{\xi_{ex}}{h^3(1)}\frac{N_1}{I_1}, \quad k_2 = \frac{N_1}{I_1}\frac{B_2 B_3}{r}, \quad T = \frac{I_1}{r}$$

Definition of symbols in the above equations

$$J_n(y) = (I_{n-1}(y) - I_n(y))/\alpha, \quad I_n = I_n(1), \quad J_n = J_n(1), \quad N_n = I_n - J_n$$

$$N_n^* = \frac{2+\alpha}{2\alpha^2} - \frac{1+\alpha}{\alpha^3}\ln(1+\alpha)(n=1), \quad \frac{2+\alpha}{2\alpha^2(1+\alpha)} - \frac{1}{\alpha^3}\ln(1+\alpha)(n=3)$$

$$\gamma = 1 + \frac{\overline{Q}}{2\overline{\lambda}}\left(\frac{d\lambda}{dQ}\right)_0$$

$$B_1 = 2(\beta\gamma - \alpha)\left(\frac{I_1 J_3}{J_1} - I_3\right) + 2\frac{I_1 I_2}{J_1} - \xi_{in} - \frac{\xi_{ex}}{h^2(1)}\left(1 - \frac{I_1}{J_1}\right)$$

$$B_2 = 3(\beta - \alpha)I_4 + \xi_{in} + \frac{\xi_{ex}}{h^3(1)}$$

$$B_3 = 2(\beta\gamma - \alpha)\left(\frac{I_1 N_3}{N_1} - I_3\right) - \xi_{in}\left(1 - \frac{I_1}{N_1}\right) - \frac{\xi_{ex}}{h^2(1)}$$

$$r = 2(\beta\gamma - \alpha)I_3 + \xi_{in} + \frac{\xi_{ex}}{h^2(1)}$$

3.3.3 Analysis method for single-degree-of-freedom translational system with leakage-flow passage of arbitrary shape

3.3.3.1 *Steady fluid force*

In the case of one-dimensional leakage-flow, steady and unsteady fluid forces can be easily analyzed for arbitrary passage shapes, such as Fig. 3.21(b), using the transfer matrix method. When a steady flow rate Q is assumed, the transfer matrix of the element connected to the entrance can be obtained. The element's outlet pressure can thus be obtained and, in turn, becomes the entrance pressure of the next element. If the

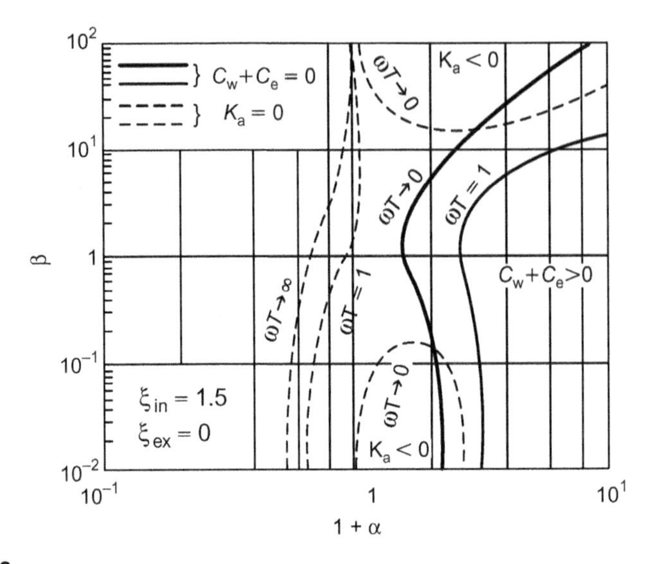

FIGURE 3.22

Region where added damping and stiffness coefficients become positive or negative when the one-dimensional tapered passage wall oscillates as a single-degree-of-freedom translating system [43].

calculation is repeated, the transfer matrix, M, of the entire flow passage from the entrance to the outlet as well as the outlet pressure, P_{ex}, can be obtained as follows:

$$\begin{bmatrix} P_{ex}/P^* \\ 1 \end{bmatrix} = M \begin{bmatrix} P_{in}/P^* \\ 1 \end{bmatrix}, \quad M = M_{ex} M_N M_{N-1} \ldots\ldots M_1 M_{in} = \begin{bmatrix} 1 & -M \\ 0 & 1 \end{bmatrix} \tag{3.31}$$

The relation between the steady flow rate and the pressure difference between the entrance and the outlet of the passage is then:

$$\overline{Q}/H_0 = \sqrt{(P_{in} - P_{ex})/\rho M} \tag{3.32}$$

The transfer matrix of each element is shown in Table 3.5.

3.3.3.2 Unsteady fluid force

The unsteady fluid force can also be analyzed using the transfer matrix method. When the transfer matrix of the j-th element is denoted by $\mathbf{m}_j$, the transfer matrix $\mathbf{m}$ of the entire flow passage from the entrance to the exit is expressed as:

$$\mathbf{m} = \mathbf{m}_{ex} \mathbf{m}_N \mathbf{m}_{N-1} \ldots \mathbf{m}_1 \mathbf{m}_{in} = \begin{bmatrix} 1 & -(i\omega A + B) & i\omega C + D & E \\ 0 & 1 & -G & 0 \\ 0 & 0 & 1 & 0 \\ 0 & 0 & 0 & 1 \end{bmatrix} \tag{3.33}$$

Table 3.5 Transfer matrices for steady flow component

Smooth flow passage element	Entrance	Outlet	Sudden section change element
$\begin{bmatrix} 1 & -\frac{1}{2}\left\{\left(\frac{1}{h_{j+1}^2}-\frac{1}{h_j^2}\right) + \beta l_j\left(\frac{1}{h_{j+1}^3}+\frac{1}{h_j^3}\right)\right\} \\ 0 & 1 \end{bmatrix}$	$\begin{bmatrix} 1 & -\frac{\xi_{in}}{2} \\ 0 & 1 \end{bmatrix}$	$\begin{bmatrix} 1 & -\frac{\xi_{ex}}{2h_{N+1}^2} \\ 0 & 1 \end{bmatrix}$	$\begin{bmatrix} 1 & -\frac{1}{2}\left(\frac{1}{h_{j+1}^2}-\frac{1}{h_j^2}+\frac{\zeta_i}{h_i^{*2}}\right) \\ 0 & 1 \end{bmatrix}$

h^*_i is the gap when ζ_i is defined.

From Eq. (3.33), the perturbation pressure and flow rate of each element (ΔP_j, ΔQ_j), as well as those of the outlet (ΔP_{N+1}, ΔQ_{N+1}), are obtained as functions of those at the entrance (ΔP_0, ΔQ_0) as:

$$\begin{aligned} &[p_{j+1}, q_{j+1}, q_w, e]^T = \mathbf{m}_{j-1}\mathbf{m}_{j-1}\ldots\mathbf{m}_1\mathbf{m}_{in}[p_0, q_0, q_w, e]^T \\ &[p_{N+2}, q_{N+2}, q_w, e]^T = \mathbf{m}[p_0, q_0, q_w, e]^T \\ &p = \Delta P/P^*, \quad P^* = \rho\overline{Q}^2/H_0^2, \quad q = \Delta Q/\overline{Q}, \quad e = \Delta H/H_0, \quad q_w = i\omega e \end{aligned} \tag{3.34}$$

where $[]^T$ indicates the matrix transpose. The transfer matrices of the different elements are shown in Table 3.6. The perturbation pressures just before the entrance and just after the exit p_0, p_{N+1} are zero. Considering the second equation of Eq. (3.34), q_0 can be obtained as a function of q_w and e as:

$$q_0 = \frac{i\omega C + D}{i\omega A + B} q_w + \frac{E}{i\omega A + B} e \tag{3.35}$$

Substituting Eq. (3.35) into the first equation of Eq. (3.34), the non-dimensional perturbation pressure distribution p_j can be obtained as a function of q_w and e. Finally, when the non-dimensional perturbation pressure is integrated numerically, the unsteady fluid force acting on the wall can be obtained as a function of q_w and e as:

$$\Delta F/P^* L = f_w \cdot q_w + f_e \cdot e \tag{3.36}$$

The added mass coefficient M_a, added damping coefficients C_w, C_e, and added stiffness coefficient K_a are defined as:

$$\begin{aligned} &M_a = -\frac{\rho L^3}{H_0}\frac{Im(f_w)}{\omega}, \quad C_w = -\frac{\rho\overline{Q}L^2}{H_0^2}Re(f_w), \quad C_e = -\frac{\rho\overline{Q}L^2}{H_0^2}\frac{Im(f_e)}{\omega}, \\ &K_a = -\frac{\rho\overline{Q}^2 L}{H_0^3}Re(f_e) \end{aligned} \tag{3.37}$$

Table 3.6 Transfer matrices for unsteady flow component

Smooth flow passage element	Entrance element	Sudden section change element
$\mathbf{m}_j = \begin{bmatrix} 1 & -(i\omega A_j + B_j) & D_j & E_j \\ 0 & 1 & -G_j & 0 \\ 0 & 0 & 1 & 0 \\ 0 & 0 & 0 & 1 \end{bmatrix}$ $A_j = \frac{l_j}{2}\left(\frac{1}{h_j} + \frac{1}{h_{j+1}}\right), \quad B_j = \beta\gamma l_j\left(\frac{1}{h_{j+1}^3} + \frac{1}{h_j^3}\right) + \frac{1}{h_{j+1}^2} - \frac{1}{h_j^2}$ $D_j = l_j\left(\frac{1}{h_{j+1}^2} + \frac{1}{h_j^2}\right), \quad E_j = \frac{3}{2}\beta l_j\left(\frac{1}{h_{j+1}^4} + \frac{1}{h_j^4}\right) + \frac{1}{h_{j+1}^3} - \frac{1}{h_j^3}$ $G_j = \begin{cases} (l_{j+1} + l_j)/2 & (j = 1, \ldots, N-1) \\ l_N/2 & (j = N) \end{cases}$	$\begin{bmatrix} 1 & -\xi_{in} & 0 & \xi_{in} \\ 0 & 1 & -l_1/2 & 0 \\ 0 & 0 & 1 & 0 \\ 0 & 0 & 0 & 1 \end{bmatrix}$ **Outlet element** $\begin{bmatrix} 1 & -\xi_{ex}/h_{N+1}^2 & 0 & \xi_{ex}/h_{N+1}^3 \\ 0 & 1 & 0 & 0 \\ 0 & 0 & 1 & 0 \\ 0 & 0 & 0 & 1 \end{bmatrix}$	$\mathbf{m}_j = \begin{bmatrix} 1 & -B_j & 0 & E_j \\ 0 & 1 & 0 & 0 \\ 0 & 0 & 1 & 0 \\ 0 & 0 & 0 & 1 \end{bmatrix}$ $B_j = \frac{1}{h_{j+1}^2} - \frac{1}{h_j^2} + \frac{\bar{\zeta}_i}{h_i^{*2}}$ $E_j = \frac{1}{h_{j+1}^3} - \frac{1}{h_j^3} + \frac{\bar{\zeta}_i \eta_i}{h_i^{*3}}$ $\eta_i = 1 - \frac{d\bar{\zeta}_i}{2\bar{\zeta}_i d\left(\frac{h_{j+1}}{h_j}\right)}\left(\frac{h_i^*}{h_{j+1}} - \frac{h_i^*}{h_j}\right)\left(\frac{h_{j+1}}{h_j}\right)$

With ΔF given by Eq. (3.36), an equation of motion for the structure similar to Eq. (3.29) can be written. The conditions of negative damping and negative stiffness can similarly be evaluated as above [44].

3.3.4 Mechanism of self-excited vibration [45]

The wall oscillation of the leakage-flow passage can induce flow resistance variation, which in turn induces oscillations of the leakage-flow velocity. In this section, the reason why a negative damping force can be easily generated is given in the case of a divergent passage [45]. This is done by considering how the flow velocity oscillations or variations affect the pressure variations in the leakage-flow passage. The second equation of Eq. (3.25) is linearized as follows:

$$\frac{\mathrm{d}p}{\mathrm{d}y} = -\left\{\frac{i\omega}{h} + \frac{2(\beta\gamma-\alpha)}{h^3}\right\}q - \frac{2}{h^2}\frac{\mathrm{d}q}{\mathrm{d}y} + \frac{3(\beta-\alpha)}{h^4}e \tag{3.38}$$

The second term in the brackets of Eq. (3.38) expresses the variation of pressure due to the flow resistance variation associated with variations in the flow rate q, while the first term expresses the pressure variation due to fluid inertia. From Eq. (3.38) we note the following regarding the various contributions to pressure variations:

1. Pressure variation due to inertia effects is proportional to $1/h$.
2. Pressure variation due to flow resistance is proportional to $1/h^3$ within the leakage-flow passage and proportional to $1/h^2$ for the entrance and the exit.

Therefore, though in the divergent passage the variation of the flow resistance and inertia force decreases with y, the inertia force can be dominant in the downstream region compared to the flow resistant related force (see Fig. 3.23).

Now, let us consider the process when the gap width increases (as passage walls move away from each other) for the divergent passage. The flow rate increases since the flow resistance of the leakage passage decreases. Near the outlet, the influence of the inertia force is dominant compared with that of the flow resistance, and the flow rate cannot increase easily. That is, pressure in the channel increases due to an effect similar to attempting to close the outlet while pumping fluid into the channel at the entrance. On the other hand, when the gap width decreases (as passage walls approach), though the flow rate decreases, the flow rate does not decrease easily near the outlet since the influence of the inertia force can be dominant. That is, pressure in the channel decreases due to an effect similar to trying to close the entrance while sucking out fluid at the outlet. Pressure increases as the gap width increases, and decreases with decreasing gap width. The increased pressure as the gap width increases acts to push the walls further apart while the suction effect as the gap width decreases acts to bring the walls even closer. The net effect then is that pressure acts as a negative damping (destabilizing) force.

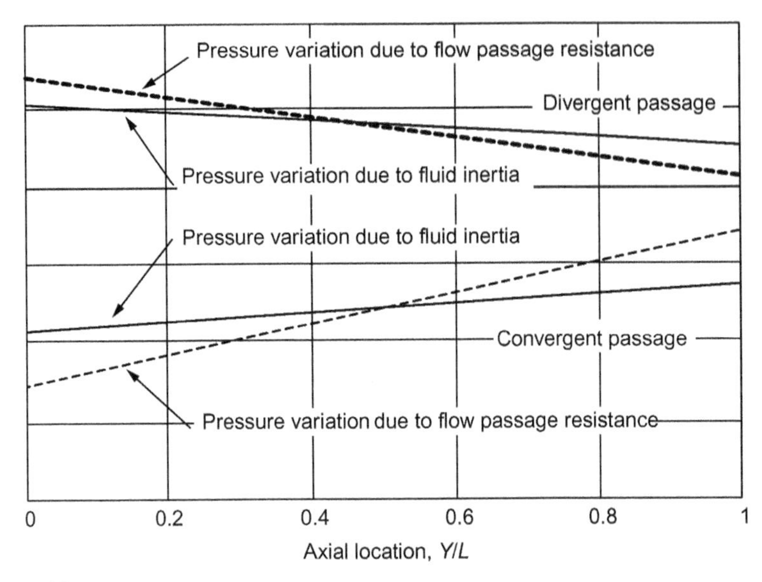

FIGURE 3.23

The distribution of variation of the flow resistance and the inertia force when the flow rate varies in the leakage passage [45].

In the case of the convergent passage, the inertia force is dominant over the flow resistant variation induced force near the entrance. Consequently, the resulting pressure variations lead to a positive damping force generated via the same effects causing the negative damping in a divergent passage. Note, that in this case, the force directions are reversed.

The foregoing is now explained using a more mathematical approach. The added damping coefficient C_e can be rewritten as follows:

$$C_e \cdot \frac{H_0^2}{\rho \bar{Q} L^2} = -\frac{k_2 T}{1+\omega^2 T^2} = -\frac{B_2(y_m - y_r)T}{1+\omega^2 T^2},$$

$$y_r = \frac{\int_0^1 \frac{2(\beta\gamma-\alpha)}{h^3(y)} y\,dy + \frac{\xi_{ex}}{h^2(1)}}{\int_0^1 \frac{2(\beta\gamma-\alpha)}{h^3(y)} dy + \xi_{in} + \frac{\xi_{ex}}{h^2(1)}}, \quad y_m = \frac{i\omega \int_0^1 \frac{1}{h(y)} y\,dy}{i\omega \int_0^1 \frac{1}{h(y)} dy} \tag{3.39}$$

where y_r and y_m are first-order moments of the inertia force and the variation of the flow resistance, respectively.

Equation (3.39) also shows that the sign (positive or negative) of C_e depends on the interrelation between the first-order moments of inertia and flow resistance forces. In addition, the condition where C_e becomes negative in the case of arbitrary-shaped passages can also be obtained from Eq. (3.39).

3.3.5 Self-excited vibrations in other cases

3.3.5.1 Rotational vibration of one-dimensional tapered passage wall

In the case of rotational (rocking) motion of the flow-passage wall, the moment around the pivot point can be calculated, from which we can obtain the added moment coefficients [43]. Though the mechanism of the self-excited vibration is similar to the mechanism in the case of translational vibration, the condition of negative damping depends on the location of the pivot point.

Figure 3.24 shows the ranges where added damping and added stiffness coefficients become positive or negative on the (α, β)-plane in the case of laminar flow, $\xi_{in} = 1.0$, $\xi_{ex} = 0$, and the non-dimensional pivot location l_f ($=L_f/L$, L_f is the pivot location) is selected to be 0, 0.5, and 0.7. The figure shows that divergence-type instability can occur in the case of the divergent passage as well as the convergent passage. Self-excited vibration can easily occur in the case of the divergent passage; however, it can also occur in the case of the convergent passage when the pivot point is upstream of the center of the passage. When the pivot point is at the entrance ($l_f = 0$), the region where divergent-type instability and self-induced vibration occur becomes limited, and the generation of divergent-type instability and self-induced vibration tends to be controlled.

3.3.5.2 Two-degrees-of-freedom translational and rotational systems and continuous systems [43,46,47]

Two-degrees-of-freedom (translational and rotational) systems and even continuous systems may also undergo the same instabilities described above for single-degree-of-freedom systems. Figure 3.25 shows the critical non-dimensional velocity $q_r(=\overline{Q}/(H_0 L\Omega)$ as a function of the area increment ratio $1+\alpha$ for a two-degree-of-freedom translational–rotational plate system [43,46]. The plate is supported at the mid-point of the passage ($l_f = 0.5$). The natural frequency of the rotational mode is close to that of the translational mode (ω_r = [rotational natural frequency]/[translational natural frequency] = 0.9) [43,46]. The figure shows that self-excited vibration can occur in the case of not only a divergent passage but also a convergent passage. Figure 3.26 presents plate vibration behavior during a cycle, showing the modal coupling of translational and rotational motions.

Figure 3.27 [47] shows the non-dimensional critical velocity for a cantilevered plate clamped at the upstream end, as a function of the parameter $\rho_f L^2/\rho_s w H_0$, where ρ_f and ρ_s are the density of the fluid and the plate; and L and w are the length and the thickness of the plate. It is seen that as the parameter $\rho_f L^2/\rho_s w H_0$ increases, higher mode instabilities can occur. This is quite similar to the tendency for a pipe conveying fluid, in which the generated mode of flutter depends on the mass ratio of the tube and the fluid in the tube.

It is concluded that self-excited vibration due to modal coupling can occur even for convergent passages, or where the pivot point of the rotation is near the entrance; note, that in this case, self-excited vibration cannot easily occur for a single-degree-of-freedom system. A simple stabilization method for this coupled

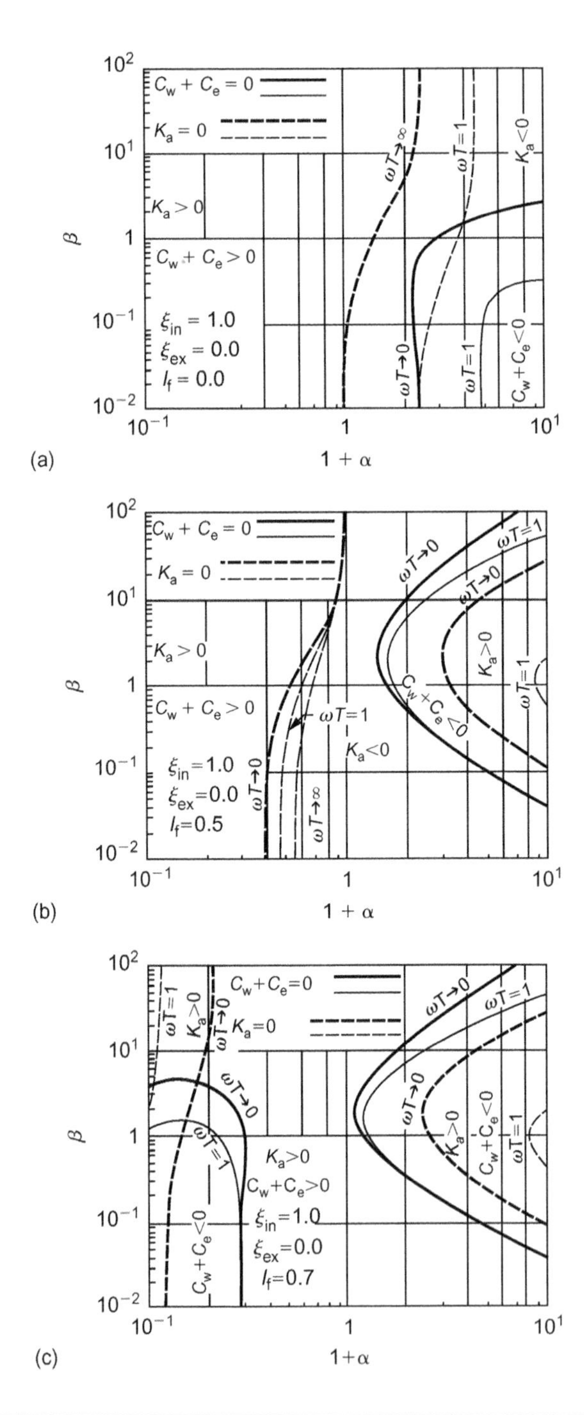

FIGURE 3.24

Region where added damping and stiffness coefficients become positive or negative for a one-dimensional tapered passage wall oscillating as a single-degree-of-freedom rotational system [43]: (a) $l_f = 0$ (the pivot point is at the entrance), (b) $l_f = 0.5$ (the pivot point is at the center of the passage), and (c) $l_f = 0.7$ (the pivot point is slightly closer to the outlet from the mid-point of the passage).

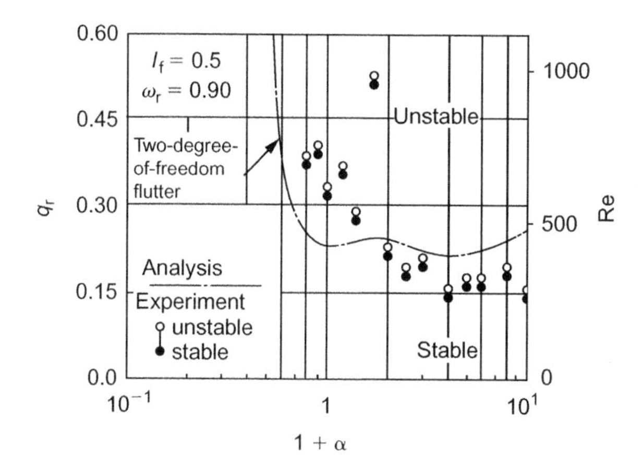

FIGURE 3.25

Critical velocity of two-degree-of-freedom translational and rotational system [46].

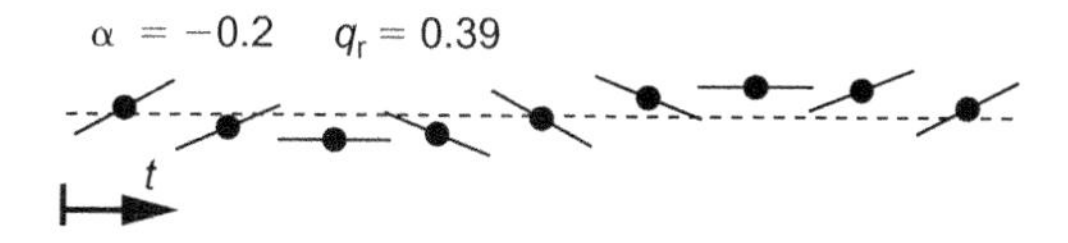

FIGURE 3.26

Plate motion during one cycle of self-excited vibration [46].

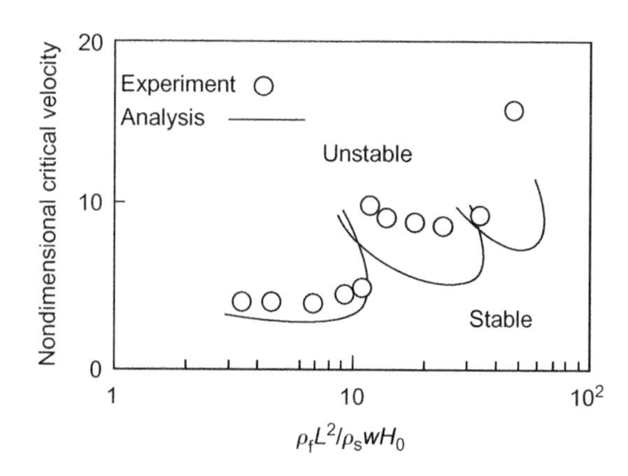

FIGURE 3.27

Critical velocity of cantilevered plate clamped at the upstream end [47].

mode instability is yet to be found. Further experimental and analytical work is needed to fully understand this instability.

Recently, the mechanism of self-excited vibration for continuous systems has been considered. As an example, the wave equation has been derived from the linearized basic equations of the leakage flow and the momentum equation of the plate. The characteristics of this equation have been investigated [45]. The phenomena are very complicated and research is expected to progress in the future.

3.3.5.3 Annular flow passage

When the structure forming an annular leakage passage vibrates as a single-degree-of-freedom system, it is effective to make the axial shape of the passage convergent to control the self-excited vibration. A study to evaluate the critical velocity quantitatively using the finite difference method has been presented by Li et al. [48].

The self-excited vibration of the thermal shield shell of a jet engine after burner, where the inner tube vibrated in a shell mode, has been reported by Ziada and Buhlmann [49]. The researchers conducted an experiment to reproduce the instability using a slightly convergent-cantilevered conical shell. They found that self-excited vibrations could be generated when the flow velocity in the annular passage exceeded a critical value, and, on the other hand, that the self-excited vibration was controlled by the flow in the inner tube. Païdoussis et al. [50] have conducted comprehensive analytical studies on the flow-induced vibration of concentric annular cylindrical shells. More recently, Kang et al. [51] have shown that flow perturbations and fluid friction strongly influence the annular-flow-induced stability behavior. Still, in order to obtain a full understanding of the underlying mechanisms of the dynamics of this complex system, further work is needed.

3.3.6 Hints for countermeasures

If the system can be identified as a single-degree-of-freedom translational or rotational system with an annular passage, both the self-excited vibration and the divergent-type instability can be controlled by changing to a convergent passage in many cases. When the flow passage geometry changes axially, self-excited vibration can be effectively controlled by moving the location of the first-order moment of the inertia force (see Eq. [3.39]) upstream of the first-order moment of the flow resistance force. In the case of a single-degree-of-freedom rotational system, self-excited vibration can be controlled when the pivot point is set at the entrance of the passage.

On the other hand, when the shape of the flow passage cannot be changed in the case of a single-degree-of-freedom system, and when the passage wall can vibrate in a beam or shell mode, increasing the natural frequency by thickening the wall and by reinforcement as well as decreasing the flow velocity are effective means to control the self-excited vibration. Note, however, that the thermal impact and thermal stress should be taken into account in the case of high temperature conditions.

The quantitative evaluation for annular-leakage flow in multi-degree-of-freedom systems or continuous systems can be conducted using numerical methods.

3.3.7 Examples of leakage-flow-induced vibration

Figures 3.20 and 3.28 show two typical components that have suffered leakage-flow-induced vibrations, as reported in the open literature. Relatively large structures such as nuclear reactor core barrels can suffer damage due to leakage-flow-induced

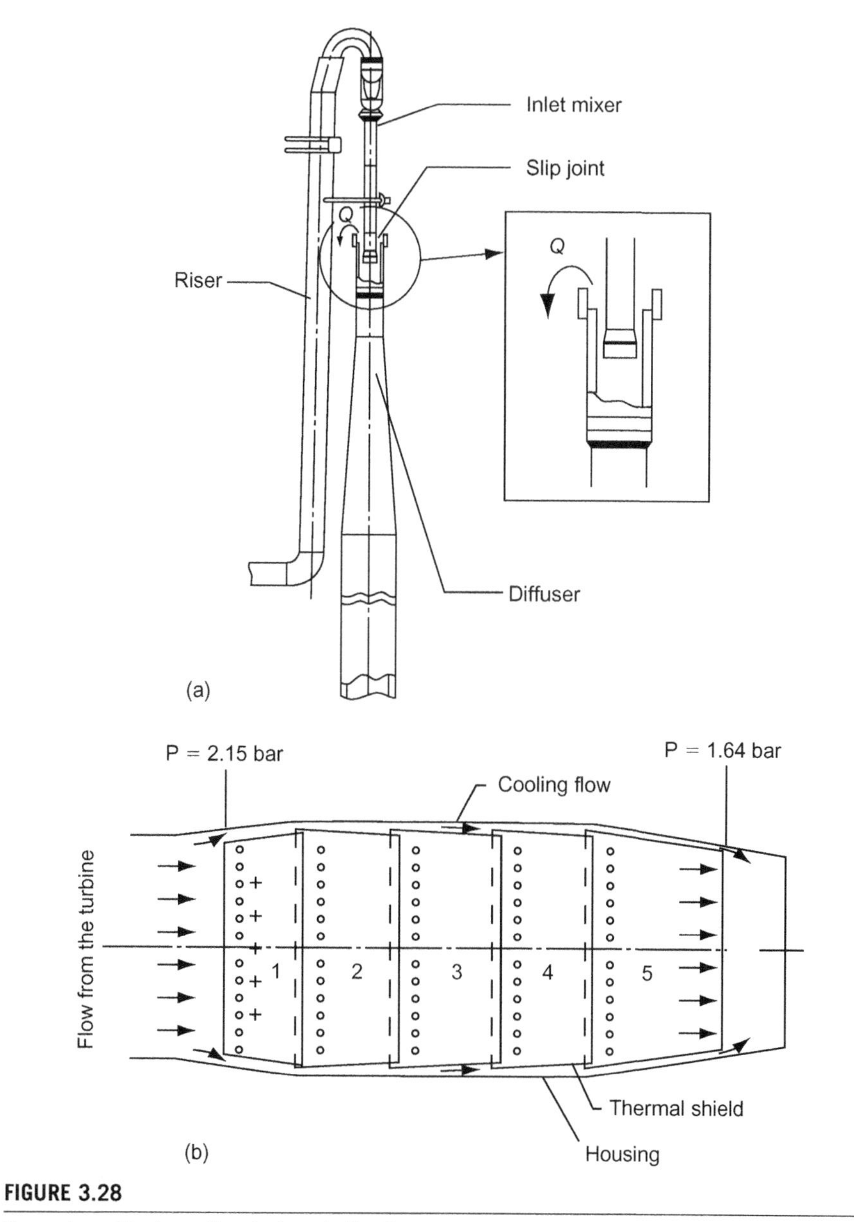

FIGURE 3.28

Examples of leakage-flow-induced vibrations.

vibration because the fluid forces induced by the leakage flow can be very large. The reported examples of Fig. 3.28 are but an indicator of possibly many more similar events in industry.

References

[1] E.P. Quinn, Vibration of Fuel Rods in Parallel Flow, GEAP-4059, General Electric Co. 1962.

[2] D.J. Gorman, An analytical and experimental investigation of the vibration of cylindrical reactor fuel elements in two-phase parallel flow, Nucl. Sci. Eng. 44 (1971) 277–290.

[3] M.J. Pettigrew, D.J. Gorman, Experimental Studies on Flow-Induced Vibration to Support Steam Generator Design, Part 1: Vibration of a Heated Cylinder in Two-Phase Axial Flow, AECL-4514, Chalk River Nuclear Laboratory, Chalk River, 1973.

[4] S.S. Chen, Flow-Induced Vibration of Circular Cylindrical Structures, Hemisphere Publishing Corp., Springer-Verlag, Washington, 1987.

[5] M.P. Païdoussis, L.R. Curling, An analytical model for vibration of clusters of flexible cylinders in turbulent axial flow, J. Sound Vib. 98 (4) (1985) 493–517.

[6] S.S. Chen, M.W. Wambsganss Jr., Response of a flexible rod to near-field flow noise, *Proceedings of the Conference on Flow-Induced Vibrations in Reactor System Components*, ANL-7685, Argonne National Laboratory, 1970.

[7] G.M. Corcos, Resolution of pressure in turbulence, J. Acoust. Soc. Am. 35 (2) (1963) 192–199.

[8] W.W. Willmarth, C.E. Wooldridge, Measurements of the fluctuating pressure at the wall beneath a thick turbulent boundary layer, J. Fluid Mech. 14 (Pt. 2) (1962) 187–210.

[9] J.M. Clinch, Measurements of the wall pressure field at the surface of a smooth-walled pipe containing turbulent water flow, J. Sound Vib. 9 (3) (1969) 398–419.

[10] H.P. Bakewell Jr., Turbulent wall-pressure fluctuations on a body of revolution, J. Acoust. Soc. Am. 43 (6) (1968) 1358–1363.

[11] H.H. Schloemer, Effects of pressure gradients on turbulent-boundary-layers wall-pressure fluctuations, J. Acoust. Soc. Am. 42 (1967) 93–113.

[12] S.S. Chen, M.W. Wambsganss, Parallel-flow-induced vibration of fuel rods, Nucl. Eng. Design 18 (1972) 253–278.

[13] M.J. Pettigrew, C.E. Taylor, Two-phase flow-induced vibration: an overview, J. Press. Vessel Technol. 116 (1994) 233–253.

[14] N. Saito, Y. Tsukuda, et al., BWR 939 Type Rod Assembly Thermal Hydraulic Tests (2) –Hydraulic Vibration Test–Tenth International Conference on Nuclear Engineering, ICONE10-22557, 2002.

[15] K. Kawamura, A. Yasuo, F. Inada, Turbulence-Induced Tube Vibration in a Parallel Steam-Water Two-Phase Flow, Fourth International Conference on Fluid-Structure Interactions, Aeroelasticity, Flow-Induced Vibration and Noise, vol. 2, ASME AD-Vol. 53, No. 2, 1997, pp. 83–92.

[16] F. Inada, K. Kawamura, A. Yasuo, A study on random fluid force acting on a tube in parallel two-phase flow, Symposium on Flow-Induced Vibrations, IMechE, 1991, pp. 379–384.
[17] D. Burgreen, J.J. Byrnes, D.M. Benforado, Vibration of rods induced by water in parallel flow, Trans. ASME 80 (5) (1958) 991–1003.
[18] J.R. Reavis, Vibration correlation for maximum fuel-element displacement in parallel turbulent flow, Nucl. Sci. Eng. 38 (1) (1969) 63–69.
[19] M.P. Païdoussis, An experimental study of vibration of flexible cylinders induced by nominally axially flow, Nucl. Sci. Eng. 35 (1) (1969) 127–138.
[20] Y.N. Chen, M. Wever, Flow-Induced Vibration in Tube Bundle Heat Exchangers with Cross and Parallel Flow, ASME Symposium on Flow-Induced Vibration in Heat Exchangers, New York, December 1970, pp. 57–77.
[21] S.S. Chen, M.W. Wambsganss, ANL-ETD-71-07 Tentative Design Guide for Calculation the Vibration Response of Flexible Cylindrical Elements in Axial Flow, Argonne National Laboratory, 1971.
[22] W.R. Hawthorne, Proc. Inst. Mech. Eng. 175 (1961) 52.
[23] M.P. Païdoussis, Dynamics of flexible slender cylinders in axial flow, part 1. Theory, J. Fluid Mech. 26 (4) (1966) 717–736.
[24] M.P. Païdoussis, Dynamics of flexible slender cylinders in axial flow, part 2. Experiments, J. Fluid Mech. 26 (4) (1966) 737–751.
[25] M.P. Païdoussis, Dynamics of cylindrical structures subjected to axial flow, J. Sound Vib. 29 (3) (1973) 365–385.
[26] M.P. Païdoussis, M. Ostoja-Starzewski, Dynamics of a flexible cylinder in subsonic axial flow, AIAA J. 19 (1981) 1467–1475.
[27] M.P. Païdoussis, Appl. Mech. Rev. 40 (2) (1987) 163–175.
[28] M.P. Païdoussis, *Trans. ASME, J. Press. Vess. Tech* 115 (1993) 2–14.
[29] T.S. Lee, J. Fluid Mech. 110 (1981) 293–295.
[30] G.S. Triantafyllou, C. Chryssostomidis, Stability of a string in axial flow, J. Energy Res. Technol. 107 (1985) 421–425.
[31] K. Washizu, *Aeroelasticity* (in Japanese), Kyoritsu Publishing Co. Ltd., 1957.
[32] R.L. Bisplinghoff, H. Ashley, R.L. Halfman, Aeroelasticity, Addison-Wesley, Cambridge, MA, 1955.
[33] Y.C. Fung, An Introduction to the Theory of Aeroelasticity, John Wiley & Sons, New York, 1955.
[34] Isogai, K. et al., Experimental Study on Transonic Flutter Characteristics of Sweptback Wing with Core Composite Plates having different Fiber Orientations, NAL-TR-827 (in Japanese), 1984.
[35] J. Nakamichi, Unsteady aerodynamics analyses around oscillating airfoils, *J. Japan Soc. Aeronaut. Space Sci.* (in Japanese) 43 (501) (1995) 549–556.
[36] H. Matsushita, Active control of flutter, J. Architect. Building Sci. (in Japanese) 109 (1351) (1994).
[37] D.S. Weaver, T.E. Unny, The hydroelastic stability of a flat plate, Trans. ASME, J. Appl. Mech. 37 (1970) 823–827.
[38] M.P. Païdoussis, S.P. Chan, A.K. Misra, Dynamics and stability of coaxial cylindrical shells containing flowing fluid, J. Sound Vib. 97 (2) (1984) 201–235.
[39] S. Ziada, E.T. Buhlmann, Model tests on shell flutter due to flow on both sides, J. Fluid Str. 2 (2) (1988) 177–196.

[40] W.K. Blake, Mechanics of Flow-Induced Sound and Vibration, Academic Press, Orlando, FL, 1986.
[41] R.D. Blevins, Flow-Induced Vibration, second ed., Van Nostrand Reinhold, New York, 1990.
[42] E. Naudascher, D. Rockwell, Practical Experiences with Flow Induced Vibrations, Springer-Verlag, Berlin, 1980, 1–81.
[43] F. Inada, S. Hayama, J. Fluid Str. 4 (1990) 395–412.
[44] F. Inada, S. Hayama, JSME Int. J. 31-1 (1988) 39–47.
[45] F. Inada, S. Hayama, Proceedings of the Seventh International Conference on Flow Induced Vibration, 2000, pp. 837–844.
[46] F. Inada, S. Hayama, J. Fluid. Str. 4 (1990) 413–428.
[47] H. Nagakura, S. Kaneko, Symposium on Flow-Induced Vibrations in Engineering Systems, Asia-Pacific Vibration Conference vol. 1, 1993, pp. 248–253.
[48] D.W. Li, S. Kaneko, S. Hayama, J. Fluid Str. 16 (2002) 909–930.
[49] S. Ziada, E.T. Buhlmann, J. Fluid Str. 2 (1988) 177–196.
[50] M.P. Païdoussis, et al., ASME Symposium on FIV-90. PVP-vol. 189, 1990, pp. 207–226.
[51] H.S. Kang, N.W. Mureithi, M.J. Pettigrew, J. Fluid Str. 35 (2012) 1–20.

CHAPTER

4 Vibrations Induced by Internal Fluid Flow

Piping systems are widely utilized to convey fluids in many industrial fields, ranging from chemical plants to biological engineering systems. The instability problem of flexible pipes conveying fluid provides a paradigm for the modeling and analysis of the instability mechanisms of fluid-structure interaction systems. Consequently, significant research on the dynamics of flexible pipes conveying fluid has been conducted.

In this chapter, the research history on dynamics of the instability phenomenon of pipes conveying fluid is presented. Evaluation methods for the vibration of pipes excited by oscillating fluid flow and two-phase fluid flow are discussed in Section 4.1. The vibration of bellows and corrugate tubes, which are major piping elements, are introduced in Section 4.2. Unstable vibrations of collapsible tubes, which have recently become important in biological engineering, are introduced in Section 4.3. Countermeasures against the different types of vibration are presented at the end of each section.

4.1 Vibration of straight and curved pipes conveying fluid

4.1.1 Vibration of pipes conveying fluid

Pipes conveying fluids are found in many industrial systems. Examples include fuel pipes in engine systems, heat transfer pipes in power generation plants, chemical plant piping, and so forth. Piping vibration problems are therefore very important in industry. The instability phenomenon of flexible pipes conveying fluid has been studied by many researchers. The stability and dynamic characteristics are now well understood. The dynamics are known to be sensitively dependent on flow velocity and support/boundary conditions.

With regard to the instability mechanism of flexible pipes conveying fluid there are two different cases: (i) unstable vibration caused by the fluid flow when the flow velocity surpasses a critical value, and (ii) vibration due to oscillating fluid flow.

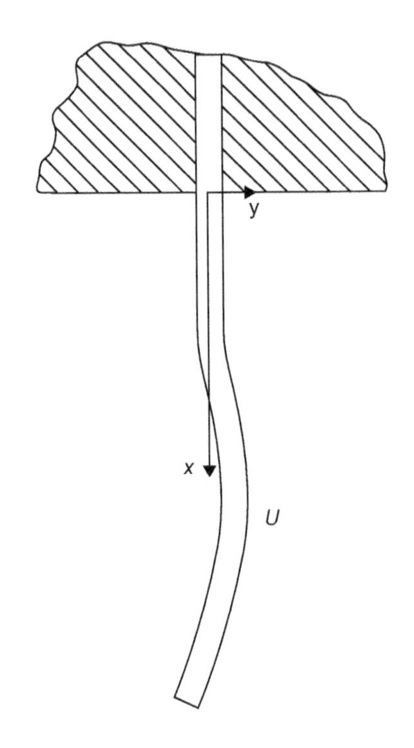

FIGURE 4.1

Flexible pipe conveying fluid and coordinate system [6]. Permission is granted by the Council of the Institution of Mechanical Engineers.

4.1.1.1 Research history

Major research on the stability and vibration of flexible pipes conveying fluid started in the 1950s in relation to the design of pipelines conveying oil. Thereafter, these studies were enhanced by the work of Benjamin [1,2]; Gregory and Païdoussis [3,4]; and Païdoussis et al. [5–13]. Many other detailed studies on the stability and dynamic behavior of flexible pipes conveying fluid have since been conducted and reported. The effects of an attached mass, damper or spring, for example, were investigated by Sugiyama et al. [14].

4.1.1.2 Modeling and stability analysis of straight pipes conveying fluid

The detailed analysis of the dynamics of straight flexible pipes conveying fluid is described by Païdoussis et al. [5–13]. In this section, the modeling and calculation methods based on these papers are introduced. The physical system analyzed is shown in Fig. 4.1. Forces and moments acting on the fluid and pipe elements, respectively, are shown in Fig. 4.2.

The system consists of a uniform pipe of length L, mass per unit length m, flexural rigidity EI, conveying fluid of mass per unit length ρA, and flowing

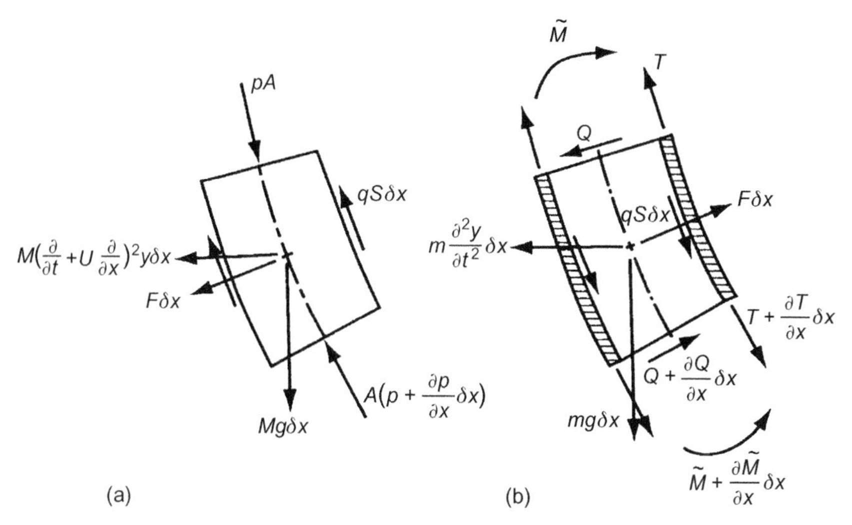

FIGURE 4.2

Forces and moments acting on elements of (a) the fluid, and (b) the pipe, respectively [6]. Permission is granted by the Council of the Institution of Mechanical Engineers.

axially with velocity U. The cross-sectional flow area is A and the fluid pressure is p. In this model, the pipe is assumed to be made of Kelvin-Voigt type visco-elastic material. The coefficient of internal dissipation due to the viscoelastic material is E^*, and the coefficient of viscous damping due to friction between the pipe with the surrounding stationary fluid medium is c.

The basic equation governing small lateral motions is:

$$E^*I\frac{\partial^5 y}{\partial t\partial x^4}+\mathrm{EI}\frac{\partial^4 y}{\partial x^4}+\left[\rho AU^2+(pA-T)|_L-\left\{(\rho A+m)g-\rho A\frac{\partial U}{\partial t}\right\}(L-x)\right]\frac{\partial^2 y}{\partial x^2}$$
$$+2\rho AU\frac{\partial^2 y}{\partial t\partial x}+(\rho A+m)g\frac{\partial y}{\partial x}+c\frac{\partial y}{\partial t}+(\rho A+m)\frac{\partial^2 y}{\partial t^2}=0 \tag{4.1}$$

This equation can be expressed in dimensionless form by defining the following dimensionless parameters:

$$\xi=x/L,\ \eta=y/L,\quad \alpha=\sqrt{\frac{I}{E(\rho A+m)}\frac{E^*}{L^2}},\quad \beta=\frac{\rho A}{\rho A+m},\quad \gamma=\frac{\rho A+m}{\mathrm{EI}}L^3 g,$$
$$\kappa=\frac{cL^2}{\sqrt{\mathrm{EI}(\rho A+m)}},\quad u=\sqrt{\frac{\rho A}{\mathrm{EI}}}UL,\quad \tau=\sqrt{\frac{\mathrm{EI}}{m+\rho A}}\frac{t}{L^2} \tag{4.2}$$

Substituting Eqs. (4.2) into Eq. (4.1), the following dimensionless equation of motion is obtained:

$$\begin{aligned}&\frac{\partial^2\eta}{\partial\tau^2}+\kappa\frac{\partial\eta}{\partial\tau}+2\sqrt{\beta}u\frac{\partial^2\eta}{\partial\tau\partial\xi}+\alpha\frac{\partial^5\eta}{\partial\tau\partial\xi^4}+\gamma\frac{\partial\eta}{\partial\xi}\\&+\left[u^2+\left\{\sqrt{\beta}\frac{\partial u}{\partial\tau}-\gamma\right\}(1-\xi)\right]\frac{\partial^2\eta}{\partial\xi^2}+\frac{\partial^4\eta}{\partial\xi^4}=0\end{aligned}\tag{4.3}$$

For a constant flow velocity, the equation of motion becomes:

$$\begin{aligned}&\frac{\partial^2\eta}{\partial\tau^2}+\kappa\frac{\partial\eta}{\partial\tau}+2\sqrt{\beta}u\frac{\partial^2\eta}{\partial\tau\partial\xi}+\alpha\frac{\partial^5\eta}{\partial\tau\partial\xi^4}+\gamma\frac{\partial\eta}{\partial\xi}\\&+\{u^2-\gamma(1-\xi)\}\frac{\partial^2\eta}{\partial\xi^2}+\frac{\partial^4\eta}{\partial\xi^4}=0\end{aligned}\tag{4.4}$$

The equation of motion is discretized using Galerkin's method. The dimensionless displacement η_n (ξ, τ) can be expressed as a superposition of a finite set of normal modes, that is,

$$\eta(\xi,\tau)=\sum_{n=1}^{N}q_n(\tau)\cdot\phi_n(\xi)\tag{4.5}$$

where q_n (τ) are the generalized coordinates, and $\phi_n(\xi)$ are the eigenfunctions of a uniform beam having the same boundary conditions as the pipe, and given by:

$$\phi_n(\xi)=(\cosh\lambda_n\xi-\cos\lambda_n\xi)+\sigma_n(\sinh\lambda_n\xi-\sin\lambda_n\xi)\tag{4.6}$$

The constants σ_n are determined by the boundary conditions.

Substituting Eq. (4.5) and $q_n=\overline{q}_n\exp(j\omega\tau)$ into Eq. (4.4), and then multiplying by $\phi_m(\xi)$, $m=1, 2, 3\ldots$ in turn, and integrating from 0 to 1, the following characteristic equation of the system is obtained by virtue of the orthogonality of the eigenfunctions:

$$\begin{aligned}&\det[\{(\Omega_n^2-\omega^2)+j(\kappa+\alpha\Omega_n^2)\omega\}\delta_{nm}+(\gamma+j2\sqrt{\beta}u\omega)b_{nm}\\&+(u^2-\gamma)c_{nm}+\gamma d_{nm}]=0\end{aligned}\tag{4.7}$$

where δ_{nm} is the Kronecker delta. The parameters Ω_n are given by $\Omega_n^2=\lambda_n^4$, and b_{nm}, c_{nm}, and d_{nm} are determined from the properties of the eigenfunctions.

For any given system, the parameters α, β, δ, and κ are known. The dimensionless complex frequencies ω_n of the system modes can be calculated for increasing values of the dimensionless flow velocity u starting with $u=0$. These complex frequencies are normally plotted in the form of Argand diagrams with dimensionless flow velocity u as the parameter.

The stability of the system is determined by the complex frequencies ω_n. The system will be dynamically stable or unstable depending on whether the imaginary component of ω_n, $\mathrm{Im}(\omega_n)$, is positive or negative. Static instability, or divergence, occurs when $\mathrm{Re}(\omega_n)$ vanishes.

4.1.1.3 Stability of straight pipes conveying fluid

Root *loci* of the complex frequencies in the case of pinned-pinned pipes with $\alpha = \kappa = \gamma = 0$ are shown in Fig. 4.3 [8], for (a) $\beta = 0.1$ and (b) $\beta = 0.5$. In

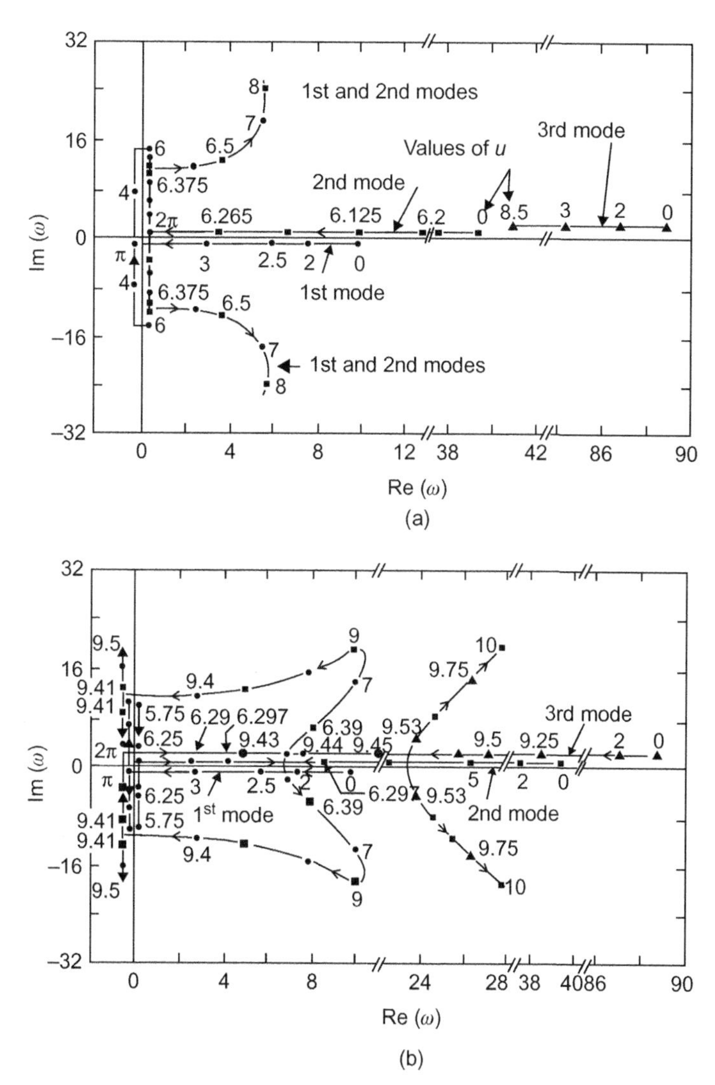

FIGURE 4.3

Root *loci* of pinned-pinned pipes with increasing flow velocity for $\alpha = \kappa = \gamma = 0$, and (a) $\beta = 0.1$ or (b) $\beta = 0.5$ [8].

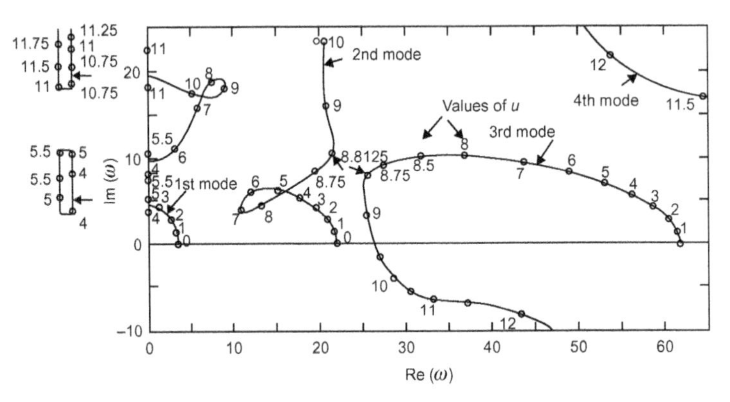

FIGURE 4.4

Root *loci* of fixed-free pipes with increasing flow velocity for $\alpha=\kappa=\gamma=0$ and $\beta=0.5$ [5].

Fig. 4.3(a), with increasing flow velocity, the frequency of the first mode decreases and eventually vanishes at the dimensionless velocity $u=\pi$. This is the first critical flow velocity for divergence. Similarly, the frequency of the second mode vanishes at $u=2\pi$. However, at a slightly higher flow velocity $u>2\pi$, the *loci* of the first and second modes coalesce on the imaginary axis and leave the axis at symmetrical points, indicating the onset of coupled-mode flutter.

In Fig. 4.3(b), the first-mode frequency vanishes at $u=\pi$. However, $u=2\pi$ does not correspond to buckling in the second mode, but it is the point where the system regains stability in its first mode. At a slightly higher flow velocity $u>2\pi$ the first- and second-mode *loci* coalesce on the real axis, once again indicating the onset of coupled-mode flutter. With increasing flow velocity u, the real part of the frequency eventually vanishes. Then, by a similar process, coupled-mode flutter occurs involving the third-mode *locus*.

In the case of fixed-free pipes, typical root *loci* of the complex frequencies with $\beta=0.5$ and $\alpha=\kappa=\gamma=0$ are shown in Fig. 4.4 [5]. In this figure, small flow velocities damp the system in all its modes and result in a reduction of the vibration frequencies. At higher flow velocities, the *locus* of at least one of the modes crosses into the unstable region, signifying that a flutter type instability occurs in this mode just above the flow velocity corresponding to a point of neutral stability. The critical flow velocity u_{cr} is the lowest flow velocity at which this instability occurs.

For the non-conservative fixed-free system, it has been found that low damping destabilizes the system. Figure 4.5 [13] shows the variation of the critical flow velocity u_{cr} versus β for progressively higher values of the viscoelastic dissipation constant α. It can be seen that, in the case of large β, the critical flow velocity decreases with the viscoelastic dissipation constant.

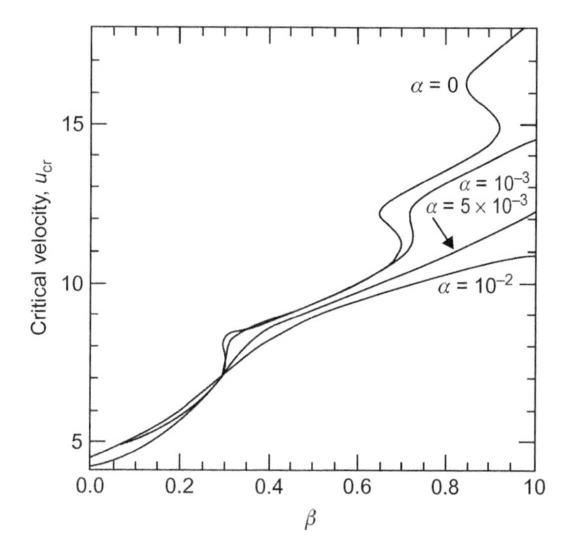

FIGURE 4.5

Variation of critical velocity of fixed-free pipes versus β for progressively higher values of the visco-elastic dissipation constant α [8].

4.1.1.4 Stability of curved pipes conveying fluid

The stability of curved pipes conveying fluid can be also studied using the approach outlined above. Detailed calculation methods and evaluation for curved flexible pipes were presented by Chen [15]. In this section, typical results obtained by Chen are discussed.

Figure 4.6 [15] shows the frequencies of a fixed-fixed curved pipe forming a 90 degree curve, as functions of the flow velocity. As the flow velocity increases, the frequencies decrease. At high flow velocities, some frequencies vanish and the system loses stability by divergence.

The complex frequencies of a fixed-free curved pipe, as functions of the flow velocity, are shown in Fig. 4.7 [15]. The effect of the flowing fluid is to reduce the frequencies of vibration and to increase the damping when the flow velocity is low. As the flow velocity increases, some roots cross the real axis and the system loses stability by flutter. The frequency of the second mode crosses the real axis at $u = 1.5$, 1.85, and 2.2; thus, the system loses stability at $u = 1.5$, regains stability at $u = 1.8$, and loses stability again at $u = 2.2$. The third and fourth modes are stable with increasing flow velocity.

4.1.1.5 Shell flutter and random vibration induced by internal fluid flow

Besides the vibration of flexible pipes discussed in the foregoing sections, flutter type instability of thin-walled flexible shells subjected to high-speed internal fluid flow was first observed by Païdoussis and Denise [16]. The flutter occurs in traveling wave modes in the circumferential direction. The critical flow

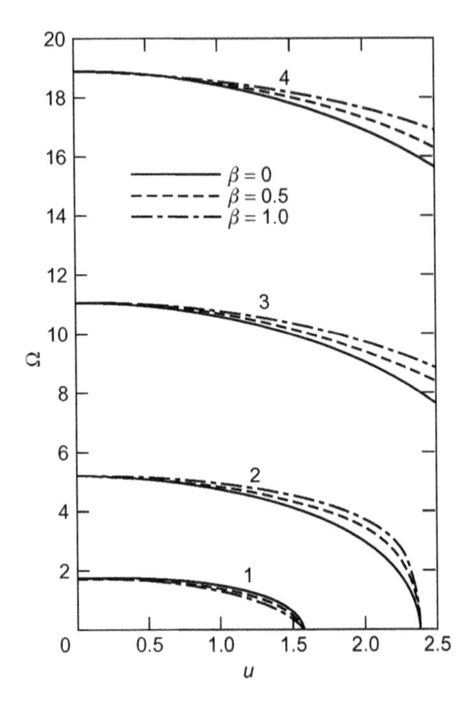

FIGURE 4.6

Variation of natural frequencies of fixed-fixed curved pipes with flow velocity u for $\tilde{\alpha} = \pi$ and $\beta = 0, 0.5, 1.0$ [15].

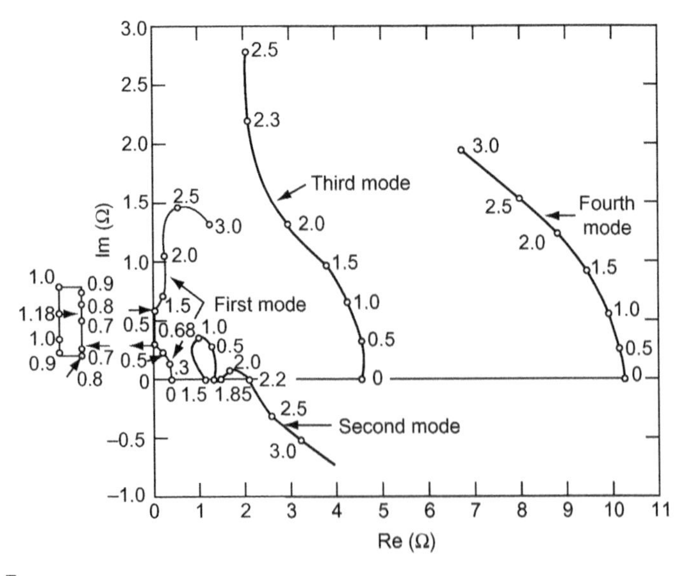

FIGURE 4.7

Root *locus* of fixed-fixed curved pipes with flow velocity u for $\tilde{\alpha} = \pi$, $\beta = 0.75$ [15].

velocity at which flutter occurs can be determined by the same method introduced above.

The turbulence generated by flow separation in short elbows should be estimated since such turbulent flow may result in shell-type flow-induced vibration. The elbow is the main source of the excitation force in piping due to the momentum variation at this location. In particular, vortex shedding at the corners of large elbows results in periodic excitation forces [17].

Based on the theory of statistical modal analysis, the power spectral density (PSD), $S_\sigma(x,f)$, of the response stress, σ, at the position x is given by the following equation:

$$
\begin{aligned}
S_\sigma(x,f) = {} & \frac{1}{16\pi^4}\sum_i\sum_j \frac{\Phi_i(x)\Phi_j(x)}{f_i^2 f_j^2 M_i M_j}\,\frac{1}{\left[\left\{1-\left(\frac{f}{f_i}\right)^2\right\}^2+4\zeta_i^2\left(\frac{f}{f_i}\right)^2\right]\left[\left\{1-\left(\frac{f}{f_j}\right)^2\right\}^2+4\zeta_j^2\left(\frac{f}{f_j}\right)^2\right]} \\
& \times\left[\left\{\left(1-\left(\frac{f}{f_i}\right)^2\right)\left(1-\left(\frac{f}{f_j}\right)^2\right)+4\zeta_i\zeta_j\frac{ff}{f_i f_j}\right\}\int_{A1}\int_{A2} S_{p1p2}{}^{c}(f)\varphi_i(x_1)\varphi_j(x_2)dA_1 dA_2\right. \\
& \left.+\left\{2\zeta_j\frac{f}{f_j}\left(1-\left(\frac{f}{f_i}\right)^2\right)-2\zeta_i\frac{f}{f_i}\left(1-\left(\frac{f}{f_j}\right)^2\right)\right\}\int_{A1}\int_{A2} S_{p1p2}{}^{Q}(f)\varphi_i(x_1)\varphi_j(x_2)dA_1 dA_2\right]
\end{aligned}
\tag{4.8}
$$

where $\Phi_i(x)$ is the i-th mode for the stress, $\varphi_i(x)$ the i-th displacement mode, f_i the i-th mode natural frequency, M_i the modal mass, and ζ_i expresses damping ratio of i-th mode. Suffixes i and j indicate the mode number. $S_{p1p2}(f)$ is the cross-spectral density of the fluctuating fluid pressures, p_1 and p_2, which act at the positions x_1 & x_2. The cross-spectral density is given by the following equation:

$$
S_{p1p2}(f) = \sqrt{Gx_1(f)Gx_2(f)}\exp\left\{-\frac{\ell}{\lambda_c}\right\}\exp\left\{-\frac{L}{\lambda_a}\right\}\exp\left\{i2\pi f\frac{L}{V}\right\} \tag{4.9}
$$

where $Gx_1(f)$ and $Gx_2(f)$ are the power spectral densities of the pressure fluctuations at points, x_1 and x_2, respectively; ℓ is the length of the circumferential arc between positions x_1 and x_2; and L the length between x_1 and x_2 along the center line of the piping. λ_c is the correlation length in the circumferential direction while λ_a is the correlation length in axial direction. V is the average flow velocity.

The pressure fluctuation power spectral density, $G(f)$, is estimated from the non-dimensional power spectral density $g(\bar{f})$ as follows:

$$
\begin{aligned}
G(f) &= \left(\tfrac{1}{2}\rho V^2\right)^2 g(\bar{f})\frac{D}{V} \\
\bar{f} &= \frac{fD}{V}
\end{aligned}
\tag{4.10}
$$

where ρ is the fluid density, D the inner diameter of the piping, and $\bar{f}$ the non-dimensional frequency. The axial correlation length λ_a and circumferential correlation length λ_c are estimated using the equations below:

$$\mathrm{Re}(\Gamma_{XY}) = \exp\left(-\frac{|\Delta x|}{\lambda}\right)\cos\left(\frac{2\pi f|\Delta x|}{U}\right) \tag{4.11}$$

$$\lambda = -\frac{|\Delta x|}{\ln(\Gamma_0) - \dfrac{f}{f_0}} \tag{4.12}$$

where $\mathrm{Re}(\Gamma_{XY})$ is real part of coherence of pressure fluctuation at the locations X,Y. Δx is the distance between X and Y. Γ_0 is the real part of the coherence at zero frequency. f_0 is the frequency where the magnitude of coherence becomes 1/e. U is average flow velocity. To obtain the circumferential correlation length, the term cos $(2\pi f|\Delta x|/U)$ is not necessary in Eq. (4.11).

The stress, σ_R, at the position x, is estimated from the power spectral density of the stress, $S_\sigma(x,f)$, as:

$$\sigma_R(x) = C_0\sqrt{\int_0^\infty S_\sigma(x,f)df} \tag{4.13}$$

where C_0 is a factor relating the RMS value to the peak value.

Measurement data indicate that the power spectral density in the pipe bend region is not constant. The pipe bend should therefore be divided into several regions and the appropriate PSD employed in each region as shown in Fig.4.8. The appropriate correlation length also needs to be estimated [18].

4.1.1.6 Hints for countermeasures

These types of vibrations may cause component failures in industry. As shown in the previous section, the vibration of the piping system will occur due to high-speed fluid flow when the pipe and its supports are flexible. Therefore the vibrations can be reduced by installing supports that increase the rigidity of the pipe and installing dampers for increased energy dissipation. Attention should be paid to the support or damper position because installing a damper may destabilize the system. A support can also destabilize the system (Sugiyama et al. [19]). These countermeasures should be applied in consideration of feasibility of the installation, cost, etc.

4.1.2 Vibration of pipes excited by oscillating and two-phase fluid flow

4.1.2.1 Research history

Forced vibrations and parametric vibrations caused by oscillating flow and two-phase fluid flow have also been reported. The unstable regions in which the parametric vibrations occur for oscillating fluid flow were examined and

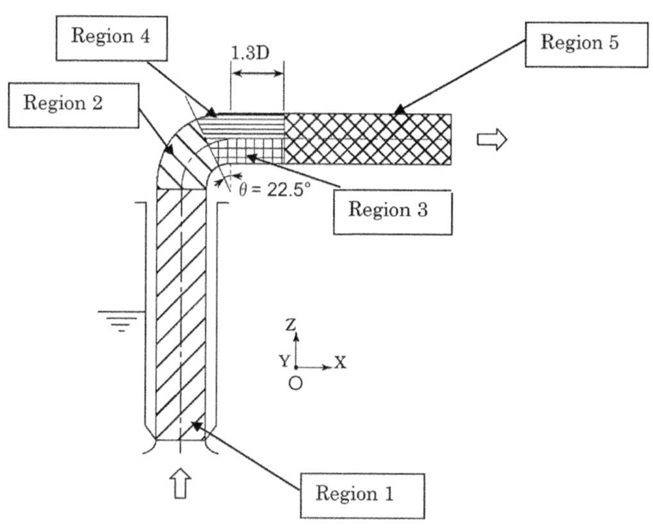

(a) Region of different types of pressure fluctuation

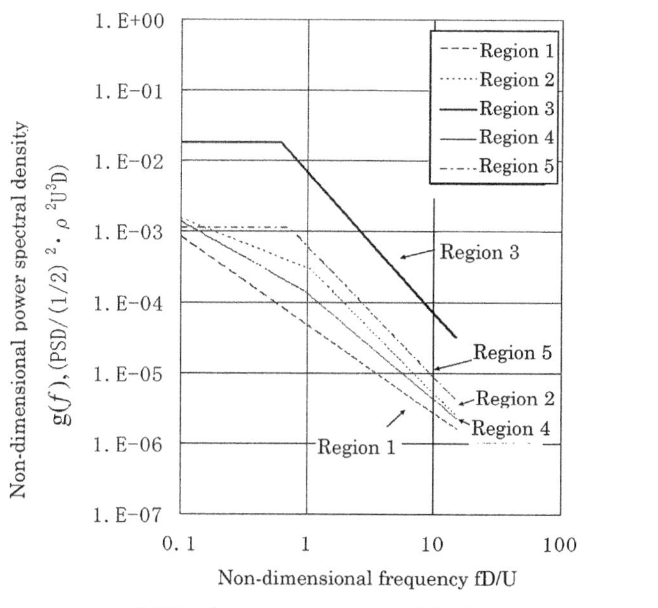

(b) Non-dimensional power spectral density

FIGURE 4.8

Pressure fluctuations at a bend: (a) Regions of different types of pressure fluctuations (b) Non-dimensional power spectral density.

reported by Païdoussis et al. [8,11,12]. The frequency response of a flexible pipe with a bend subjected to oscillating fluid flow was reported by Hayama and Matumoto [20].

In this section, the modeling and calculation methods presented by Hayama and Matumoto [20] are introduced. This is followed by a discussion of the oscillating-flow-induced vibration response of a flexible pipe with a bend.

4.1.2.2 Frequency response of pipes with bends

An analytical model of the pipe of length L with oscillating fluid flow is illustrated in Fig. 4.9 [20]. In this analytical model, a curved section is installed at the end of the pipe. The detailed configuration of the bend with radius R_b and subtended angle θ_b is shown in Fig. 4.10.

The fluid forces generated at the bend ($x = L$) can be expressed as:

$$F_{bY} = -\rho A u_e^2 \sin\theta_b - \rho A R_b (1 - \cos\theta_b)\frac{\partial u_e}{\partial t}$$
$$F_{bX} = \rho A u_e^2 (1 - \cos\theta_b) + \rho A R_b (1 - \sin\theta_b)\frac{\partial u_e}{\partial t} \tag{4.14}$$

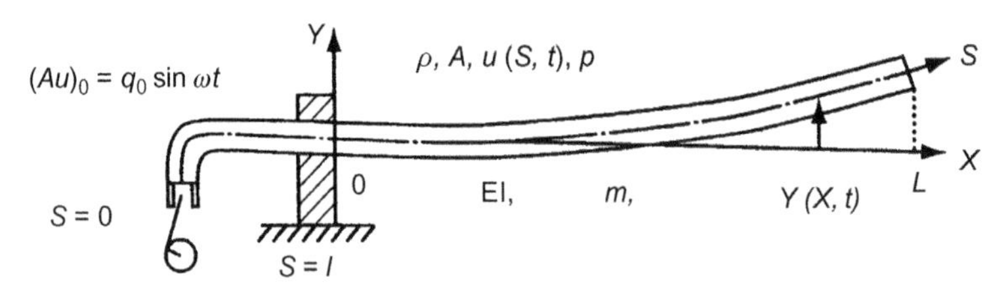

FIGURE 4.9

Pipes with harmonically oscillating fluid flow and its coordinate system [20].

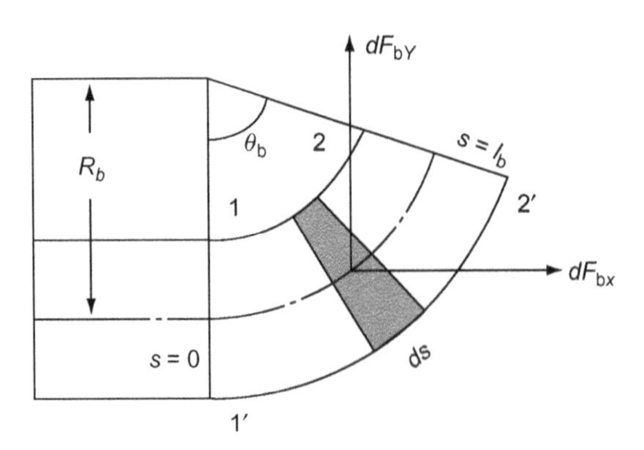

FIGURE 4.10

Bent pipe and coordinate system [20].

where ρ is the fluid density, A is the cross-sectional flow area, and u_e is the flow velocity at the bend.

The basic equation of motion of the pipe excited by these fluid forces is:

$$(m+\rho A)\frac{\partial^2 Y}{\partial t^2}+\mathrm{CI}\frac{\partial^5 Y}{\partial t\partial X^4}+\mathrm{EI}\frac{\partial^4 Y}{\partial X^4}-F_{\mathrm{b}X}\frac{\partial^2 Y}{\partial X^2}=F_{\mathrm{b}Y}\delta(X-L) \tag{4.15}$$

where m is the mass per unit length, C the damping coefficient, EI the flexural rigidity of the pipe, and $\delta(X-L)$ the Dirac delta function.

By using the modal analysis method, the governing equation with respect to the modal coordinate Z_n becomes:

$$\begin{aligned}&\ddot{Z}_n+2\zeta_n\dot{Z}_n+\{1-\varepsilon_1 C_{1\mathrm{b}X}\mu\cos(\mu\tau-\alpha)-\varepsilon_2 C_{2\mathrm{b}X}\sin^2(\mu\tau-\alpha)\}Z_n\\&=-\varepsilon_1 C_{1\mathrm{b}Y}\mu\cos(\mu\tau-\alpha)-\varepsilon_2 C_{2\mathrm{b}Y}\{1-\cos(2\mu-2\alpha)\}/2\end{aligned} \tag{4.16}$$

where $C_{1\mathrm{b}X}$, $C_{2\mathrm{b}X}$, $C_{1\mathrm{b}Y}$, and $C_{2\mathrm{b}Y}$ are determined from the properties of the eigenfunctions.

In the above governing equation, the major dimensionless parameters defined using the n-th natural frequency ω_n, the frequency of the oscillating flow velocity ω, and the maximum flow velocity $u_{\max}$ are:

$$\begin{aligned}&\zeta_n=C\omega_n/2E, && \mu=\omega/\omega_n, && \tau=\omega_n,\\&\varepsilon_1=\rho AL\omega_n u_{\max}/P_c, && \varepsilon_2=\rho A u_{\max}^2/P_c, && P_c=\lambda_n^2\mathrm{EI}/L^2\end{aligned}$$

From Eq. (4.16), it is found that the excitation forces are generated by reaction forces due to the acceleration of the fluid in the pipe and spatial momentum change. The former excitation forces due to the acceleration of the fluid have the same frequency as that of the oscillating fluid flow, while the latter excitation forces due to the spatial momentum change, vary at twice the frequency of the oscillating fluid flow.

A typical experimental result of the frequency response of the pipe, which has a bend of $\theta_b=90$ degrees at the end, is shown in Fig. 4.11 [20]. It is seen that resonance occurs in the system when $\mu=\omega/\omega_n=1$ and $\mu=1/2$, and that unstable vibrations due to parametric excitation may occur when $\mu=2$ and the fluid has a high density or the pipe has a low buckling load. The parametric excitation is generated by the oscillating axial force due to the spatial momentum change of the fluid flow at the bend. The same vibrations introduced above may occur in deformed pipes, but the vibration amplitude at resonance does not usually become large, hence the deformation of the pipe generally remains small.

4.1.2.3 Hints for countermeasures

The vibrations discussed in this section will normally occur due to the oscillating fluid flow and pressure pulsation in the piping system. Therefore, the piping vibration can be reduced by various methods: installing a pulsation (oscillation) suppression device that reduces the pulsation (oscillation) itself, installing an

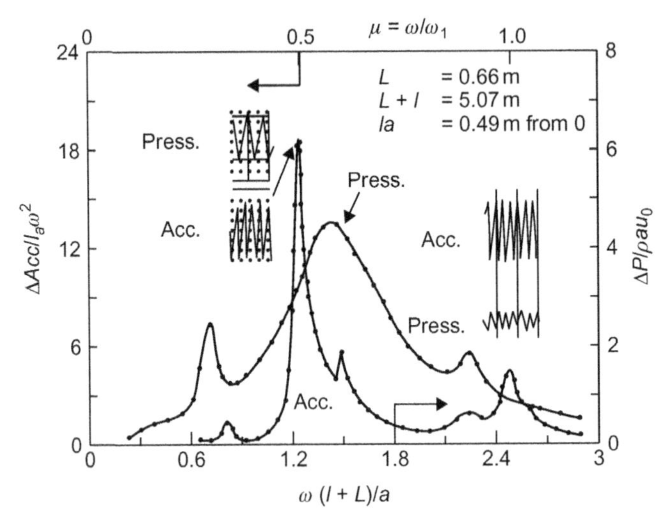

FIGURE 4.11

Frequency response of pipe with harmonically oscillating fluid flow for $\theta_b = 90$ degrees [20].

orifice that adds damping, adjusting the piping length to change the resonance frequency, and installing a damper (structural damping) that dissipates the vibration energy. These countermeasures should be applied in consideration of feasibility of the installation, cost, etc.

4.1.3 Piping vibration caused by gas–liquid two-phase flow

4.1.3.1 Summary

Piping conveying two-phase flow sometimes suffers extreme vibration for specific two-phase flow conditions. The piping can sometimes not be supported sufficiently to avoid flow-induced vibration due to the need to reduce thermal stress caused by thermal expansion of the piping. For this reason, piping vibration due to two-phase flow-induced excitation forces sometimes inevitably occurs.

4.1.3.2 Two-phase flow regimes

Gas–liquid two-phase flow in piping has various flow patterns, corresponding to different conditions, different flow rates, physical properties, etc. as shown in Figs. 4.12 and 4.13. Piping vibration due to gas–liquid two-phase flow has characteristic excitation forces corresponding to these flow patterns. The flow patterns are classified as follows:

1. *Bubbly flow*: mainly liquid flow with dispersed bubbles; the gas flow rate is very low and the flow-induced forces are small.

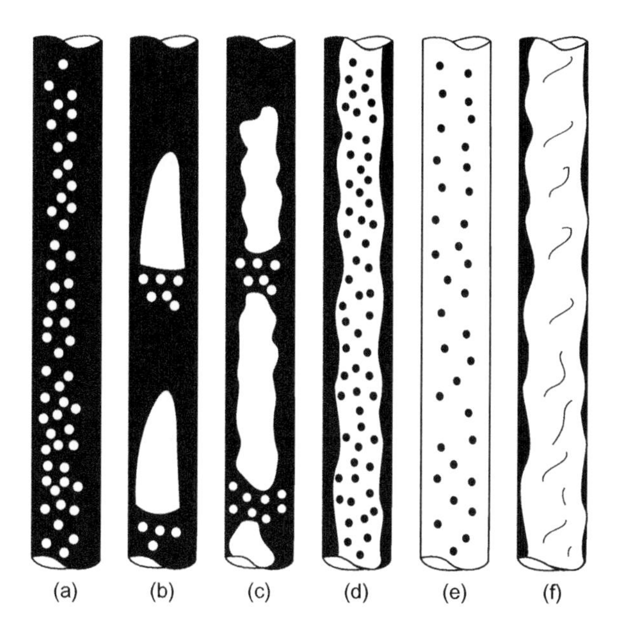

FIGURE 4.12

Flow patterns in vertical piping: (a) bubble flow, (b) slug flow, (c) froth flow, (d) annular mist flow, (e) mist flow, and (f) inverse annular flow.

2. *Plug flow and slug flow*: large bubbles move in the piping intermittently in the case of plug flow. For slug flow, large bubbles, having the shape of artillery shells that almost fill the pipe, and liquid slugs flow alternately. Two-phase flow turbulence is low for both flow patterns. When a liquid slug, which has a high density, passes through the elbow portion of the piping, excitation forces are generated. These forces can cause piping vibration.
3. *Froth flow*: liquid slugs contain many bubbles in the case of froth flow. The shape of the gas bubble is distorted and the flow is highly turbulent. Flow-induced forces are also large because of the high turbulence.
4. *Mist flow*: the main flow is gas flow. Liquid droplets flow in the piping like mist. This flow is basically the opposite of bubbly flow. Flow-induced forces are therefore small.
5. *Annular flow*: this flow consists of liquid film flow adhering to the inner wall of the piping, and gas flow in the center of the piping. The liquid film flow is not completely stable. Piping vibration sometimes occurs due to turbulence in the liquid film flow.
6. *Stratified flow and wavy flow*: gas–liquid two-phase flow is stratified into the upper side gas and the lower side liquid in the case of horizontal piping. If the interfacial surface of gas and liquid is not wavy, the flow is called a stratified flow. If the interfacial surface is wavy, it's called a wavy flow.

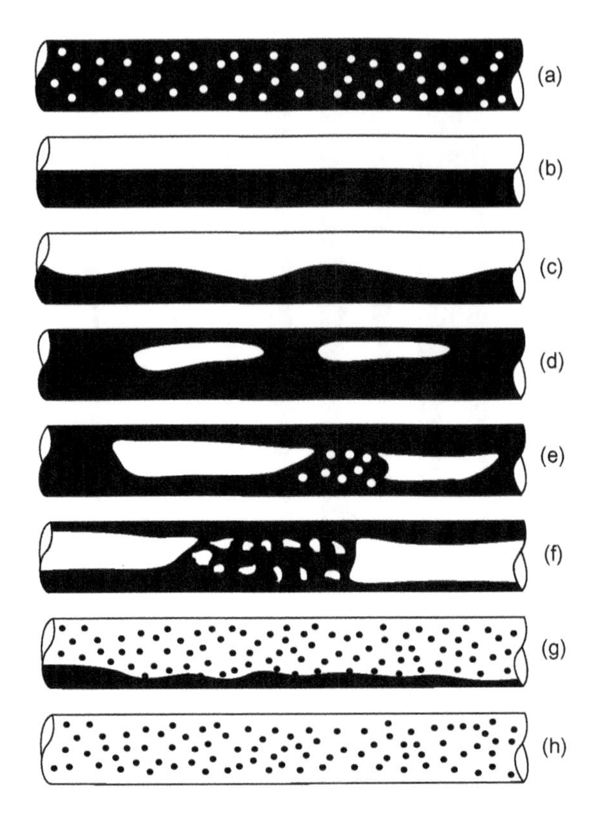

FIGURE 4.13

Flow patterns in horizontal piping: (a) bubbly flow, (b) stratified flow, (c) wavy flow, (d) plug flow, (e) slug flow, (f) froth flow, (g) annular mist flow, and (h) mist flow.

In the case of slug and froth flows, piping vibration occurs due to flow-turbulence excitation forces and excitation forces caused by passage of liquid slugs through the elbow portion of the piping.

4.1.3.3 Evaluation method

For evaluation of piping vibration caused by gas–liquid two-phase flow, specifying or estimating flow patterns is of primary importance because the characteristics of the excitation force depend on the two-phase flow patterns. Many maps to estimate flow patterns from flow conditions, flow rate, and physical properties exist. For example, one of the maps is the Weisman map shown in Figs. 4.14 and 4.15 [21,22]. The values on horizontal and vertical axes can be obtained from the superficial liquid flow velocity V_{SL} $(=(1-\alpha)U_L)$, the gas flow velocity V_{SG} $(=\alpha U_G)$, and correction factors (Table 4.1 [22]). Piping must be designed in order to avoid flow patterns which have large excitation forces, for instance, froth flow and slug flow, by using these flow pattern maps.

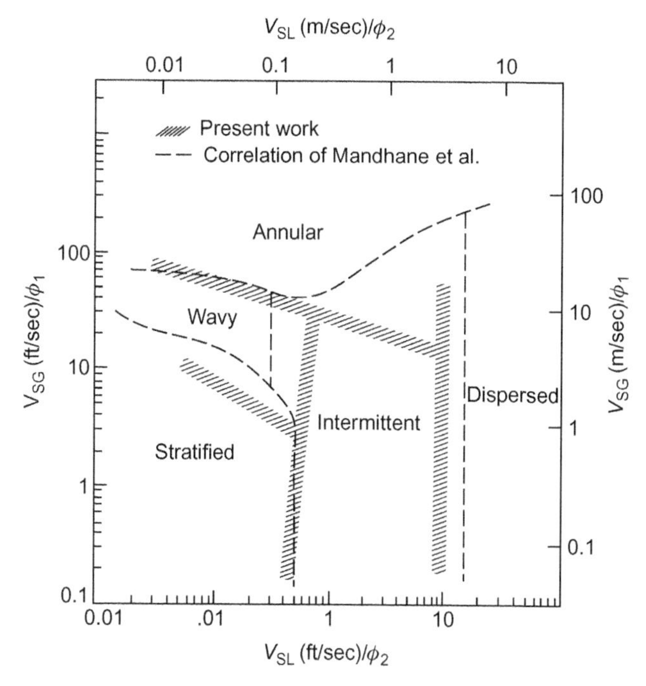

FIGURE 4.14

Flow pattern map for horizontal piping [21].

Two-phase flow-induced vibration phenomena are explained by using vibration test results of simply supported horizontal piping conveying air–water two-phase flow as shown in Fig. 4.16. For low air volume fraction, the plug flow pattern occurs. When the volume ratio of air is large, the flow pattern changes to slug flow. In the case of these flow patterns, the variance of the measured vibration strain in the horizontal piping is shown in Fig. 4.17 [23]. Piping vibration generally increases with two-phase flow velocity. For slug flow, vibration amplitudes are less dependent on two-phase flow velocity although vibration amplitudes are large. For plug flow, vibration amplitudes strongly depend on two-phase flow velocity although vibration amplitudes are small. At the points A, D, and H in Fig. 4.17, piping vibration becomes relatively large in comparison with the other points. At these points, vibration amplitudes are large because of the occurrence of parametric vibrations.

4.1.3.4 Case examples and methods to reduce piping vibration

Few examples of vibration of piping conveying two-phase flow are published in the open literature. However, many cases of piping vibration must occur. In the design of piping conveying two-phase flow, stable flow patterns such as annular

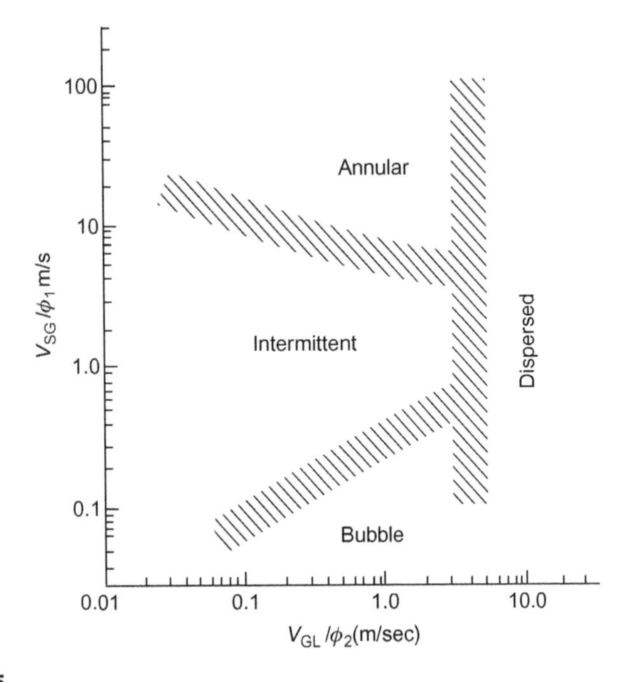

FIGURE 4.15

Flow pattern map for vertical piping [22].

Table 4.1 Conversion coefficients of flow pattern map [22]

Flow orientation		φ_1	φ_2
Horizontal, vertical, and inclined flow	Transition to dispersed flow	1.0	$(\rho_L/\rho_{SL})^{-0.33}\ (D/D_s)^{0.16}\ (\mu_{SL}/\mu_L)^{0.09}\ (\sigma/\sigma_s)^{0.24}$
	Transition to annular flow	$(\rho_{SG}/\rho_G)^{0.23}\ (\Delta\rho/\Delta\rho_s)^{0.11}\ (\sigma/\sigma_s)^{0.11}\ (D/D_s)^{0.415}$	1.0
Horizontal and slightly inclined flow	Intermittent-separated transition	1.0	$(D/D_s)^{0.45}$
Horizontal flow	Wavy-stratified transition	$\left(\frac{Ds}{D}\right)^{0.17}\left(\frac{\mu_G\rho_{SG}}{\mu_{SG}\rho_a}\right)^{1.55}\left(\frac{\Delta\rho\sigma_s}{\Delta\rho_s\sigma}\right)^{0.69}$	1.0
Vertical and inclined flow	Bubble-intermittent transition	$(D/D_s)^n\ (1-0.65\cos\alpha)$ $n=0.26\ e^{-0.17(V_{SL}/V^S_{SL})}$	1.0

s denotes standard conditions. $D_s = 1.0$ *in* $= 2.54$ *cm,* $\rho_{SG} = 0.0013$ *kg/l,* $\rho_{SL} = 1.0$ *kg/l,* $\mu_{SL} = 1$ *centipoise,* $\sigma_s = 70$ *dynes/cm, and* $V^S_{SL} = 1.0$ *ft/s* $= 0.305$ *m/s.*

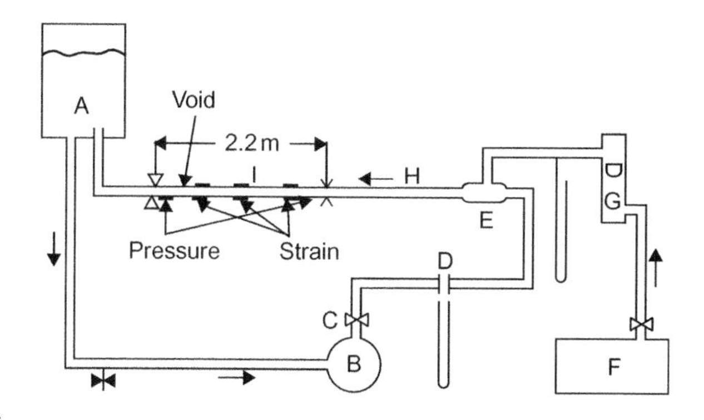

FIGURE 4.16

Test equipment for piping vibration caused by gas–liquid two-phase flow [23].

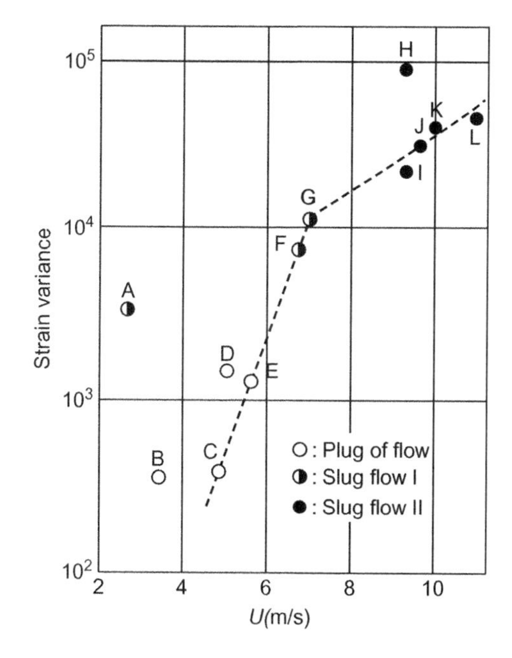

FIGURE 4.17

Piping vibration due to gas–liquid two-phase flow [23].

flow are preferable in order to avoid piping vibration. However, in the case of vertical piping conveying upward flow, the flow patterns tend to be unstable, for example slug flow, because pressure decreases as the two-phase fluid flows upward [24]. For froth flow or slug flow patterns, extreme piping vibration sometimes occurs. In this case, piping repeats free vibrations non-periodically at its natural frequency because two-phase flow-induced forces are random [24]. Once large piping vibrations occur due to two-phase flow, it is usually difficult to change the flow pattern as a countermeasure. Some support structures can be attached to the piping, properly taking thermal stress into consideration, in order to avoid piping vibration. It is important to avoid two-phase flow during start-up and shut-down conditions.

4.2 Vibration related to bellows

In general, bellows are used in piping systems requiring flexibility due to thermal deformation. Unfortunately, this added flexibility also makes them susceptible to vibration. When the internal flow velocities are sufficiently high, significant flow-induced vibrations can develop and lead to fatigue failure of the bellows. In this section, an outline of the excitation mechanisms of flow-induced vibration of bellows caused by internal flow is presented. Calculation methods to determine natural frequencies of the bellows, which are important when implementing countermeasures to avoid bellows vibration, are described. Single bellows, double bellows (Fig. 4.18) and corrugated pipes are considered. The internal flow considered is single-phase flow.

4.2.1 Vibration of bellows

4.2.1.1 Mechanism of flow-induced vibration

Flow-induced vibration of bellows due to internal flow (Fig. 4.18) takes place when the frequency of periodic fluctuations of free shear layers created by the bellows convolutions resonates with the mechanical natural frequency of the bellows, as illustrated in Fig. 4.19. Therefore, to eliminate flow-induced vibration of the bellows due to internal flow it is necessary to avoid the resonance between the frequency of periodic fluctuation of free shear layers and the mechanical natural frequency of the bellows.

4.2.1.2 Other vibration mechanisms

Bellows sometimes generate loud noise when the free shear layer oscillating frequency resonates with an acoustic natural frequency of the piping system, including the bellows [25]. This type of resonance should, as far as possible, be avoided.

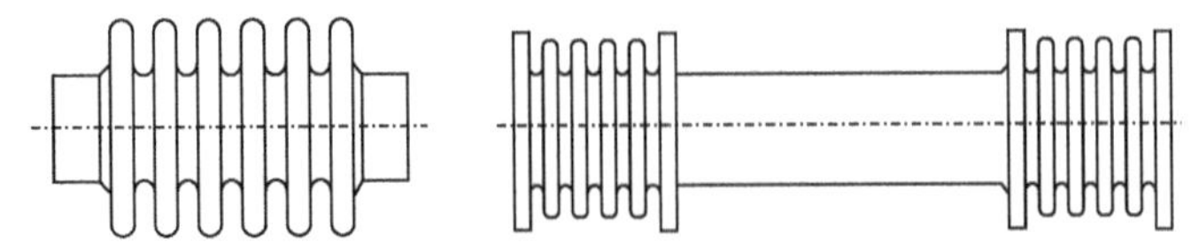

FIGURE 4.18

Typical sketch of single and double bellows.

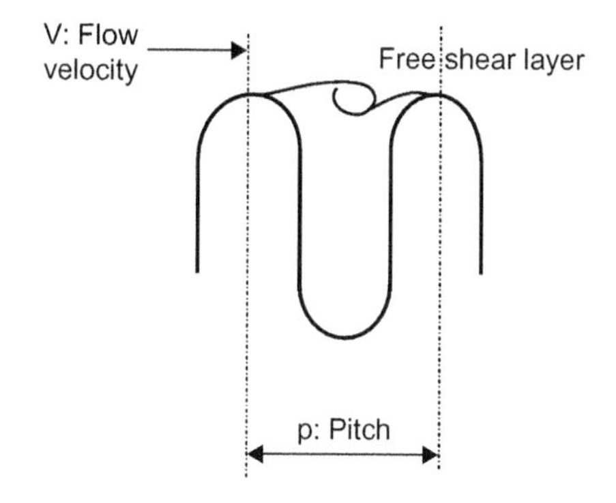

FIGURE 4.19

Free shear layer created by bellows.

When bellows are installed into a line in which pressure pulsations exist, parametric vibration of the bellows can occur if the pulsation frequency is twice the mechanical natural frequency of the bellows. In this situation, a pulsation suppression device is required to reduce the vibration level of the bellows.

4.2.1.3 History of studies on bellows vibration

4.2.1.3.1 Studies on mechanical natural frequencies of bellows

Jakubauskas (1999) [26] showed that the mechanical natural frequency of bellows can be calculated accurately by considering fluid added mass due to motion of the convolutions under vibration in addition to the bellows' mass as indicated in the Expansion Joint Manufacturers Association (EJMA) [27] standard. Morishita et al. (1989) [28] applied a Timoshenko beam model and showed that the effect of rotary inertia of the bellows is significant for the determination of the natural frequency of transverse vibrations of a single bellows. Jakubauskas and Weaver [29] verified the study implemented by Morishita et al. and presented a calculation method for the determination of the natural frequencies of transverse vibration of the bellows taking into account the fluid added mass. Based on a similar theory, Jakubauskas and Weaver [29] also developed a calculation method for double-bellows natural frequencies.

4.2.1.3.2 Studies on bellows Strouhal numbers

Weaver and Ainsworth [30] and Jakubauskas and Weaver [29,31] found that the Strouhal number for bellows away from an upstream elbow is 0.45 while the Strouhal number just downstream of a 90 degree elbow is 0.58.

4.2.1.4 Evaluation method for the flow-induced vibration of bellows

The excitation force acting on bellows due to internal flow varies periodically with the frequency f_v given by:

$$f_v = \mathrm{St}\frac{V}{p} \tag{4.17}$$

where V is the mean flow velocity through the bellows, p the convolution pitch, and St the Strouhal number.

Vibration amplitudes of the bellows increase when the frequency of free shear layer fluctuations resonates with the mechanical natural frequency of the bellows. To avoid flow-induced bellows vibration, it is necessary to separate the frequency of fluctuation of the free shear layer from the mechanical natural frequency of the bellows. In order to achieve this, the mean flow velocity through the bellows V should be less than the critical velocity V_c expressed by Eq. (4.18) (in general, $V < 0.75V_c$ is recommended).

$$V_c = \frac{fp}{\mathrm{St}} \tag{4.18}$$

It is necessary to accurately predict the natural frequency of the bellows in order to calculate the critical velocity. Calculation methods for both the axial and transverse natural frequency of a single bellows and the transverse natural frequency of double bellows are described. The bellows' dimensions used in the natural frequency calculations are shown in Figs. 4.20 and 4.21. The nomenclature used in this chapter is given in Table 4.2.

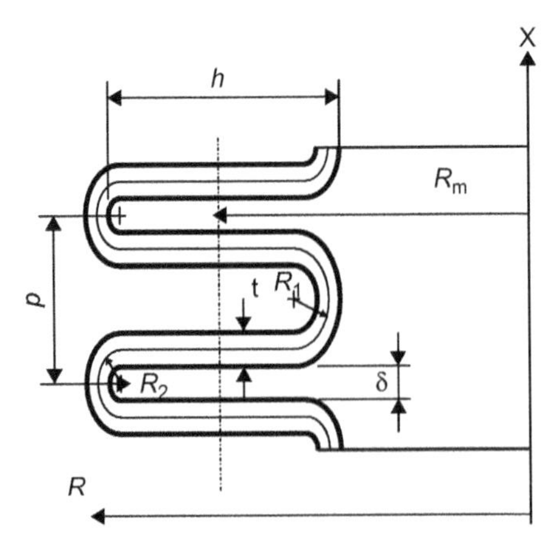

FIGURE 4.20

Details of bellows geometry.

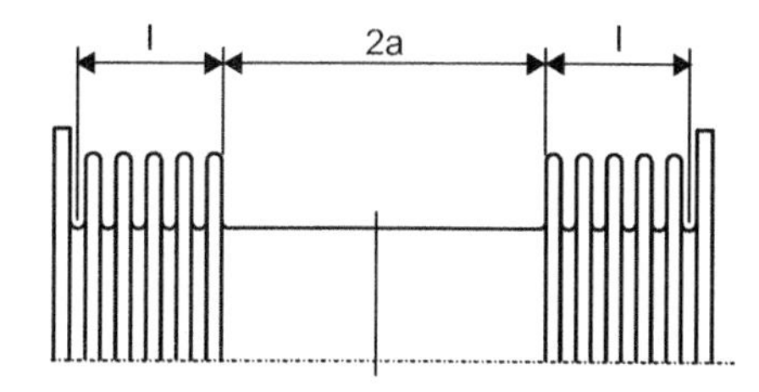

FIGURE 4.21

Dimensions of double bellows.

Table 4.2 Nomenclature

Symbol	Description	Symbol	Description
A_{ik}	*i*th constant for mode *k*	P	Fluid pressure
a	Half length of connecting pipe	p	Convolution pitch
E	Modulus of elasticity	R	Mean radius of connecting pipe
F	Bellows natural frequency	R_1	Root convolution radius
f_k	Bellows natural frequency in *k*th mode	R_2	Crown convolution radius
H	Total convolution height	R_m	Mean radius of bellows convolutions
I	Moment of inertia	t	Convolution thickness
J_p	Mass moment of inertia of half connecting pipe about pivot	V	Mean flow velocity
K	Axial stiffness of bellows per half convolution	V_c	Critical velocity
k_s	Lateral support stiffness	X_k	Bellows mode shape in *k*th mode
l	Live length of bellows	α_{f2k}	Integral function for bellows distortion fluid added mass in *k*th mode for transverse vibration
M_s	Lateral support effective mass	α_{f2kA}	Integral function for bellows distortion fluid added mass in *k*th mode for axial vibration of single bellows
m_b	Bellows mass per unit length	δ	Width of convolution
m_f	Fluid added mass per unit length	μ	Added mass coefficient
m_p	Connecting pipe mass per unit length	ρ_f	Fluid density
m_{tot}	Bellows mass including fluid added mass	ρ_b	Bellows material density
		ρI	Effective mass moment of inertia of bellows and contained fluid

4.2.1.4.1 Fluid added mass for axial vibration of single bellows

Figure 4.22 shows vibration mode shapes for single bellows. Considering the axial bellows vibration [26], the total fluid added mass m_f is assumed to consist of three components: the first associated with rigid body motion of the convolutions in the axial direction m_{f1A}, the second associated with the convolution distortion m_{f2A}, and the third associated with return flow in the central area of the cross-section of the bellows m_{f3A}. The total fluid added mass m_f can be used in a calculation procedure prepared by the EJMA Standard to obtain more accurate natural frequencies of the bellows. The fluid added mass per unit length can be expressed by the following equations:

$$m_{f1A} = 2\pi \frac{hR_m}{p}(2R_2 - t)\rho_f \tag{4.19a}$$

$$m_{f2A} = \alpha_{f2kA}\mu R_m^3 \rho_f \tag{4.19b}$$

$$m_{f3A} = \frac{8R_m h(2R_2 - t)}{(2R_m - h)^2 p} m_{f1} \tag{4.19c}$$

$$\alpha_{f2kA} = 1.85\left(\frac{k}{l}\right)^2 p \quad (k\text{:mode})$$

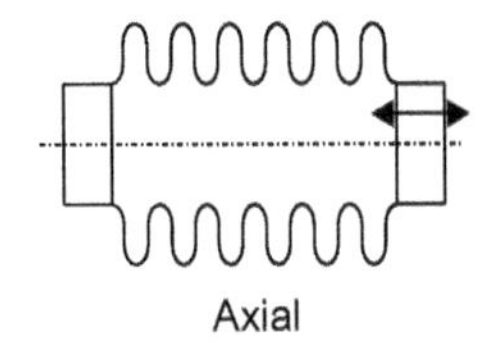

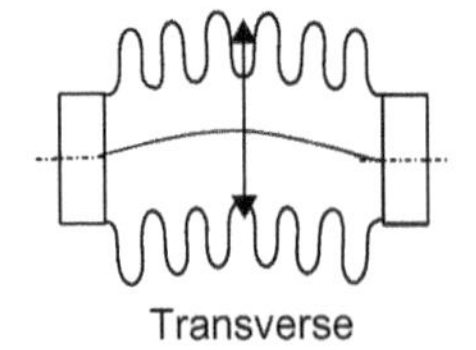

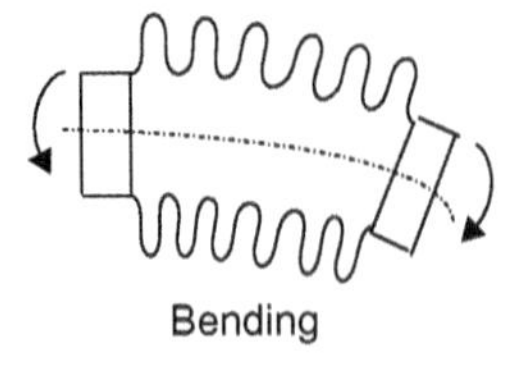

FIGURE 4.22

Vibration modes of single bellows.

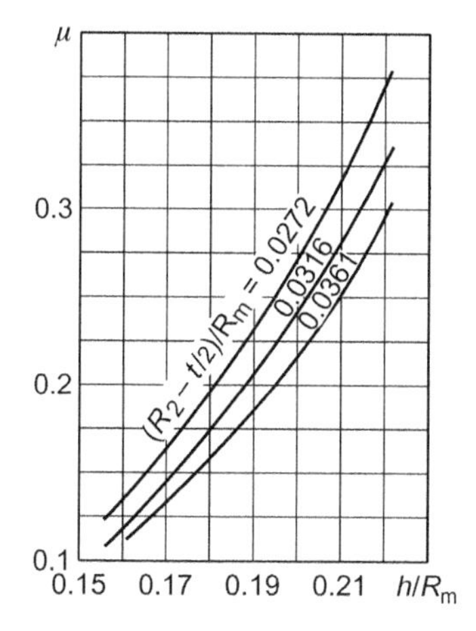

FIGURE 4.23

Half-convolution added mass coefficient for axial vibration [26,29].

The added fluid mass coefficient μ is shown in Fig. 4.23.

The fluid added mass per unit length m_f is thus determined by the following equation:

$$m_f = (m_{f1A} + m_{f2A} + m_{f3A}) = \left\{ 2\pi \frac{hR_m}{p}(2R_2 - t)\left[1 + \frac{8R_m h(2R_2 - t)}{(2R_m - h)^2 p}\right] + \alpha_{f2k}\mu R_m^3 \right\}\rho_f \tag{4.20}$$

4.2.1.4.2 Fluid added mass and natural frequency of transverse vibration of single bellows

Transverse vibration modes of single bellows are shown in Fig. 4.22. The equation to calculate the mechanical natural frequency of the bellows transverse vibration [29] based on the Timoshenko beam model is expressed considering bellows rotary inertia and fluid added mass due to bellows distortion. The total mass per unit length m_{tot} of the bellows is given by the following equation:

$$m_{tot} = (m_f + m_b) \tag{4.21}$$

$$m_b = \frac{4\pi R_m}{p}(h + 0.285p)t\rho_b \tag{4.22}$$

$$m_f = (m_{f1} + m_{f2})$$
$$= \left\{ \pi \left(R_m - \frac{h}{2} + \frac{2hR_2}{p} \right)^2 + \alpha_{f2k} \mu R_m^3 \right\} \rho_f \quad (4.23)$$
$$m_{f1} = \pi \left(R_m - \frac{h}{2} + \frac{2hR_2}{p} \right)^2 \rho_f$$
$$m_{f2} = \alpha_{f2k} \mu R_m^3 \rho_f$$
$$\alpha_{f2k} = 0.066 \frac{A_{1k}^2}{l^4} \left(R_m - \frac{h}{2} \right)^2 p$$

where m_b is the bellows mass per unit length and m_f the fluid added mass per unit length. The fluid added mass m_f is composed of the mass of fluid per unit length of the bellows m_{f1} and the convolution distortion component m_{f2}. The added mass coefficient μ is presented in Fig. 4.24.

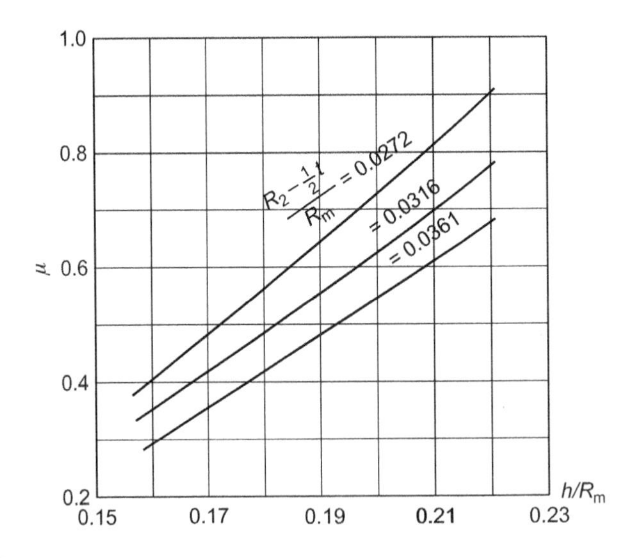

FIGURE 4.24

Half-convolution added mass coefficient for transverse vibration [29,31].

The equation of motion for bellows transverse vibrations is:

$$EI_{\mathrm{eq}}\frac{\partial^4 w}{\partial x^4}+m_{\mathrm{tot}}\frac{\partial^2 w}{\partial t^2}-\rho I\frac{\partial^4 w}{\partial x^2 \partial t^2}+P\pi R_{\mathrm{m}}^2\frac{\partial^2 w}{\partial x^2}=0 \qquad (4.24)$$

where EI_{eq} is the equivalent bending stiffness, w the transverse displacement, and P the fluid pressure.

Considering the boundary conditions of the bellows, the k-th transverse mechanical natural frequency of the bellows, f_{k}, is given by the following equation:

$$f_{\mathrm{k}}=\frac{R_{\mathrm{m}}}{4\pi l^2}A_{1\mathrm{k}}\left[\frac{kp-4\pi l^2 PA_{2\mathrm{k}}}{m_{\mathrm{tot}}+\frac{\rho_1 I}{l^2}A_{4\mathrm{k}}}\right]^{1/2} \qquad (4.25)$$

$$\rho I=\pi R_m^3\left(\left(\frac{2h}{p}+0.571\right)t\rho_b+\frac{h}{p}(2R_2-t)\rho_f\right)$$

The coefficients $A_{1\mathrm{k}}$, $A_{2\mathrm{k}}$, $A_{4\mathrm{k}}$ for the first four modes of single bellows are given in Table 4.3.

4.2.1.4.3 Transverse natural frequencies of double bellows

Figure 4.25 shows the vibration mode shapes for double bellows. Based on the theory outlined in part (b) above, natural frequencies of transverse and rocking vibrations of bellows [31] are determined by the equations below. The added mass coefficient is shown in Fig. 4.24. The constants $A_{1\mathrm{k}}$, $A_{2\mathrm{k}}$, $A_{3\mathrm{k}}$, $A_{4\mathrm{k}}$, and $A_{5\mathrm{k}}$ are presented in Table 4.4 for lateral modes and Table 4.5 for rocking modes (Figs. 4.26–4.31).

$$f_{\mathrm{k}}=\frac{R_{\mathrm{m}}}{4\pi l^2}A_{1\mathrm{k}}\left[\frac{kp-4\pi l^2 PA_{2\mathrm{k}}+(4k_{\mathrm{s}}l^3A_{3\mathrm{k}}/R_{\mathrm{m}}^2)}{m_{\mathrm{tot}}+(\rho_1 IA_{4\mathrm{k}})/l^2+BA_{5\mathrm{k}}}\right]^{1/2} \qquad (4.26)$$

Table 4.3 A_{ik} constants for transverse modes [29]

Mode no. *k*	**1**	**2**	**3**	**4**
$A_{1\mathrm{k}}$	22.37	61.67	120.9	199.9
$A_{2\mathrm{k}}$	0.02458	0.01211	0.00677	0.00374
$A_{4\mathrm{k}}$	12.30	46.05	98.91	149.4

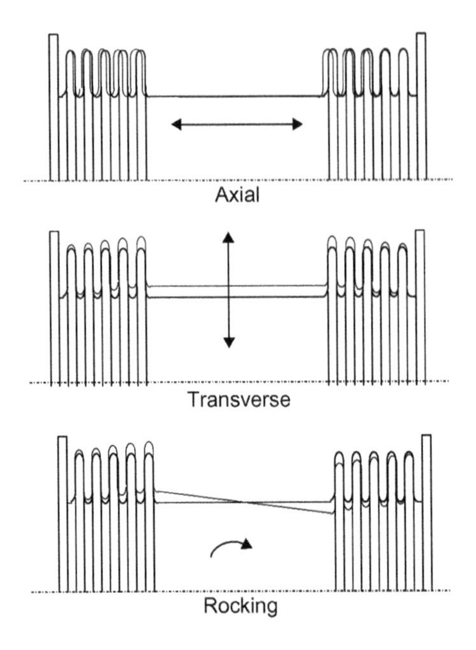

FIGURE 4.25

Vibration modes of double bellows.

Table 4.4 A_{ik} constants for lateral modes [31]

Mode #	Without transverse support			With transverse supports
A_{ik}	1	2	3	1
A_{1k}	5.650	Fig. 4.26	Fig. 4.27	5.62
A_{2k}	0.10	0.0249	0.0119	0.0998
A_{3k}	0	0	0	0.0826
A_{4k}	3.193	Fig. 4.26	Fig. 4.27	3.152
A_{5k}	2.656	Fig. 4.26	Fig. 4.27	2.611

$$m_{\mathrm{tot}} = (m_{\mathrm{f}} + m_{\mathrm{b}})$$

$$m_{\mathrm{b}} = \frac{4\pi R_{\mathrm{m}}}{p}(h + 0.285p)t\rho_{\mathrm{b}}$$

$$m_{\mathrm{f}} = \left\{\pi\left(R_{\mathrm{m}} - \frac{h}{2} + \frac{2hR_2}{p}\right)^2 + \alpha_{\mathrm{f2k}}\mu R_{\mathrm{m}}^3\right\}\rho_{\mathrm{f}}$$

Table 4.5 A_{ik} constants for rocking modes [30]

Mode #	Without transverse support			With transverse supports
A_{ik}	1	2	3	1
A_{1k}	Fig. 4.28	Fig. 4.29	Fig. 4.30	Fig. 4.31
A_{2k}	Fig. 4.28	Fig. 4.29	Fig. 4.30	Fig. 4.31
A_{3k}	0	0	0	Fig. 4.31
A_{4k}	3.08	Fig. 4.29	Fig. 4.30	3.073
A_{5k}	Fig. 4.28	Fig. 4.29	Fig. 4.30	Fig. 4.31

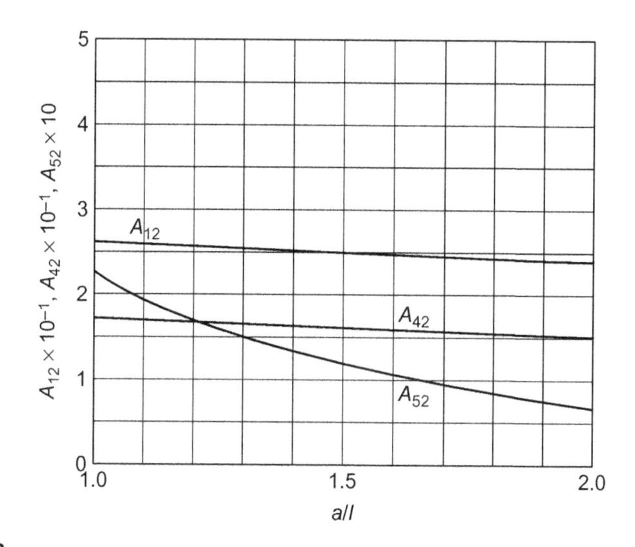

FIGURE 4.26

Coefficients for second lateral mode; no lateral support [31].

$$\alpha_{f2k} = 0.066 \frac{A_{1k}^2}{l^4} \left(R_m - \frac{h}{2} \right)^2 p$$

$B = [M_s + (m_p + m_f)a]/l$: for transverse modes

$B = J/(a^2 l)$: for rocking modes

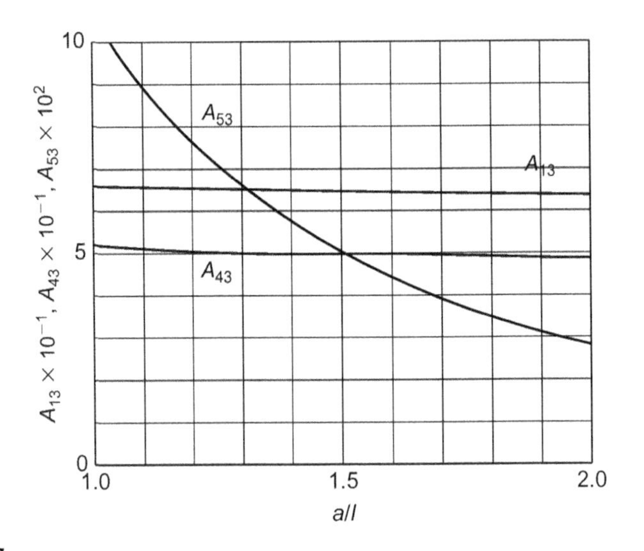

FIGURE 4.27

Coefficients for third lateral mode; no lateral support [31].

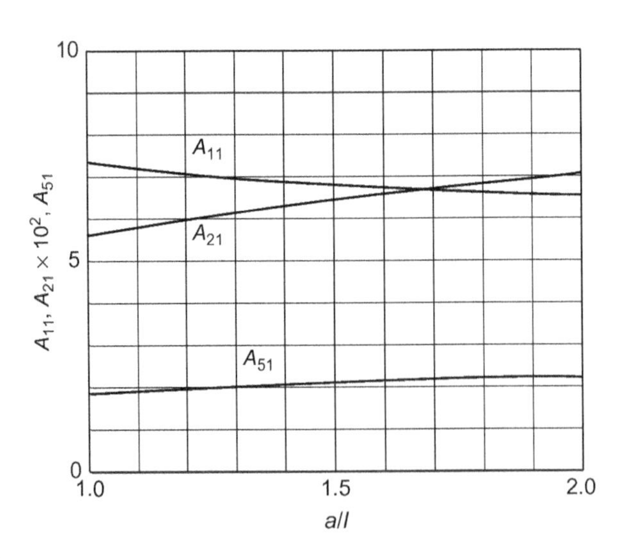

FIGURE 4.28

Coefficients for first rocking mode; no lateral support [31].

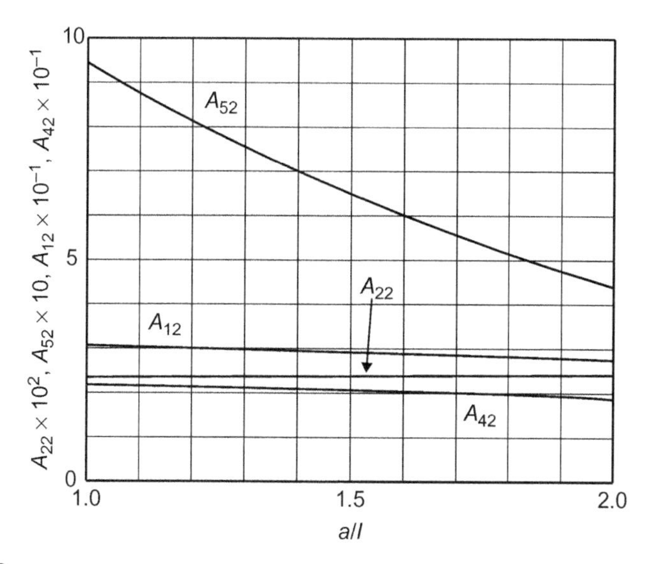

FIGURE 4.29

Coefficients for second rocking mode; no lateral support [31].

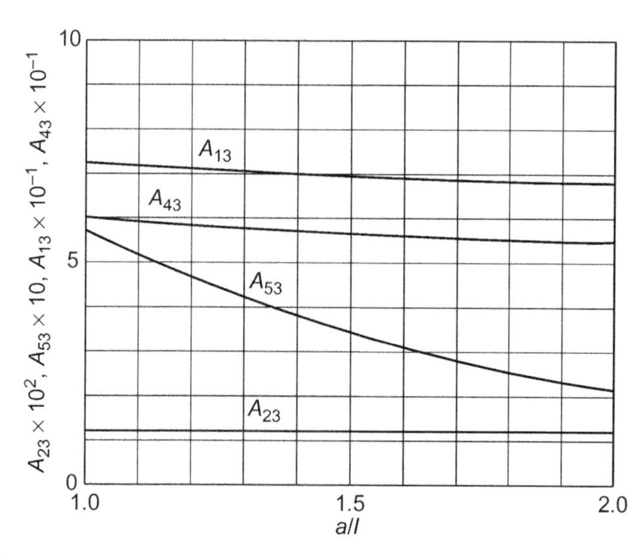

FIGURE 4.30

Coefficients for second rocking mode; no lateral support [31].

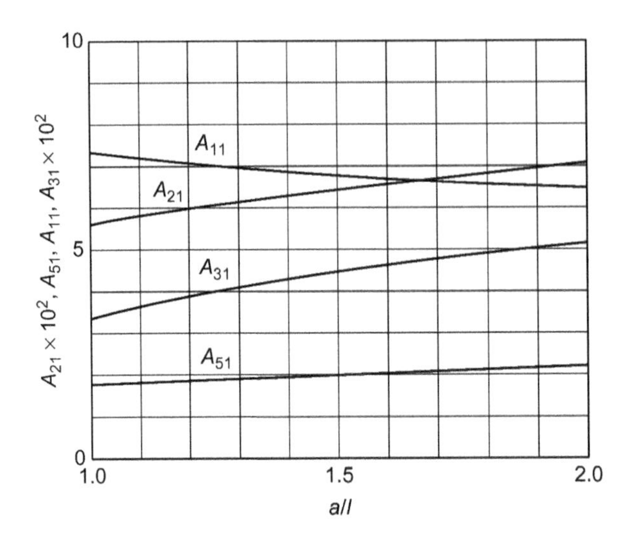

FIGURE 4.31

Coefficients for first rocking mode with lateral support [31].

$$J_p = \frac{(m_p + m_f)a^3}{3} + \frac{(2m_p + m_f)aR^2}{4} + M_s a^2$$

$$\rho I = \pi R_m^3 \left(\left(\frac{2h}{p} + 0.571 \right) t\rho_b + \frac{h}{p}(2R_2 - t)\rho_f \right)$$

4.2.2 Hints for countermeasures and examples of flow-induced bellows vibrations

4.2.2.1 Hints for countermeasures

Countermeasures to avoid bellows vibrations due to fluid flow are described next. When the mean flow velocity through the bellows is roughly higher than 0.75 times the critical velocity, vibrations can be expected. Countermeasures include the following:

1. Changing the flow rate so that the mean flow velocity through the bellows falls below the critical velocity.
2. Addition of a straight length section between an elbow and the bellows, when the bellows are installed downstream of an elbow and the straight length is not long enough.
3. Shifting the mechanical natural frequency of the bellows so that the critical velocity may be higher than 1.33 times the mean velocity of the bellows.

4. Eliminating the creation of free shear layers from the bellows convolutions by installation of sleeves within the bellows. The sleeve thickness should be determined carefully considering flow conditions. Corrosion should also be considered because stagnation regions are created between sleeves and bellows.

4.2.2.2 Examples of flow-induced bellows vibration and component failure

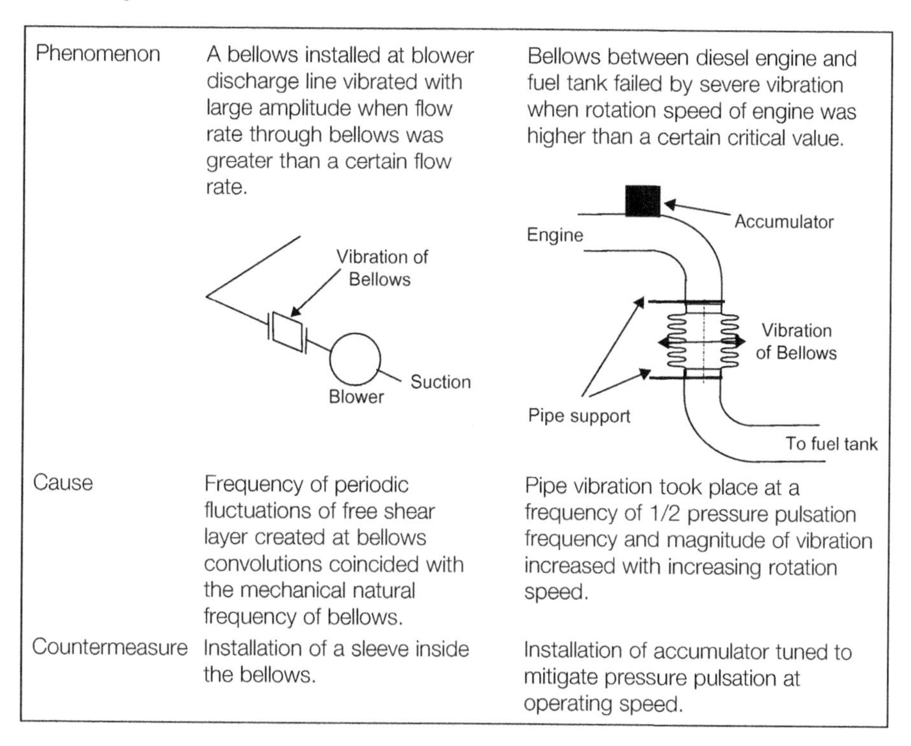

Phenomenon	A bellows installed at blower discharge line vibrated with large amplitude when flow rate through bellows was greater than a certain flow rate.	Bellows between diesel engine and fuel tank failed by severe vibration when rotation speed of engine was higher than a certain critical value.
Cause	Frequency of periodic fluctuations of free shear layer created at bellows convolutions coincided with the mechanical natural frequency of bellows.	Pipe vibration took place at a frequency of 1/2 pressure pulsation frequency and magnitude of vibration increased with increasing rotation speed.
Countermeasure	Installation of a sleeve inside the bellows.	Installation of accumulator tuned to mitigate pressure pulsation at operating speed.

4.3 Collapsible tubes

4.3.1 Summary

The term collapsible tube is generically used for a tube with soft thin walls that can be easily be bent and collapsed. The cross-section of the tube may change when exposed to internal–external pressure difference or flow variance. By the interaction of pressure variations and tube deformation, self-excited vibration will occur with strongly non-linear dynamics. Collapsible tubes show similar characteristics to internal organs in the human body, such as blood vessels, vocal cords, urethras, and lung bronchi. Since the tube behavior has similarities to the phenomena observed in natural and artificial organs, many related studies have been conducted in the field of bioengineering. Other examples are heart–lung machines, Korotkoff sounds heard in blood-pressure measurements, sound generation in vocal cords, and snoring.

4.3.2 Self-excited vibration of collapsible tubes

4.3.2.1 Historical background

The dynamical behavior of collapsible tubes has been investigated for more than 100 years [32]. In most of the previous studies, simple experimental models have been utilized, for instance, the model of Fig. 4.32 [32]. In this system, the tube is connected to identical rigid pipes at both upstream and downstream ends, and confined in a pressurized condition by the surrounding fluid, where the pressure is controlled by the liquid level or injected gas. A reservoir tank is connected to the upstream end of the tube, and the flow rate in the tube is controlled by a valve at the downstream end. Based on this system, complicated dynamics of the collapsible tubes have been investigated both experimentally and analytically.

4.3.2.2 Evaluation methods

The typical behavior of the collapsible tube model can be expressed by a characteristic curve which relates the flow rate and the pressure difference between the two ends of the tube (ΔP), shown in Fig. 4.33 [34]. For low flow rates, ΔP increases linearly since the pressure loss in the tube is considerably large due to the collapsed tube shape and the resulting narrow flow path. Beyond a certain flow rate, the tube recovers a circular cross-sectional form on the upstream side causing ΔP to decrease, leading to what is termed negative resistance. When the collapsed region is eliminated from the whole tube, ΔP shows a gentle increase with flow rate. The

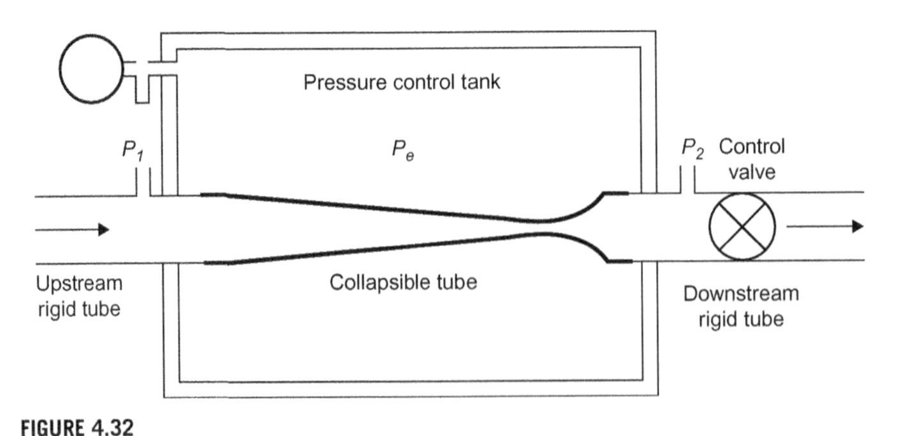

FIGURE 4.32

Typical collapsible tube experimental system [33].

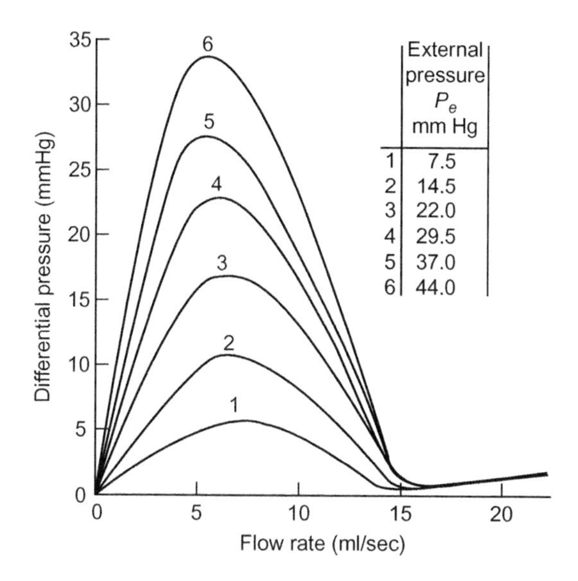

FIGURE 4.33

Flow rate–pressure characteristic curve [34].

larger the external pressure, the larger is the peak in the characteristic curve before the negative resistance region. ΔP, however, shows almost consistent behavior after the recovery of the tube shape, independently of external pressure conditions.

This characteristic curve can be quantitatively reproduced by analytically solving the following equations [35]: the momentum equation of a simplified one-dimensional model for the tube internal flow, the continuity equation of tube cross-sectional area, the equilibrium equation of the force acting on the tube wall segment, and flow resistance characteristics due to the tube transformation, which is the so-called tube law. The tube law is expressed as an empirical relationship between the collapsed ratios of cross-section and the transmural (internal minus external) pressure of the tube, as shown in Fig. 4.34 [36].

In the experimental system described above, self-excited vibrations with large amplitudes will occur in a non-circular cross-section state, mainly in the negative resistance region on the characteristic curve. Several mechanisms have been proposed to explain the onset of these vibrations. These include a periodic shift between subcritical and supercritical flow in hydraulic jump, unsteady flow separation at the collapsed region, flow instability of viscous fluid, and instability in the interaction of the tube characteristics and the rigid elements connected to the tube ends. In order to explain this mechanism theoretically, the collapsible tube has been treated as a lumped parameter system or a one-dimensional distributed parameter system in previous studies.

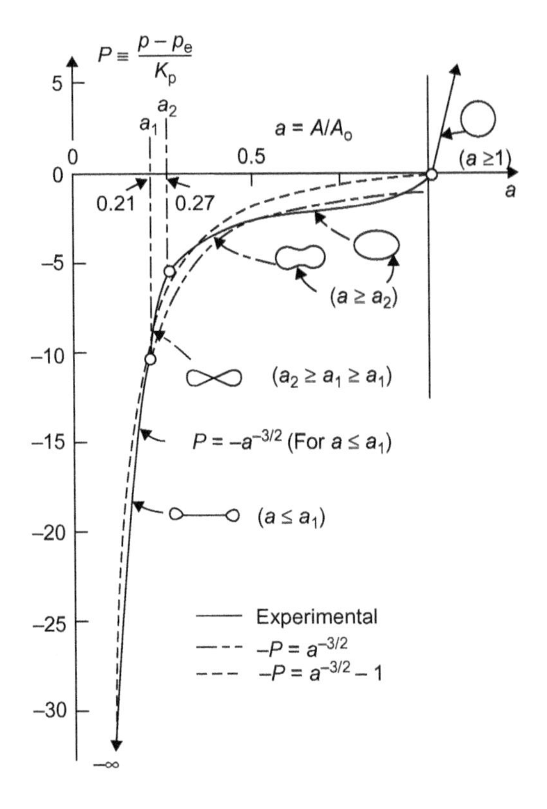

FIGURE 4.34

Tube law [36].

As a lumped parameter system model, Hayashi et al. [37] divided the tube into three elements: upstream region, collapsed region, and downstream region. Then, by adding a rigid pipe element at both ends, they introduced basic equations for the resulting five regions. With the constant conditions of upstream pressure and external pressure, the basic equations were linearized and transformed into a characteristic frequency equation. Stable conditions were determined employing the Routh-Hurwitz criterion. Stability criteria were superposed on the characteristic curve (see Fig. 4.35) and shown to qualitatively explain the experimental results.

Treating it as a distributed continuous system, Hayashi et al. [38] have gathered the aforementioned four basic equations of the simplified one-dimensional model, and additional momentum equations of the fluid, in the rigid pipe at the upstream and downstream ends of the tube. These six equations were discretized and solved numerically under transient conditions to obtain the dynamic characteristics. The results qualitatively predicted the time variance of cross-section area, flow rate, and pressure.

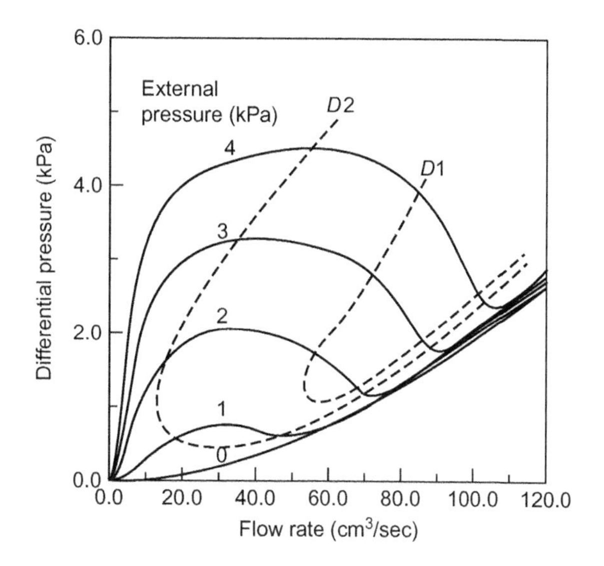

FIGURE 4.35

Unstable region in the characteristic curve [37].

Although many other studies have been conducted, a systematic theory to explain this complicated phenomenon has not been developed.

4.3.3 Key to prevention of collapsible-tube vibration

Studies on collapsible tubes have existed for nearly a century. However, these mainly dealt with simplified experimental models such as shown in Fig. 4.32. There have been a few investigations which modified or analyzed the phenomena in actual internal organs and artificial organs. Although stability analysis methods may be applied to avoid self-excited vibrations in a simplified experimental system, the methods are not developed enough to evaluate actual practical engineering or biological systems. Concrete preventive measures depend on future work.

References

[1] T.B. Benjamin, Proc. Roy. Soc. A 206 (1961) 457–486.

[2] T.B. Benjamin, Proc. Roy. Soc. A 206 (1961) 487–499.

[3] R.W. Gregory, M.P. Païdoussis, Proc. Roy. Soc. A 293 (1966) 512–527.

[4] R.W. Gregory, M.P. Païdoussis, Proc. Roy. Soc. A 293 (1966) 528–542.

[5] M.P. Païdoussis, Mechanical Engineering Research Laboratories Report MERL 69-3, Department of Mechanical Engineering, McGill University, Montreal, Quebec, Canada, 1969.
[6] M.P. Païdoussis, J. Mech. Eng. Sci. 12 (2) (1970) 85–103.
[7] M.P. Païdoussis, J. Mech. Eng. Sci. 12 (4) (1970) 288–300.
[8] M.P. Païdoussis, N.T. Issid, J. Sound Vib. 33 (3) (1974) 267–294.
[9] M.P. Païdoussis, J. Mech. Eng. Sci. 17 (1) (1975) 19–25.
[10] M.P. Païdoussis, G.X. Li, J. Fluids Struct. 7 (1993) 137–204.
[11] M.P. Païdoussis, C. Sundararajan, J. Appl. Mech. 42 (1975) 780–784.
[12] M.P. Païdoussis, N.T. Issid, J. Appl. Mech. 43 (1976) 198–202.
[13] C. Semler, H. Alighanbari, M.P. Païdoussis, J. Appl. Mech 65 (1998) 642–648.
[14] Y. Sugiyama, et al., *Trans. JSME* (in Japanese) 54 (498, C) (1988) 353–356.
[15] S.S. Chen, J. Appl. Mech. 40 (2) (1973) 362–368.
[16] M.P. Païdoussis, J.P Denise, J. Sound Vib. 20 (1) (1972) 9–26.
[17] K. Hirota, et al., Flow-induced vibration of a large-diameter elbow piping in high Reynolds number range; random force measurement and vibration analysis, Proceedings of FIV2008, 2008.
[18] FBR Safety Unit, Proposed guideline of flow-induced vibration evaluation for hot-leg piping of primary cooling system in sodium-cooled fast reactor, JAEA-Res. (2011) 2011–2021.
[19] Y. Sugiyama, Y. Tanaka, T. Kishi, H. Kawagoe, J. Sound Vib. 100 (2) (1985) 257–270.
[20] S. Hayama, M. Matumoto, *Trans. JSME* (in Japanese) 52 (476, C) (1986) 1192–1197.
[21] J. Weisman, et al., Int. J. Multiphas. Flow 5 (6) (1979) 437–462.
[22] J. Weisman, S.Y. Kang, Int. J. Multiphas. Flow 7 (3) (1981) 271–291.
[23] F. Hara, *Trans. JSME* (in Japanese) 42 (360, C) (1976) 2400–2411.
[24] H. Matsuda, *Piping Technol.* (in Japanese) 40 (6) (1998) 28–35.
[25] Y. Nakamura, N. Fukamachi, Sound generation in corrugated tubes, Fluid Dyn. Japan Society for Fluid Mechanics Res. 7 (1991) 255–261.
[26] V.F. Jakubauskas, Added fluid mass for bellows expansion joints in axial vibrations, Trans. ASME, J. Press. Vess. Technol. 121 (1999) 216–219.
[27] Standards of the Expansion Joint Manufacturers Association, Inc., seventh ed., EJMA, 1998.
[28] M. Morishita, N. Ikahata, S. Kitamura, Dynamic analysis methods of bellows including fluid-structure interaction, Metallic Bellows and Expansion Joints-1989, ASME PVP-Vol.168, (1989) 149–157.
[29] V.F. Jakubauskas, D.S. Weaver, Transverse vibrations of fluid filled bellows expansion joints, Symposium on Fluid Structure Interaction, Aeroelasticity, Fluid-Induced Vibration and Noise, ASME AD-Vol. 53-2, vol. 2, (1997) 463–471.
[30] D.S. Weaver, P. Ainsworth, Flow induced vibrations in bellows, Trans. ASME, J. Press. Vess. Technol. III (1989) 402–406.
[31] V.F. Jakubauskas, D.S. Weaver, Transverse natural frequencies and flow induced vibrations of double bellows expansion joints, J. Fluid Struct. 13 (1999) 461–479.
[32] F.P. Knowlton, E.H. Starling, J. Physiol. 44 (1912).
[33] C. Cancelli, T.J. Pedley, J. Fluid Mech. 157 (1985) 375–404.
[34] W.A. Conrad, IEEE Trans. Bio-Medical Eng. BME-16 (4) (1969) 284–295.

[35] S. Hayashi, R. Toyoda, K. Sato, *Trans. JSME* (in Japanese) 57 (534, B) (1991) 1–7.
[36] A.H. Shapiro, Trans. ASME J. Biomedical Eng. 99 (1977) 126–147.
[37] S. Hayashi, M. Tanba, T. Hayase, *Trans. JSME* (in Japanese) 60 (579, B) (1994) 46–51.
[38] S. Hayashi, T. Hayase, H. Kawamura, *Trans. JSME* (in Japanese) 62 (594, B) (1996) 146–153.

CHAPTER

5 Vibration Induced by Pressure Waves in Piping

Flow in piping generally has random frequency characteristics. However, a particular frequency component may become dominant when a change of flow interacts with fluid machinery such as compressors and pumps, as well as equipment such as valves, branch pipes, and perforated plates. When the particular frequency matches the acoustic natural frequency of any equipment, oscillations are strongly amplified. The associated pressure pulsations excite the piping and/or equipment, which may result in damage.

Flow oscillation in piping has been addressed by a research subgroup and a workshop of the Japan Society of Mechanical Engineers (JSME) [1]. However, the phenomenon has never been discussed comprehensively. This chapter presents an outline of the physical phenomenon, research history, calculation, and evaluation methods. The effect on equipment when pressure waves occur in piping is discussed. Some case examples and countermeasures based on the latest information are also presented.

5.1 Pressure pulsation in piping caused by compressors

5.1.1 Summary

Compressors and blowers installed in chemical plants can efficiently transport and pressurize fluids as well as transfer power. Their classification by type is shown in Table 5.1. This chapter describes the flow oscillations (pressure pulsations) in piping and the vibration of piping caused by suction/discharge flows in such fluid machinery. In addition, some cases of automobile engines are introduced.

5.1.2 Explanation of the phenomenon, and the history of research/evaluation

5.1.2.1 What is the pressure pulsation phenomenon?

Specific flow oscillations occur in piping connected to compressors or blowers. These oscillations can be categorized into two groups: oscillations caused by pressure pulsations in the fluid machinery and the piping system; and

Flow-Induced Vibrations.

Table 5.1 Classification of compressors and blowers

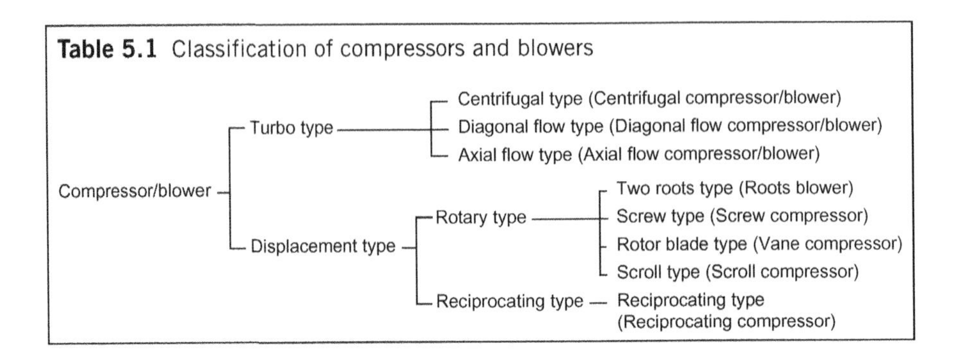

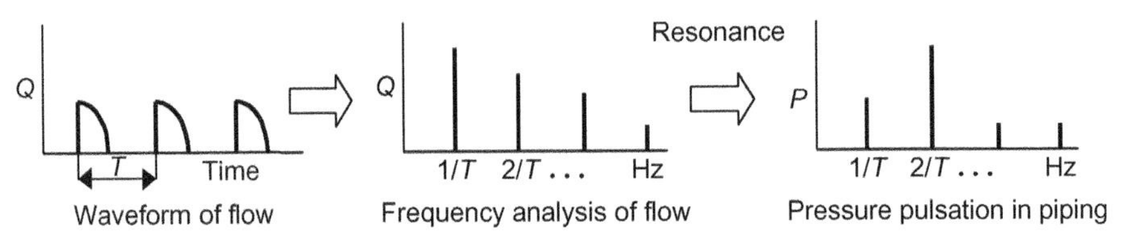

FIGURE 5.1

Suction/discharge flow and pressure pulsation in reciprocating compressors.

oscillations caused by the change of momentum (centrifugal action) at pipe bends, etc. Only the former is discussed here as the latter seldom cause any problem in gas systems. Refer to Section 4.1.2 for a discussion on the latter.

Displacement-type compressors, such as reciprocating compressors, cause large flow fluctuations because of the intermittent suction/discharge flow, as shown in Fig. 5.1. The resulting pressure pulsations contain many harmonic components of the rotational speed. The reciprocating compressor rotational speed is generally low and the frequency of the pressure pulsation is 100 Hz or less in many cases. This pressure pulsation is generally too weak to cause any problem. However, it may be amplified due to coincidence with an acoustic natural frequency of the piping. The resulting resonance can induce severe piping vibration and/or malfunction of instruments, compromising plant operation. As a familiar example, the exhaust gas discharged from an automobile exhaust pipe may have large flow fluctuations. Compressors having a series of blades, such as turbomachinery, cause flow fluctuation at the blade passing frequency (blade number × compressor rotational speed) and its harmonic components. The frequency of the pulsations is generally several hundred hertz, often resulting in severe noise problems.

5.1.2.2 Research history

Axial-type and centrifugal-type turbomachinery do not cause large pressure pulsations due to their different compressing and blowing mechanisms. There has been little work done on pressure pulsation in these machines. On the other hand, for displacement-type compressors, such as reciprocating compressors, pressure pulsation has been the subject of research over the years since the suction/discharge flow fluctuations are relatively large. Among the displacement-type compressors, the discussion below focuses on the reciprocating type, for which a lot of research has been carried out with the goal of regulating the maximum allowable level of pulsation. For turbomachinery, when the flow fluctuation is known, the pressure pulsation can be evaluated using the same techniques developed for piping systems connected to reciprocating compressors.

In September 1952, the Southern Gas Association (SGA) in the USA organized the pulsation research committee and launched a project on pulsation analysis. This was the beginning of systematic research on pulsation in piping connected to reciprocating compressors. The Southwest Research Institute (SwRI) was selected as an external research agency and achieved good results in pulsation analysis using an analog computer for the first time in the autumn of 1954. They started calculations for practical use in the summer of 1955, and have undertaken pulsation analysis on a commercial basis around the world.

In the 1960s, the American Petroleum Institute Standard 618 (API Standard 618) imposed the requirement that pulsation analysis be done at the design stage of process plants in order to control pulsations to an appropriate level. The API Standard 618 had specified that pulsation analysis be done using analog simulation until the third edition revised the standard in February 1986. This was based on the background that the digital computer at the time had difficulty performing pulsation analysis considering the interaction between reciprocating compressors and piping which is a requirement of the API Standard 618. Based on an analogy between equations for fluid flow and those for an electric circuit, the analog simulation method [2,3] evaluates the pulsation in piping by simulating the fluid characteristic with an L-C-R circuit on the assumption that pressure and flow in the fluid system are equivalent to voltage and current in the electric circuit. However, the third edition of API Standard 618, revised in 1986, removed the restriction on the kind of computer to be used and allowed the use of digital computers for pulsation analysis. In addition to this revision of the API Standard 618, along with the remarkable progress in the performance of digital computers and achievements in research efforts on pulsation in piping in the late 1960s, the mainstream of pulsation analysis in piping in Japan moved to digital simulations and today, pressure pulsation in piping can be calculated with sufficient accuracy for design purposes.

5.1.3 Calculation and evaluation methods

5.1.3.1 Calculation method

5.1.3.1.1 Calculation method for pressure pulsations in piping

Generally, the transfer matrix method, developed and employed in the field of mechanical vibration and acoustics, is extensively used in pulsation analysis. This method has been used since the beginning because the acoustic natural frequency in piping can be obtained theoretically. However, the analysis may be difficult when the piping system is complicated due to bypass-lines or in the presence of a piping network. In this case, alternative calculation/analysis methods, such as the stiffness matrix method, the modal analysis method, and the finite element method, have been developed and used in computations for practical use. The internal fluid is considered as a one-dimensional flow when the pipe diameter is less than 1/5 of the pulsation wavelength [4] in the above analysis methods. When such an assumption is not applicable, the boundary element method, which is used in the field of acoustics, can be employed effectively. Table 5.2 shows the calculation/analysis methods that are currently used in calculations for practical use. The impedance method [31] that uses the ratio of the pressure fluctuation to the flow fluctuation and the pressure pulsation analysis method [32] that applies the nodal analysis method for an electric circuit network are not listed. These methods are rarely used in pulsation analysis today. The transfer matrix method is described below.

5.1.3.1.2 Transfer matrix method

The transfer matrix method [5–14] is the most extensively used method in pulsation analysis. In this method, the piping system is divided into several piping elements such as a pipe, a side branch, a volume (vessel), and resistance (orifice or valve). The 2×2 transfer matrix M is then formulated to relate the pressure fluctuation p and the flow fluctuation q between the inlet and outlet of each piping element, as expressed below:

$$\begin{Bmatrix} p_{\text{out}} \\ q_{\text{out}} \end{Bmatrix} = [M] \begin{Bmatrix} p_{\text{in}} \\ q_{\text{in}} \end{Bmatrix} \tag{5.1}$$

The subscripts 'in' and 'out' indicate conditions at the inlet and outlet of the piping element, respectively. On the assumption that pressure and flow vary sinusoidally, the transfer matrix is formulated by substituting assumed periodic solutions in both the equation of motion and the continuity equation, and then solving the equations in consideration of boundary conditions. Table 5.3 shows the transfer matrices when damping is not considered. For the total piping system, the 2×2 transfer matrix M_{T} is formulated as Eq. (5.2), by multiplying the transfer matrices of all the piping elements.

$$\begin{Bmatrix} p_{\text{end}} \\ q_{\text{end}} \end{Bmatrix} = [M_n][M_{n-1}][M_{n-2}]\cdots[M_2][M_1] \begin{Bmatrix} p_{\text{strt}} \\ q_{\text{strt}} \end{Bmatrix} = [M_{\text{T}}] \begin{Bmatrix} p_{\text{strt}} \\ q_{\text{strt}} \end{Bmatrix} a \tag{5.2}$$

The subscripts 'strt' and 'end' in Eq. (5.2) represent conditions at the starting point and end point of a piping system, respectively. Frequency equations in which determinants of the transfer matrix M_{T} for the total system equal zero are obtained by giving boundary conditions (open or closed) at the starting point and the end point of the piping system. The acoustic natural frequency is calculated by solving the frequency equations. The pressure fluctuations and flow fluctuations at each point are determined according to the following procedure, using

Table 5.2 Calculation/analysis methods for pressure pulsations

Calculation method	Outlines, etc.	Key equation
Transfer matrix method References [5–14]	• This is the most extensively used method to calculate the pressure pulsation response in piping in the frequency domain • The acoustic natural frequency in piping can be calculated theoretically • This analysis may be difficult when the piping system is complicated with bypass lines and/or piping network	$\begin{Bmatrix} p_{\text{out}} \\ q_{\text{out}} \end{Bmatrix} = [M]\begin{Bmatrix} p_{\text{in}} \\ q_{\text{in}} \end{Bmatrix}$
Stiffness matrix method Reference [15]	• This is a frequency domain method, which improved on the transfer matrix method so that complicated piping systems can be analyzed • It is named after the methodology where the relation between flow and pressure is formulated similarly to the stiffness matrix in structural analysis	$\begin{Bmatrix} q_{\text{in}} \\ q_{\text{out}} \end{Bmatrix} = [A]\begin{Bmatrix} p_{\text{in}} \\ p_{\text{out}} \end{Bmatrix}$
Modal analysis method References [16–21]	• This method applies modal superposition, often used in vibration analysis, to the pressure pulsation analysis related to a complicated piping system • The natural frequencies and the modal functions are calculated using the transfer matrix method. The pressure pulsation is calculated by the time history analysis. The computation time of this method is shorter compared with direct numerical integration	$q = \sum_{i=1}^{\infty} \psi_i(x)\cdot y_i(t)$ $p = \sum_{i=1}^{\infty} \phi_i(x)\cdot z_i(t)$
Finite element method References [22–25]	• This was developed as the pulsation analysis method for complicated piping systems	$[M]\dot{q} + [C]q + [K]\int q\,dt = p$

(Continued)

Table 5.2 (Continued)

Calculation method	Outlines, etc.	Key equation
	• The method models the fluid as a one-dimensional problem similar to longitudinal vibration analysis, formulates a whole system matrix similarly to the structural analysis, and calculates the pressure pulsation. Both frequency and time domain analyses can be performed	
Boundary element method	• This is a frequency domain method that is applied in cases where the internal fluid cannot be analyzed as a one-dimensional flow. Example applications include large tanks and silencers	$[H]\{p\} = [G]\{q\}$
References [26–30]	• Computation time can be reduced when the system under analysis is sufficiently large compared with the pulsation wavelength	

p, pressure fluctuation; q, flow fluctuation.

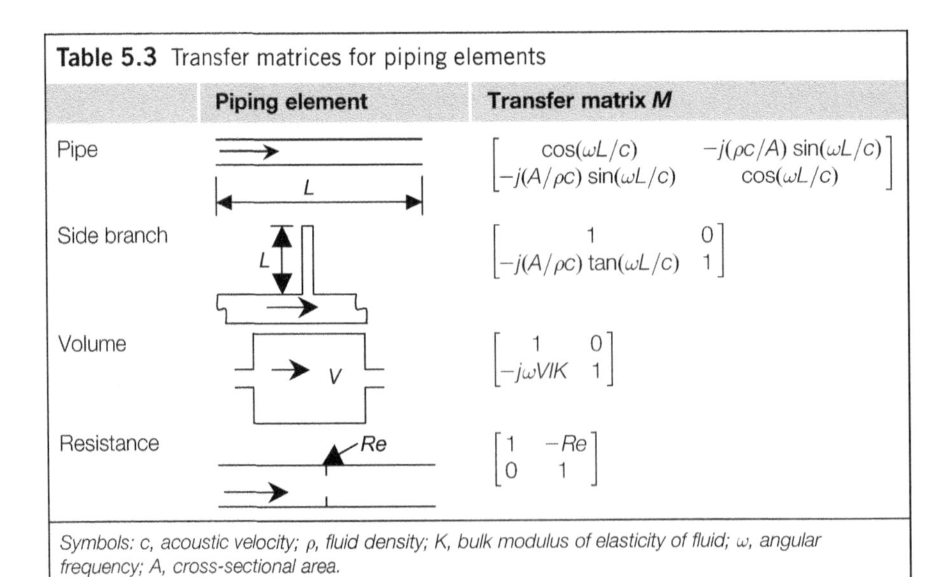

Table 5.3 Transfer matrices for piping elements

	Piping element	Transfer matrix M
Pipe		$\begin{bmatrix} \cos(\omega L/c) & -j(\rho c/A)\sin(\omega L/c) \\ -j(A/\rho c)\sin(\omega L/c) & \cos(\omega L/c) \end{bmatrix}$
Side branch		$\begin{bmatrix} 1 & 0 \\ -j(A/\rho c)\tan(\omega L/c) & 1 \end{bmatrix}$
Volume		$\begin{bmatrix} 1 & 0 \\ -j\omega V/K & 1 \end{bmatrix}$
Resistance		$\begin{bmatrix} 1 & -Re \\ 0 & 1 \end{bmatrix}$

Symbols: c, acoustic velocity; ρ, fluid density; K, bulk modulus of elasticity of fluid; ω, angular frequency; A, cross-sectional area.

Eq. (5.2), in which the starting point and the end point of the piping system are related by the transfer matrix:

1. Assuming, for instance, that a compressor is installed at the starting point, apply an appropriate flow excitation at this point.
2. Enter the boundary condition at the end point (e.g. the pressure fluctuation is zero for an open end, the flow fluctuation is zero for a closed end).
3. Calculate the pressure fluctuation at the starting point, and the pressure fluctuation or the flow fluctuation at the end point.
4. Calculate the pressure and flow fluctuations at any point of the piping system based on the above results.

The transfer matrix is formulated for a given excitation frequency ω. Hence, for a piping system connected to a reciprocating compressor, the calculation of pressure pulsations must be performed for every frequency corresponding to an integral multiple of the compressor rotational speed.

5.1.3.1.3 Evaluation of the fluid damping force in piping

The evaluation of the fluid damping force in piping is quite critical, because it affects the calculation accuracy for pulsations in piping. For a fluid damping adopted for calculating the pulsations in piping, Binder [33] proposed an experimental formula; however, the calculation accuracy was not satisfactory and clearly further research is required.

Fluid damping in plant piping is mainly due to pressure loss at a reducer section, a bend (elbow), an orifice, or friction within pipes. It is expressed as a non-linear function proportional to the second power of velocity. In the vicinity of a compressor, in particular, the oscillating flow velocity is added to the average flow velocity. Since the calculation is quite complicated when using the non-linear equation, Hayama et al. [34–36] studied an equivalent linearization method, which calculates a linear damping where the lost energy is equivalent to that dissipated in one cycle of the pulsation due to the non-linear damping. This equivalent linearization method is currently used extensively. It has been proved that the calculation accuracy for the pulsation in piping is satisfactory even in the case where an average flow exists. The equivalent damping coefficient [35] is presented in Table 5.4.

5.1.3.1.4 Determination of flow excitation by compressors

The suction/discharge flow variation in compressors (i.e. the oscillating flow) is derived based on the assumption that the pressure on the suction (discharge) side is constant and the valve opens (closes) ideally. The method for reciprocating compressors is outlined below, since the oscillating flow is easily determined from mechanical specifications [37]. For other displacement-type machines such as Roots blowers and screw compressors, the oscillating flow can be determined in the same way if the suction (discharge) opening geometry is known. Note,

Table 5.4 Equivalent damping coefficient [35]

Equivalent damping coefficient K_1 of the basic wave component $K_1 = \|u_1\|k_1$	(Reference) Equivalent linear equation with a term of the second power of velocity $\|u\|u \cong K_0 \cdot U + K_1 \cdot u_1 \sin \omega_1 t$ $(u = U + u_1 \sin \omega_1 t)$
where, $k_1 = \frac{4}{\pi}\beta \sin^{-1}\beta + \frac{4}{3\pi}(\beta^2 + 2)\sqrt{1-\beta^2}$	$(0 \leq \beta \leq 1)$
$k_1 = 2\beta$	$(1 < \beta)$ $(\beta = U/\|u_1\|)$

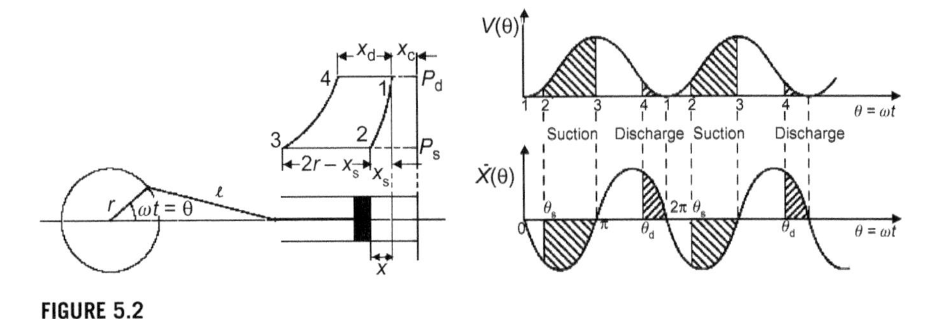

FIGURE 5.2

Suction/discharge flow waveform in reciprocating compressors [37].

however, that the oscillating flow cannot be determined uniquely for turbomachinery such as centrifugal-type compressors.

A schematic diagram of a reciprocating compressor and its suction/discharge flow rate time history are shown in Fig. 5.2. The change of cylinder volume V and the volume velocity $\dot{X}$ in the figure are expressed as follows:

$$
\begin{aligned}
V(\theta) &= Ax_c + Ar\{(1+\varepsilon/4) - \cos\theta - \varepsilon\cos 2\theta/4\} \\
\dot{X}(\theta) &= -Ar\omega(\sin\theta + \varepsilon\sin 2\theta/2)
\end{aligned}
\tag{5.3}
$$

where A is the cylinder cross-section area, r the crank radius, ω the crank rotation angular velocity, $\varepsilon = r/l$ (l being the connecting rod length), θ the crank angle, and x_c the top dead center clearance. In Fig. 5.2, the crank angle is expressed as follows, when the suction and the discharge valves open:

$$
\begin{aligned}
\theta_s &= \cos^{-1}\left(2\delta_1/\left(1+\sqrt{1+2\varepsilon\delta_1}\right)\right) \quad \text{where } \delta_1 = (1+\varepsilon/2 - x_s/r) \\
\theta_d &= \cos^{-1}\left(2\delta_2/\left(1+\sqrt{1+2\varepsilon\delta_2}\right)\right) \quad \text{where } \delta_2 = (1+\varepsilon/2 - x_d/r)
\end{aligned}
\tag{5.4}
$$

Assuming that the piston volume velocity $\dot{X}$ in Eq. (5.3) is the suction/discharge flow rate, the pressure pulsations in the piping can be analyzed via a Fourier series decomposition of the shaded area in Fig. 5.2 as an oscillating flow in the frequency domain analysis, or by using it directly in a time history analysis.

5.1.3.1.5 Remarks on calculating and evaluating pressure pulsations

Flow interaction with the compressor and piping system. Pressure pulsations in the piping are calculated using the compressor suction/discharge flow rates as a forced excitation, on the assumption that the pulsations do not affect the compressor flow. For reciprocating compressors, however, the API Standard 618 [38] recommends taking into consideration the interaction between the fluid and the compressor and piping system in the case of a certain level of compressor capacity. Accordingly, Matsuda et al. [17–19], Kato et al. [24,25,39], and Fujikawa [40] developed an appropriate analysis technique, which is now in practical use. Fujikawa [40], and Kato et al. [39] reported the conditions that are likely to cause the fluid interaction, with some case examples for reciprocating compressors. These studies clarified some particular conditions to improve the calculation accuracy further by taking into consideration the fluid–structure interaction; for instance, when a large pressure pulsation occurs around a compressor. As for displacement-type compressors, since fluid on the suction side and fluid on the discharge side are separated by the compressor, the pressure pulsations in the piping may be calculated at each side independently. For turbomachinery, on the other hand, a particular pressure pulsation mode (surging) may occur, e.g. in a blower-piping system depending on the relationship between the blower characteristic and the pipe restriction characteristic, as well as the blower installation position [41]. The underlying mechanism is described in Section 5.2.

Influence of compressor valves. In order to accurately estimate the flow excitation force by a compressor, the dynamic characteristics of the compressor valves are critical. Some results, from several parametric studies, were reported related to the movement of reciprocating compressor valves and the influence on pulsations in piping [42–45].

Change of the fluid acoustic velocity. In industrial plants, the process gas quality may differ slightly between the estimated value at the design stage and the actual value in practical operation. In such a case, the acoustic resonance frequencies change due to the difference in the gas acoustic velocity; the calculated values may therefore differ from the actual values. In order to avoid this risk, SwRI adopts the maximum value of pressure pulsations as the calculation result while changing the gas acoustic velocity (practically, the excitation frequency) about $\pm 5\%$ of the design value.

Effect of two or more compressors. In many cases in industrial plants, two or more independent compressors are installed in the same piping system. The API Standard 618 [38] recommends taking this into consideration for reciprocating compressors. As a calculation method for pressure pulsations when two reciprocating compressors operate simultaneously, Fujikawa et al. [46] proposed a frequency domain analysis. The analysis converts the relative phase of the two compressors into the in-phase and the out-of-phase parts for every harmonic component of the flow excitation force, to solve for the pressure pulsations for the two cases. The larger response for every frequency is adopted for safety.

Calculation method for piping vibration excited by pressure pulsations. Pressure pulsations in piping induce excitation forces. Piping sections excited by pressure pulsation are shown in Fig. 5.3. The forces are always applied to the piping sections as shown in the figure. Large vibrations occur when the pressure pulsation frequency matches the mechanical natural frequency of the piping. Only a few researchers have focused on the steady-state response of the piping under pressure pulsation excitation. Hayama and Matsumoto [48] described a concept for the fluid force at pipe bends. Wakabayashi et al. [49] and Tanaka and Fujita [47,50] reported good results in comparison with calculation results and experimental data for piping vibration related to reciprocating compressors and pumps, respectively. For compressors, the pulsations in piping can be considered as an external excitation force to the piping since the internal fluid is a gas, and it is not necessary to take into consideration the interaction between the pulsations and the pipe vibration. Basically, the steady-state response of the piping vibration is obtained by applying pressure pulsations to the piping sections, in consideration of the phase, as an excitation force as shown in Fig. 5.3.

5.1.3.2 Evaluation method

5.1.3.2.1 Maximum allowable level of pulsation

Evaluation criteria for pressure pulsations in piping are only specified for reciprocating compressors in API Standard 618. Since no criterion is specified for the other displacement-type compressors and turbomachinery, the evaluation work for pulsations is quite difficult. The same evaluation criteria as those for reciprocating compressors may be applied to other types of compressors when the pulsation

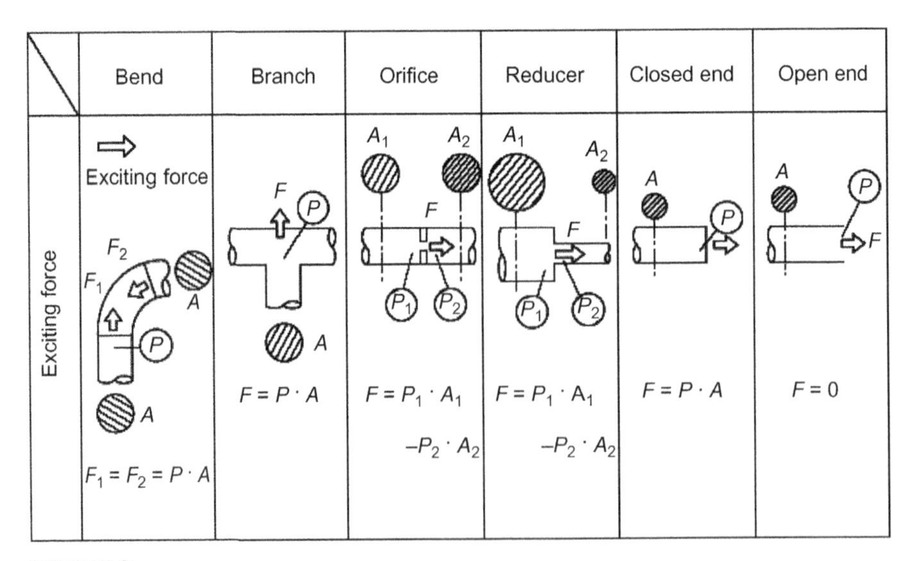

FIGURE 5.3

Piping sections excited by pressure pulsations [47].

frequencies in piping are about 100 Hz or less. On the other hand, when the frequency range is beyond 200 Hz and up to several hundred hertz, the pulsations should be treated as noise.

The maximum allowable level of pulsation in piping is shown below. This is the level specified in API Standard 618 (fourth edition) [38] for reciprocating compressors. In this standard, the evaluation criteria are classified into three design approaches based on compressor power and the discharge pressure as shown in Fig. 5.4. The maximum allowable level of pulsation for each design approach is expressed using the equations given in Table 5.5.

5.1.3.2.2 Maximum allowable level of piping vibration

The maximum allowable level of piping vibration has not been specified in most publications. The design approach 3 defined in API Standard 618 [38] for reciprocating compressors requires analyzing the piping vibration. However, it only specifies that mechanical vibrations should not cause a cyclic stress level in excess of the endurance limits of the piping material. No maximum allowable level is specified. The API standard states that the peak-to-peak cyclic stress shall not exceed 179.2 MPa (26,000 psi) for carbon steel pipes at an operating temperature below 371 °C (700 °F). Some companies have independently established their own maximum allowable level of vibration as a company standard. In the 1970s, SwRI published an evaluation criterion for piping vibration related to reciprocating compressors and pumps [51], which is most useful as a primary diagnosis of the main piping at plant sites. SwRI's evaluation criteria for the piping vibration cover the frequency range up to 300 Hz, which can be used satisfactorily in evaluating the mechanical vibration of main piping (refer to Fig. 5.5).

5.1.4 Hints for countermeasures

5.1.4.1 How to reduce pressure pulsations

Pulsations in piping reach a high level due to acoustic resonance. Therefore, understanding the acoustic characteristics is essential and fundamental for pulsation reduction. Calculation formulae for acoustic resonance frequencies are given in Table 5.6.

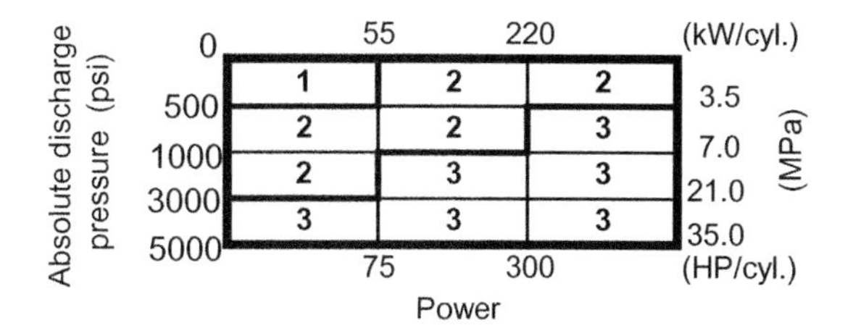

FIGURE 5.4

Selection criteria for API design approach number [38].

Table 5.5 Maximum allowable level of pulsation specified in API Standard 618 [38]

Design approach number	Maximum allowable level of pulsation	Symbol	Remarks
1	$P_1 = 4.1 / (P_L)^{1/3}$	P_1: Maximum allowable level of pressure pulsation ratio (%) (=Peak-to-peak pressure pulsation amplitude/P_L) P_L: Average mean absolute line pressure (bar)	• Maximum allowable level at the line side of pulsation suppression devices (snubber) • No allowable level in piping is specified
2	$P_1 = (a/350)^{1/2}\{400/(P_L \cdot ID \cdot f)^{1/2}\}$	P_1: Maximum allowable level of pressure pulsation ratio (%)(=Peak-to-peak pressure pulsation amplitude/P_L) P_L: Average mean absolute line pressure (bar) ID: Inside diameter of line pipe (mm) f: Pulsation frequency (Hz) = (RPM/60) $\cdot N$ RPM: Machine speed (revolutions per minute) $N = 1, 2, 3, \ldots$ a: Speed of sound (m/s)	• Allowable level for piping systems • This design approach requires simulating the pulsation in consideration of dynamic iteration between reciprocating compressor and piping systems
3	Same as design approach 2	Same as design approach 2	• Same as design approach 2 • This design approach requires calculating the mechanical natural frequency of reciprocating compressor and piping, and avoiding the coincidence of the mechanical natural frequency and the pressure pulsation frequency • If not avoiding, this design approach requires calculating the piping vibration responses excited by pressure pulsations

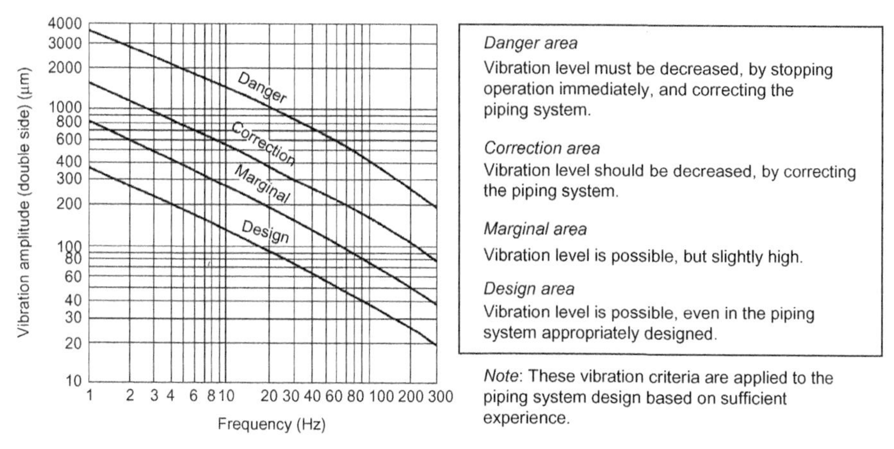

FIGURE 5.5

SwRI's evaluation criteria for piping vibration [52].

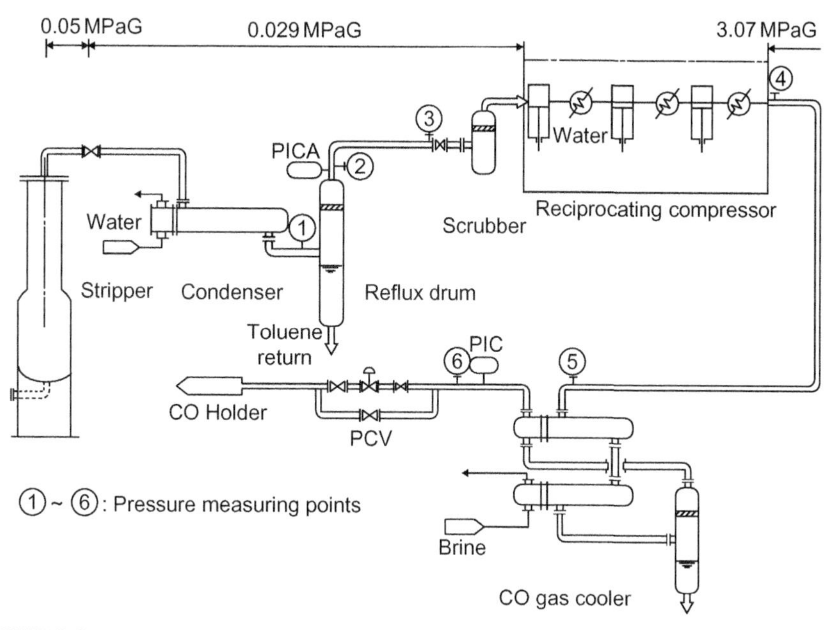

FIGURE 5.6

Reciprocating compressor: schematic diagram of plant [23].

Pulsations in piping can be reduced by various methods: installation of an orifice that adds fluid damping, adjusting piping length to change the resonance frequency, installation of a pulsation suppression device (snubber) that reduces the pulsation itself, and installation of a 1/4 wavelength branch pipe that aims to reduce a

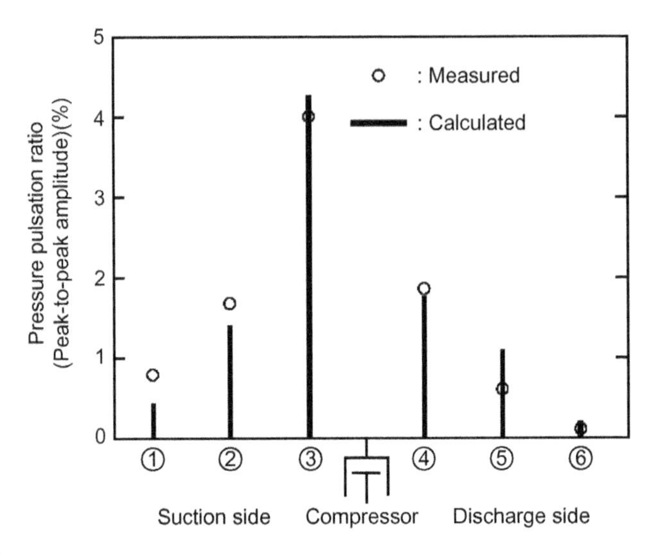

FIGURE 5.7

Comparison between measured and calculated pressure at each point [23].

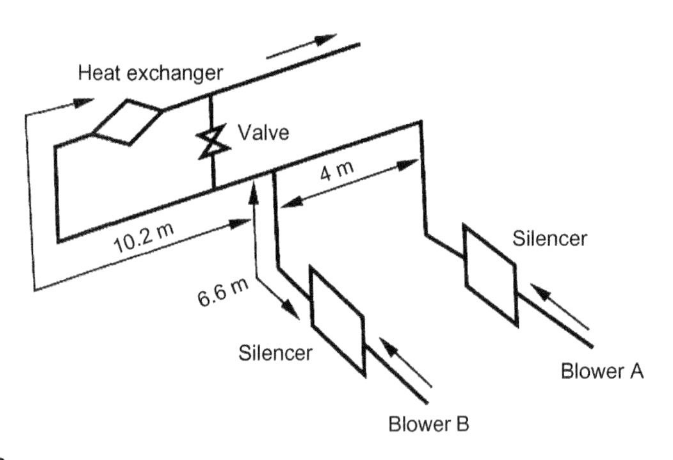

FIGURE 5.8

Roots blower: schematic diagram of plant [63].

particular frequency component. These should be applied in consideration of viability of the installation and cost. The solution methods are listed in Table 5.7 [52].

5.1.4.2 Methods to reduce piping vibration

Large piping vibrations are often caused by coincidence between the pulsation frequency and the mechanical natural frequency of the piping. It is essential to avoid this resonance condition. Adding damping to the piping by means of a hydraulic

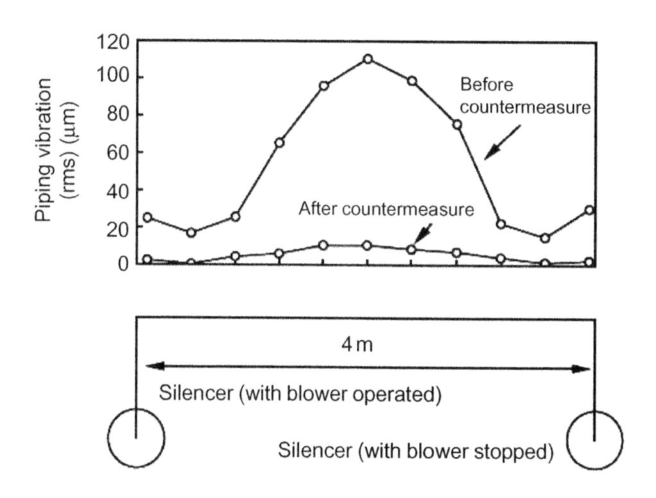

FIGURE 5.9

Result of vibration field data (effect of countermeasure) [63].

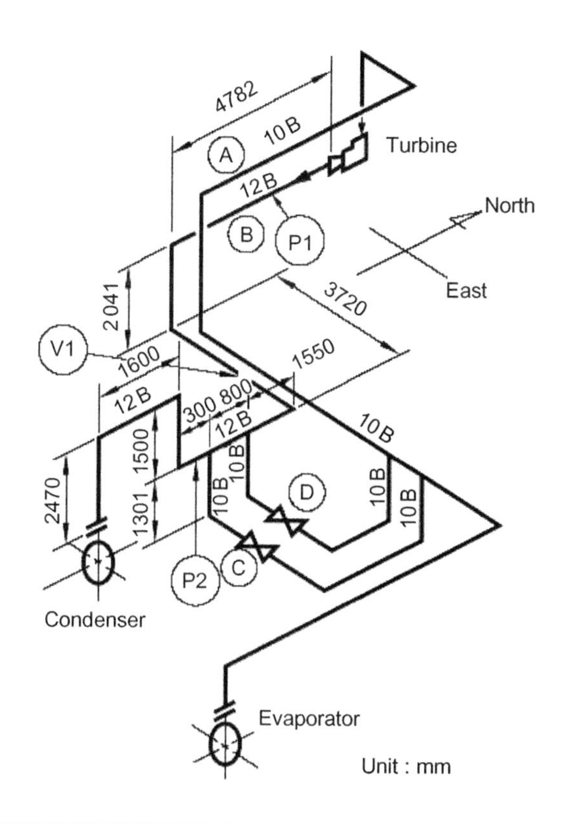

FIGURE 5.10

Turbine: schematic diagram of plant [64].

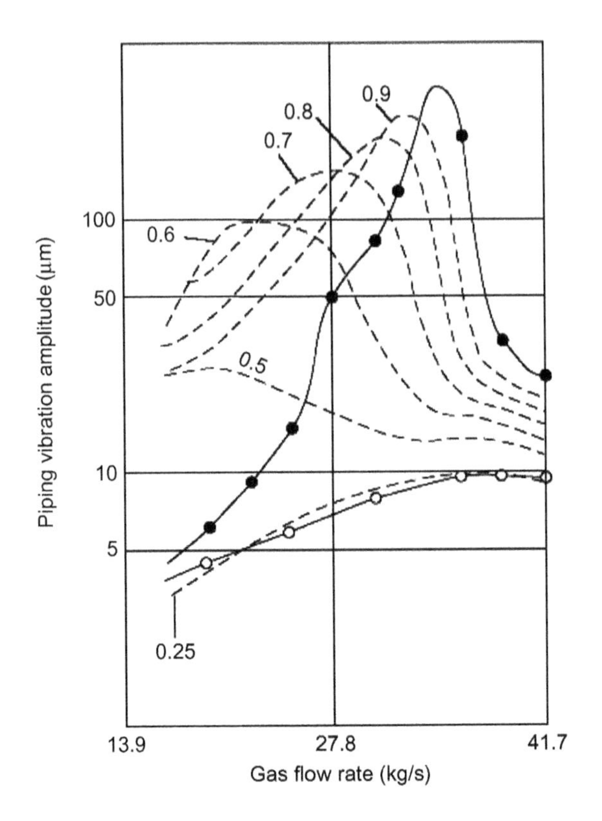

FIGURE 5.11

Result of piping vibration field data (before [●] and after [○] countermeasure) [64].

cylinder or a dynamic damper can be effective. However, it should be noted that piping vibration tends to be larger when the piping support is too weak; supports are often designed to be weak in order to avoid thermal stresses. It is also important to reinforce and/or support drains and valves subjected to pulsation forces. Various methods to reduce piping vibration are outlined in Table 5.8 [58].

5.1.5 Case studies

Some case studies of pressure pulsation problems which occurred in actual plants are shown in Table 5.9. Generally, displacement-type compressors have larger flow variations than turbomachinery. There are, therefore, many examples of problems of pressure pulsations in piping due to these compressors. For turbomachinery, on the other hand, there are few examples because fundamentally, the compression mechanism in turbomachinery does not cause any pressure pulsation. Noise [67], swirling stall, and surging are, however, often experienced in turbomachinery.

These problems will be discussed in a subsequent section.

Table 5.6 Acoustic resonance frequencies

Type of resonance	Acoustic resonance frequency f_0
Pipe (1) Open – Open or Closed – Closed 1/2 wavelength resonance	$f_o = (c/2L)n$ $n = 1, 2, 3, \ldots$
(2) Open – Closed 1/4 wavelength resonance	$f_o = (c/4L)(2n-1)$ $n = 1, 2, 3, \ldots$
(3) Helmholtz resonance	$f_o = (c/2\pi)(A/LV)^{1/2}$
(4) Volume – Restriction – Volume resonance (low pass filter)	$f_o = (c/2\pi)\{(A/L)\cdot(1/V_1 + 1/V_2)\}^{1/2}$

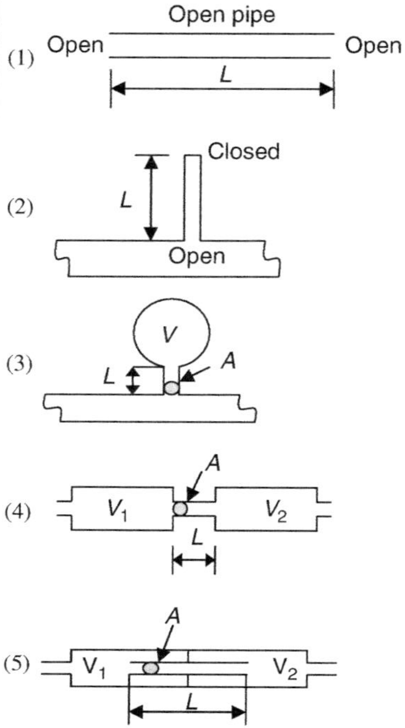

(1) Open pipe: Open, Open, L

(2) Closed, Open, L

(3) V, A, L

(4) V_1, V_2, A, L

(5) V_1, V_2, A, L

c: Acoustic velocity, *V*: Volume

A: Cross-section area

L: Pipe length (add an open-end correction length of 0.4*d* [for every open end, *d*: pipe diameter])

Table 5.7 Methods to reduce pressure pulsation

	Measures	Description/feature	Remarks
1.	Installing an orifice [53,54]	• Adds damping to reduce pressure pulsation • This is the simplest countermeasure to reduce pulsation with minimum change of piping • Installing at an appropriate location is effective also for the change of acoustic velocity in a broad range	• Install at the location where flow fluctuation is large • Installing at the inlet or outlet of a large volume is most effective (inlet/outlet of snubber) • Note that the pressure loss is quite large. Usually, an opening area ratio of 1/4 is adopted
2.	Adjusting piping length	• In order to avoid acoustic resonance, adjust piping length to change the acoustic natural frequency • The 1/4 wavelength resonance in a safety valve piping or a closed branch with a gate block valve can be avoided effectively	• Pay attention when the acoustic velocity changes over a wide range • Pay attention to piping flexibility
3.	Changing piping diameter	• For reducing pressure pulsation, use a piping of larger diameter at the location where pressure pulsation is large, in order to increase the acoustic volume/capacitance	• For the location where size increase of piping is predicted, consider required space at design stage • Pay attention to piping flexibility
4.	Installing a side branch (1/4 wavelength pipe) and a resonator	• Installing the dynamic damper that aims at a particular frequency is sometimes considered • Breaking the resonance mode in main piping reduces the pressure pulsation	• Install at the antinode of pressure pulsation • Pay attention to the effect when the acoustic velocity changes over a wide range • Adding damping is also effective [55] • Pay attention to the supports to prevent vibration since pulsations may be large in a branch pipe
5.	Increasing surge drum size (snubber) volume	• Theoretically, a larger volume reduces pressure pulsations more • This can be the most effective measure to reduce pressure pulsation	• Install as close to compressor as possible • It is important to consider a snubber having a large enough volume in the planning stage (refer to API Standard 618 [38])

(*Continued*)

Table 5.7 (Continued)

	Measures	Description/feature	Remarks
6.	Installing second (additional) drum	• Install the second drum to complement the snubber volume • Use as a low pass filter, combining with snubber	• Delivery period may be a problem in emergency case • It may be quite expensive
7.	Installing a pulsation damper and a snubber internal	• It works as a low pass filter • Use as a low pass filter by installing an internal to a snubber	• Pay attention to material selection when the fluid is corrosive gas • Maintenance is required. Expensive • Generally, pressure loss is large

5.2 Pressure pulsations in piping caused by pumps and hydraulic turbines

5.2.1 Outline

In this section, the phenomena related to steady pressure pulsations that occur in pipeline systems including pumps and hydraulic turbines are discussed. The first discussion is focused on positive displacement pumps and turbo-pumps, which are used to transport, supply, and pressurize liquids, particularly water in various plants. Discussion on Francis and Kaplan turbines for hydroelectric power plants follows. Furthermore, pipelines, tanks, and other equipment which accompany pumps or turbines are examined. The causes of pulsation are divided roughly into two kinds, namely forced vibrations and self-excited vibrations. In either case, pulsations will do damage to machines or plants directly and may be the cause of secondary damage by propagation of vibrations and noise to the surroundings. However, since solution methods vary depending on the cause of pulsations, it is important to first identify the cause.

Pulsations in simple pipelines that do not include pumps or hydraulic turbines were discussed in Section 4.1, while pulsations caused by flow velocity fluctuations were presented in Section 5.1.

5.2.2 Explanation of phenomena

5.2.2.1 Piping section

As a method for the numerical simulation of steady pulsation in a piping section, the transfer matrix method is most commonly used. Amplitudes of pressure P and flow rate Q are taken as components of the state vector. Ignoring damping effects, the following equation relates quantities at two arbitrary points:

$$\begin{bmatrix} P(x) \\ Q(x) \end{bmatrix} = \begin{bmatrix} \cos(\omega x/c) & -j(\rho c/A)\sin(\omega x/c) \\ -j(A/\rho c)\sin(\omega x/c) & \cos(\omega x/c) \end{bmatrix} \begin{bmatrix} P(0) \\ Q(0) \end{bmatrix} \tag{5.5}$$

Table 5.8 Methods to reduce piping vibration

	Measures	Description/feature	Note for practice
1.	Adding support to main piping	• In order to avoid mechanical resonance, determine the interval between piping supports so that the mechanical natural frequency of the piping is sufficiently away from the pulsation frequency in the piping • Install the piping support to avoid the out-of-plane vibration • For the piping support, take thermal stress into consideration	• For a piping system around reciprocating compressors, the mechanical natural frequency of piping should be 6 times or more of the rotational speed of reciprocating compressors • Note that the out-of-plane stiffness at a bend section is weaker than the stiffness at a straight section • A piping support that is sufficiently flexible to thermal stress and sufficiently stiff relative to vibration is used in a practical design [56]
2.	Adding damping to main piping	• Install a hydraulic cylinder and a dynamic damper to the section where the piping vibration is strong, in order to add damping to the piping [56,57]	• Some maintenance work may be required against time-dependent changes in hydraulic cylinder or dynamic damper
3.	Reinforcing drain, vent, etc.	• In order to enhance the stiffness, use a boss with a thicker wall than usual for the small size branch pipe such as a drain or a vent • Reinforcing with a brace is effective to prevent vibration	• Pay attention to the branch pipe with instruments such as pressure gauges, since it is likely to vibrate
4.	Supporting valves	• For heavy equipment such as valves, support from the foundation and/or a structure with sufficient stiffness	• The valves' supports should be designed sufficiently strong, since pulsation always has an impact on them

where A, ρ, c, and ω are the inner cross-sectional area of the pipe, the fluid density, the pressure wave propagation velocity, and the angular frequency, respectively.

Because the pressure wave propagation velocity c strongly depends on the state of the fluid, caution is required. When the fluid is a liquid, the elastic deformation of the pipe cannot be neglected. The following well-known equation, in

Table 5.9 Example cases

	Machinery type	Vibration/pulsation problems	Possible causes	Measures
1.	Reciprocating compressor: gas recovery plant in an ironworks [23]. Refer to Figs 5.6 and 5.7	• Malfunction of instruments (pressure gauges)	• Acoustic resonance with a frequency of 2 times the compressor rotational speed • The maximum allowable level of pulsation to the pressure gauge was as severe as 3%	• The possible problem was avoided by changing the pressure gauge location
2.	Reciprocating compressor: petroleum refinery [59]	• Bending vibration of piping • The piping vibrated 900 μm with a frequency of 2 times the reciprocating compressor rotational speed	• A large pulsation (17 Hz) at a frequency 2 times the reciprocating compressor rotational speed occurred	• 18 orifices were installed. Their diameter ratio to the pipe diameter was 1/2 • The piping vibration was decreased from about 900 to 350 μm
3.	Reciprocating compressor: reactor [60]	• Vibration of the reactor and the structure • The pressure pulsation in the piping connected to the reactor was 0.92 psi, and the vertical vibration amplitude of the structure was 261 μm	• The gas blown into the bottom of the reactor from a reciprocating compressor flowed intermittently due to pressure pulsation; the forced excitation affected the reactor vertically	• An orifice was installed, and the structure was reinforced • The pressure pulsation was decreased from 0.92 to 0.19 psi, and the piping vibration was decreased from 261 to 86 μm

(Continued)

Table 5.9 (Continued)

	Machinery type	Vibration/pulsation problems	Possible causes	Measures
4.	Reciprocating compressor: drains and vents in a piping system [61,62]	• High harmonics pulsations occurred when measuring pressure fluctuation using the drain and/or the vent	• A 1/4 wavelength resonance in a branch pipe such as a drain	• An orifice was installed at the inlet of the drain, which area ratio was 1/4. Note that installing an orifice may decrease the expected pressure fluctuation in the measurements
5.	Roots blower: direct reduction steel plant [29]	• Pressure pulsations with a component of 4 times (the fundamental 29 Hz) and 12 times (third order, 87 Hz) the rotational speed occurred	• Intermittent discharge flow from the Roots blower	• A large-sized silencer (2.2 m diameter, 6 m height) was installed
6.	Roots blower: chemical plant [63]. Refer to Figs. 5.8 and 5.9	• Radial (shell mode) vibration of piping • The blower outlet piping vibrated 100 μm radially peak-peak	• The piping was excited by the acoustic resonance in piping	• A 1/4 wavelength pipe (side branch) was installed between the roots blower and the silencer, which was tuned at 55 Hz (pulsation frequency)
7.	Expansion turbine: chemical plant [64]. Refer to Figs. 5.10 and 5.11	• Bending vibration of piping • Piping vibration occurred at the partial load operation (80%) of the turbine • The piping vibrated with the fourth acoustic natural frequency of 31 Hz, and amplitude of 200 μm or more	• Possibly caused by an unstable flow due to interference of the swirling flow at the turbine outlet with the secondary flow at a piping bend section	• Installing a honeycomb (discharge angle coefficient: 0.25) at the turbine outlet eliminated the vibration

8.	Automobile diesel engine [65]	• A bellows supplying fuel to the engine vibrated several centimeters when the engine was operated above a certain rotational frequency	• Parametric vibration caused by a periodic variation of the bellows stiffness due to internal pressure pulsations	• An accumulator that was tuned to the pressure pulsation frequency at the driving rotational speed range was installed • The pressure pulsation was decreased to about l/20 of the original level
9.	Truck engine [66]	• In the driver's compartment of a 4-ton truck, a noise around 100, 250, and 400 Hz occurred when the engine was operated at 2000 rpm or more	• The peak components of the noise spectrum at the air intake were at the same as frequencies • The design change of the air cleaner and an air intake duct caused the elimination of insertion loss of the air intake system	• The insertion losses of the target frequency were improved about 20 dB by installing a resonator to both the air intake duct and the air cleaner

which the elastic deformation of the pipe is taken into account, is used in such cases [68,69]:

$$c = 1/\sqrt{\rho(1/K + d/Eh)} \tag{5.6}$$

where K is the bulk modulus of the fluid, E the Young's modulus of the pipe, d the average diameter of the pipe, and h the pipe wall thickness.

If air bubbles are mixed in the liquid, the velocity of the pressure wave varies significantly with the void fraction because of air compressibility. The wave velocity in this case is given by:

$$c = 1/\sqrt{\{\alpha_1/p + (1-\alpha_1)/K + d/Eh\}\{\alpha_1\rho_g + (1-\alpha_1)\rho\}} \tag{5.7}$$

where α_1, p, and ρ_g are the void fraction, the mean pressure, and the density of the gas, respectively [70]. Measured values of c are presented in Fig. 5.12 [71]. Within the same pipeline, the value of c may vary along the pipe length. For example, the propagation velocity of the pressure wave in a suction pipe usually differs from that in a discharge pipe.

5.2.2.2 Forced pulsations caused by positive displacement pumps

Most positive displacement pumps are powerful sources of pulsations. Pulsations in the suction and discharge pipes synchronize with each other. However, in many cases, it is not necessary to take into consideration the coupling between pulsations in the suction and discharge pipes because a pressure wave cannot propagate through a positive displacement pump. A positive displacement pump acts as a source with known capacity.

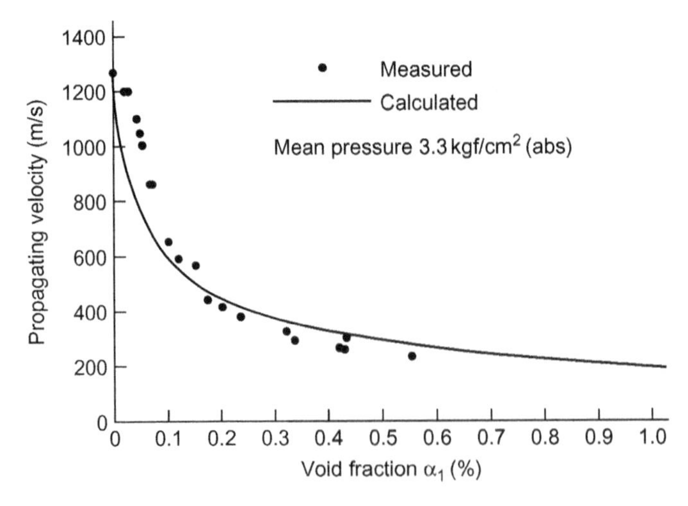

FIGURE 5.12

Influence of void fraction on pressure wave velocity in water pipes [71].

Let us consider a discharge pipe as an example. Let $t_{ij}(x)$ and L_d be elements of the transfer matrix and the length of the discharge pipe, respectively. The subscript d is given to the quantities at the outlet of the pipe and the subscript p to the quantities at the pump outlet. If the outlet end of the pipe is open, $P_d = 0$. This leads to the relation:

$$P_p = -Q_p \cdot t_{12}(L_d)/t_{11}(L_d)$$

Thus, when the discharge flow rate from the pump Q_p is known, the pressure and flow rate amplitudes at an arbitrary point can be obtained from the following equation:

$$\begin{Bmatrix} P(x) \\ Q(x) \end{Bmatrix} = \begin{bmatrix} t_{11}(x) & t_{12}(x) \\ t_{21}(x) & t_{22}(x) \end{bmatrix} \begin{Bmatrix} -t_{12}(L_d)/t_{11}(L_d) \\ 1 \end{Bmatrix} Q_p \tag{5.8}$$

The discharge flow from an actual positive displacement pump is intermittent. The pressure in the discharge pipe therefore contains many higher harmonic components. Equation (5.8) is valid for a single frequency. It is therefore necessary to check for resonance not only at the fundamental frequency but also at higher harmonics. Some accidents caused by resonance at higher harmonic frequencies have been reported. In many actual cases, two or more piston pumps are operated in parallel to reduce pulsation, as shown in Fig. 5.13.

The API Standard 674 [72] has detailed regulations on reciprocating pumps. The API standards are compiled data of the knowledge and experiences that have been accumulated by manufacturers and users to date. In the chapter on basic design, the minimum life of equipment, including auxiliary machines, is specified as 20 years. Furthermore, it is prescribed that machines must be designed and manufactured so that they can be operated continuously for at least 3 years.

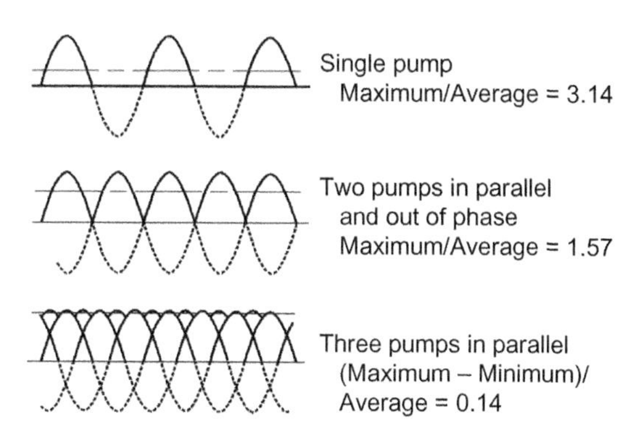

FIGURE 5.13

Illustration of pulsation reduction by means of parallel operation of multiple pumps.

5.2.2.3 Forced pulsations caused by turbo-pumps

5.2.2.3.1 The primary features of forced pulsations

Turbo-pumps may act as sources of forced pulsations. The frequency f (Hz) of forced pulsations is generally related to the rotational speed N (rpm) as follows:

$$f = mN/60$$

where m is the harmonic order.

An example of the frequency spectrum is shown in Fig. 5.14 [73]. Z is the number of impeller blades. Pulsations of order $m = Z$, $2Z$, and so on are inevitably generated, even if the pump is manufactured and operated ideally. However, pulsations of order $m = 1$, 2, etc. may also be generated due to various causes (e.g. eccentric rotation of the pump shaft). A pressure wave can propagate through a turbo-pump. Thus, pulsations in the suction and discharge pipes may be coupled. It is therefore necessary to treat the pulsation as a phenomenon of the whole pipeline, including both the suction and discharge pipes.

5.2.2.3.2 Piping system natural frequency

It is necessary to determine the resonance frequency (i.e. the natural frequency of the whole piping system). If the pump characteristic can be assumed linear, the transfer matrix between both ends of the pump can be written as follows [74]:

$$\begin{Bmatrix} P \\ Q \end{Bmatrix}_{\text{out}} = \begin{bmatrix} G_{11} & G_{12} \\ G_{21} & G_{22} \end{bmatrix} \begin{Bmatrix} P \\ Q \end{Bmatrix}_{\text{in}} + \begin{Bmatrix} P \\ Q \end{Bmatrix}_{\text{S}} \tag{5.9}$$

This formulation is known as a two-port model. The last term on the right-hand side contains the amplitudes of the pressure and the flow rate due to pump excitation. Currently, the matrix elements $G_{11} \sim G_{22}$ can be determined only by

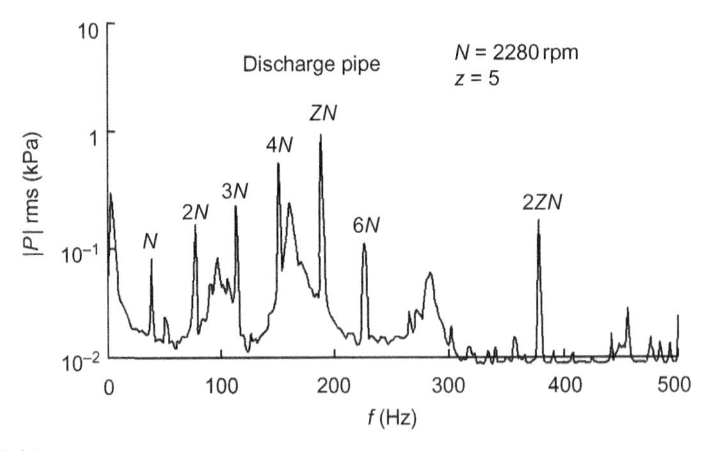

FIGURE 5.14

An example of the frequency spectrum of pulsations caused by a pump [73].

measurement at every excitation frequency. However, if they are known, the transfer characteristics of a pump can be expressed in a form including the mass effect and the resistance effect. In a specific example, the simplification $\{P_S, Q_S\}^T = 0$ in Eq. (5.9) is adopted in order to identify the matrix elements and to calculate the natural frequencies of the pipeline [73,75]. It is then assumed that the pulsation source is placed at the upper suction end.

The analytical model of the pump can be replaced by a simpler one [73]. For example, the pump can be replaced by a pipe element with an equivalent length and an equivalent diameter. In this method, the transfer matrix is used to determine the model parameters. It is assumed that the pulsation source is placed at one end as shown in Fig. 5.15. The equivalent length l_{eq} and diameter d_{eq} corresponding to the pump are determined so as to match the amplitude distribution of pressure and flow rate obtained experimentally.

5.2.2.3.3 Flows in pumps and pulsations

The effect of a centrifugal pump as a source of pulsations is closely related to the flow field within the pump. This means that the strength of the pulsations strongly depends on the shape and dimensions of the pump. It is therefore important to consider these parameters at the design stage. In the review paper [76] the effects of shapes and dimensions of pump components are summarized. In the paper, more than 50 recent references are listed. Based on this work, the items considered helpful for pump design are presented in Tables 5.10 and 5.11.

5.2.2.4 Self-excited pulsations of pipelines containing turbo-pumps

The kinetic energy of a flow can be converted into pulsation energy through various mechanisms. Some pulsations can be identified as self-excited vibrations of

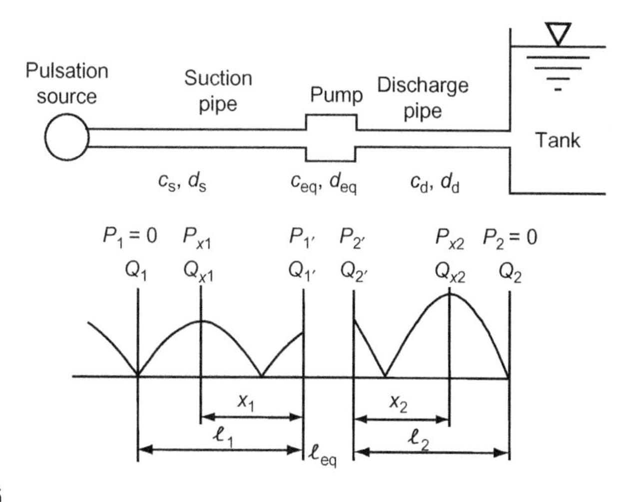

FIGURE 5.15

Pipe element model equivalent to a pump [73].

Table 5.10 Characteristics of noise generation mechanisms in centrifugal pumps [76]

Mechanism of noise generation	Flow/pressure field characteristics	Parameters controlling design
Impeller blade and volute tongue interaction	Impingement of wake flow on volute tongue	Lip clearance, impeller blade trailing edge, and volute tongue geometry
	Oscillation in stagnation pressure near volute tongue	Lip clearance, blade and volute tongue geometry
	Unsteady flow recirculation and separation at tongue surface	Lip clearance, volute tongue geometry (tip radius)
Impeller blade and flow interaction	Discontinuity in pressure field between discharge and suction sides of impeller blade	Impeller blade trailing edge geometry
	Local flow separation on impeller blade surface	Smooth surface finish
Non-uniform outflow from impeller	Flow and stagnation pressure oscillation at pump discharge	Impeller blade geometry
	Jet flow impingement on volute tongue	Lip clearance, impeller blade, and volute tongue geometry
Vortex interaction with flow	Extended trains of vortex structures near volute tongue and blade (may coincide)	Impeller blade trailing edge, impeller blade, and volute tongue geometry

the fluid columns in pipelines. In general, the frequency of self-excited vibrations has no special relation with the rotating speed of the pump. It is, instead, closely related to the natural frequency of the system.

A classic type of self-excited vibration in pipeline systems containing a turbopump is surging. The illustrative model of single-degree-of-freedom surging by Fujii [77] is described next. The model system (Fig. 5.16) includes a pump and a tank having pressure head H and a cross-sectional area A. Suppose that the characteristic of the system is as shown in Fig. 5.17. The average flow rate at the operating point (i.e. a position of equilibrium) is set at Q_0. Using the notation $Q_P - Q_0 = q$, the characteristic equations of the system are as follows:

$$H + m\frac{dq}{dt} = f(q),\ A\frac{dH}{dt} = q$$

where the coefficient m is related to the inertia of the system, and $f(q)$ is the characteristic of the pump near the operating point. Eliminating H from these two equations, the following equation is obtained:

$$m\frac{d^2q}{dt^2} - a(q)\frac{dq}{dt} + \frac{1}{A}q = 0 \tag{5.10}$$

Table 5.11 Effect of pump design on pressure pulsation [76]

Design parameter	Recommendations		Comments
	Hydraulic design	Pulsation control	
Lip clearance (in percent of impeller radius)	5–10%	~10%	Large clearance may lead to excessive flow circulation and strong pulsation May have significant impact on pump performance
Volute tongue geometry	Tip radius 6.3–25 mm	Large tip radius Inclined cutwater edge (3D design)	Elimination of the effect of flow separation Prolong the time of blade/cutwater passage
Number of impeller blades		High number of impeller blades.	Select number of blades to avoid system acoustic resonance
Twisting and staggering of impeller blades	Twist angle 15–22º		Prolong the time of blade/cutwater passage
Impeller blade trailing edge	Thickness must be sufficient to sustain hydraulic load	Profile to reduce wake width	Marginal effect on pump performance
Pump volute geometry	Uniform and smooth flow passage	Increase internal impedance	High impedance pump must be placed near pressure node

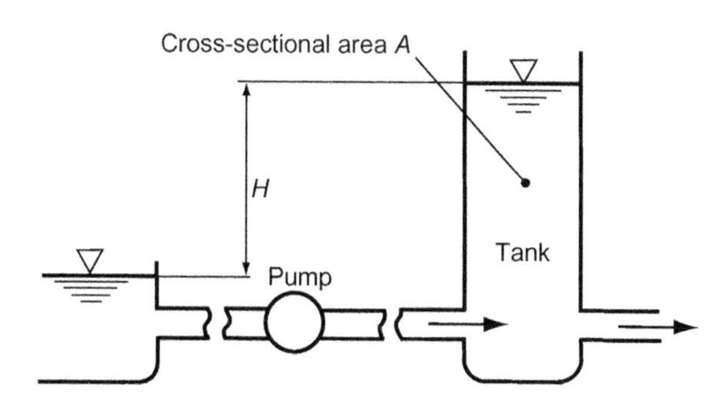

FIGURE 5.16

Pipeline system including a pump and a tank.

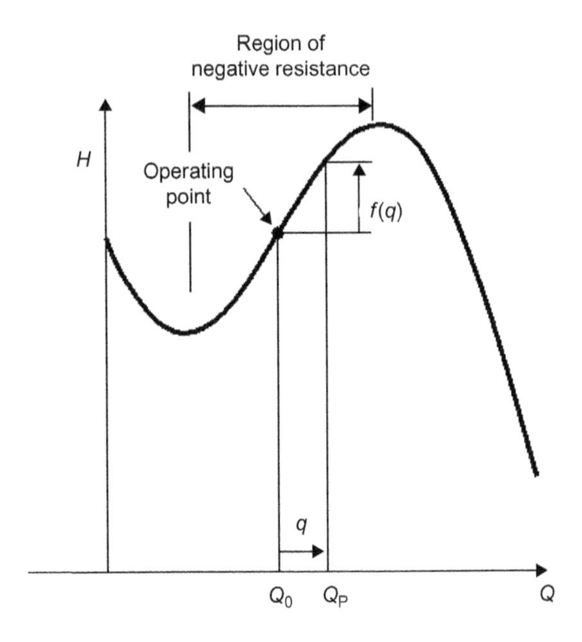

FIGURE 5.17

Example of a pressure head versus flow rate curve for a pump having a negative resistance characteristic.

where $a(q) = \mathrm{d}f(q)/\mathrm{d}q$. Small oscillations expressed by the solution of the differential equation (5.10) around the point of $q = 0$ grow if $a(q) > 0$ and decay if $a(q) < 0$. In other words, if the operating point of the system is located in a positive slope region to the right of the Q–H curve, the operating point is unstable. In such a case, large self-excited oscillations can occur. If pulsation occurs, L, the energy per cycle, will be accumulated in the system. L can be estimated by the following equation:

$$L = \oint a(q) \frac{\mathrm{d}q}{\mathrm{d}t} \mathrm{d}q$$

Depending on whether L is positive or negative, pulsations grow or are attenuated. If $L = 0$, the pulsation reaches a stationary state and a limit cycle is traced on the Q–H plane.

Whether pulsations are due to surging is judged by whether or not the operating point of the system is in the positive slope region of the Q–H curve. However, even if the operating point is within this region, surging may not necessarily occur. Surging may occur easily in a pipeline in which a tank or an accumulator is located downstream of the pump and a resistive component, such as a flow-regulating valve, is located further on downstream. This means that surging requires a potential energy accumulating element corresponding to the term q/A in Eq. (5.10). When the fluid contains many air bubbles, the bubbles can play the same role as installed accumulators. More detailed research has been done on surging [78].

Self-excited vibrations other than surging can also occur. For example, it has been reported that self-excited pulsations occurred in the low flow rate region of a high-pressure multi-stage pump [79,80]. Since the operating point of the system was located in the portion of the Q–H curve having a negative slope, it was determined to be a type of self-excited vibration different from surging. This case may be analyzed as follows. A pipeline is divided into several pipe elements. A model for analysis is developed using the transfer matrix. Laplace transformation of the transfer matrix is carried out next. Thus, the elements of the matrix contain the operator s. A frequency equation is obtained from the boundary conditions at both ends of the pipeline. Values of s which satisfy the frequency equation are obtained by a numerical calculation. s represents a complex natural frequency and has the form $s = \alpha + j\beta$. The system is unstable when $\alpha > 0$. The imaginary part of s, that is, β, is the angular natural frequency.

In the low-frequency region in which the wavelength of the pressure wave is much longer than the pump size, Eq. (5.9), which expresses the transfer characteristics of the pump portion, can be simplified as follows:

$$\begin{Bmatrix} P \\ Q \end{Bmatrix}_{\text{out}} = \begin{bmatrix} 1 & Z_P \\ 0 & 1 \end{bmatrix} \begin{Bmatrix} P \\ Q \end{Bmatrix}_{\text{in}} \tag{5.11}$$

In this equation, it is assumed that $\{P_S, Q_S\}^T = 0$. Z_P is called the pump impedance. It is identified by vibration tests and the transfer matrix of the line system. The real and imaginary parts of Z_P, $\text{Re}(Z_P)$ and $\text{Im}(Z_P)$, are the pump resistance and the pump inertance, respectively. It has become clear through numerical experiments that the stability of the system is governed mainly by the pump resistance $\text{Re}(Z_P)$. The system is usually stable since the pipeline has positive resistance. However, if $\text{Re}(Z_P)$ surpasses the pipeline resistance and if the value of α becomes positive, pulsations will arise. Figure 5.18 shows an example of the measured relationship between the pump resistance and the rotating speed [79]. Detailed information on measurement methods for pump characteristics, including Z_P, can be found in the Refs. [75,80,81].

5.2.2.5 Vibrations and pulsations caused by suction of vortices into turbo-pumps

When the suction pipe entrance of a centrifugal pump is placed in a tank with a free surface, a whirlpool or vortex in the tank may be sucked into the pipe [82,83]. A whirlpool is accompanied by a core of air extending from the free surface; air is inhaled by the pump when the core is deep. Even in the case where only an underwater vortex arises, low pressure in the vortex core causes a kind of cavitation by dissociating dissolved air from water. The dissociated air may be

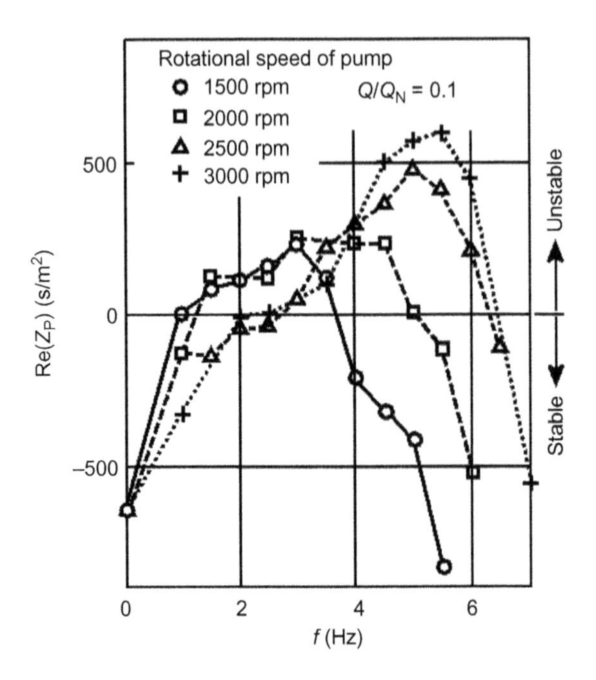

FIGURE 5.18

Relation between pump resistance and rotational speed [79].

inhaled into the pipe. The vibration or pulsation problems below are caused by this phenomenon (refer to Fig. 5.19):

1. Swirling flow which arises in a suction pipe and contains many air bubbles causes an unbalanced torque to the pipe. This can be a cause of structural vibrations and noise.
2. An air column is interrupted by the turbo-pump impeller. This causes a change of the fluid force acting on the impeller. Resulting vibrations and pressure pulsations may damage the impeller and piping [84,85].
3. In some cases, flow may vary significantly and seem almost intermittent [83,84,86].

The countermeasure to the foregoing problems is to prevent whirlpools and air inhalation. To achieve this, the following proposals have been put forward:

1. The clearances surrounding the suction pipe should be small.
2. Installation of a cross-shaped guide vane at the front or the inside of the suction pipe is recommended.
3. A free surface near the suction port in a tank should be avoided. For this purpose, a roof inclined on the water surface is effective.
4. A raft-like floating structure installed around the pipe is effective to prevent the formation of a whirlpool.

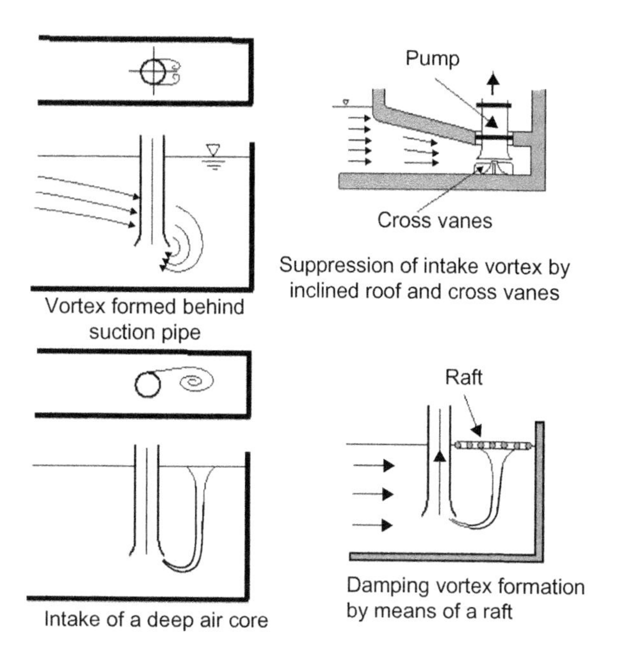

FIGURE 5.19

Suppression of intake vortex.

5.2.2.6 Other types of pump vibrations

When a loss of coolant accident occurs in the pump and pipeline system for heat transport in a nuclear reactor, the void fraction will rise because of reduction of suction pressure. This introduces two-phase flow instability. It is believed that the observed pump instability, intermittence of the flow due to separation of air bubbles, density wave, and acoustic sound resonance are factors related to this phenomenon. There is continuing research on the interactions of these factors [87]. A method for analyzing a primary cooling water loop system that contains a pressurized water reactor, a steam generator, and a primary cooling water pump using transfer matrices has also been reported [88]. It is supposed in the analysis that the flow in the cooling water loop system is almost single-phase flow but partially two-phase. The transfer matrices used in the analysis are different from those described thus far.

If cavitation occurs in the flow through a pump, strong vibrations will occur in the pump and, in the worst case, may result in breakage. Cavitation is strongly dependent on the properties of the fluid conveyed through the pump. It may be mentioned that the cause of the Japanese H-II rocket launch failure in November, 1999 was cavitation-induced failure of a fuel system pump.

5.2.2.7 Hydraulic turbines

In Francis and Kaplan turbines, strong vibrations can occur in the main parts of the turbine and the draft tube.

5.2.2.7.1 The blade passing frequency

The wakes behind the guide vanes produce a non-steady flow in Francis turbine runners [89]. This unsteady flow causes pressure pulsations and stress fluctuations in the runner blades of all Francis-type turbines. Stress concentration may result depending on the shape of each part of the turbine rotor; the result may be fatigue failure and cracking. The dominant frequency is the rotational frequency multiplied by the number of guide vanes. Higher frequencies caused by shedding of Kármán vortices are also observed. The latter depends strongly on the cross-sectional shape of the trailing edge of the guide vane. The results of examination show that the trailing edge must not be cut off at right angles to the flow direction; ideally the angle should be about 45 degrees. It should also be noted that a semicircle shape is perhaps the worst geometry of the trailing edge from the viewpoint of prevention of vibrations (see Fig. 5.20).

5.2.2.7.2 Vibrations and pulsations in hydraulic turbine draft tubes

Vibrations and pressure pulsations in hydraulic turbine draft tubes may arise under partial load operation [90]. In general, the central and circumferential parts

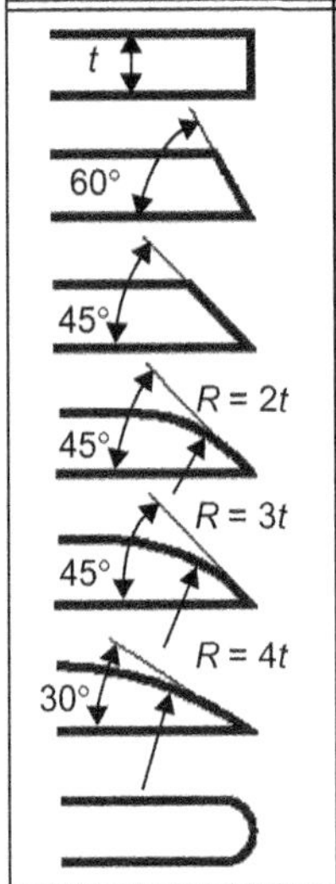

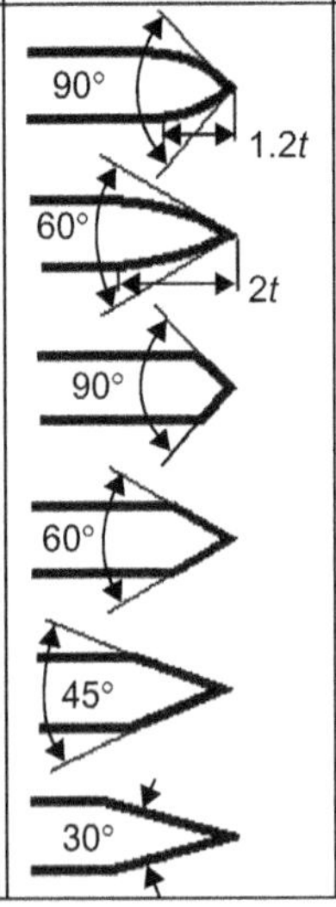

Trailing edge	A	B	Trailing edge	A	B
t	100 (100)	100	90°, 1.2t	(0)	
60°	(48)		60°, 2t	(0)	
45°	38 (20)	112	90°	190 (230)	96
45°, $R = 2t$	3 (0)	131	60°	380 (360)	83
45°, $R = 3t$	0	149	45°	43	117
30°, $R = 4t$	0	181	30°	0	159
	(260)				

The frequency may be calculated by the following approximate formula:

$$f = 0.19 \frac{B}{100} \frac{v}{(t + 0.56)} \text{ (Hz)}$$

v = water velocity (m/s), t = plate thickness in mm

B = constant depending on geometry (see table).

FIGURE 5.20

Influence of trailing edge on vibrations (case of plane blade in parallel flow) [90].

of the flow that enter a draft tube from a runner, swirl as a forced vortex and a free vortex, respectively. The vortex core has very low pressure; thus, air cavities appear. On the other hand, the draft tube acts as a diffuser by causing downstream pressure recovery. Since the pressure rise is large in the core when a swirling flow enters the draft tube, a dead water core appears. As a result, cavities coil around the perimeter of the dead water core. Thus, a spiral-formed cavitation appears. When the swirling flow accompanied by cavitation rotates, associated pressure fluctuations may also cause structural vibration. This is the so-called draft tube surging.

The cavity cores may grow along the axis of the draft tube downstream, and they may disappear rapidly near the elbow portion of the draft tube because of increasing pressure. This causes intense pressure fluctuations. As a cure, the taper angle of the draft tube entrance portion is enlarged so that the static pressure in the vortex core may rise quickly downstream. This is effective in eliminating cavitation and suppressing pressure pulsations.

5.2.2.7.3 Self-excited pulsations

There are cases where pressure pulsations in pipelines containing a hydraulic turbine may be regarded as a self-excited vibration. An example is given in [91] where fluctuations of pressure and electric power output were observed in low output states of a hydraulic turbine–electric generator system. When the guide vane opening is small, the circumferential component of the flow into the runner becomes large in comparison with the radial component. The flow alternately separates and re-attaches to the runner blades. As a cure, the guide vanes were opened and closed asynchronously. In the low output state, only a certain guide vane was opened and all the others closed. The direction of the flow into the runner can then be stabilized (see Fig. 5.21).

It has also been reported that another kind of self-excited pulsation based on the characteristic of the check valve with a flexible diaphragm occurred in a hydraulic turbine–pump–pipeline system for irrigation [92] (Fig. 5.22).

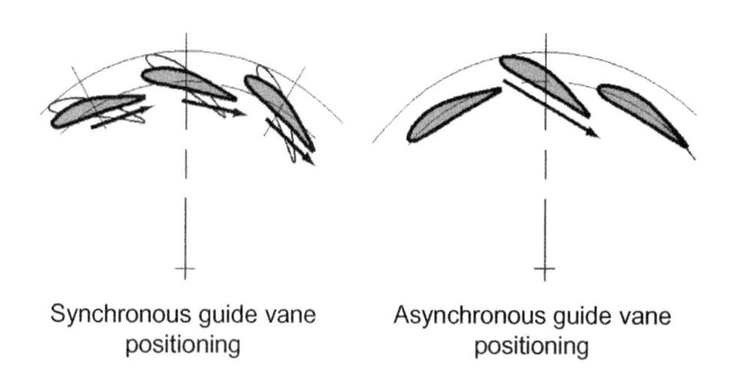

FIGURE 5.21

Synchronous and asynchronous guide vane positioning at partial load [91].

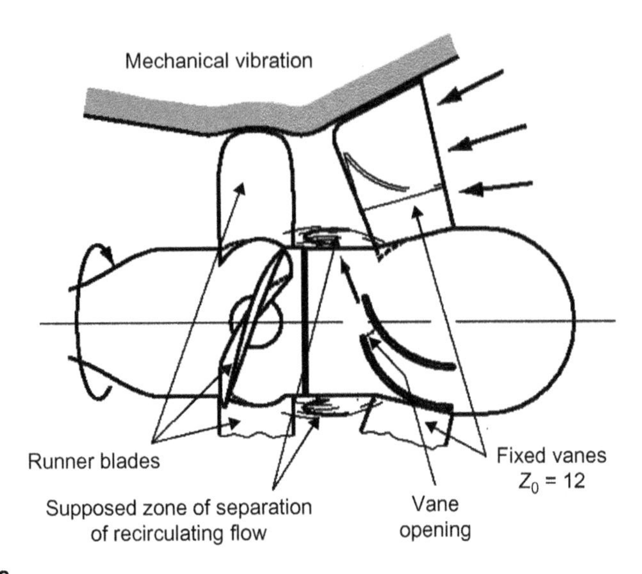

FIGURE 5.22

Vibration of an axial flow turbine with fixed guide vanes [93].

5.2.3 Vibration problems and suggested solutions

If pulsations are due to forced vibrations, the first option is to avoid resonance. If possible, the operation condition such as the rotation speed of the pump should be changed. Installation of an air chamber, an accumulator or a side branch, and revision of the number of centrifugal pump blades are possible solutions accompanied by design changes. However, installation of an air chamber may cause surging at low flow rates. Moreover, since the vortices produced at the connection of a side branch with the main pipe may cause new vibration and pulsations, installation of a side branch needs caution. In small-scale pipelines such as in buildings, pipes may be made of flexible materials to suppress pulsation or to prevent vibration transmission. A silencer may also be inserted [94]. A silencer consists of a series of expansion chambers and choking plates, much like an internal combustion engine muffler.

In the case of self-excited vibration, the fundamental solutions involve changing the system parameters to displace the operating point into the stable region. In order to increase the apparent flow rate, measures such as moving a component to the suction side from the discharge side are sometimes taken. One of the cures for surging problems is to make the Q–H curve slant downward to the right. Another solution is to install resistance components such as valves downstream of the pump. The latter cure prevents the occurrence of a negative resistance domain in the system.

Other examples of problems and solutions, in addition to the cases illustrated so far, are summarized in Table 5.12.

Table 5.12 Examples of problem cases for pumps and hydraulic turbines

	Machine	Phenomenon	Causes	Cure
1.	Three reciprocating pumps in parallel operation [95]	When the temperature in the suction pipe reached 200°C or higher, unusual vibrations occurred	Pressure fluctuation by rapid collapse of steam bubbles and of gas bubbles which are separated from water and resonance	Accumulator was formed by enclosing nitrogen gas in upper part of the volume bottle
2.	Swash plate axial piston pump for hydraulic excavator [96]	Rapid increase of pressure pulsations near 2000 rpm	Simulations indicated resonance	Installation of a side branch
3.	Vertical mixed flow sewer pump [97]	Large pulsations accompanied by a water column separation in piping	Resonance of the pulsations with torsional vibrations of the shaft connecting motor and pump	In some cases, piping length was changed; in others, the radius of the shaft was increased
4.	Centrifugal pump for waste treatment facility [98]	Vibration accompanied by unusual noise occurred in the tank installed on suction-side	Resonance of liquid column in the tank with pulsations	The number of impeller blades was changed
5.	Multi-stage centrifugal pump feeding water to boiler [99]	Vibration in which the ZN component is dominant	Forced vibration by the fluid force generated in interference of the impeller blade and the guide vanes	The angle of the fourth-stage guide vanes was changed by 20 degrees in the anti-rotation direction from the third-stage guide vanes

(Continued)

Table 5.12 (Continued)

	Machine	Phenomenon	Causes	Cure
6.	Multi-stage centrifugal pump [100]	Vibration at 80% of shaft speed	Force on an impeller caused by swirling flow	Reduction of the eccentricity between diffuser/blower and impeller Installation of a swirl breaker
7.	Vertical mixed flow pump [101]	Vibration with a broad band spectrum	Cavitation at the valve placed downstream of the pump	A perforated-plate orifice was installed downstream of the valve so that the pressure at the valve increased
8.	Axial-flow hydraulic turbine [93]. (see Fig. 5.12)	Vibration of housing and noise with irregular frequency of 140–176 Hz	It was presumed that flow separates in downstream portion of the guide vanes	The guide vanes were shortened
9.	Francis turbine [102]	Large pressure fluctuations	A cavitation core grows towards the elbow where it disappears	The taper suction pipe entrance was increased

5.3 Pressure surge or water hammer in piping system

5.3.1 Water hammer

The term water hammer is used synonymously to describe transient flow or pressure surge in liquid conveying pipelines. Use of the term 'water hammer' is customarily restricted to water.

Figure 5.23 schematically shows the propagation of a pressure wave caused by instantaneous stoppage of liquid flow at a downstream inline valve. The positive pressure wave front at the valve propagates at the speed of sound towards the upstream reservoir, where it is reflected. Providing that the reservoir pressure is unchanged, there results an unbalance condition at the upstream reservoir at the instant of arrival of the positive pressure wave. The liquid starts to flow backward and a wave front of negative pressure forms and propagates downstream toward the valve. Thus, the pressure wave travels repeatedly between the valve and the upstream reservoir and gradually attenuates due to frictional damping. Downstream of the valve, when the pressure drops to the vapor pressure of the liquid, vaporization occurs and the vapor void volume or vaporized region grows. This phenomenon is referred to as liquid column separation and leads to a sudden pressure rise when the vapor void collapses. Such a pressure surge occurs not only when a valve closes but also due to pump trips and pump start-ups.

5.3.2 Synopsis of investigation

Although there is an example of pipeline transportation and water supply facility constructed in the northern ancient city of Habuba Kabira in Syria circa 3500–3000 BC, it was during the renaissance era that DA Vinci used the simple formula for flow rate as a product of velocity and pipe sectional area [103]. In the middle of the nineteenth century, unsteady liquid flows and pressure waves were mathematically studied using the elastic theory of liquids. Menabrea [104] and

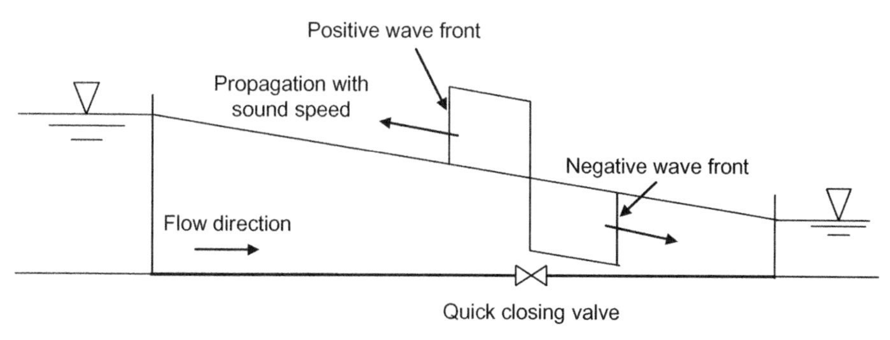

FIGURE 5.23

Wave propagation due to instantaneous valve closure.

Joukowski [105] proposed that the pressure rise due to instantaneous valve closure can be expressed by the equation:

$$J = \rho \cdot c \cdot \Delta V \tag{5.12}$$

where J is the pressure rise due to the change of fluid velocity ΔV; ρ is the fluid density, and c the wave speed. Allievi [106] applied the method for the slow valve closure to render the assumption of linear change of flow at valve. Until the early 1930s, the mathematical model derived by integration of Joukowsky's equation ignoring pipe friction was used. From the early 1930s to the early 1960s a graphical method taking account of pipe friction, iteratively, with the model above was utilized by plotting on the head-velocity diagram [107]. With the development of high-performance computers, these methods are no longer used but they are useful for understanding the mechanism of water hammer.

There are several numerical computation methods, including the method of characteristics (MOC), the semi-implicit two-step Lax–Wendroff method (2-LW) [108], the fully implicit method [109], and the finite element method [110,111]. MOC is the most commonly used method for fluid transient analysis, especially for liquid or water hammer analysis. The advantage of MOC is the ease of handling complex boundary conditions such as pumps and compressors. Also, MOC guarantees a numerically stable solution providing that the Courant condition is satisfied [108]. Depending on the nature of the problem, other methods may be used. For example, 2-LW can be applied to solve impulsive shock wave problems and column separation problems [112]. A fully implicit method is advantageous when solving slow transient problems that may occur in a gas pipeline [113].

5.3.3 Solution methods

5.3.3.1 Method of characteristics (MOC)

This section describes MOC [114–116] as a calculation method for transient flow in liquid pipelines, its application to column separation problems, and practical aspects of frequency-dependent friction.

The equations governing unsteady fluid flow in pipelines consist of two partial differential equations of momentum and continuity. A general solution to the partial differential equations is not available. However, these hyperbolic-type partial differential equations can be mathematically transformed by MOC into two ordinary differential equations that are valid along particular characteristic lines C^+ and C^-, in time and space [117]. These two ordinary differential equations can be recast into the following finite difference equations:

$$C^+: c \cdot (u_P - u_R) + g \cdot (h_P - h_R) - g \cdot u_R \cdot \Delta t \cdot \sin\alpha + \frac{f \cdot u_P \cdot |u_R|}{2 \cdot D} \cdot c \cdot \Delta t = 0 \tag{5.13}$$

$$C^{-}: c\cdot(u_P - u_S) - g\cdot(h_P - h_S) + g\cdot u_S\cdot\Delta t\cdot\sin\alpha + \frac{f\cdot u_P\cdot|u_S|}{2\cdot D}\cdot c\cdot\Delta t = 0 \quad (5.14)$$

where f is the Darcy–Weisbach friction factor, g the gravitational acceleration, c the wave speed, α the pipe slope and u the fluid velocity. The piezometric head h is defined by $h = p/\gamma + z$, where γ is the unit weight of the fluid, z the elevation of the pipe above a reference datum, D the pipe diameter, and Δt time step.

Figure 5.24 shows a schematic visualization of the characteristic lines C^+ and C^- along which the corresponding finite difference equations are valid. Each characteristic line physically represents a forward or backward propagating wave; $dx/dt = c + u$ for C^+ and $dx/dt = c + u$ for C^-. Equations (5.13) and (5.14) can be solved at a grid point P to obtain the two unknowns of u_P and h_P.

The MOC for a given time step, requires interpolations of (u_R, h_R) at R and (u_S, h_S) at S since the slopes of C^+ and C^- are different depending on local fluid velocity and local wave speed. The Courant condition $\Delta t < \Delta x/(c + u)$ is necessary to obtain a stable numerical solution.

The unknowns u_P and h_P in a pipe can be explicitly calculated; however, a specific boundary condition is necessary at the pipe ends where only one of the equations C^+ or C^- is available. The simplest boundary condition is a reservoir having constant liquid level for which the boundary condition is: h_P = constant. Substituting this condition into Eqs. (5.13) or (5.14), the unknown u_P at the reservoir can be obtained.

One of the most complicated boundary conditions is that of a pump, where four unknowns, that is $(u_P, h_P)_{SUC}$ on the suction side and $(u_P, h_P)_{DIS}$ on the discharge side must be solved for using the following four equations: (i) the continuity equation between suction and discharge sides, (ii) pump characteristic curve of flow-head at rated speed, (iii) the wave equation (5.13) at suction side, and (iv) the wave equation (5.14) at discharge side. Calculation of the behavior during power failure requires three additional equations besides the pump characteristic curve of flow-head at rated speed; these are (i) the

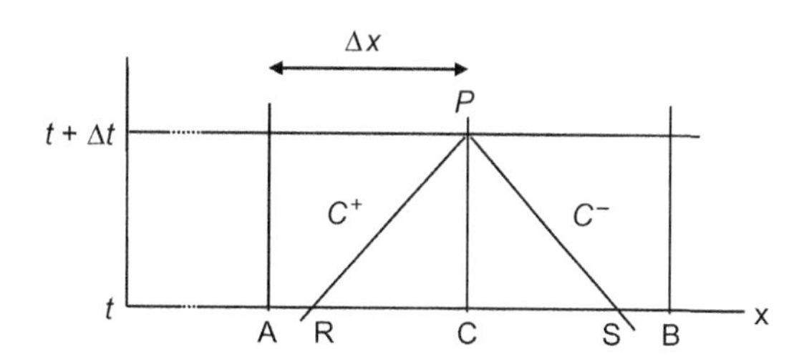

FIGURE 5.24

Characteristic line of MOC on x–t plane.

relation between the shaft torque and rotating speed, (ii) homologous relations on total characteristic of flow-head, and (iii) total characteristic of flow-torque [114].

The wave speed varies depending on gas entrainment in the liquid, as expressed by Eq. (5.7) in Section 5.2.2. This is true even with very small amounts of gas present in the form of bubbles. Also, the wave speed strongly depends on the elasticity of the pipe wall. For example, the wave speed in thin flexible wall pipes is remarkably reduced when compared to that in open space.

The solutions of Eqs. (5.13) and (5.14) precisely represent the pressure rise caused by the effect of line pack in long pipelines due to frictional pressure gradient as well as the pressure rise due to Joukowsky's equation (5.12), providing that sufficient discretization is implemented. However, in extreme cases, such as highly viscous fluids, the first-order accuracy of the friction term of Eqs. (5.13) and (5.14) may cause an unstable solution. A stability study [118] shows that the spatial discretization Δx must satisfy $f \cdot \Delta x \cdot u/(2D \cdot c) \ll 1$ in order to have a stable solution.

5.3.3.2 Column separation

When the pressure drops to the vapor pressure level, the liquid vaporizes. This condition is often met downstream of valves closing quickly and after pump trips [119]. Vapor bubbles, which may be physically dispersed homogeneously or collected into single or multiple void spaces, grow when the downstream liquid column flows downward by its inertial force. The most common concern in analyzing this type of liquid vaporization is prediction of the magnitude of the pressure rise when the vapor bubbles collapse. If the pipeline is a single line and the aim of the analysis is focused on knowing the pressure rise at the first collapse, rigid liquid column theory [120] can be used to predict the reverse velocity of the downstream liquid column and the subsequent Joukowsky pressure rise when the void collapses. However, practically, it is recommended to use MOC with an appropriate column separation model, such as the discrete-vapor-cavity model and the discrete free-gas cavity model [114,121].

In the discrete-vapor-cavity model, the pressure at each grid point is kept constant at the vapor pressure when the pressure drops to this value. With this condition, Eqs. (5.13) and (5.14) yield the upstream velocity u_{Pu} and downstream velocity u_{Pd}, respectively, at the grid point P. The vapor volume V_P is given by:

$$V_P = V_C + \Delta V = V_C + \{\psi \cdot (u_{Pd} - u_{Pu}) + (1 - \psi) \cdot (u_{Cd} - u_{Cu}\} \cdot A \cdot \Delta t \qquad (5.15)$$

where A is the pipe sectional area and ψ is a weighting factor. A value of $\psi = 1$ gives an implicit and stable solution. Although this method is easily implemented and reproduces the essential features of the physical phenomenon, numerical oscillations or unrealistic pressure spikes appear when vaporization occurs at multiple grid points in the pipeline. The source of the oscillation is wave reflections from multi-cavities. Reliable results are normally obtained for systems having a clearly defined isolated cavity position.

Most water and crude oil pipelines contain small amounts of free gas in the fluid. Such cases require the consideration of gas volume expansion when the pressure decreases and nearly equals the vapor pressure. The discrete free-gas cavity model lumps the same amount of free gas at each grid point. If α_N is the void fraction at the atmospheric pressure P_N for a constant mass of free gas distributed at each grid point, the equation for the lumped free gas yields the following Eq. (5.16) at each time step:

$$V_P = (P_N \cdot \alpha_N \cdot A \cdot \Delta x/\gamma)/(h_P - z - h_v) \tag{5.16}$$

Here, Δx is the length between neighboring nodes (i.e. $\Delta x = L/N$, where L is pipe length and N the number of equal pipe segments). The four unknowns u_{Pu}, u_{Pd}, h_P, and V_P can be obtained by solving Eqs. (5.13)–(5.16). This model shows good agreement with experimental results for bubbly flow in which air is dispersed in the continuous liquid phase. It is noted that in spite of using the pure liquid wave speed between the nodes, the model yields realistic wave speeds in two-phase flow. For pure liquid, a void fraction of 10^{-7} at standard atmospheric pressure is typically used [121].

More precise numerical models start with the governing equations for two components of free gas and pure liquid [112,122]. It is assumed that there is little or no slip between the two components in the homogeneous bubbly mixture for low void fraction. The partial pressure of the free gas represents an unknown pressure instead of the piezometric head h_P. The wave speed of the homogeneous two-phase mixture is applied. Again, MOC can be used for integration of the governing equations. However, a variable wave speed depending on void fraction may affect the accuracy of interpolation. In this case, the two-step Lax–Wendroff method is recommended [112].

5.3.3.3 Frequency-dependent friction

The mechanism of shear stress generation in pipe walls in transient flow is quite different from that in steady flow. The velocity near the wall responds in phase with the pressure gradient since the frictional force dominates relative to the inertia force, whereas the acceleration is in phase with the pressure gradient in the central portion of the pipe since the inertial force is dominant here. This means that the friction term of Eqs. (5.13) and (5.14) based on velocity-squared will not accurately model transient losses [123–127].

Uchida [128] derived a solution of the frequency-dependent friction factor for sine-wave pulsation in laminar flow, and Zielke [123] used a weighting function applied to the time history of unsteady laminar flow. Kagawa et al. [126] developed a fast computation method to calculate the weighting function. Extensive efforts have been made to develop sophisticated methods for turbulent flow

[129,130]. Fortunately, however, quasi-steady modeling is normally sufficiently accurate during the early phase of a transient. The frequency-dependent friction is important for high-frequency pulsative flows and laminar flows often observed in highly viscous fluids.

5.3.3.4 Design practice and criteria

The basic design of a liquid piping system requires consideration and study of pressure surge. Typical examples are summarized in Table 5.13.

Various design codes and standards specify the limitation on surge pressure in a piping system. Since pressure surge is categorized as an occasional load, the allowable surge pressure is normally greater than the design pressure of the system. For example, ASME/ANSI B31.3 for the chemical and petrochemical plant piping systems specifies a maximum allowable surge pressure of 1.33 times the design pressure of the system. For long-distance pipelines, ASME/ANSI B31.4 specifies 1.1 times the design pressure. It should be noted that each code and standard has a unique definition of the allowable stress, which leads to different definitions of the allowable surge pressure.

Table 5.14 shows typical data necessary for surge analysis of a piping system. Initial pressure and flow balance are first determined based on the boundary conditions before transient analysis can be implemented.

5.3.4 Countermeasures

Table 5.15 summarizes typical countermeasures for water hammer or pressure surge. Numerical computations are necessary for the detailed design of the countermeasures.

Figure 5.25 shows a flow chart for valve stroking analysis most frequently encountered by engineers.

5.3.5 Examples of component failures

Numerous examples of surge-related problems including structural coupling problems [135,136] have been reported in the literature. The following examples are typical cases.

5.3.5.1 Example-1: Power failure of pump

Figure 5.26 shows a schematic representation of the system [137]. The ship loading system consisting of 48 in $\times$ 6.5 km pipeline transporting crude oil from pump station to tank farm on the top of a hill. It was reported that a sudden power failure of the pump station caused severe damage of the anchor supports approximately 5 km away from the station. Investigations showed that the propagation of

Table 5.13 Typical examples of pressure surge

Event	Descriptions
On–off valve: Quick closure	• Quick closure of valve causes pressure rise that is at least higher than Joukowski's pressure given by Eq. (5.12) if effective stroke time of the valve is less than the pipeline period of $2 \cdot L/c$ • The pressure rise is proportional to the initial fluid velocity • Ship loading lines with quick-release coupler valve is the typical example for the study
On–off valve: Quick opening	• Quick opening of valve causes severe pressure rise if there is a vapor cavity downstream of the valve. Sudden opening of the valve pressurizes the vapor cavity and collapses it leading to severe pressure rise • Such situations are likely to occur at the stage of priming and re-start of plant facilities
Pump station: Power failure	• Power failure of motor or turbine driver causes sudden reduction of discharge pressure, and causes column separation if the pressure reduces below the vapor pressure • Subsequent rejoining pressure rise due to backpressure occurs when the vapor column instantly closes or collapses
Pump station: Start-up	• Sudden increase of discharge flow and pressure due to pump start-up sometimes causes collapse of downstream vapor cavities • Inappropriate air release from air valves at downstream high point also causes severe pressure rise
Check valve: Reverse flow	• In pump station, power failure induces sudden closure of discharge check valve due to reverse flow. This causes a large shock force on the check valve, pipe support, and pump nozzle • If pumps are in parallel operation and one pump trips, downstream high pressure causes large reverse flow until the check valve closes • The shock force is proportional to the reverse flow velocity • Dynamic characteristics determine the reverse velocity, and should be considered for the selection of the check valve
Lift-type safety valve: Blow-down	• In liquid systems, long lead line from branch point to safety valve sometimes causes large pressure oscillation [131] and structural vibration when the safety valve blows • If the valve with long lead line installed at discharge of the pump having droopy head-flow characteristics and the set point of the valve is near the discharge pressure, such a possibility becomes high

Table 5.14 Necessary data for pressure surge analysis

Item	Contents
Piping data	Configuration of pipe network; pipe lengths; outside diameter and wall thickness; elevation; Young's modulus of pipe material; pipe friction factor; design pressure
Control valve data	Cv-valve opening curve, stroke speed, closing time
Check valve data	Maximum Cv, dynamic characteristics (deceleration versus reverse velocity)
Pump data	Head-flow characteristic curve; torque-flow characteristic curve; rated and operation conditions (flow, head, speed, torque); rotational inertia (GD2); starting torque curve of driver; design pressure of case
Tank data	Inside diameter, liquid surface level, gas pressure at liquid surface, gas volume and polytropic exponent in case of closed-type tank
Fluid properties	Bulk modulus, density, viscosity, vapor pressure, void fraction of dissolving gas
Process data	Initial pressure and flow balance

a negative pressure wave formed a vaporized two-phase zone near the tank farm, and column separation and subsequent rejoining pressure caused a shock force on the support system. It is stressed that the reduced wave speed in two-phase flow makes the wave traveling time between the elbows large, causing large energy input into the support system.

5.3.5.2 Example-2: Operational change of pump station

Figure 5.27 shows a schematic description of the event [138]. In a cooling water network system (200–500 mm inside diameter piping and a few hundred meters in total pipe length), one pump failure and the subsequent automatic start-up of a stand-by pump to recover the system pressure injured a worker near a high elevation point (see Fig. 5.27). Column separation and subsequent rejoining pressure dislocated the piping system around the high elevation point and hit the worker. In this case, the temporal suction bypass operation made the situation worse, as the pressure at the high elevation point decreased, falling below the vapor pressure when the pump tripped.

5.3.5.3 Example-3: Unexpected valve opening

Figure 5.28 gives a schematic description of the event [139]. The system injects ammonia vapor into vapor tank through pressure control valve PCV that flushes 16 barA ammonia liquid into 4.5 barA vapor. A shutdown operation of the system required both valves PCV and HV to be closed and liquid ammonia filled

Table 5.15 Typical countermeasures and cautions

Countermeasures	Explanation and caution
Valve stroking [132]	• Most on–off valves have large flow coefficient or Cv value. Therefore, flow starts to reduce typically at 30% to 10% opening in closure action. Such effective stroke depends on valve type • Maximum pressure rise due to valve stroking can be determined by $P_{Surge} = \mathrm{Min}(J, J \cdot T_p/T_{eff})$, where J is Joukowski's pressure, T_p is pipeline period of $2 \cdot L/c$, and T_{eff} is effective stroke time, which corresponds to typically 10% to 30% of full stroke time • Bi-linear stroke is effective to reduce pressure rise, where valve closes from 30% to 10% in the early stages of stroking
Surge tank	• One purpose is to provide free surface near surge source and minimize pipeline period. In the case of high-pressure line, closed-type surge tank with gas pressure control system is necessary • Another purpose is to provide liquid into the main line to prevent column separation [106]. This design concept has often been used in the past until good prediction methods to calculate rejoining pressure were established
Air valve	• The high elevation point in the piping system sometimes reaches the vapor pressure after shutdown of the facility. Re-start operation collapses the vapor void and may cause severe pressure rise • Installation of air valve at this point prevents sudden closure or collapse of the void • Numerical analysis is necessary to determine flow coefficient of the air valve [133]
Lift-type safety valve	• As described in Table 5.13, lift-type safety valve requires caution for installation • Moreover, lift speed of this type of valve is not high enough to relieve high-velocity surge wave
Relief valve	• Rubber-type quick relief valve is effective to relieve high-velocity surge wave • Slop tank and re-injection system to main line is required
Quick close check valve [134]	• Split-disk-type and nozzle-type check valves have good dynamic characteristics to minimize reverse flow velocity Swing-type check valve is not recommended for this purpose • For protection of pump casing against slamming force at discharge check valve, installation of bellows between pump nozzle and check valve is effective
Damped check valve [115]	• Another approach against sudden check valve closure is to apply the concept of slow valve closure by using a damped check valve • In general, dynamic characteristics for this type of valve are not available. Experimental data are essential for real design [135] • Application of motor-operated valve is effective instead of this type of check valve function. Control sequence linked to pump failure is necessary

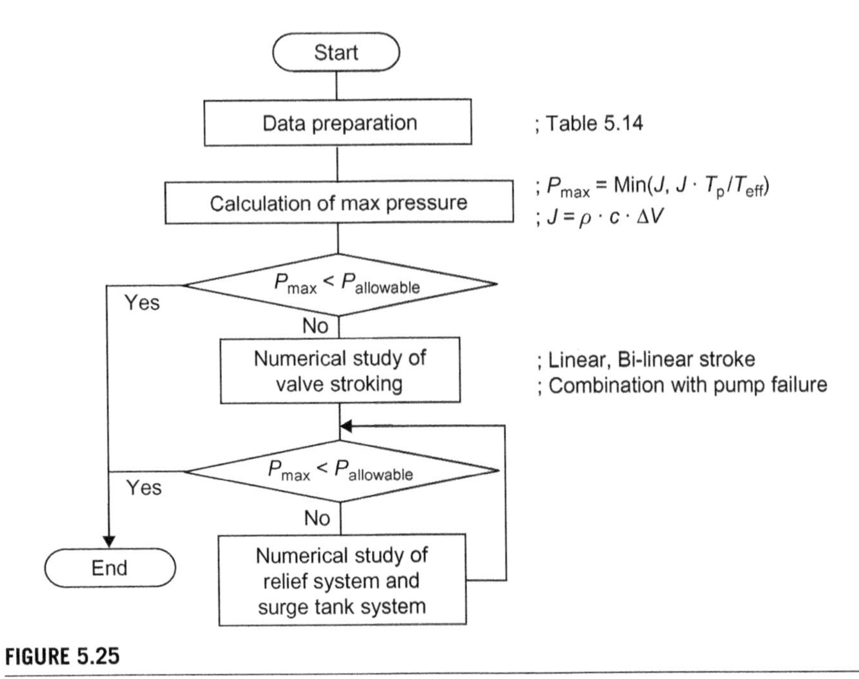

FIGURE 5.25

Flow chart for valve stroking analysis.

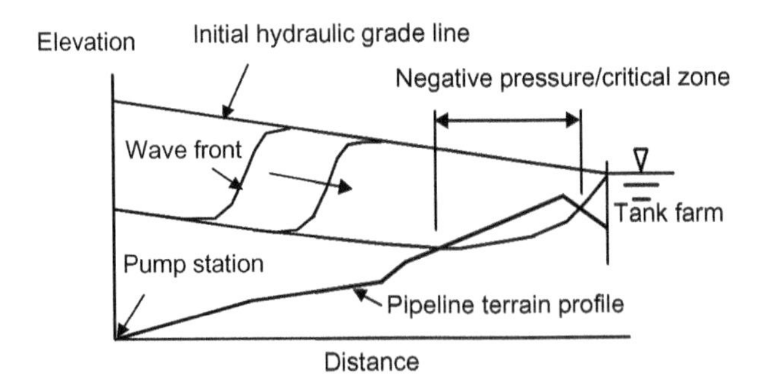

FIGURE 5.26

Support damage due to negative pressure wave.

between the valves. The re-start operation required opening of the valve HV first. This re-start operation caused sudden blow-out of ammonia liquid from the safety valve near HV and injured the operator. Leakage of a small amount of liquid ammonia through valve PCV resulted in ammonia vaporization and accumulation in the pipe between the valves, then sudden opening of valve HV caused collapse of the vapor column and severe rejoining pressure rise.

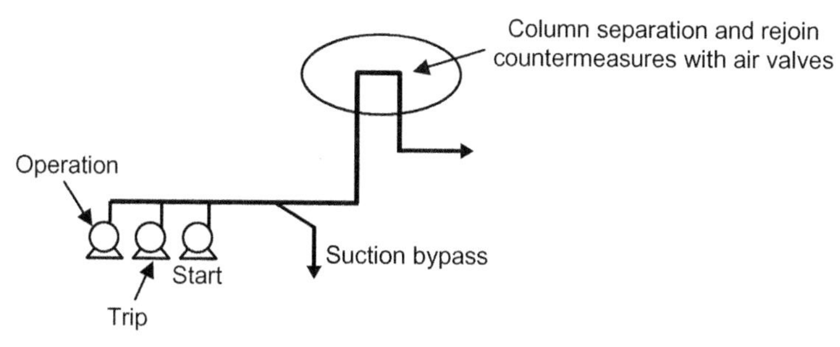

FIGURE 5.27

Pipe dislocation due to operational change in pump station.

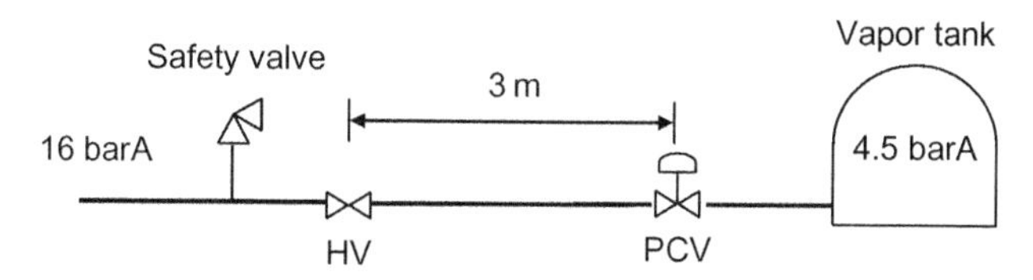

FIGURE 5.28

Unexpected blow-out from safety valve caused by re-start operation.

5.4 Valve-related vibration

Vibrations may occur in valves used to adjust pressure and flow rate in piping systems, as well as valves used for the piston/cylinder of reciprocating compressors. The vibration mechanisms depend on the valve type and the flow direction (upward or downward) in the piping system. Valves are a root cause of vibrations in piping systems. In this section, valve vibrations due to flow and coupling between valve vibration and flow behavior are discussed.

5.4.1 Valve vibration

5.4.1.1 Outline

When a valve is set at a small opening and the fluid force due to the resulting leakage flow is large, leakage-flow-induced vibrations and turbulence-induced vibrations occur. These may lead to damage to the valve itself.

5.4.1.2 Self-excited valve flutter

Flutter, similar to that encountered for aircraft wings, can occur with butterfly valves (Fig. 5.29) for some valve opening conditions (refer to Section 3.2.2). Increasing the stiffness of the valve can generally raise the critical flow velocity. Flutter can also be avoided by operating with a large opening. An additional valve or introduction of an orifice can also suppress flutter.

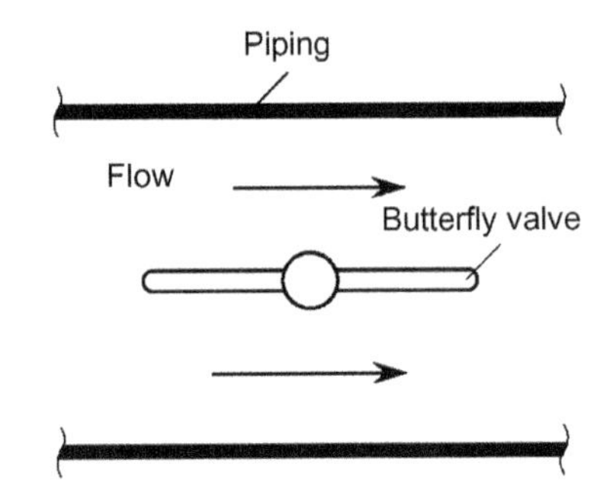

FIGURE 5.29

Schematic view of butterfly valve.

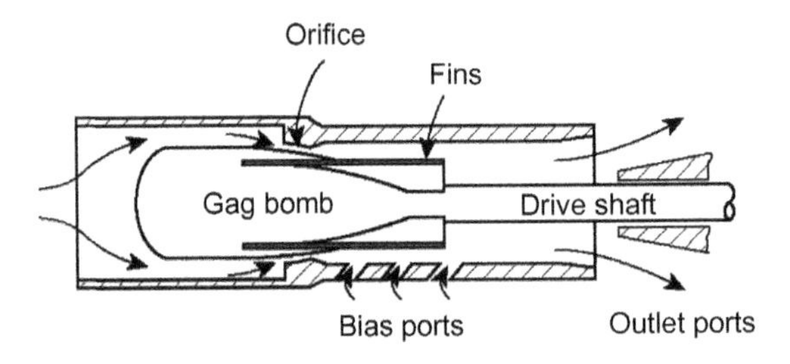

FIGURE 5.30

Self-excited vibration of a valve having an annular clearance [140].

5.4.1.3 Leakage-flow-induced vibration

5.4.1.3.1 Outline

Consider the valve shown in Fig. 5.30. Displacement of the gag bomb changes the leakage-flow rate through the orifice. The change in flow rate in turn modifies the fluid force acting on the gag bomb at the orifice in a feedback loop, resulting in vibration. In the case of divergent channel geometries, or inlet local pressure loss geometries, the resultant negative fluid damping force can lead to self-excited vibrations [141].

5.4.1.3.2 Evaluation method

Flutter-type or divergence-type instability can often be avoided by employing convergent channel geometries. Divergent geometries, in addition to being destabilizing, have the added disadvantage of increased turbulence excitation forces. These geometries should therefore be avoided.

5.4.1.4 Turbulence-induced valve vibration

5.4.1.4.1 Overview

Flow turning around the valve can result in turbulence generation which leads to valve vibration. In partially open steam valves, turbulence due to choked flow at

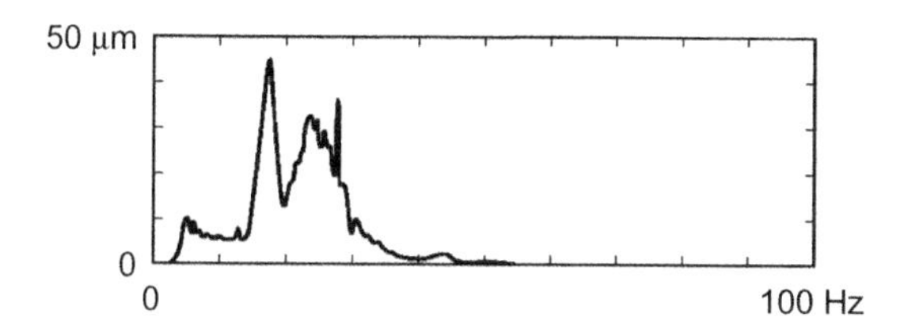

FIGURE 5.31

Frequency spectrum of valve vibration caused by cavitation [143].

the throat may occur, resulting in vibration of not only the valve itself, but also piping downstream of the valve. Furthermore, resonant vibrations can also occur when vortex shedding within the valve occurs at a frequency close to the valve natural frequency [142].

In valves used to control liquid flow, acceleration of the flow at the valve throat may lead to cavitation. The random pressure fluctuations resulting from the ensuing bubble collapse may also cause significant valve vibration [143]. Figure 5.31 shows an example of the vibration response spectrum during cavitation-induced vibration.

5.4.1.4.2 Evaluation methods

Since vibration is induced by flow turbulence, the result is a broadband response with a peak at the valve natural frequency. Intense cavitation occurs when the cavitation coefficient is below $1 \sim 1.5$, thus:

$$C_v = \frac{P - P_v}{\frac{1}{2}\rho U^2} < 1 \sim 1.5 \tag{5.17}$$

In Eq. (5.17), P and U are the pressure and the velocity, respectively, ρ is the fluid density, and P_v the saturation vapor pressure of the fluid.

To reduce vibrations, two possible countermeasures may generally be considered: reducing the turbulence excitation force or increasing the stiffness and damping of the valve elements. In the case of vortex shedding induced resonance, the solution is to change the natural frequency of the valve.

To eliminate cavitation, piping design should be reviewed to avoid valve installation at low-pressure locations. Alternatively, multiple valves may be installed, thereby reducing flow velocity and eliminating cavitation.

5.4.2 Coupled vibrations involving valves, fluid and piping

5.4.2.1 Summary

In systems where a valve is connected to piping, self-excited vibrations involving valve-fluid-piping coupling sometimes occur, even when the valve alone is stable against self-excited vibrations. These types of coupled vibrations often

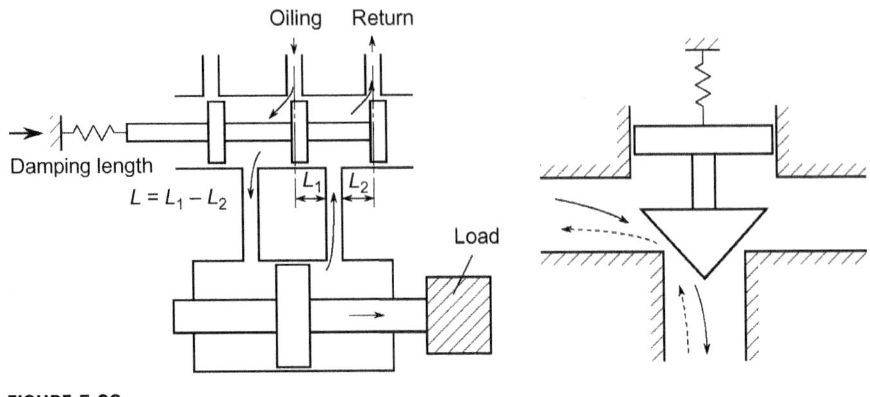

FIGURE 5.32

Examples of valves to be discussed.

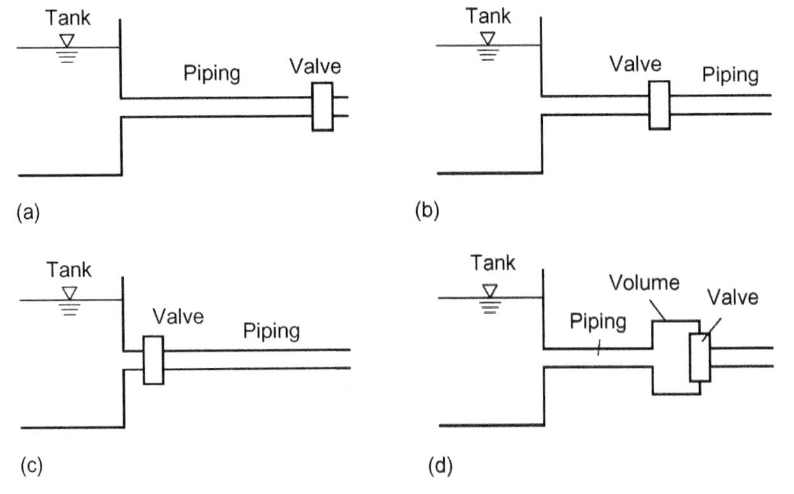

FIGURE 5.33

Examples of valve position relative to piping: (a) valve located at exit of piping, (b) valve located at middle of piping, (c) valve located at piping inlet, and (d) valve and adjacent volume.

occur in plants, hydraulic structures, and high-pressure oil hydraulic piping systems. Many kinds of valve-piping coupled systems with spool valves and poppet valves have been studied, as shown in Figs. 5.32 and 5.33.

5.4.2.2 Vibration phenomena and history of research and evaluation

Weaver [140] showed that self-excited vibration occurs due to the delay of pressure pulsations in the piping relative to the valve vibration when the valve, which closes as the fluid starts to flow, is connected to the piping systems. Fujii [144] showed that coupled valve-piping self-excited vibration occurs if the natural frequency of the piping system is lower than that of the valve and the flow velocity through the valve reduces as upstream pressure increases. Maeda and Fujii [144,145] studied the case where the valve opening lift increases as upstream pressure increases. Fujii [144] showed that self-excited vibrations occur if the natural frequency of the piping system is higher than that of the valve in this case.

Ezekiel and Ainsworth [146,147] studied the stability of valve-piping systems in cases where a spool valve is connected downstream of the piping. Spool valves are stable against self-excited vibrations if the damping length L (defined in Fig. 5.32) is positive. However, self-excited vibrations of the valve-piping system can occur in the case of a specific relationship between the natural frequency of the valve and piping system even if the valve itself is stable to self-excited vibration.

Valve vibration for valves which close or open as fluid starts to flow is the subject of the next section.

5.4.2.3 Self-excited vibration of flow-restriction valves

5.4.2.3.1 Time-delay-induced vibrations

Self-excited vibrations are known to occur in flow-restriction valves [140], such as that shown in Fig. 5.34, at small apertures. The linear force equilibrium condition for the valve is given by:

$$kx_0 = \rho g \, \Delta H A_v \tag{5.18}$$

where k is the valve structural rigidity, x_0 the equilibrium displacement, ρ the fluid density, ΔH the hydrostatic head, and A_v the effective cross-sectional area. The valve will remain closed when kx_0 is smaller than the pressure head force on the right, and open when kx_0 is larger than this term.

As the valve opens or closes, the flow through the valve reacts by accelerating or decelerating. However, due to fluid inertia, the flow response is delayed

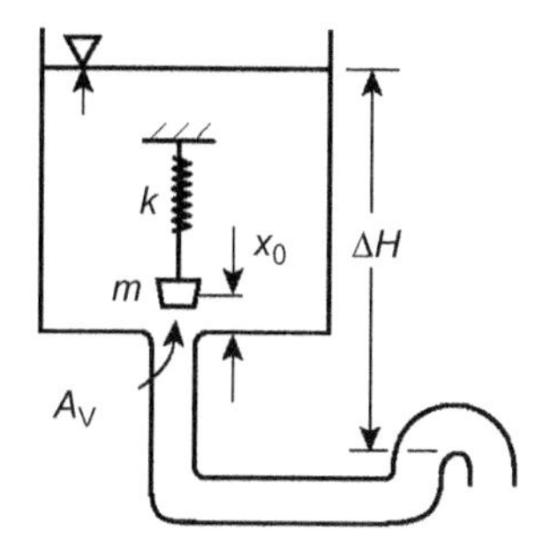

FIGURE 5.34

Valve that closes as flow increases [140].

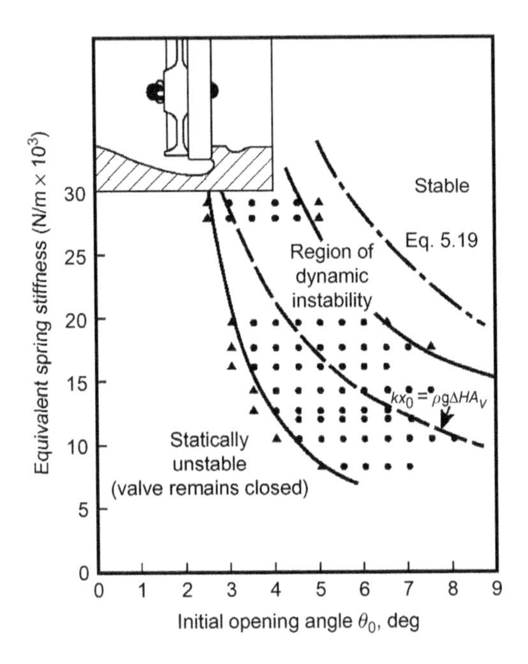

FIGURE 5.35

Condition for self-excited vibration of check valve [140].

relative to the valve displacement. If, because of this time lag, the resulting fluid force amplifies the valve motion, self-excited vibration results.

In Fig. 5.35, the condition for occurrence of self-excited vibrations for a directional control check valve is shown, based on the relation between the equivalent spring stiffness and the initial opening angle θ_0.

An example of the hysteresis effect due to fluid inertia is shown in Fig. 5.36. The time delay is strongly influenced by the magnitude of the fluid inertia force. The higher the fluid density, and the larger the piping, the larger is the fluid inertia force.

A large inertia force leads to long period high-amplitude vibrations while smaller inertia forces result in short period low-amplitude vibrations.

Based on Kolkman's linear analysis, a simple stability criterion has been developed. The stability boundary for self-excited vibration is:

$$kx_0 < 2\rho g\ \Delta HA\left(\frac{m}{\rho AL} + 1\right) \approx 2\rho g\ \Delta HA \tag{5.19}$$

where m is the structural mass. To minimize the valve aperture between the boundaries for permanently open and permanently closed valve conditions, the rigidity k should be large. By ensuring that the selected rigidity also meets the stability condition of Eq. (5.19), self-excited vibration will be avoided.

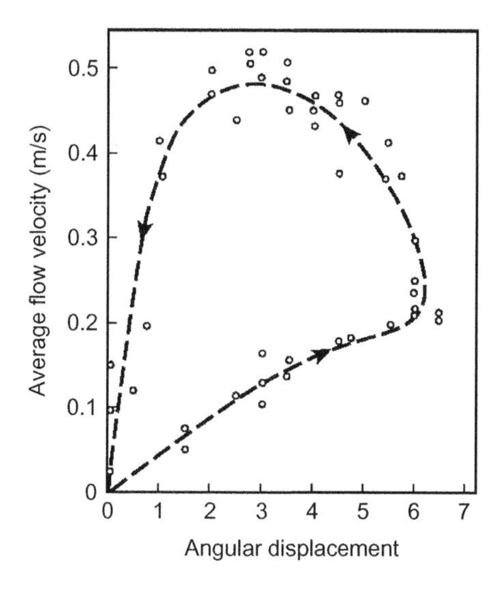

FIGURE 5.36

Hysteresis caused by fluid inertia [140].

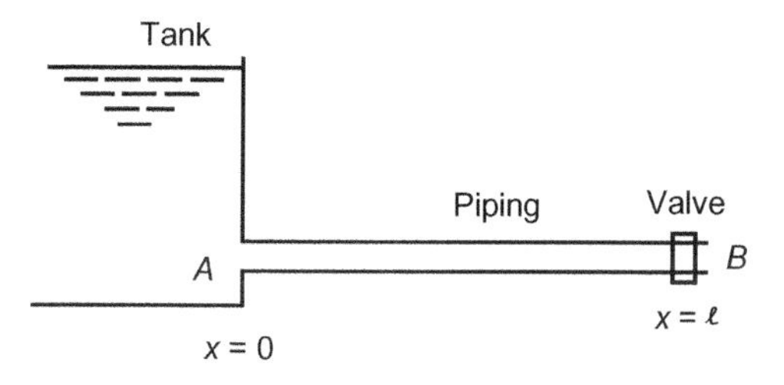

FIGURE 5.37

System consisting of piping, valve, and constant pressure tank [148].

Notice that the right hand side of Eq. (5.19) corresponds to twice the force needed for static equilibrium given in Eq. (5.18).

5.4.2.3.2 Self-excited vibration related to coupling between pressure pulsations in piping and valve vibration

The stability of a system which consists of a constant pressure tank, piping, and valve as shown in Fig. 5.37 is described in this section [148]. Boundary conditions at the inlet and exit of the piping and equilibrium points are shown in Fig. 5.38. P is the pressure and V the flow velocity. If a normal valve is located at the exit, P is proportional to the square of the flow velocity. P_s is the pressure in the tank; P_0 and V_0 at the intersection point of the two curves are the pressure and flow velocity in the piping at the stable condition. When the pressure in

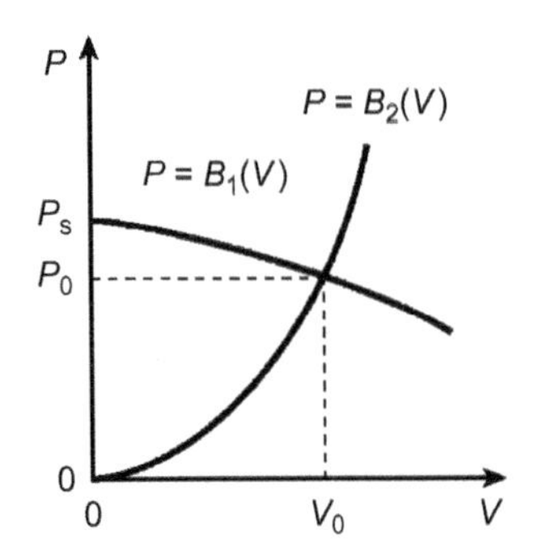

FIGURE 5.38

Boundary condition and equilibrium position [148].

the piping becomes $P_0 + p$ and the velocity $V_0 + v$, the fluctuations p and v can be expressed as follows:

$$p = F(t - x/a) + f(t + x/a) \tag{5.20}$$

$$v = (1/\rho a)\{F(t - x/a) - f(t + x/a)\} \tag{5.21}$$

where ρ is the fluid density, and a the wave propagation velocity. The pressure boundary conditions near the equilibrium point (V_0, P_0) are linearized as follows:

$$P = -b_1 v \ (b_1 > 0, \text{ inlet end}) \tag{5.22}$$

$$P = b_2 v \ (b_2 > 0, \text{ outlet end}) \tag{5.23}$$

The following equations are obtained by taking the reflection condition of the wave at the inlet end and the exit end into consideration:

$$\begin{aligned} F(t) &= R_1 R_2 F(t - 2\ell/a) \\ R_1 &= (1 - b_1/\rho a)/(1 + b_1/\rho a)\} \\ R_2 &= (1 - b_2/\rho a)/(1 + b_2/\rho a) \end{aligned} \tag{5.24}$$

$|R_1 R_2|$ is the growth rate when the wave makes a round-trip in the piping. If $|R_1 R_2| < 1$, the system is stable. In the case of the equilibrium condition shown in Fig. 5.38, the system is stable because $b_1 > 0$, $b_2 > 0$, $|R_1| < 1$, and $|R_2| < 1$. Instability of a valve which closes as the pressure increases, as shown in Fig. 5.39(a), may be explained as follows. The clearance h between the valve and the valve seat is given by:

$$h = h_m - A_b P/k \tag{5.25}$$

where h_m is the maximum clearance, A_b the valve area over which the pressure acts, and k the spring constant. If the valve opening area is proportional to h

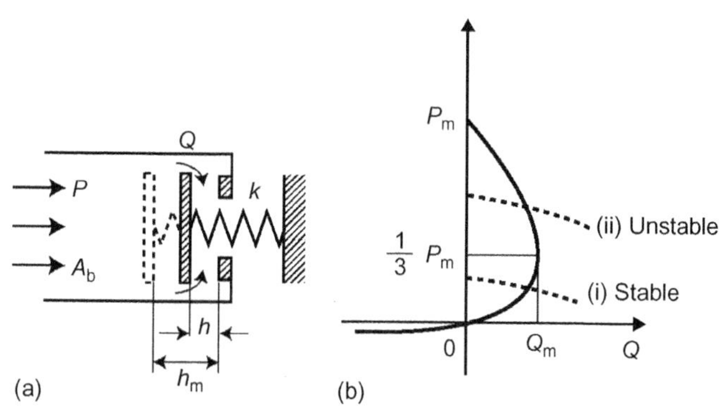

FIGURE 5.39

Auto-valve in pipeline system [148]: (a) valve which closes at high pressure, and (b) corresponding pressure-flow rate characteristic.

and the flow rate coefficient is constant, the flow rate Q can be expressed as follows:

$$Q = C(P_m - P)\sqrt{P} \tag{5.26}$$

Equation 5.26 is shown as a solid line in Fig. 5.39(b), which relates the pressure P to the flow rate Q. In the case where this valve is located at the exit end, the system is stable if the pressure at the equilibrium point is less than $P_m/3$, as shown by line (i) in Fig. 5.39(b). On the other hand, if the source pressure is high and the pressure at the equilibrium point is greater than $P_m/3$, as shown by line (ii), $b_2 < 0$ in Eq. (5.24) and $|R_1| > 1$. If b_1 is small, $|R_1R_2| > 1$, making the system unstable. Fluid forces are obtained from the pressure difference across the valve shown in Fig. 5.39(a). For spool valves and poppet valves fluid forces are obtained using momentum theory [149].

5.4.2.4 Self-excited vibration of valves which open due to flow

5.4.2.4.1 Self-excited vibration of the auto-valve: effect of compressibility-induced time delay [145]

Diesel engine fuel injection valves sometimes undergo self-excited vibrations. Consider the valve and valve chamber type shown in Fig. 5.40. The chamber pressure P_s is assumed constant and the exit pressure $P_2 = 0$. If the valve displacement x is considered sufficiently small, the flow rate Q through the valve is given by:

$$Q = C_d \pi \, dx \sin \alpha \sqrt{\frac{2P}{\rho}} \tag{5.27}$$

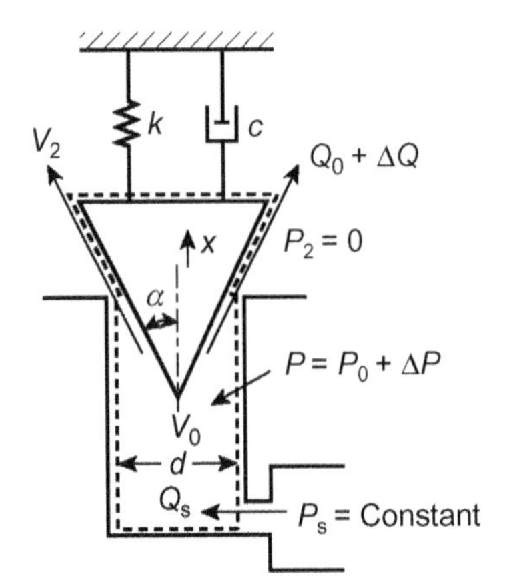

FIGURE 5.40

Valve model [145].

where C_d is the valve capacity coefficient. Linearization of Eq. (5.27) about the equilibrium point (x_0, P_0) leads to the following equation:

$$\frac{\Delta Q}{Q_0} = \frac{\Delta x}{x_0} + \frac{\Delta P}{2P_0} \tag{5.28}$$

Applying the continuity equation to the valve chamber, where Q_s is the volume flow rate into the chamber and K the fluid bulk modulus, leads to the relations:

$$\Delta Q_s = \Delta Q - \frac{V_0}{K}\frac{d(\Delta P)}{dt} \tag{5.29}$$

$$Q_s = C_s\sqrt{\frac{2(P_s - P)}{\rho}} \tag{5.30}$$

$$\Delta Q_s = \frac{-C_s}{\sqrt{2\rho(P_s - P_0)}}\Delta P \tag{5.31}$$

The valve equation of motion may be written by taking Eqs. (5.28), (5.29), and (5.31) into consideration. Defining the valve effective area as A, the equation of motion in the Laplace domain is:

$$\begin{gathered} ms^2\tilde{X} + cs\tilde{X} + k\tilde{X} = A\tilde{P} \\ \tilde{P} = \left[\frac{-C_s}{Q_0\sqrt{2\rho(P_s - P_0)}} - \frac{1}{2P_0} + \frac{V_0}{Q_0 K}s\right]^{-1}\frac{\tilde{X}}{x_0} \end{gathered} \tag{5.32}$$

where $\tilde{X}$ and $\tilde{P}$ are the Laplace transforms of the displacement x and pressure P, respectively.

From Eq. (5.32), the variation of the net pressure acting on the valve is delayed (relative to the valve motion) due to fluid compressibility in the valve chamber. This may result in a negative damping force. The Routh–Hurwitz criterion can be used to evaluate the occurrence of self-excited vibration. Generally, self-excited vibration occurs when x_0 and c are small. Operating the valve at a small opening or adding damping can eliminate the self-excited vibration.

5.4.2.4.2 Self-excited vibration related to coupling of pressure pulsations in piping and valve vibration

Fujii [144] shows that self-excited vibration occurs when the natural frequency of the valve is lower than that of the pressure pulsations by an analysis similar to the method described in Section 5.4.2.3.

5.4.3 Problem cases

Problem cases related to valves are numerous. Some examples are presented below.

5.4.3.1 Vibration of penstock in Yatsusawa electric power plant

In the Yatsusawa electric power plant [150], violent, periodic, and continuous pressure pulsations occurred in a penstock when the operation of the waterwheel was stopped and the gate valve closed. According to the investigation results, the gate valve had actually remained slightly open at a few centimeters before the fully closed position because of a damaged gasket. There was therefore a small amount of leakage flow from the waterwheel guide vanes even when the vanes were fully closed. Self-excited vibrations related to coupling between the valve and pressure pulsations in the piping occurred because the gate valve reduced the flow rate through the valve as the pressure difference increased. Self-excited vibration did not occur after the gasket was fixed and the gate valve was fully closed.

5.4.3.2 Self-excited vibration of agricultural channel decompression valve

The flow must be controlled in order to keep a specific constant pressure in the piping as well as a constant flow rate at the end of an irrigation canal. An automatic response type of decompression valve, shown in Fig. 5.41, is used to adjust and control pressure in the piping. When the sensitivity of the decompression valve was adjusted so that it could respond quickly against external perturbations, self-excited vibrations related to coupling between valve vibration and pressure fluctuation in the piping occurred. The self-excited vibration can be avoided by weakening the response sensitivity of the decompression valve [151,152].

5.4.3.3 Self-excited vibration of automobile power steering system valve

Vibration with noise sometimes occurs when a power steering system, such as shown in Fig. 5.42, is activated quickly [153]. The power steering system consists

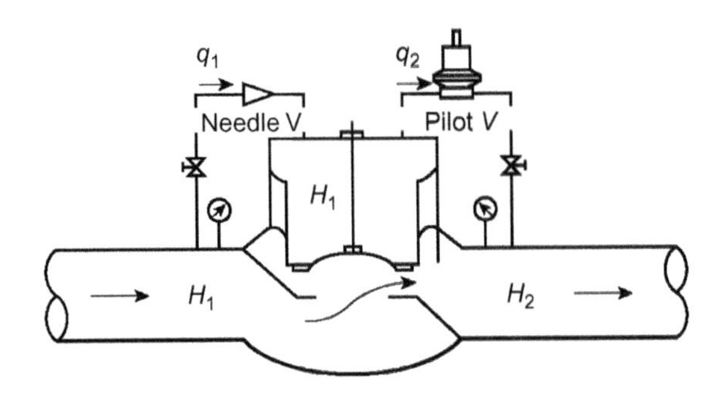

FIGURE 5.41

Automatic response-type decompression valve [151].

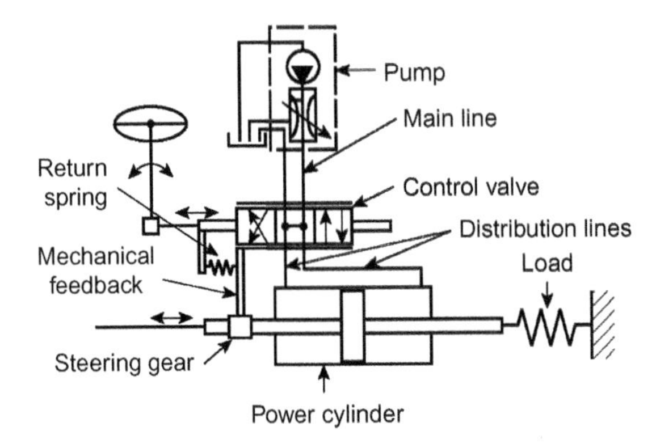

FIGURE 5.42

Schematic view of power steering system [153].

of a pump, steering gear, a control valve, and the power cylinder. The steering gear connects the steering wheel (the input) to the power cylinder piston (the output) mechanically. The valve controls the pressure of the oil supply into the power cylinder based on the deviation between the steering wheel and the piston displacements. For this system, self-excited vibrations related to coupling between the valve and piping occur when the steering wheel starts to be returned toward the straight position. Countermeasures include lowering the sensitivity of the control valve, selecting an adequate main line length, shortening the distribution line length, and reducing the piston mass.

5.4.3.4 Resonance vibration and malfunction of steam pipe pressure switch

During plant start-up, the small opening condition of a steam valve resulted in high-speed choked flow at the throat. As a result, a high-frequency (several hundred hertz) acoustic mode was excited in the large-diameter piping downstream of the valve (Fig. 5.43). High-amplitude pressure pulsations were propagated far

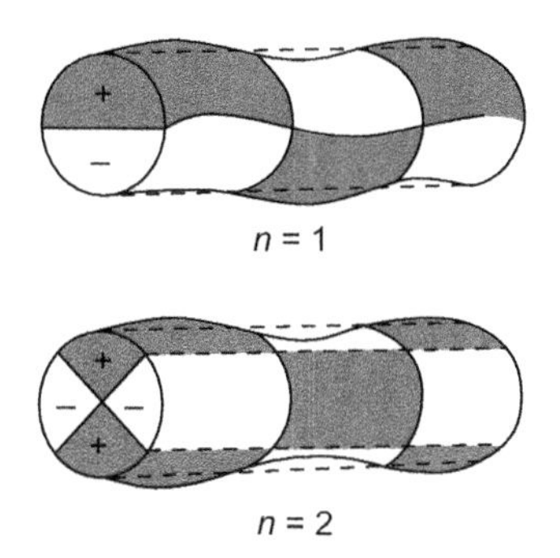

FIGURE 5.43

Examples of coupled vibration of pipe structural modes and high-order acoustic modes.

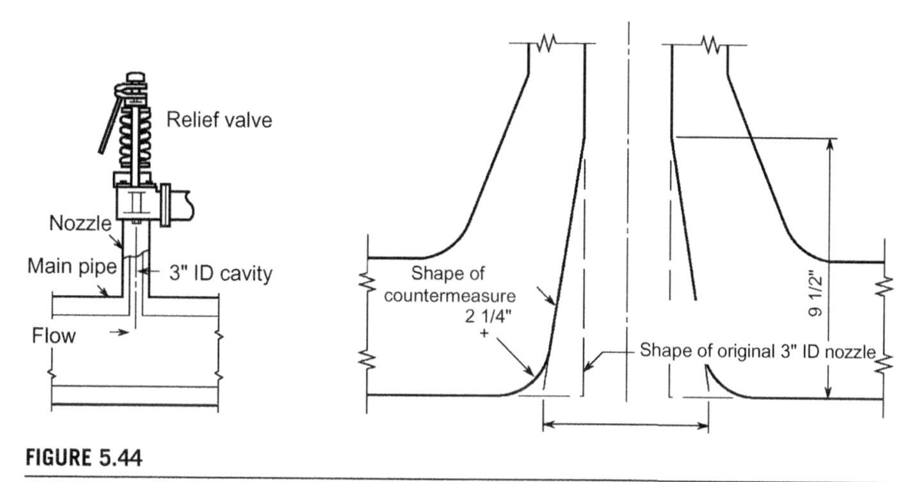

FIGURE 5.44

Cavity tone at relief valve and preventive countermeasure [154].

downstream to the location of the steam pipe pressure switch. This switch had a natural frequency close to the acoustic mode frequency. The switch failed, causing plant shutdown. As a counter measure, the pressure switch was moved to a location less susceptible to pressure fluctuations.

5.4.3.5 Cavity tone excitation of relief valve

A strong one-quarter (1/4) wavelength cavity tone developed in the relief valve, shown in Fig. 5.44, which was located four diameters downstream of a bend in a steam pipe [154]. Large-amplitude vibrations and acoustic noise resulted.

Two possible solutions to the problem are:

1. Modification of the valve nozzle shape as shown in Fig. 5.44. The modified nozzle shape eliminates stable vortex shedding, which consequently eliminates the cavity tone excitation.

2. Installation of a relief valve far downstream (nine to ten diameters) of the bend, thus avoiding the high-level turbulence closer to the bend.

5.4.3.6 Vibration of high-pressure bypass control valve

The cage of the steam valve shown in Fig. 5.45 cracked due to resonant vibrations. As the valve opens or closes, the acoustic peak frequency decreases or increases, respectively. It was found that the peak frequency under certain conditions leads to resonance with the cage.

The underlying mechanism is believed to be the interaction between shear flow separated from the valve seat and the cage via upstream feedback and self-amplification of the resulting pressure fluctuations [155].

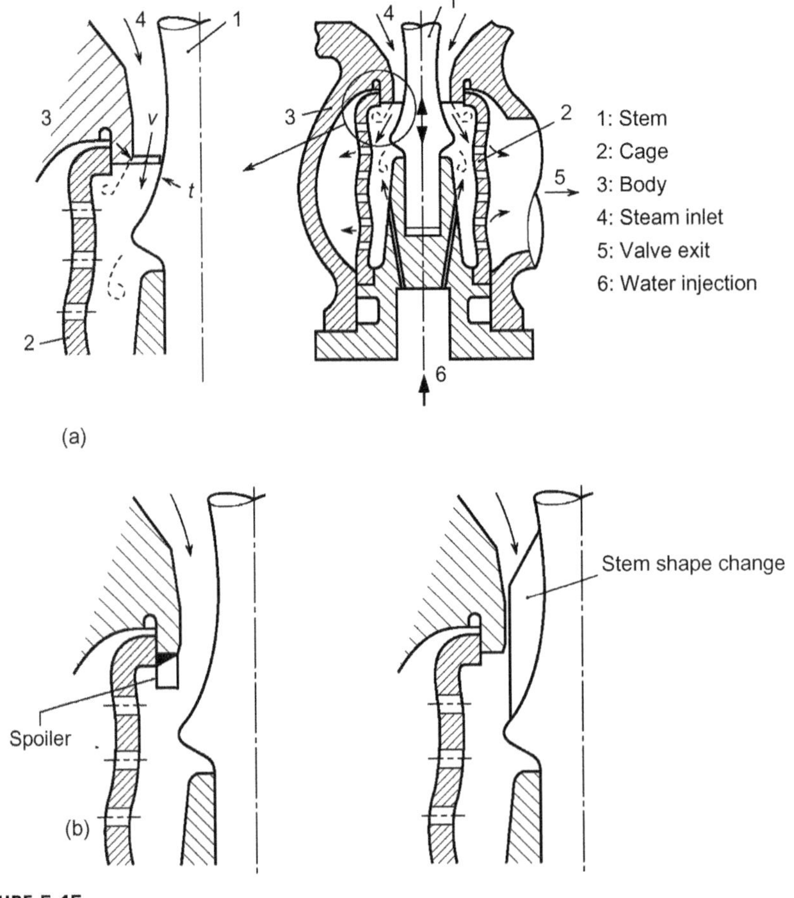

FIGURE 5.45

(a) Flow in the high-pressure bypass control valve and, (b) vibration countermeasure [155].

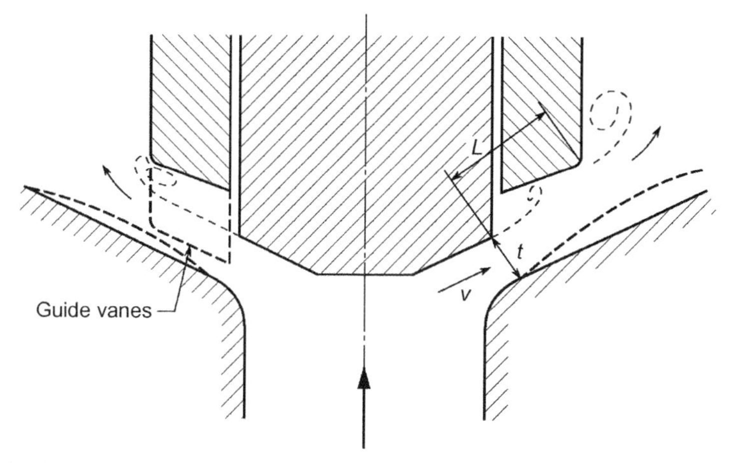

FIGURE 5.46

Flow and countermeasure in turbine control valve [155].

Possible solutions include installation of a spoiler near the flow separation and/or modification of the shape of the valve stem as shown in Fig. 5.45. These have the effect of increasing flow turbulence, thus weakening the flow correlation at the attachment point.

5.4.3.7 Vibration of high-pressure turbine control valve

The valve shown in Fig. 5.46 generated high acoustic noise which also resulted in damage to the piston ring between the stem and the shroud. It is believed that flow separating from the valve tip interacted with the shroud, resulting in large pressure fluctuations [155]. Resonance of these pressure fluctuations with one of the valve cavity modes then led to high acoustic noise. The problem was solved by introducing guide vanes that hindered flow in the circumferential direction.

5.4.4 Hints for countermeasures against valve vibration

Depending on valve type, countermeasures associated with different vibration mechanisms exist. In general, increasing valve structural stiffness and damping has a positive effect against vibration.

Vibration often occurs under conditions of small valve aperture. It is therefore important to correctly select a valve that will operate with the appropriate aperture within the range of the design flow conditions. This is an important consideration since transient start-up conditions and partial load operating conditions can be easily overlooked.

In the case of vibration problems with a single-stage valve, the installation of an orifice or additional valves can resolve the problem by reducing the pressure drop per stage.

5.5 Self-excited acoustic noise due to flow separation

5.5.1 Summary

When an obstacle or acoustic reflector is placed downstream of a free shear layer, discrete vortices are formed, giving rise to flow noise and structural vibration. This phenomenon, which is well known as cavity tone, hole tone, or self-excited vibration of orifice plates, is frequently observed in power and process plant piping systems. Figure 5.47 illustrates impinging jet configurations which produce self-sustained flow oscillations [156].

Large-scale boundary-layer separation at the pipe inlet or flow control valves causes low-frequency vibrations. High-rate pressure-reducing devices generate high-frequency pressure fluctuations downstream. These are possible causes of leakage and fatigue fracture in pipelines.

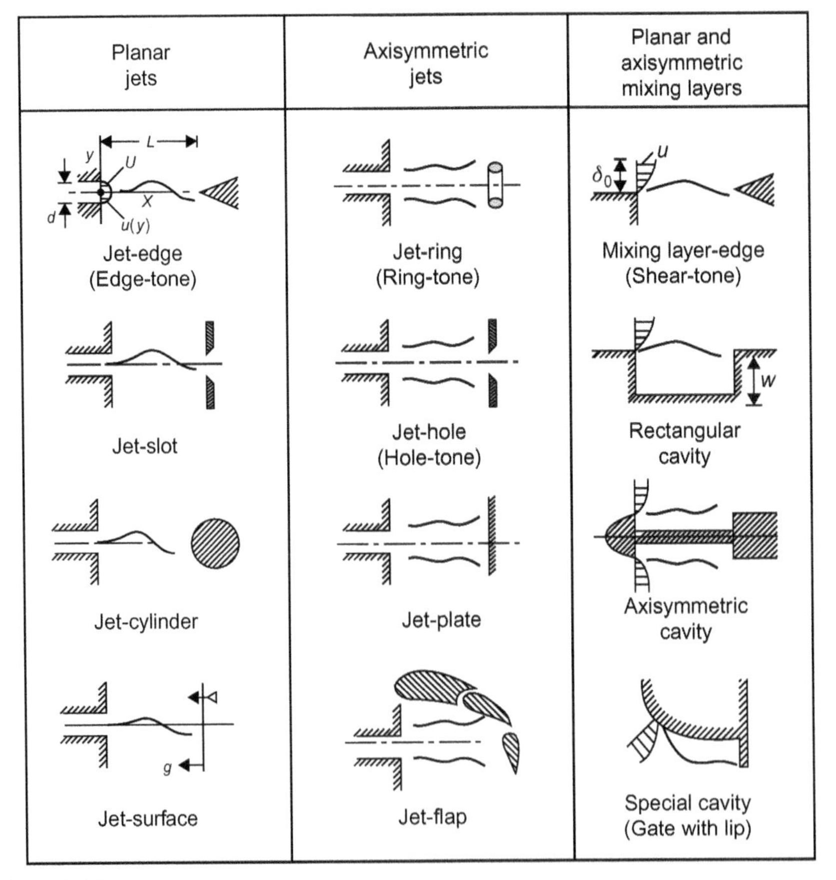

FIGURE 5.47

Basic configurations of shear-layer and impingement-edge geometries that produce self-sustained oscillations [156].

5.5.2 Outline of excitation mechanisms

5.5.2.1 Cavity tone

Figure 5.48 depicts a two-dimensional flow, with free stream velocity U, over a cavity having length l and depth h. A sound pulse generated by the shear layer impinging at the trailing edge of the cavity propagates upstream and controls the flow separation from the leading edge. The resulting periodic vortex in the shear layer causes a pure tone, which is also called a cavity tone or cavity noise. The pure tone frequency f is expressed by the modified Rossiter equation [157, 158]:

$$f = \frac{U}{l} \frac{n - \alpha}{\frac{M}{\sqrt{1 + (\kappa - 1)M^2/2}} + \frac{U}{U_c}} \tag{5.33}$$

where n is an integer associated with the number of vortices inside the cavity (mode number or stage number), α an experimental constant, M the free stream Mach number, κ the specific heat ratio of the fluid, and U_c the convective speed of the vortices. As discussed in Chapter 2, the non-dimensional frequency, known as the Strouhal number, is defined as:

$$St = \frac{fL}{U} \tag{5.34}$$

where U scales the variation of velocity in the length scale L.

Since the convective speed of the disturbances in the cavity U_c has a significant role, the Strouhal number based on the free stream velocity U and the longitudinal cavity length l is influenced not only by the geometry of the cavity but also by the characteristics of the boundary layer, such as momentum and displacement thicknesses, Reynolds number and Mach number. Figure 5.49 shows the variation of the Strouhal number $St = fl/U$ with the normalized longitudinal cavity length l/h in the case of turbulent flow at the separation point and low Mach number [159]. The Strouhal number increases with increasing cavity length and gradually approaches to a constant value. In Ref. [160] it is reported that for $M > 0.2$,

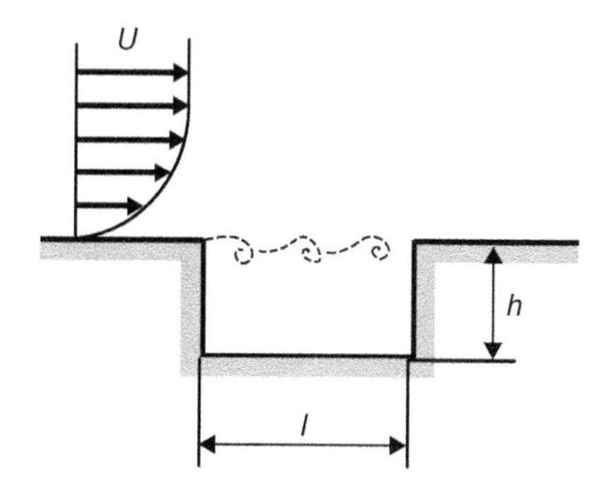

FIGURE 5.48

Flow over a two-dimensional cavity.

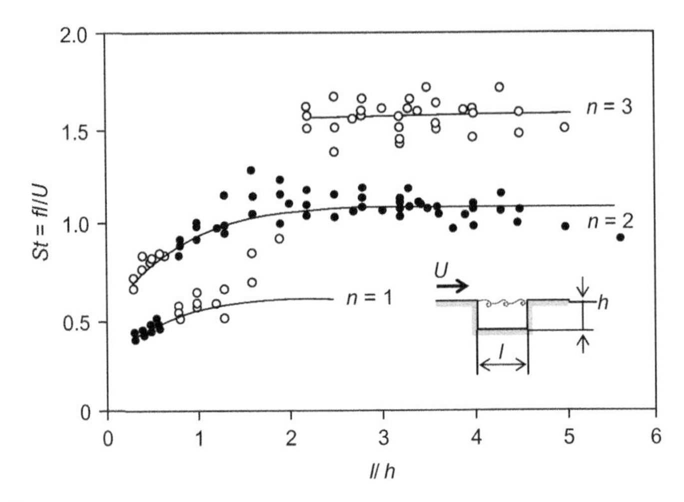

FIGURE 5.49

Variation of organized-oscillation Strouhal number with cavity length for fluid-dynamic oscillations in a two-dimensional cavity [159].

predictions of Eq. (5.33) are consistent with the experimental data when $U_c/U = 0.6$ and $\alpha = 0.25$.

For engineering applications the cavity tone may be generated in three-dimensional cavity geometries as shown in Fig. 5.50 [158]. In the presence of resonators the pressure oscillations in the cavity increase in magnitude as the resonance condition is approached. For a typical geometry the acoustic mode natural frequencies f_r are given by:

$$\left.\begin{aligned} f_r &= (c/2\pi)\sqrt{A/VH} && \text{for Helmholtz resonator} \\ &= ic/4h\ (i = 1, 3, 5, \ldots) && \text{for deep cavity} \end{aligned}\right\} \tag{5.35}$$

where c is the speed of sound, A the cross-sectional area of the opening, V the volume within the resonator, H the depth of the opening plus a correction factor equal to approximately 0.8 times the diameter of the opening [161], and h the cavity depth.

The sound pressures when the vortex shedding frequency coincides with the acoustic natural frequency are examined for a Helmholtz resonator and a rectangular deep cavity in Refs. [162,163]. Closed side branches are treated as a variation of a deep cavity. The rounding off effect of rectangular side-branch entrances on acoustic resonance is investigated in Ref. [164]. Rounding off the upstream corner of the side branch increases the flow velocity required to generate peak acoustic resonance while having a relatively minor effect on the magnitude of the peak. The Strouhal numbers, corresponding to the quarter wave resonance frequency based on $l + r_u$ (r_u = upstream entrance radius) are, however, independent

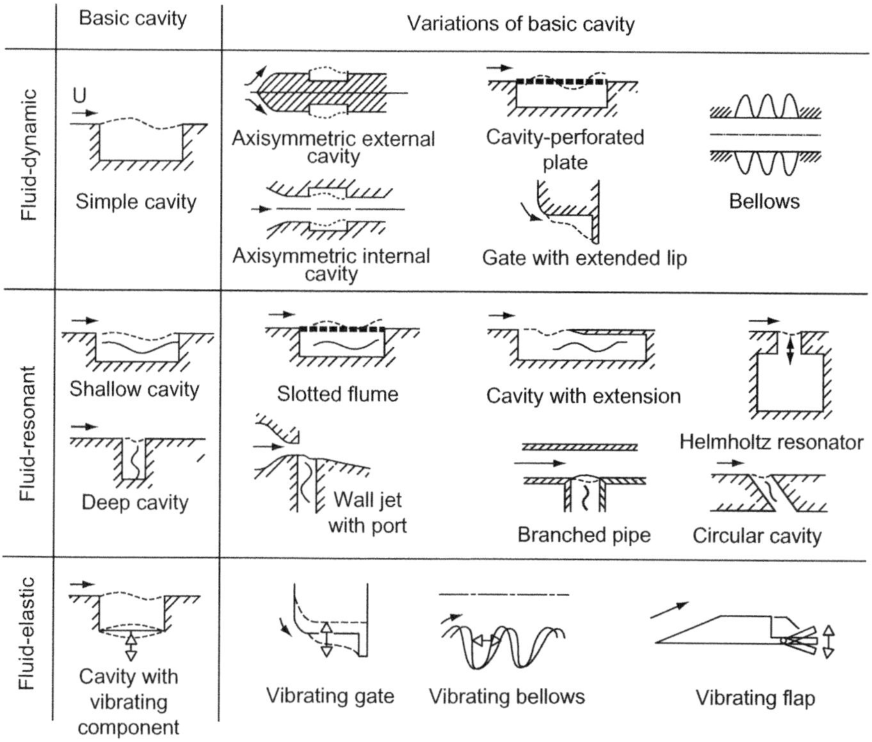

	Basic cavity	Variations of basic cavity		
Fluid-dynamic	U Simple cavity	Axisymmetric external cavity Axisymmetric internal cavity	Cavity-perforated plate Gate with extended lip	Bellows
Fluid-resonant	Shallow cavity Deep cavity	Slotted flume Wall jet with port	Cavity with extension Branched pipe	Helmholtz resonator Circular cavity
Fluid-elastic	Cavity with vibrating component	Vibrating gate	Vibrating bellows	Vibrating flap

FIGURE 5.50

Matrix categorization of fluid-dynamic, fluid-resonant, and fluid-elastic types of cavity oscillations [158].

of r_u. On the other hand, increasing the downstream radius substantially reduces the maximum acoustic pressure. The frequency at the peak acoustic pressure is relatively unaffected by the change in downstream radius. In order to apply the experimental data obtained for rectangular side branches to circular side branches, the rectangular side-branch width l is replaced by the effective diameter $d_e = 4l/\pi$ [165].

For piping systems involving two branches in close proximity, as shown in Fig. 5.51, the pressure pulsation amplitudes at resonance increase with a decrease of spacing l or the angle θ between the branches [166]. Amplitudes of acoustic pulsations at the resonance of single, tandem, and coaxial arrangements increase in this order. As the diameter ratio of the side branch relative to the main pipe is increased, the maximum amplitude at resonance increases and the resonance range shifts to higher Strouhal numbers [167]. For a side branch installed downstream of an elbow, the critical Strouhal number based on the average velocity is

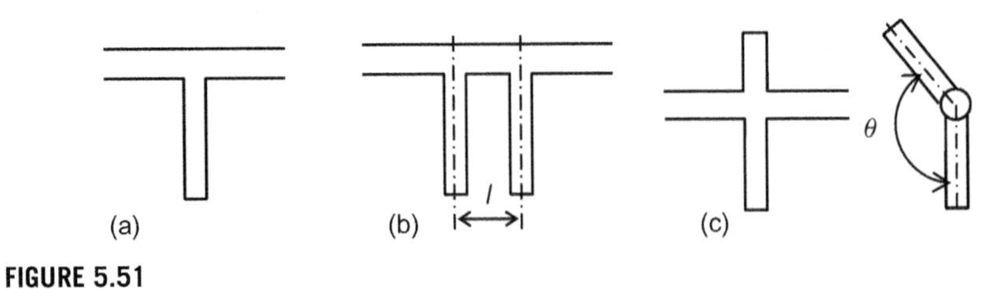

FIGURE 5.51

Configurations of closed side branches: (a) single branch, (b) tandem branches, and (c) coaxial branches.

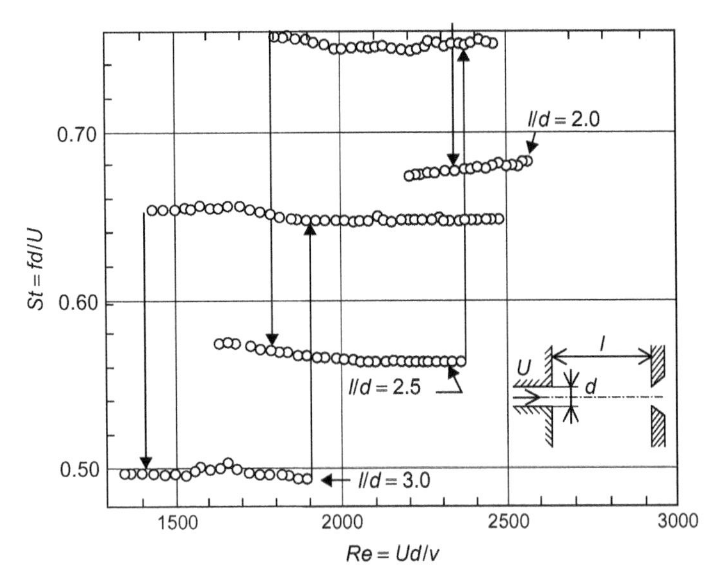

FIGURE 5.52

Variations of Strouhal numbers with Reynolds number for the hole-tone system [168].

higher when the branch is at the outer side of the elbow as compared to the inner side of the elbow. This is because the local velocity at the outer side of the elbow is higher than average [167].

5.5.2.2 Hole tones

When an axisymmetric jet impinges on a flat plate with a hole normal to the jet axis, a self-excited sound called a hole tone is generated. In this case, vortex rings are formed. As can be seen from Fig. 5.52 the Strouhal number $St = fd/U$ (d = nozzle exit diameter) takes almost constant value regardless of the Reynolds number $Re = Ud/v$ (v = kinematic viscosity of fluid) [168]. However, when the Reynolds number exceeds a critical value, a jump in the Strouhal number occurs and the mode number or stage number shifts to a new one. A strong hysteresis is observed between the cases of increasing and decreasing velocity. The Strouhal number decreases with an increase in the spacing l, but tends to approach a constant value independent of l when l instead of d is used as the characteristic length. When a cavity is formed by a side wall between the nozzle and the downstream orifice plate it frequently operates as a Helmholtz resonator [169].

Two closely spaced orifice plates installed in a pipeline induce similar tones [170–174]. Because of the presence of multiple acoustic natural frequencies, the vortex shedding frequency sequentially locks into higher or lower frequencies as the velocity is varied over a wide range. The pressure pulsation amplitude is maximum at the velocity where the natural vortex shedding frequency coincides with natural acoustic mode frequency. Using smoke visualization, it has been confirmed that distinct vortex rings are formed between the two orifice plates during lock-in. The number of vortices corresponds to the mode number [171].

5.5.2.3 Self-excited vibration of perforated plate and orifice plate

A pure tone may be produced from a perforated plate or orifice plate installed in a pipe even if no obstacle is placed downstream. The vortex ring from the perforated plate or orifice plate again acts as an energy source. Figure 5.53 shows the frequency f and amplitude $|p|$ of the pressure fluctuations as functions of mean velocity v at the perforated plate hole, where d is the hole diameter, m the number of holes and h the plate thickness [175]. The occurrence of lock-in is significantly affected by the flexural rigidity of the perforated plate. The lock-in velocity range, the frequency, and amplitude of the pressure fluctuations during lock-in, depend on the pipe length both upstream and downstream of the perforated plate. The Strouhal number based on the plate thickness and mean velocity at the perforated plate hole takes the value 0.9, 1.8, or 2.7 for mode number $n = 1, 2, 3$, respectively, regardless of the plate thickness, as long as no lock-in occurs. A pure tone may also be generated at an orifice plate with a single hole [176]. Flow visualization shows that the vortex ring spacing approximately equals the plate thickness, one half the plate thickness and one-third the plate thickness for $n = 1, 2, 3$, respectively [177].

5.5.2.4 Low-frequency vibration due to flow separation

A recirculation cell, due to large-scale flow separation at a pipe inlet or inside a flow control valve, frequently causes low-frequency vibrations. This phenomenon seems to be associated with instability of the pressure inside the cell due to volumetric flow change, but the excitation mechanism remains unclarified.

5.5.2.5 Acoustic fatigue of piping systems

Steam or gas flow through high-rate pressure-reducing devices such as safety valves, depressurizing valves and restriction orifices frequently produces severe

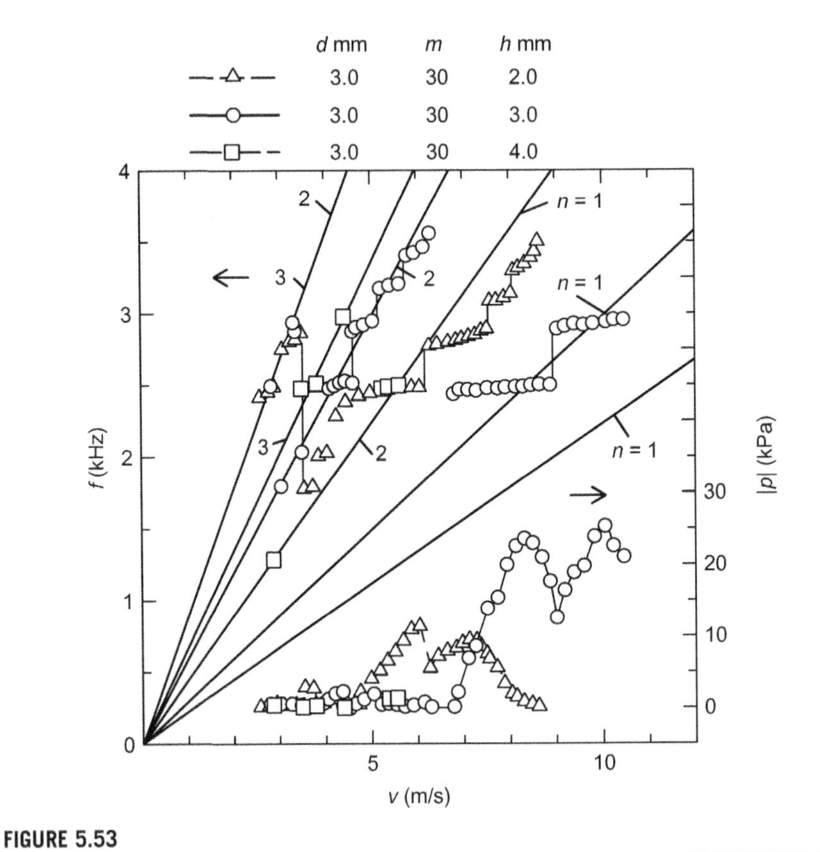

FIGURE 5.53

Frequencies and amplitudes of pressure fluctuations for different perforated plate thickness [175].

acoustic vibration causing piping system fatigue failure. The broad-banded random excitation differs from the single frequency self-excited excitations discussed above. Since the pipe wall thickness is relatively thin, the risk of acoustic fatigue generally increases with increasing pipe diameter.

As shown in Fig. 5.54, the diametral acoustic modes and corresponding circumferential structural modes must be focused on if the frequency is higher than about 500 Hz. Guidelines for designing piping systems against acoustic fatigue have been proposed, based on the data of practical pipes [178,179]. Recently, numerical analysis of resonant stress amplitudes in pipe walls during coincidence with the acoustic diametral modes has also been conducted [180,181]. Figure 5.55 illustrates the limit values of the energy parameter $M\Delta p$ as a function of pipe geometry parameter D/t, where M is the downstream Mach number, Δp the pressure drop across the pressure reducing station, D the pipe internal diameter, and t

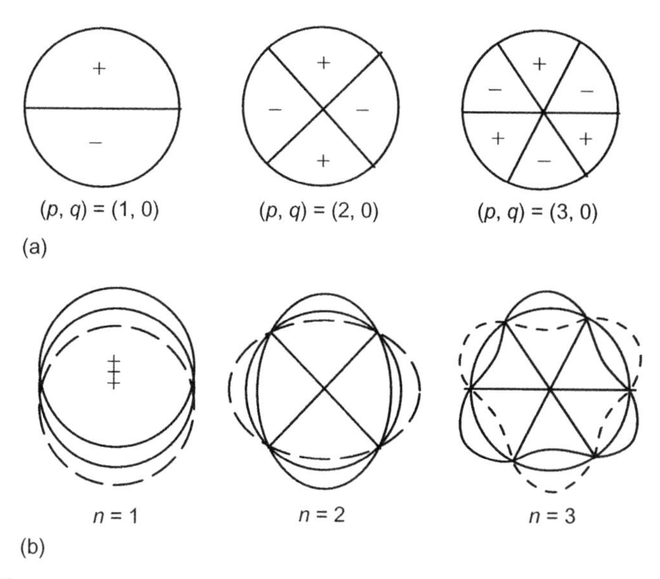

FIGURE 5.54

Examples of diametral acoustic modes and corresponding circumferential structural modes with matching wave numbers: (a) acoustic modes, and (b) structural modes.

the pipe wall thickness [180]. Figure 5.55 is available as a primary screening tool for piping system design.

5.5.3 Case studies and hints for countermeasures

5.5.3.1 Cavity tone

In order to attenuate the cavity tone oscillation amplitude, modifications of the rectangular cavity geometry, such as trailing- and leading-edge ramps, and offsets are effective [158,182–184]. The effectiveness of an unequally spaced groove of a labyrinth seal at a centrifugal pump [185] and a decrease in the cavity volume of a multi-stage restrictor at a flow control valve [186] has been reported.

5.5.3.2 Hole tone and the self-excited vibration of perforated plate and orifice plate

Suppression of the flow separation at the hole entrance by a rounded-off or sloping upstream corner is effective. In the case of two closely spaced orifice plates installed in a pipe, the countermeasure must also consider the downstream orifice. Extreme vibrations due to impinging of an annular jet from a high-pressure bypass valve were attenuated by the attachment of a perforated plate with non-sharp edge holes just downstream of the valve plug [187].

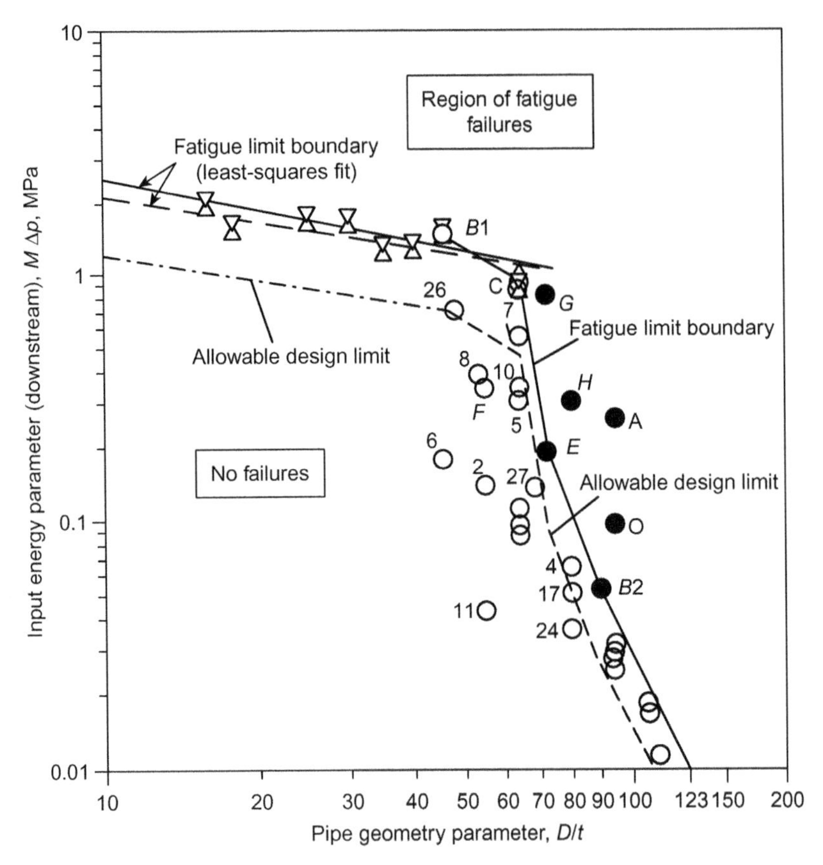

FIGURE 5.55

Fatigue limit diagram, for piping systems subjected to internal acoustic loading, relating acoustic input energy parameter to pipe geometry parameter [180].

5.5.3.3 Low-frequency vibration due to flow separation

Sudden changes in cross-sectional area and flow direction must be avoided to prevent flow separation. Low-frequency piping vibration occurred during the start-up of a large natural gas compression facility. Severe vibrations at 6 Hz were generated upstream of the throttling globe valve when the valve was opened more than 70%. A valve-plug design change eliminated the formation of the hydraulic instability and piping vibration. As can be seen from Fig. 5.56, which shows the numerical flow simulation inside the valve before the countermeasures were realized, a large recirculation cell due to flow separation was clearly forming upstream of the valve plug [188].

Considerable vibrations at 5 Hz in a pipe system between a flashing vessel and heat exchangers resulted in leakages and subsequent shutdown of a chemical

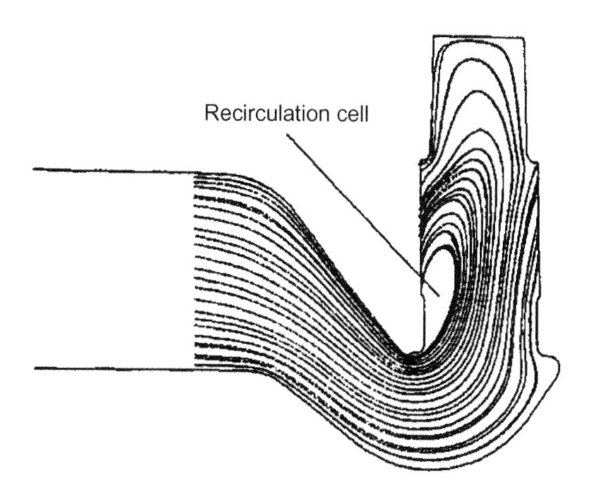

FIGURE 5.56

Pathlines in the symmetry plane of a standard plug at 100% stroke [188].

plant [189]. The modification, consisting of installation of a perforated pipe section between the outlets, was very effective in preventing flow separation from the pipe wall at the outlets of the flashing vessel.

5.5.3.4 Acoustic fatigue of piping system

Since increasing the local velocity at a pressure-reducing device causes severe pressure fluctuations with high frequencies, adopting a cage valve or a multi-stage pressure-reducing device is effective for reducing the power of the excitation source. Reinforcement of the pipe structure, for instance using circumferential stiffener rings, is also an option.

References

[1] JSME Report on Flow-induced Vibration in Mechanical Engineering, (P-SC10, 1980); JSME Research Subgroup Report on the Unsteady Characteristic and the Analyses of Flow in Fluid Machinery/Piping System (P-SC104, 1989); JSME Material on Various Problems in Piping System Including Fluid Machinery (the 243rd workshop, 1965); JSME Material on Fundamentals and Practice of the Flow-induced Vibration (the 505th workshop, 1980); JSME Material on Basics and the Latest Topics of the Flow-induced Vibration (the 674th workshop, 1988); and, JSME Material on Various Troubles Caused by Flow, (No. 00-43 workshop, 2000) (in Japanese).

[2] J.V. Hughes, J.M. Sharp, ASME Paper 56-A-200, 1956.

[3] C.R. Sparks, J.C. Wachel, Hydrocarb. Process. (July 1997) 183–189.

[4] JSME, *Mechanical Engineering Handbook* (in Japanese), JSME, 1991, pp. A5–134.

[5] O. Taniguchi, *Vibration Engineering Handbook* (in Japanese), Yoken-do, Japan, 1983. p. 1063

[6] T. Abe, T. Fujikawa, N. Ito, Bull. JSME 13 (59) (1970) 678.

[7] T. Sakai, S. Saeki, Bull. JSME 16 (91) (1973) 54.
[8] T. Sakai, S. Saeki, *Trans. JSME (Part 1)* (in Japanese) 38 (309) (1972-5) 1007.
[9] T. Sakai, S. Saeki, *Trans. JSME (Part 1)* (in Japanese) 39 (319) (1973-3) 862.
[10] T. Munakata, M. Ohba, T. Toyoda, H. Wada, Technical Report of the Ishikawajima-Harima Heavy Industries, Ltd. (in Japanese), 15, 3, 1975-5, p. 311.
[11] Y. Mohri, S. Hayama, *Proc. JSME* (in Japanese) 49 (439, C) (1983-3) 351.
[12] M. Nishimura, S. Fukatsu, K. Izumiyama, M. Hasegawa, *Trans. JSME* (in Japanese) 54 (504, C) (1988-8) 1740.
[13] M. Nishimura, S. Fukatsu, K. Izumiyama, M. Hasegawa, *Trans. JSME* (in Japanese) 54 (504, C) (1988-8) 1747.
[14] Y. Mohri, S. Hayama, *Trans. JSME* (in Japanese) 49 (439, C) (1983-3) 351.
[15] K. Yamada, Y. Noda, *Fluid Eng.* (in Japanese) 10 (10) (1974) 647–656.
[16] S. Hayama, et al., Preprint of the JSME (in Japanese), 740, 3, 1974, p. 135.
[17] H. Matsuda, S. Hayama, *Trans. JSME* (in Japanese) 51 (563, C) (1985-3) 515.
[18] H. Matsuda, S. Hayama, *Trans. JSME* (in Japanese) 52 (481, C) (1986-9) 2365.
[19] H. Matsuda, S. Hayama, ASME PVP, vol. 154, Book No. H00469, 1989, p. 17.
[20] H. Matsuda, S. Hayama, *Trans. JSME* (in Japanese) B (940-26) (1994-7) 609.
[21] H. Matsuda, S. Hayama, ASME PVP, vol. 310, Book No. H00973, 1995, p. 117.
[22] S. Hayama, Preprint of the JSME (in Japanese), 730, 14, 1973-10, pp. 113–116.
[23] T. Fujikawa, M. Kurohashi, M Kato, M. Aoshima, H. Yamamura, KOBELCO Technol. Rev. 37 (1) (1987) 59.
[24] M. Kato, Y. Inoue, T. Fujikawa, M. Aoshima, *Proc. JSME* (in Japanese) 52 (481, C) (1986-1) 2375–2381.
[25] M. Kato, E. Hirooka, Y. Inoue, S. Sato, *Trans. JSME* (in Japanese) 58 (554, C) (1992-10) 2907–2911.
[26] C.A. Brebbia, S Walker, Boundary Element Techniques in Engineering, Newness-Butterworths, London, 1980 (Translated jointly by N. Kamitani, M. Tanaka, K. Tanaka, Fundamentals and Application of Boundary Element Method, Baihukan, 1981.)
[27] T. Tanaka, T. Fujikawa, T. Abe, H. Utsuno, *Trans. JSME* (in Japanese) 50 (453, C) (1984-5) 848–856.
[28] T. Tanaka, T. Fujikawa, T. Abe, H. Utsuno, *Trans. JSME* (in Japanese) 50 (460, C) (1984-12) 2356–2363.
[29] T. Tanaka, H. Utsuno, T. Masuda, N. Kanzaki, *Trans. JSME* (in Japanese) 53 (491, C) (1987-7) 1443–1449.
[30] S. Suzuki, M. Imai, S. Ishiyama, *Trans. JSME* (in Japanese) 52 (473, C) (1986) 310–317.
[31] E.B. Wyle, V.L. Streeter, Fluid Transients, McGraw Hill, USA, 1978.
[32] Y. Mohri, S. Hayama, *Proc. JSME* (in Japanese) 55 (509, C) (1989-1) 52–57.
[33] R.C. Binder, J. Acoust. Soc. Am. 15 (1) (1943) 41–43.
[34] S. Hayama, et al., Bull. JSME 20 (146) (1977) 955.
[35] S. Hayama, et al., *Proc. JSME* (in Japanese) 45 (392) (1979-4) 422.
[36] Y. Mori, S. Hayama, *Proc. JSME* (in Japanese) 52 (481) (1986-9) 59.
[37] JSME, Research Subgroup Report on Pressure Pulsations in Piping Systems caused by Reciprocating Compressors, P-SC105, 1989.
[38] API (American Petroleum Institute) Standard 618, fifth ed., December 2007.
[39] M. Kato, E. Hirooka, K. Murai, T. Fujikawa, Symposium on Flow-Induced Vibration and Noise, ASME Winter Annual Meeting, vol. 6, 1992, pp. 237–244.

[40] T. Fujikawa, *Proc. JSME* (in Japanese) 57 (533, C) (1991-1) 148–153.
[41] O. Taniguchi, *Vibration Engineering Handbook* (in Japanese), Yoken-do, Japan, 1976. p. 1033.
[42] M. Costagliola, Trans. ASME, *J. Appl. Mech.* 14 (4) (1950) 415.
[43] J.F.T. Maclaren, et al., Inst. Mech. Eng. 4 (1974) 9–17 Conference Publication.
[44] M. Kato, M. Kurohashi, T. Fujikawa, M. Aoshima, *Trans. JSME* (in Japanese) 54 (505, C) (1988-9) 2148.
[45] H. Matsuda, S. Hayama, *Trans. JSME* (in Japanese) 54 (506, C) (1988-10) 2465.
[46] T. Fujikawa, M. Kato, M. Ito, H. Nomura, *Trans. JSME* (in Japanese) 55 (512, C) (1989-4) 904–909.
[47] M. Tanaka, K. Fujita, *Trans. JSME* (in Japanese) 53 (487, C) (1987-3) 591.
[48] S. Hayama, M. Matsumoto, *Trans. JSME* (in Japanese) 52 (476, C) (1986-4) 1192.
[49] K. Wakabayashi, A. Arai, J. Yamada, *Trans. JSME* (in Japanese) 63 (605, B) (1997-1) 231–236.
[50] M. Tanaka, K. Fujita, *Trans. JSME* (in Japanese) 53 (491, C) (1987-7) 1363.
[51] J.C. Wachel, C.L. Bates, Hydrocarb. Process. (October 1976) 152.
[52] H. Matsuda, *Safety Engineering* (in Japanese) 34 (3) (1995) 197.
[53] T. Abe, Preprint of the *JSME* (in Japanese) 200 (1968-9) 73.
[54] K. Mitsuhashi, Y. Kodera, Preprint of the *JSME* (in Japanese) 750 (14) (1975-10) 13.
[55] T. Kanda, *Trans. JSME* (in Japanese) 55 (512, C) (1989-4) 910–915.
[56] H. Ohashi, *Fluid Machinery Handbook* (in Japanese), Asakura-Shoten, Japan, 1998. p. 535.
[57] The JSME, *Vibration Damping Technology* (in Japanese), Yoken-do, Japan, 1998. p. 220.
[58] H. Matsuda, T. Yamaguchi, *Piping Technology* (in Japanese) 347-27 (8) (1985-7) 66.
[59] The JSME D&D 98, V_BASE forum (in Japanese), No. 98-8 (II), 1998-8, pp. 32–34.
[60] The JSME D&D 98, V_BASE forum (in Japanese), No. 98-8 (II), 1998-8, pp. 35–36.
[61] H. Matsuda, S. Hayama, *Trans. JSME* (in Japanese) 53 (496, C) (1987-12) 2510.
[62] H. Matsuda, S. Hayama, Asia-Pacific Vibration Conference, 93/Symposium on FIVES, Proceeding, vol. 1, 1993-11, p. 213.
[63] The JSME V_BASE forum (in Japanese), No. 920-65, 1992, pp. 22–23.
[64] M. Kato, Y. Sakamoto, Y. Nosaka, *R&D Tech. Rep. Kobe Steel* (in Japanese) 42 (4) (1992) 55–58.
[65] The JSME D&D 96, V_BASE forum (in Japanese), No. 96-5 (II), 1996-8, pp. 86–87.
[66] The JSME D&D 97, V_BASE forum (in Japanese), No. 97-10-3, 1997-7, pp. 14–15.
[67] The JSME material on Today's Vibration and Noise in Fluid Machinery (the 636th workshop) (in Japanese), 1986, pp. 65–70.
[68] D.J. Korteweg, Ann. Phys. Chem. 5 (1878) 525–542.
[69] O. Simin, Proc. Water Works Ass. (1904) 35.
[70] Y. Mori, et al., *Trans. JSME* (in Japanese) 39 (317) (1973) 305.
[71] T. Kobori, et al., *Hitachi Hyouron* (in Japanese) 37 (10) (1955) 33.
[72] API STANDARD 674, Second ed., American Petroleum Institute, 1995.
[73] M. Sano, *Trans. JSME* (in Japanese) 49 (440B) (1983) 828.
[74] G. Rzentkowski, S. Zbrojia, J. Fluid. Struct. 14 (2000) 529–558.

[75] H. Kawata, et al., *Trans. JSME* (in Japanese) 55 (514) (1989) 1584.
[76] G. Rzentkowski, Flow Induced Vibration, 328, *ASME PVP*, 1996, pp. 439–454.
[77] S. Fujii, *Trans. JSME* (in Japanese) 13 (44) (1947) 184.
[78] H. Kusama, et al., *Trans. JSME* (in Japanese) 20 (89) (1954) 15.
[79] H. Kawata, et al., *Trans. JSME* (in Japanese) 52 (480) (1986) 2947.
[80] H. Kawata, A study on the pulsating phenomena in centrifugal pumps and pipelines, Doctoral dissertation (in Japanese), The University of Tokyo.
[81] H. Kawata, et al., *Trans. JSME* (in Japanese) 55 (514) (1986) 1590.
[82] Y.N. Chen, IAHR/UTAM Symposium, Karlsruhe, 1979, pp. 265–278.
[83] R.E. Elder, IAHR/UTAM Symposium, Karlsruhe, 1979, pp. 285–286.
[84] J.C. Wang, et al., IAHR/UTAM Symposium, Karlsruhe, 1979, pp. 333–335.
[85] C. Wei-Yih, A.B. Rudavsky, IAHR/UTAM Symposium, Karlsruhe, 1979, pp. 337–339.
[86] H. Ohashi, *The Fluid-Machinery Handbook* (in Japanese), Asakura Publishing, 1998, 147.
[87] G. Rzentkowski, EAM, pp. 525–537.
[88] S. Benedek, J. Sound Vib. 177 (3) (1994) 337–348.
[89] H. Brekke, Seventh International Conference Pressure Surge, BHC Group, 1996, pp. 399–415.
[90] See p. 241 of Ref [87].
[91] A.H. Glattfelder, et al., IAHR/UTAM Symposium, Karlsruhe, 1979, pp. 293–297.
[92] H. Brekke, Eighth International Conference Pressure Surge, BHR Group 2000, pp. 599–610.
[93] O. Eichler, IAHR/UTAM Symposium, Karlsruhe, 1979, pp. 240–249.
[94] A. Bihhadi, K.A. Edge, in: Ziada, Staubi (Eds.), Flow Induced Vibration, Balkema, Rotterdam, 2000, pp. 607–613.
[95] JSME, The collection of the draft papers presented in the 69th National Conference of the JSME (in Japanese), No. 910-62, vol. D, 1991, p. 16.
[96] JSME, The collection of data of the vibration and noise problems presented in the v_BASE forum of the JSME (in Japanese), No.99-7 III, 1999, p. 29.
[97] JSME, The collection of the draft papers presented in the ordinary general meeting of JSME, No. 920-17, vol. D, 1992, p. 664.
[98] JSME, The collection of the draft papers presented in the 69th National Conference of the JSME (in Japanese), No. 930-63, vol. G, 1993, p. 156.
[99] JSME, The collection of data of the vibration and noise problems presented in the v_BASE forum of the JSME (in Japanese), 1997, p. 57.
[100] JSME, The collection of the draft papers presented in the 69th National Conference of the JSME (in Japanese), No. 920-17, vol. D, 1992, p. 642.
[101] JSME, The collection of data of the vibration and noise problems presented in the v_BASE forum of the JSME (in Japanese) 1992, p. 11.
[102] T. Kubota, H. Aoki, IAHR/U Symposium, Kahrsruhe, 1979, pp. 279–284.
[103] G. Garbrecht, Fifth International Conference on Pressure Surges, BHRA, 1986, pp. ix–xx.
[104] L.F. Menabrea, Note sur les effects de choc de l'eau dans les conduits, C. R. Acad. Sci. (Paris) 47 (1858) 221–224.
[105] N. Joukowsky, Waterhammer, Procs. AWWA 24 (1904) 341–424.

[106] L. Allievi, Theory of Water-Hammer, Translated by E. Halmos, Rome, Typography Riccardo Garroni, 1925.
[107] J. Parmakian, Waterhammer Analysis, Dover, 1963.
[108] P.J. Rorche, Computational Fluid Dynamics, Hermosa, 1985. p. 250.
[109] M. Amein, et al., *J. Hydraul. Div.*, ASCE 101 (HY6) (June 1975) 717–731.
[110] N.D. Katopodes, E.B. Wylie, Symposium on Multi-dimensional Fluid Transients, ASME, New Orleans, LA, December 1984, pp. 9–16.
[111] H.H. Rachford, E.L. Ramsy, SPE 5663, ASME, Dallas, September 1975.
[112] C. Kranenburg, *J. Hydraul. Div.* ASCE 100 (HY10) (October 1974) 1383–1398.
[113] T. Kiuchi, Int. J. Heat Fluid Flow 15 (5) (October 1994).
[114] B.E. Wylie, V.L. Streeter, Fluid Transients in Systems, Prentice-Hall, Englewood Cliffs, NJ, 1993.
[115] A.R.D. Thorley, Fluid Transients in Pipeline System, D. & L. George Ltd., 1991. p. 242.
[116] J. Yokoyama, Suigeki Nyuumon, Nishin Syuppan (in Japanese), 1980.
[117] A.H. Shapiro, The Dynamics and Thermodynamics of Compressible Fluid Flow, 2, Krieger, FL, 1983. p. 972.
[118] G.G. O'Brien, et al., J. Math. Phys. 29 (1951).
[119] J.T. Kephart, *Trans. ASME*, Ser. D 83 (3) (September 1961) 456–460.
[120] T. Tanahashi, et al., *Trans. JSME* (in Japanese) 35-2 (279) (1969) 2217–2226.
[121] G.A. Provoost, E.B. Wylie, Fifth International Symposium on Column Separation, Germany, September 1981.
[122] E.B. Wylie, Third International Conference on Pressure Surges, BHRA, March 1980, pp. 27–42.
[123] W. Zielke, *J. Basic Eng.*, ASME, *Ser. D* 90 (1) (1968) 109–115.
[124] A.K. Trikha, *J. Fluid Eng.*, ASME 97 (March 1975) 97–105.
[125] J.L. Achard, G.M. Lespinard, J. Fluid Mech. 113 (1981) 283–298.
[126] T. Kagawa, et al., *JSME Proc.* (in Japanese) 921 (1983)No. 83-0063.
[127] S. Oomi, et al., *Trans. JSME* (in Japanese) 47-424-B (1981) 2282.
[128] S. Uchida, ZAMP 7 (1956) 403–422.
[129] S. Oomi, et al., *Trans. JSME* (in Japanese) 50-457-B (1984) 1995–2003.
[130] S. Hayama, et al., *Trans. JSME* (in Japanese) 45 (1979) 422–432.
[131] G. Taylor, Seventh International Conference Pressure Surge, BHRA, 1996, pp. 343–362.
[132] N. Nakata, et al., *Ebara Jihou* (in Japanese) 114 (1980) 39–43.
[133] A.C.H. Kruisbrink, Proceedings of the Third International Conference on Valves and Actuators, 1990, pp. 137–150.
[134] A.R.D. Thorley, Proceedings of the Fourth International Conference on Pressure Surges, BHRA, 1983, pp. 231–242.
[135] K.K. Botros, B.J. Jones, O. Roorda, Fluid Structure Interaction," ASME PVP, vol. 337, No. 27-1, 27-2, 1996, pp. 241–264.
[136] D.C. Wiggert, A.S. Tijsseling, *Appl. Mech. Rev.*, ASME 54 (2002) 455–481.
[137] A.B.D. Almeida, Fifth International Conference Pressure Surge, BHRA, 1986, pp. 27–34.
[138] F.A. Locher, J.S. Wang, Seventh International Conference Pressure Surge, BHRA, 1996, pp. 211–223.
[139] M. Hamilton, G. Taylor, Seventh International Conference Pressure Surge, BHRA, 1996, pp. 15–27.

[140] D.S. Weaver, E. Naudascher, D. Rockwell (Eds.), Practical Experiences with Flow-Induced Vibrations, Springer-Verlag, 1980, pp. 305–319.
[141] Inada, Hayama, *Proc. JSME C* (in Japanese) 53 (1986) 933–939.
[142] JSME D&D Conf. /V-BASE No.139.
[143] JSME D&D Conf. /V-BASE No.76.
[144] S. Fujii, *Proc. JSME* (in Japanese) 18 (73) (1952) 182–184.
[145] T. Maeda, *Proc. JSME* (in Japanese) 35 (274) (1969) 1285–1292.
[146] F.W. Ainsworth, Trans. ASME 78 (4) (1956) 773–778.
[147] F.D. Ezekiel, Trans. ASME 80 (4) (1958) 904–908.
[148] O. Taniguchi, *Vibration Engineering Handbook* (in Japanese), Yoken-do, Japan, 1976, pp. 1068–1071.
[149] T. Takenaka, E. Urata, *Proc. JSME* (in Japanese) 71 (599) (1968-12) 1684.
[150] S. Fujii, *Sci. Mach.* (in Japanese) 1 (11) (1949).
[151] K. Cho, Transactions of the Faculty of Agriculture of Kagoshima University, vol. 39, 1989, pp. 273–286.
[152] K. Cho, et al., Transactions of the Japanese Society of Irrigation, Drainage and Reclamation Engineering (in Japanese), vol. 135, 1988-6, p. 91.
[153] T. Matsunaga, et al., Proceedings of Symposium on Fluid Power, Japan Society of Fluid Power System, 1991, pp. 125–128.
[154] M.D. Berstein, et al., Flow-Induced Vibration, 154, *ASME PVP*, 1989. p. 155.
[155] S. Ziada, et al., J. Fluid. Struct Vol. 3 (1989) 529–549.
[156] D. Rockwell, E. Naudascher, Annu. Rev. Fluid Mech. 11 (1979) 67–94.
[157] J.E. Rossiter, RAE Tech. Rep. No. 64037 (1964) and Reports and Memoranda No. 3438, 1964.
[158] D. Rockwell, E. Naudascher, Trans. ASME, J. Fluid. Eng. 100 (1978) 152–165.
[159] D. Rockwell, Trans. ASME, J. Fluid. Eng. 99 (1977) 294–300.
[160] K.K. Ahuja, J. Mendoza, NASA Contractor Report, Final Report Contract NAS1-19061, Task 13, 1995.
[161] J.S.W. Rayleigh, The Theory of Sound, Vol. II, Dover, 1945, pp. 180–183.
[162] F.C. DeMetz, T.M. Farabee, AIAA Paper 77-1293, 1977.
[163] S.A. Elder, T.M. Farabee, F.C. DeMetz, J. Acoust. Soc. Am. 72 (2) (1982) 532–549.
[164] D.S. Weaver, G.O. Macleod, Proceedings of the Symposium on Flow-Induced-Vibration, ASME, PVP, Vol. 389, 1999, pp. 291–297.
[165] J.C. Bruggeman, Doctoral Dissertation, Technische Universität Eindhoven, Eindhoven, The Netherlands, 1987.
[166] S. Ziada, E.T. Bühlmann, IMechE, C416/009, 1991, pp. 435–444.
[167] S. Ziada, S. Shine, J. Fluid. Struct. 13 (1) (1999) 127–142.
[168] R.C. Chanaud, A. Powell, J. Acoust. Soc. Am. 37 (5) (1965) 902–911.
[169] T. Morel, Trans. ASME, J. Fluid. Eng. 101 (1979) 383–390.
[170] F.E.C. Culick, K. Magiawala, J. Sound Vib. 64 (3) (1979) 455–457.
[171] H. Nomoto, F.E.C. Culick, J. Sound Vib. 84 (2) (1982) 247–252.
[172] R.E. Harris, D.S. Weaver, M.A. Dokainish, Proceedings of the International Conference on Flow Induced Vibrations, 1987, pp. 35–59.
[173] P.N. Ramoureux, D.S. Weaver, IMechE, C416/093, 1991, pp. 303–312.
[174] JSME, The JSME D&D 96, v_BASE forum (in Japanese), No. 96-5 (II), 1996, pp. 82–83.

[175] M. Sano, JSME Int. J. C 41 (1) (1998) 30–36.
[176] K. Okui, R. Yamane, F. Mikami, E. Takekoshi, *Turbomachinery* (in Japanese) 20 (10) (1992) 636–641.
[177] K. Okui, R. Yamane, F. Mikami, E. Takekoshi, *Turbomachinery* (in Japanese) 21 (2) (1993) 87–91.
[178] V.A. Carucci, R.T. Mueller, ASME Paper No. 82-WA/PVP-8, 1982.
[179] K.J. Marsh, P.J. van de Loo, G. Spallanzani, R.W. Temple, Concawe Report, No. 85/52, 1985.
[180] Eisinger, F.L. and Francis, J.T., Proceedings of the Symposium on Flow-Induced Vibration, ASME PVP, Vol. 389, 1999, pp. 393–399.
[181] I. Hayashi, T. Hioki, H. Isobe, Preprint of JSME Conf. (in Japanese), No. 01-12, 2001, pp. 116–119.
[182] S. Ethembabaoglu, Division of Hydraulic Engineering, University of Trondheim, Norwegian Institute of Technology, 1973.
[183] M.E. Franke, D.L. Carr, AIAA Paper 75-492, 1975.
[184] X. Zhang, X.X. Chen, A. Rona, J.A. Edwards, J. Sound Vib. 221 (1) (1999) 23–47.
[185] Y. Handa, K. Toyonaga, *Turbomachinery* (in Japanese) 18 (4) (1990) 220–224.
[186] JSME, The JSME D&D 96, v_BASE forum (in Japanese), No. 96-5 (II), 1996, pp. 80–81.
[187] JSME, The JSME Conf., v_BASE forum (in Japanese), No. 024-1, 2002, pp. v11–v12.
[188] P.J. Schafbuch, T. McMahon, T. Kiuchi, Proceedings of the Symposium on Fluid–Structure Interaction, Aeroelasticity, Flow-Induced Vibration and Noise, ASME, AD, vol. 53, No. 2, 1997, pp. 507–516.
[189] E. van Bokhorst, F. Goos, H.J. Korst, J.C. Bruggeman, Proceedings of the Symposium on Fluid–Structure Interaction, Aeroelasticity, Flow-Induced Vibration and Noise, ASME, AD, vol. 53, No. 2, 1997, pp. 533–541.

CHAPTER

Heating-Related Oscillations and Noise

6

In the past, many acoustic oscillations and noise-related problems in combustors, chemical plants, etc. have been reported. Many unsolved problems remain because most cases are complicated phenomena often involving interaction between different mechanisms.

In this chapter, we discuss acoustic oscillations and noise caused by heating. This includes combustion-driven oscillations, oscillations due to steam condensation, as well as boiling-related flow-induced vibrations. The mechanisms of oscillation and noise generation, evaluation methods and countermeasures are presented, based on work currently reported in the literature.

6.1 Acoustic oscillations and combustion noise

6.1.1 Introduction

Loud noise caused by combustion is often encountered in plants where boilers, heaters, furnaces, turbines, and heat exchangers are used. It is very important to suppress this noise due to the resulting effect on the environment or plant operation.

Combustion noise may be classified into two types [1], combustion-driven oscillations, and combustion roar. Combustion-driven oscillations are a type of self-excited vibration, resulting from the interaction between acoustic and heating systems. Oscillations occur at a single frequency, or at most a few frequencies, as shown in the spectrum of Fig. 6.1. Similar oscillations may be caused by a heater or a cooler located in a tube. The principal mechanism is illustrated in Fig. 6.2, where a heater controlled by acoustic oscillations (in a feedback loop) is located in a Helmholtz tube. The Helmholtz tube can be modeled by a mass representing the throat and a spring representing the compressibility of the gas in the volume below, so that free oscillations may occur. If the power supplied to the heater is synchronized with the velocity of the fluid mass within the throat, as shown in Fig. 6.2, the mass is accelerated by the thermal expansion of the heated gas, resulting in amplification of the acoustic oscillations. The oscillations occur at the natural frequency of the system and are amplified only when the gas is heated at

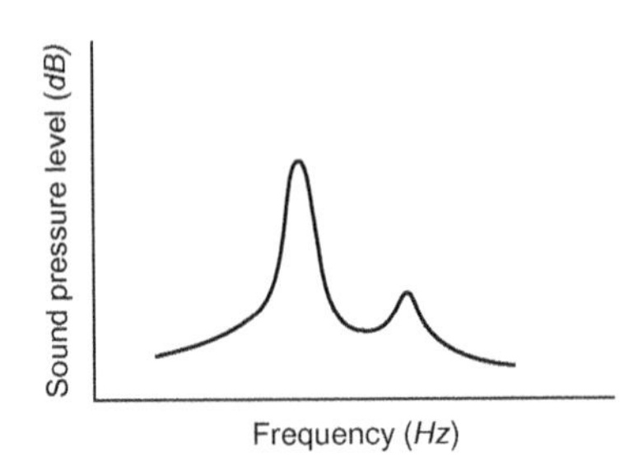

FIGURE 6.1

Spectrum of combustion-driven oscillations.

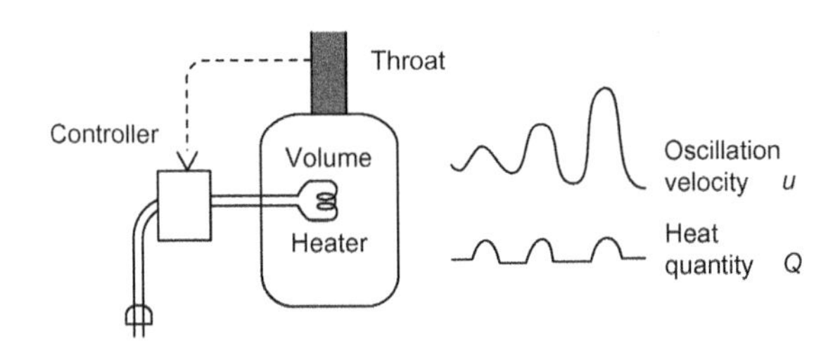

FIGURE 6.2

Generation of heat-driven oscillations.

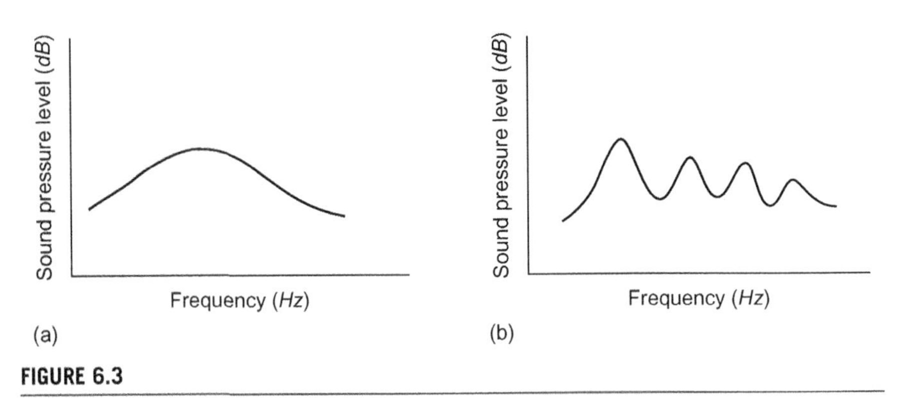

FIGURE 6.3

Spectrum of combustion roar: (a) open flame, and (b) flame in a chamber.

the correct phase. If the phase of the power supply is reversed, the oscillations are damped out.

Combustion roar is a type of forced oscillation excited by turbulence, generated by heat fluctuations. Since combustion is a random process, the heat release has randomly fluctuating components. The resulting thermal expansion excites the air surrounding the flame, which produces sound. The sound spectrum has a wide band (Fig. 6.3(a)) when the flame is in the open free field. However, when the combustion occurs in a furnace or an enclosed chamber, dominant peaks may appear due to resonances, as shown in Fig. 6.3(b).

This section deals with combustion driven oscillations first, followed by combustion roar.

6.1.2 Combustion-driven oscillations

6.1.2.1 Research history and evaluation criteria

In 1777, Higgins observed that sound was generated by a hydrogen burner when located in a tube, as shown in Fig. 6.4[2]. In 1850, Sondhauss investigated the phenomenon of sound generation by heat in a tube with one end closed and the other open. Rijke also performed a similar experiment using a tube with both ends open [3]. Because of their contributions, the closed-open tube and the open-open tube with a heater are called the Sondhauss tube and the Rijke tube, respectively. Bosscha and Rises found that a cooled wire net also generates sound similarly to a Rijke tube [4]. These heat-induced sounds are also called singing flames or gauze tones. The mechanism of sound generation is the interaction between the acoustic system and the heating system. Rayleigh indicated in his book *The Theory of Sound* [5] that sound generation is amplified if heat is supplied when density is high, lost when density is low, and damped out under the opposite conditions. In his book *The World of Sound*, Bragg [6] states that the acoustic oscillations grow if heat is gained during the expansion phase or lost during the compression phase.

The following three explanations have been proposed for the underlying excitation mechanism.

6.1.2.1.1 Explanation based on temperature change around a heater

This idea is mainly valid for the Rijke tube with mean flow. Air column oscillations may occur in the apparatus shown in Fig. 6.5. For example, sound at a frequency of 10 Hz is generated when the heater is located approximately 4.3 m

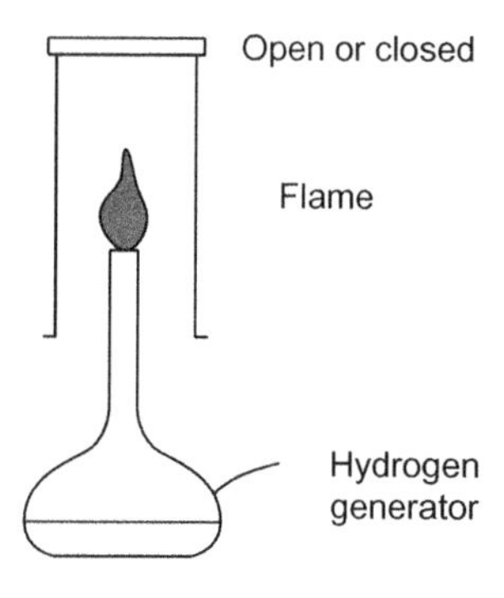

FIGURE 6.4

Higgins's experiment.

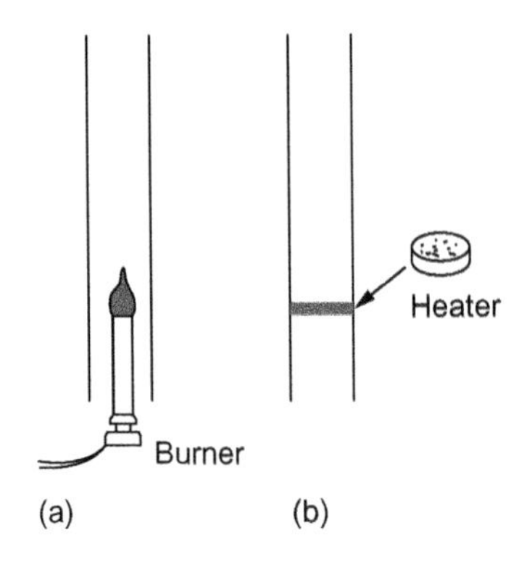

FIGURE 6.5

Rijke tube heated by (a) burner and (b) heater.

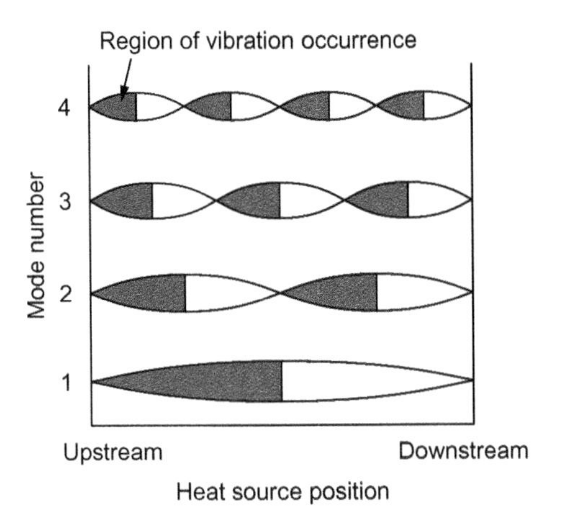

FIGURE 6.6

Heater positions susceptible to oscillations for different pressure modes.

from the upstream end of a 17 m long tube. Higher frequency sound at 600 Hz is generated for a heater position of about 7.5 cm from the end of a 30 cm long tube [6]. The frequency of the sound is the natural frequency of the acoustic system. Occurrence of acoustic oscillations is dependent on the position of the heater and the vibration mode shape. The vibration mode is excited when the heat source is located between the node and the antinode of the pressure mode shape in the flow direction. Conversely, the same mode is suppressed when the heat source is located between an antinode and a node as shown in Fig. 6.6. The mechanism discovered by Saito [7] is illustrated in Fig. 6.7. Figure 6.7(c) is a schematic of the temperature distribution near the heat source located as shown in Fig. 6.7(b). The temperature is affected by the vibration velocity. When the gas-flow speed

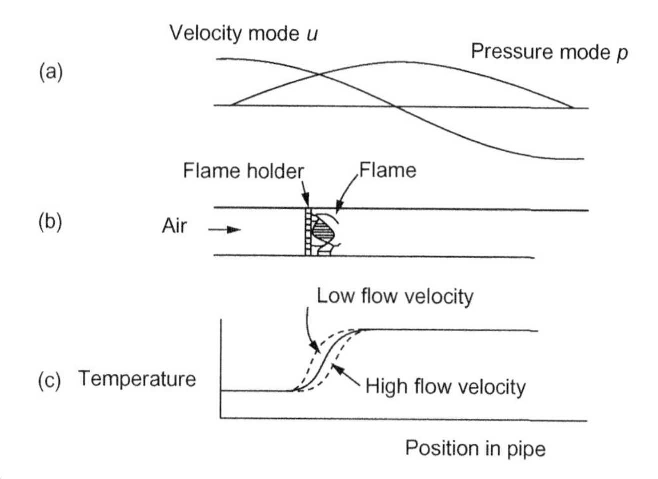

FIGURE 6.7

Mechanism of heat-driven air column vibration [7].

increases the gas temperature decreases, as shown by the lower dotted line. On the other hand, the temperature increases when the gas speed decreases. This means that the temperature increases in phase with the expansion process for the fundamental acoustic mode, which results in the occurrence of acoustic oscillations. Madarame [8] presented a quantitative analysis of this phenomenon. He carried out experiments using the apparatus shown in Fig. 6.8 and confirmed the existence of the unstable regions in Fig. 6.6. Madarame also developed a calculation method for the destabilization energy.

Yamaguchi [9] performed tests using an apparatus similar to that depicted in Fig. 6.8. It is reported that the acoustic oscillations are unlikely to occur when the length l_1 or l_2 is short, as seen in Fig. 6.9. The Japan Burner Research Committee [10] performed some experiments for countermeasures by using a small size rig similar to Fig. 6.8, as well as a practical-sized rig, and demonstrated that installing a resonator or reducing the upstream volume is effective. In addition, they developed a computer program for computation of a design criterion, which is presented later.

6.1.2.1.2 Explanation based on thermo-acoustic theory

Katto et al. [11] performed experiments using a Sondhauss tube to investigate the heating-induced oscillations. The mechanism of oscillation in a Sondhauss tube may be explained by thermo-acoustic theory. This approach was introduced by Swift [12] based on results by many researchers, including Carter, Feldman, Kramer, and Rott. When the closed end of a tube is at a higher temperature than the open end, acoustic oscillations may occur. In this condition, shown in Fig. 6.10, the oscillating fluid mass moves to the closed end and then it is heated by the wall at the closed end. Next, the fluid mass loses heat during the expansion

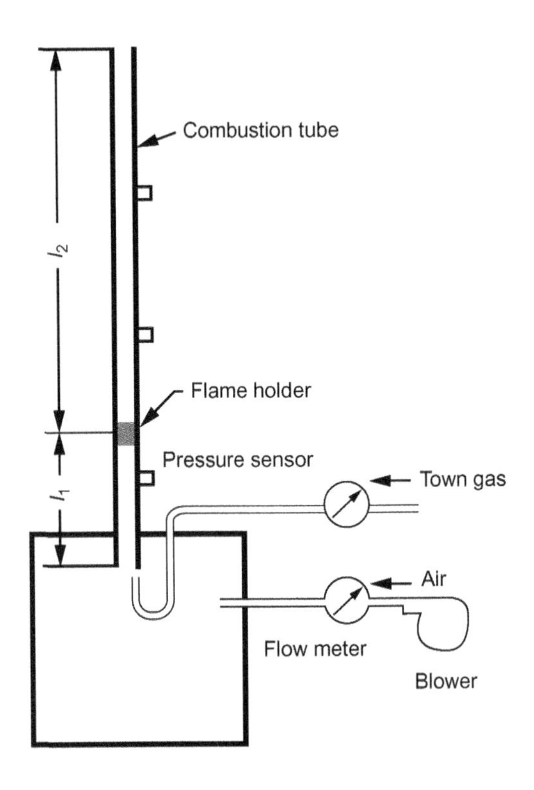

FIGURE 6.8

Acoustic vibration test apparatus [8,9].

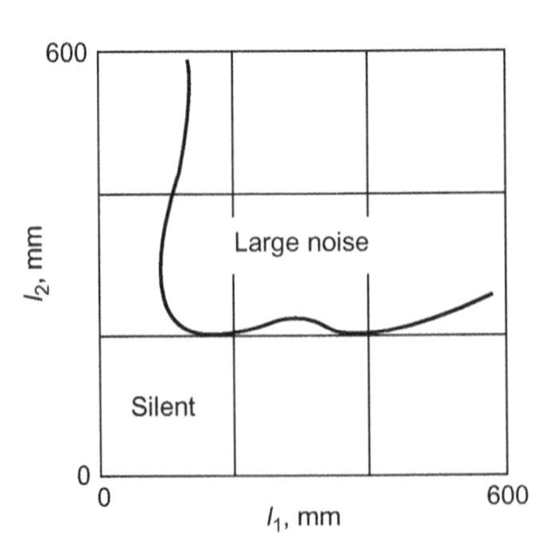

FIGURE 6.9

Vibration occurrence zone [9].

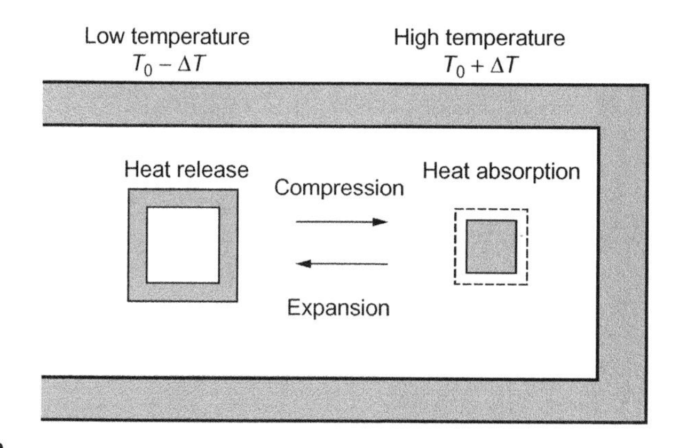

FIGURE 6.10

Mechanism of thermo-acoustic oscillations [12].

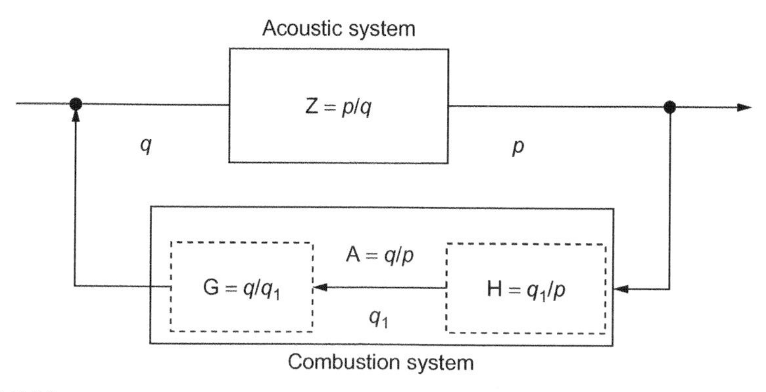

FIGURE 6.11

Feedback system.

process as it moves to the open end. This process repeats and leads to thermo-acoustic oscillations. These oscillations are also known to occur in thermal engines.

6.1.2.1.3 Explanation by feedback theory

Feedback theory is often used to explain combustion-driven oscillations. Referring to Fig. 6.11, Z is the transfer function of the acoustic system relating the pressure p or velocity u to the volume change rate q due to heat release. 'A' represents the transfer function of the combustion system between q and p or q and u (the figure illustrates the case of pressure feedback). It is well-known that a system is unstable and oscillations occur if the loop gain $|ZA|$ is greater than unity and the phase angle greater than 180 degrees. This condition is easily satisfied at the resonance frequency. The effect of combustion on temperature change has also been investigated. Jones [13] considered the fact that the fuel supply was affected by pressure, and a time lag existed in the combustion process. Merk [14] and Putnam and Dennis [15] stated that heat convection at the walls was influenced by velocity and that heat supply lagged the velocity by about 45 degrees, while Carrier [16] reported a phase lag of $3\pi/8$ rad (67.5 degrees).

For pressure feedback, the following mechanism may be considered. At low pressure the fuel supply increases because of the increased pressure difference, and generates heat with a time lag due to the travel time to the flame position. If the phase lag becomes larger than 3/4 of a period, the system gains heat in the expansion process, leading to oscillations. Fig. 6.11 shows a pressure feedback model [17]. The transfer function A is divided into G and H, where $H = q_1/p$ is the characteristic of volume change rate versus pressure, and $G = q/q_1$ is the response at the flame position. The functions G and H are expressed as complex numbers with amplitude and phase. Their values can be obtained experimentally or theoretically, based on the model shown in Fig. 6.12. A typical example is shown in Fig. 6.13. In this model, it takes some time for the fuel ejected from the nozzle to travel the distance x to the flame location prior to combustion. The diffusion-combustion time lag is considered larger than in the premixing case, while the solid fuel time lag is larger than the corresponding liquid fuel lag. There is, however, little quantitative information available on the different time lags.

In the velocity feedback theory, combustion is influenced by the oscillation velocity. For example, in atomizing combustion, the heat release is larger when the inlet speed is high. The mechanism shown in Fig. 6.7 is a type of velocity feedback.

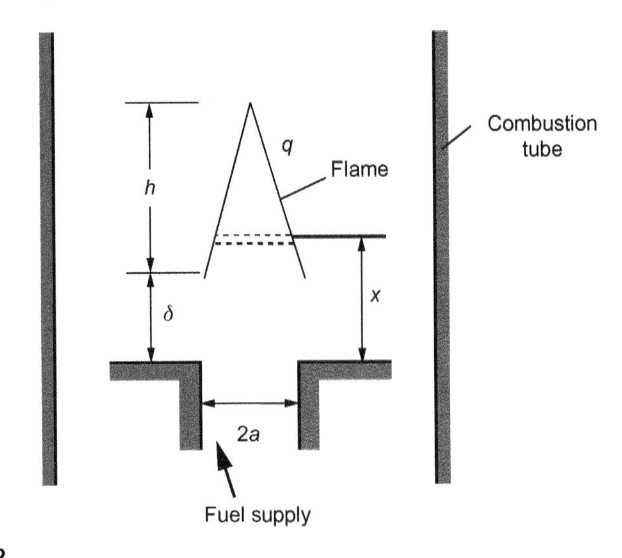

FIGURE 6.12

Time lag by fuel transportation.

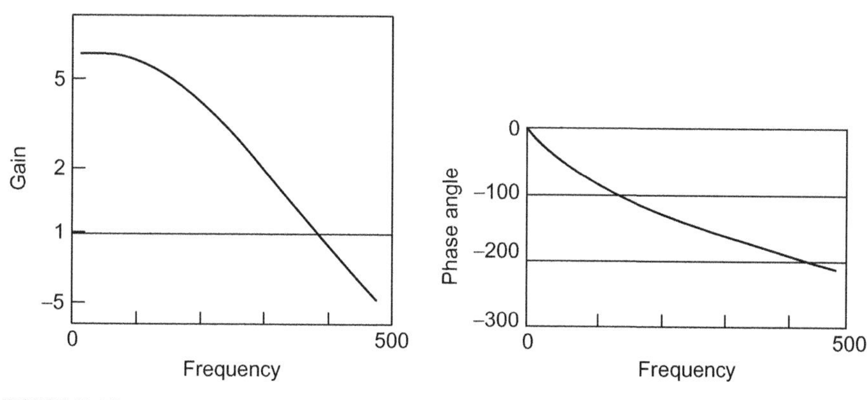

FIGURE 6.13

Transfer function $G = q/q_1$.

In addition to the three typical mechanisms presented above, other mechanisms may exist [18–21]. Furthermore, the phenomena observed in actual physical systems may be the result of a combination of different mechanisms. Many results are available [22–61].

6.1.2.2 Evaluation methods

It is difficult to accurately predict the occurrence of combustion oscillations. However, some evaluation methods have been proposed and are used in practice.

6.1.2.2.1 Rayleigh's criterion [1]

Rayleigh considered the work done on the system during one cycle of oscillation, and wrote the following equation for the energy change, assuming that the volume change is proportional to the heat release rate:

$$\Delta E = \oint pq \, \mathrm{d}t \propto \oint p\dot{Q} \, \mathrm{d}t \propto \dot{Q}_0 p_0 \cos \theta \tag{6.1}$$

Here, $\dot{Q}$ is the heat release rate, p the pressure, θ the phase angle, and the subscript 0 indicates amplitude. When $\Delta E > 0$, oscillations occur. For example, suppose the displacement x in Fig. 6.2 is given by $x = A \sin \omega t$, then the velocity is $u \equiv \dot{x} = A\omega \cos \omega t$ and the pressure $p \propto -A \sin \omega t$. If $Q \propto u$, that is $\dot{Q} \propto \dot{u} = -A\omega^2 \sin \omega t$, then:

$$\Delta E = \oint p\dot{Q} \, \mathrm{d}t \propto \omega^2 A^2 \oint \sin^2 \omega t \, \mathrm{d}t > 0$$

This means the system is unstable and oscillations occur. In Rayleigh's criterion, estimation of the phase angle θ is difficult. Furthermore, mean flow effects are neglected.

6.1.2.2.2 Madarame's criterion

Madarame [22] developed a calculation method for the energy input during one cycle in the Rijke tube, shown in Fig. 6.8, assuming that the heating rate Q can be expressed as:

$$Q(\tau) = \Theta\sigma^{m+1}\tau^{m}e^{2\sigma\tau}/m \tag{6.2}$$

and representing the pressure p, velocity u, and temperature T as:

$$u = u_0 + u_1 e^{i\omega t} \quad p = p_0 + p_1 e^{i\omega t} \quad T = T_0 + T_1 e^{i\omega t} \tag{6.3}$$

The steady term T_0 is calculated first, followed by the fluctuation term T_1 Then, the destabilizing energy per cycle can be obtained as follows;

$$\Delta E = A\beta \iint \mathrm{R_e}(-ip_1 e^{i\omega t})\frac{\partial}{\partial t}\mathrm{R_e}(T_1 e^{i\omega t})dx\,dt = \frac{\pi}{\omega}p_1 u_1 A\beta\Theta \mathrm{Re}(G) \tag{6.4}$$

Here, τ is the time lag from the start of firing, Θ the adiabatic combustion temperature, while σ and m are combustion parameters. A is the cross-sectional area of the tube, β the thermal expansion coefficient, and ω the angular frequency. Re(G) indicates the real part of G.

The Japan Burner Research Committee developed an evaluation method for practical combustion furnaces based on Madarame's theory. First, the natural frequencies and mode shapes are calculated using the finite element method where the chamber system is modeled using one-dimensional wave propagation theory. The energy coefficient Re(G) is calculated numerically. Once the energy coefficient is determined, Eq. (6.4) is used to determine the destabilizing energy ΔE. For practical evaluation it is recommended to use the growth ratio (negative damping ratio) ζ_h, defined as:

$$\zeta_h = \Delta E/(4\pi E) \tag{6.5}$$

where E is the total energy of the mode considered. If $\zeta_h > \zeta$ (the damping ratio of the system), the oscillation amplitude grows. For example, Table 6.1 shows results of an evaluation carried out for the mid-sized practical burner system shown in Fig. 6.14. The natural frequency and the growth ratio for each mode are

Table 6.1 Growth ratio for each mode

Mode	Frequency (Hz)	Growth ratio (ζ_h)	Stability evaluation
1	19.8	−5.02	O
2	45.8	−0.23	O
3	65.8	1.22	X
4	80.2	0.04	Δ
5	167.2	−2.38	O

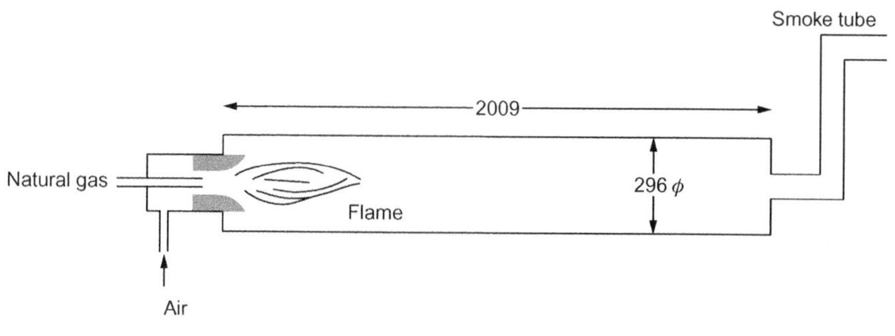

FIGURE 6.14

Mid-sized furnace used in experiments.

presented in Table 6.1. A large and positive ζ_h value indicates that oscillations in the corresponding mode are easily generated. The symbol X indicates cases where large oscillations occurred, while Δ indicates low-amplitude oscillations. No oscillations were observed in the experiments for the case indicated by O. This makes physical sense since ζ_h is negative, indicating stability.

6.1.2.2.3 Evaluation by feedback theory

Feedback theory [17] can be employed if the transfer functions of the acoustic system Z and the combustion system A are known. If the loop gain of AZ is larger than unity and the phase angle greater than 180 degrees, that is:

$$|AZ| > 1 \quad \text{and} \quad \arg(AZ) > 180 \text{ degrees} \tag{6.6}$$

then oscillations will occur. It is important to have the appropriate transfer functions of the heating system for this method.

6.1.2.2.4 Eisinger's criterion

Eisinger [24] proposed the following criterion for the Sondhauss tube and Rijke tubes shown in Fig. 6.15, based on many field cases:

$$(\log \xi)^2 = 1.52(\log \alpha - \log \alpha_{\min}) \tag{6.7}$$

where $\xi = (L - l)/l$, and $\alpha = T_h/T_c$. L is the length from the open end to the closed end, and l the distance from the open end to the heat source. T_h and T_c are the temperatures at the heat source and the open end, respectively. The constant $\alpha_{\min} = 2.14$ is the value at the most unstable position of the source, which is $\xi = 1(l = L/2)$. Fig. 6.16 is a plot of Eq. (6.7). The region above the curve is unstable. In this criterion, the effect of the mean flow is not considered. The criterion also does not agree with Madarame's theory. However, it is practical and easy to use.

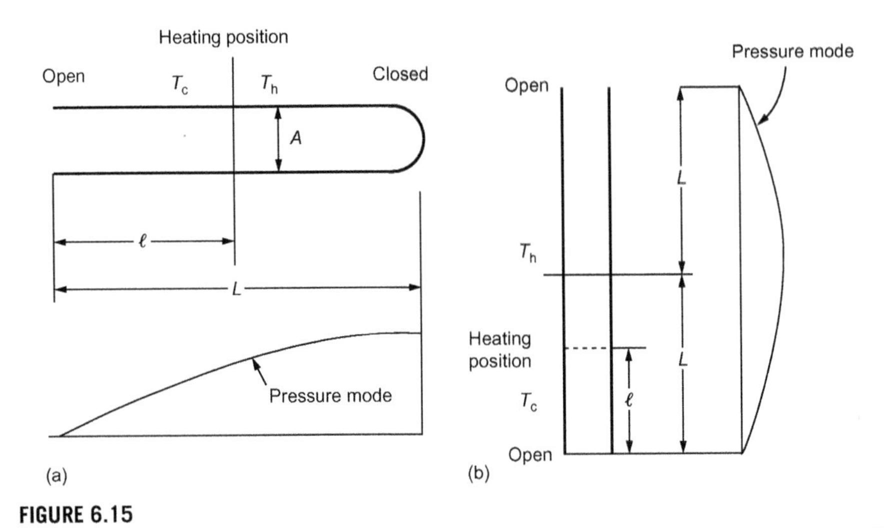

FIGURE 6.15

Sondhauss and Rijke tubes.

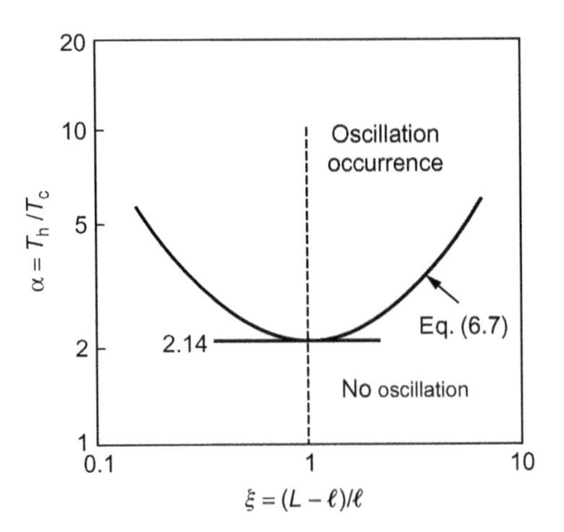

FIGURE 6.16

Stability criterion chart.

6.1.2.3 Examples of combustion oscillation problems and hints for countermeasures

Because combustion-driven oscillations are self-excited, increasing system damping is effective in reducing the destabilization force effect. The effective driving forces can also be reduced by changing mode shapes.

The following countermeasures are usually considered for structural or combustion oscillation conditions:

1. Installing resonators, such as Helmholtz resonators, or side-branch tubes.
2. Using an orifice to increase damping.

3. Changing the length or the diameter of the chamber or duct, making holes in the chamber, and inserting a baffle plate.
4. Reducing the upstream volume of the fuel or air supply line.
5. Stabilization of the flame by adjusting the swirler, changing nozzle position, or direction of nozzle jet angle.
6. Lengthening the flame and changing the flame position.
7. Modifying the fuel/air ratio, hence, reducing thermal load (reduction of heat release).

Many problem cases are reported in the literature [25–33]. Here, two typical cases are presented.

6.1.2.3.1 Heat-induced vibration in a steel plant blast furnace

Fig. 6.17(a) shows a section view of a blast furnace. Air and fuel gases are introduced through the burner port at the bottom and combustion gases travel through the dome-shaped section and the accumulator. When the fuel flow was increased during operation, large-amplitude pressure oscillations occurred in the furnace and structural vibrations of the piping became very large. The frequency measured near the burner position was 7.0–7.5 Hz. The pressure amplitude was small at the dome and large at the burner port, corresponding to the fundamental mode,

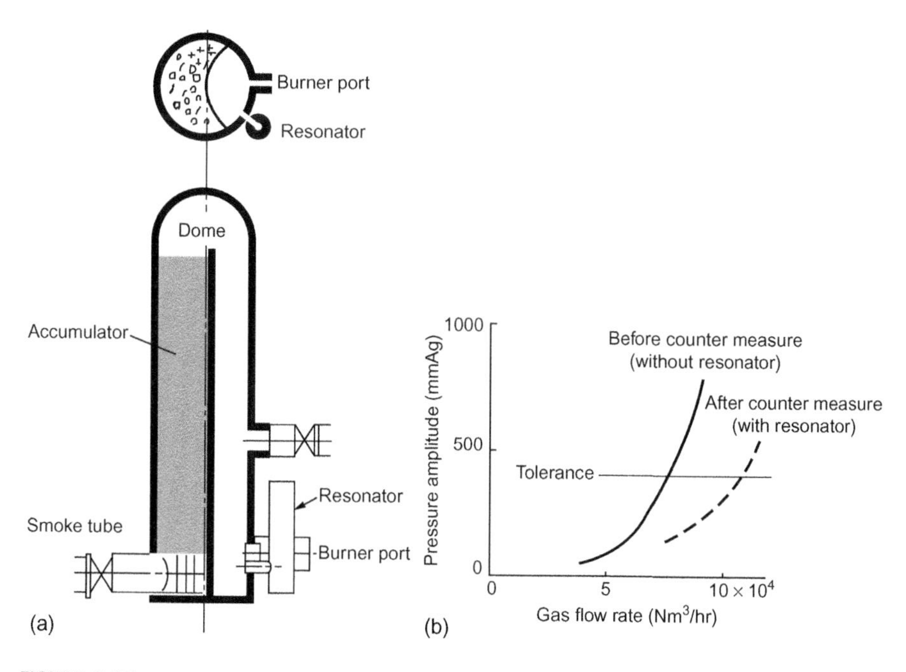

FIGURE 6.17

Blast furnace: (a) furnace configuration, and (b) pressure amplitude versus fuel flow rate.

with 1/4 wavelength. The solid line in Fig. 6.17(b) shows the magnitude of the pressure against the fuel flow rate.

The following countermeasures were attempted:

1. Introduction of uniform flow at the burner.
2. Redesign of the flame-holder equipment.
3. Installation of a Helmholtz resonator.
4. Introduction of high-pressure loss at the burner.
5. Reinforcement of the structures.

Among these countermeasures, the resonator was the most effective. The resonator position is shown in Fig. 6.17(a). The effectiveness of the resonator is shown by a broken line in Fig. 6.17(b). With this countermeasure, the vibration amplitude was reduced to within the allowed range for normal operation [27,30].

6.1.2.3.2 Vibration in a water tube boiler

In the small boiler depicted in Fig. 6.18, combustion oscillations occurred and severe flame fluctuations were observed at the condition of low air ratio combustion [30]. The stable zone was limited to a narrow air ratio zone. It was concluded

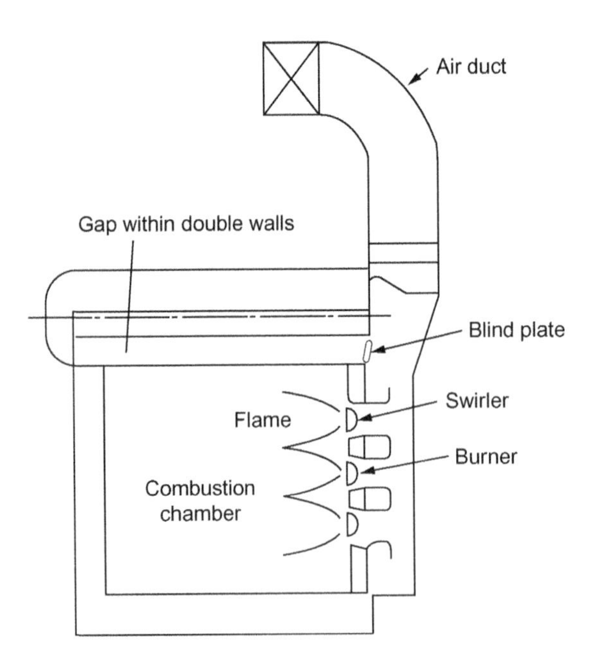

FIGURE 6.18

Water-tube-type boiler.

from frequency and phase measurements that the phenomenon was that of organ-type oscillations. Many countermeasures were tried, among which the following were found to be effective:

1. Blocking off the space between the air duct and gap within the double walls using blind plates, which reduced the upstream volume of the burner, thereby changing the acoustic mode.
2. Changing the swirler shape from a blade type to a flat type.
3. Change of nozzle, including reduction of the number of jets with wide jet angle.

These countermeasures successfully eliminated the combustion oscillations.

6.1.3 Combustion roar

6.1.3.1 Research history and evaluation

Combustion roar is a type of forced acoustic oscillation caused by fluctuating heat release. The sound level may reach 80 to 100 dB in boiler burners and 120–140 dB in jet engine combustion. Thomas and William [62] demonstrated the mechanism of roar noise. Causing ignition at the center of a soap bubble inflated by premixed fuel with air, the flame radius was found to increase with increasing combustion volume. The fluid flame accelerates to generate the sound pressure, which is expressed as:

$$P = \rho/(4\pi l)\mathrm{d}^2V/\mathrm{d}t^2$$

where ρ is the gas density, V the flame volume, and l the distance from source to observation point.

The work of many researchers on combustion roar, for example, Smith and Kilham [63], Hurle [64], Strahe [65], Shivashankara [66], and others, has been discussed by Suzuki [67]. In Japan, Kotake [68], Urie and Sato [69], Kazuki et al. [70], as well as the JSME research committee [1] have studied and reported on combustion roar. Other work is also reported in Refs. [71–94].

Flow in a flame normally becomes turbulent in practical burners, resulting in randomly fluctuating combustion. This non-uniform combustion becomes a monopole sound source. Kazuki et al. [70] state that roar consists of local roar, tip combustion roar, and jet sound, based on observation of a turbulent Bunsen burner. In a flame such as that shown in Fig. 6.19, turbulence generated by the instability of the shear layer around the burner is amplified by combustion when it goes through the flame zone and propagates upward, accompanying the local roar. On the other hand, the flow in the central part is ignited at random by turbulence from the flame tip, causing tip roar. In addition, gas flow is accelerated by thermal expansion, producing noise by shear action.

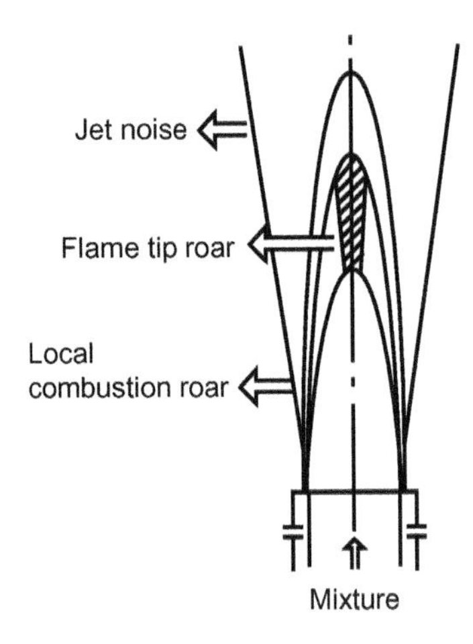

FIGURE 6.19

Turbulent flame.

6.1.3.2 Evaluation methods

The combustion roar sound power has been estimated by many researchers. Generally, the power P is dependent on the velocity U, the burner size D, and the combustion speed S. P can be expressed in the form:

$$P \propto U^{\alpha} S^{\beta} D^{\gamma}$$

For the various types of flames the following relations have been proposed:

Thomas (cubic flame)	: $P \propto S^2 D$
Smith (premixed flame)	: $P \propto U^2 S^2 D^2$
Strahe (spreading flame)	: $P \propto U^2 S^2 D^2$
Strahe (wrinkle flame)	: $P \propto U^3 S^3 D^3$
Kotake, Hatta (diffusion flame)	: $P \propto U^4 D^3$
Shivashankara (premixed flame)	: $P \propto U^{2.68} S^{1.35} D^{2.84}$
Muthukrishnan (flame holder)	: $P \propto U^{2083} S^{1.89} D^{2.77}$

Kanabe et al. [73] proposed the following equation based on experiments with a premixed propane burner with air under 'open air' conditions:

$$P = 1.65 \times 10^{-3} Re^{1.25} D^{0.5} S^{2.16}$$

where Re is the Reynolds number. Kotake states that the sound power is proportional to U^{α}, where α is approximately 2 for premixed combustion, and approximately 4 for diffusion flames [1]. Roar has a wide-band spectrum.

However, some dominant frequencies may arise in combustion chambers or ducts due to resonance effects.

6.1.3.3 Hints for countermeasures and practical examples

The occurrence of roar is inevitable during combustion. However, the following ideas may be considered for its suppression:

1. For the combustion chamber or burner:
   Stabilization of the flame, reduction of the gas flow velocity, use of a multi-burner, increasing the flame length.
2. For soundproofing:
   Laying of absorbent material, using enclosed resonators and reinforcement.
3. Avoidance of resonance:
   Avoid multiple acoustic reflections between parallel walls in ducts or chambers.

Figure 6.20 is an example of a low-noise-type burner [74,75] developed for reducing roar in which absorbent material is used and the air passage is enclosed.

Noise level specifications are very low for home equipment. Figure 6.21(a) is a boiler for home use. In the initial design, a peak in the noise spectrum was observed. A slant plate, as shown in Fig. 6.21(b), was installed at the exhaust duct in order to avoid resonance. Its effectiveness was demonstrated.

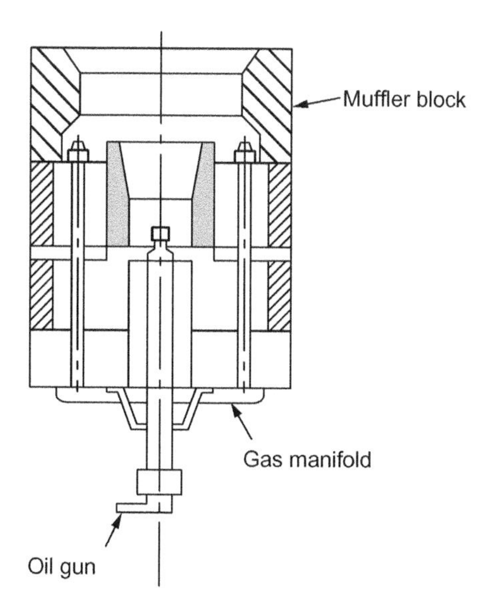

FIGURE 6.20

Low-noise burner.

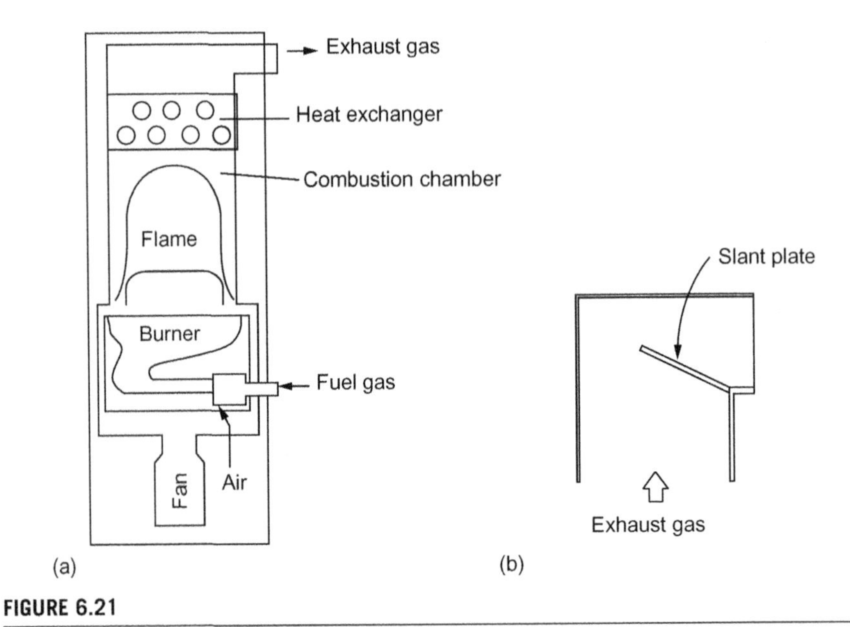

FIGURE 6.21

Home use boiler: (a) structure of the boiler, and (b) a slant plate for countermeasure.

6.2 Oscillations due to steam condensation

6.2.1 Introduction

In steam-water two-phase flow lines expansion/condensation of the steam may cause pressure oscillations, vibrations, and damage to piping and electrical instruments. Examples include flow in the export steam condensation system of a marine boiler, and flow in the suppression pool and feedwater piping of nuclear power plants. Significant research has been conducted to clarify the underlying mechanism and preventive measures.

Research has been conducted focusing on emergency cooling piping systems in nuclear power plants; for example, the full-scale experiments known as 'the Marviken project, [95] in Sweden (and the same type of experiments in Japan, USA, Germany, and Italy), detailed model experiments on emergency cooling piping systems in nuclear power plants [96], noise reduction of steam condensation in cooling water [97], and experimental research on heat transfer with steam condensation [98]. Additionally, there are several applications utilizing the condensation oscillations (e.g. degreasing and washing by steam condensation [99] and cooling control of high-temperature metal by condensation oscillations [100]).

6.2.2 Characteristics and prevention

Pressure oscillations caused by condensation (condensation oscillations) can be classified into chugging, two-phase flow oscillations in piping, and steam

Table 6.2 Condensation-induced oscillations

Class	Chugging	Two-phase flow oscillations in piping	Steam hammer
Cause	1. Inflow of steam to cooling water 2. Inflow of cooling water to steam flow	Water/steam two-phase flow in piping system	Trapping of steam in cooling water
Characteristics	1. Flow oscillation by time-lag between steam inflow rate and condensation rate 2. Flow oscillation by relation among the inertia, condensation rate and pressure fluctuation	Flow oscillation by sudden change in condensation rate caused by sudden change of cooling water flow rate	Water hammer by sudden condensation of trapped steam
Affected system	1. Boiler export steam condensation system 2. Nuclear power plant suppression pool	Water/steam two-phase piping system	Boiler and nuclear power plant feedwater line
Preventive method	1. Raise cooling water temperature Suppress condensation by mixing noncondensable gas Make outflow pattern random using many different-diameter holes 2. Raise cooling water temperature Increase steam flow rate	Enlarge piping diameter Avoid sudden increase of the cooling water flow rate	Enlarge piping diameter Avoid accumulation in piping Introduce slope in piping system Install drain trap at lower elevation of piping system Mix air to reduce condensation speed

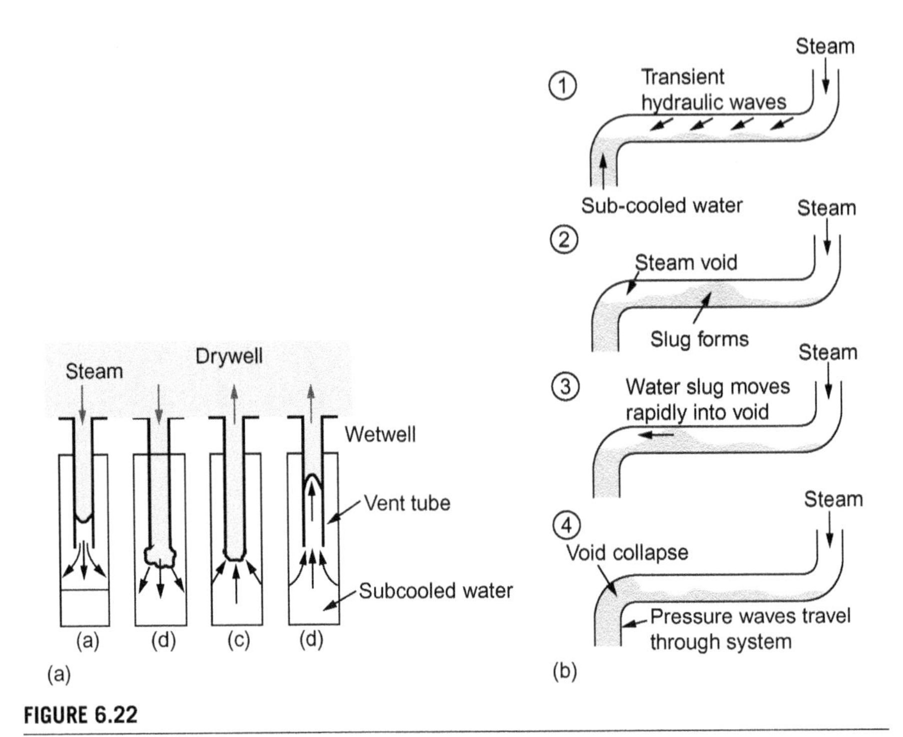

FIGURE 6.22

Schematics of (a) chugging, and (b) steam hammer [104].

hammer [101,102]. Table 6.2 and Fig. 6.22 show the characteristics of condensation oscillations, affected systems, and preventive methods. Chugging and steam hammer result in large-amplitude pressure oscillations. Although two-phase flow in piping generally results in relatively small amplitudes, sustained large-amplitude system oscillations may occur in the case of a sudden increase of the cooling water flow rate.

6.2.3 Examples of practical problems

1. *Experiment on feedwater piping system for nuclear power plant*[103]: in 1989, at the Loviisa nuclear power plant (Finland), a steam hammer experiment at the T-connection of a feedwater piping system was conducted when replacing piping damaged by corrosion. Steam hammer caused by the condensation of the steam trapped in the cooling water occurred during plant start-up. The piping structure was modified to suppress the pressure amplitude. The pressure amplitude was successfully reduced below the endurance limit.
2. *Pressurized water reactor* (PWR) *feedwater piping system*[104]: in 1973, at the Indian Point No. 2 PWR steam generator (USA), a horizontal piping

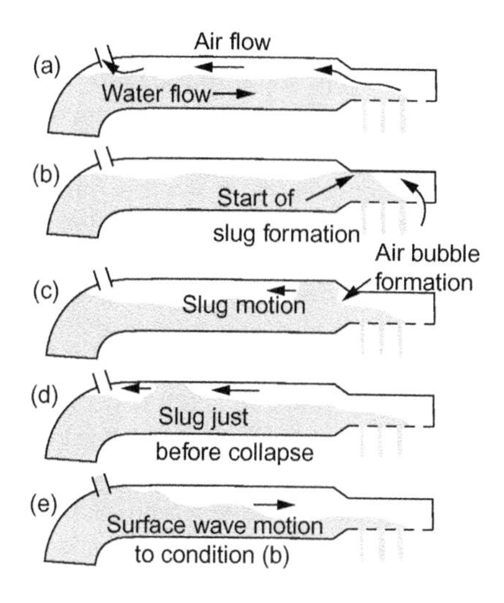

FIGURE 6.23

Steam hammer in feedwater piping [104].

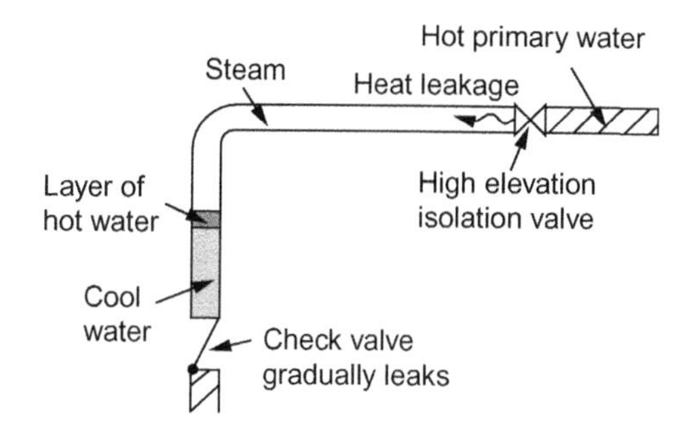

FIGURE 6.24

Steam hammer at emergency cooling system [104].

section was damaged when the feedwater was supplied by an auxiliary feedwater system because of the shut-off of the main feedwater system due to a turbine trip. The US Nuclear Regulatory Commission (NRC) confirmed that the cause was steam hammer by conducting (a) a scale-model experiment (Fig. 6.23), (b) an analysis, and (c) a field test. The pressure amplitude was reduced by redesigning the piping systems.

3. *Boiling water reactor* (BWR) *emergency cooling system*[104]: in the US, 10% of the snubbers of the emergency cooling system of a BWR (Fig. 6.24) were damaged during inspection. The cause was thought to be the following: (a) water accumulated at the check valve installed at the vertical straight piping

section and temperature stratification was formed by the high-temperature steam in the piping; (b) the temperature stratification was disturbed and broken by the pump start-up and the low-temperature water came in contact with the high-temperature steam; and (c) steam hammer occurred because of the rapid condensation of the steam at the contact surface. These phenomena were confirmed by a flow transient analysis (using the Hytran code) and structural analysis of the piping (using the PIPSYS code).

4. *Boiler feedwater pump suction line in thermal power plant* [104] (Fig. 6.25): at a thermal power plant in England, an emergency relief valve ((d) in Fig. 6.25) was damaged by steam hammer. In the piping, an emergency feedwater tank (at atmospheric pressure and low temperature, (c)) was installed at an elevation higher than the main hot feed-tank (a) (having high pressure and high temperature). The level of the emergency reserve feed-tank (c), had been designed such that its pressure equalled that of the main tank side at the position of the relief valve, and the relief valve had been designed to work when the water level of the main hot feed-tank (a) was low. During many years of operation, the emergency relief valve (d) worked effectively to control any increase in pressure and decrease in water level in the main feed-tank. However, in this case, the water rose up the piping to the emergency reserve feed-tank when the pressure in the main tank surpassed the design pressure. The flushed steam then came in contact with the low-temperature water, causing steam hammer.

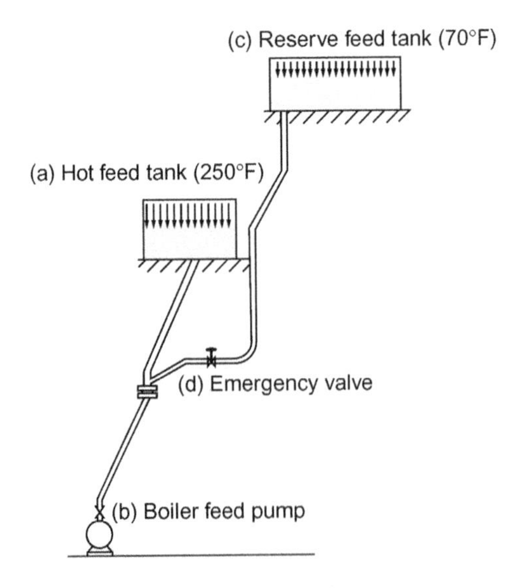

FIGURE 6.25

Feedwater piping in boiler [104].

6.3 Flow-induced vibrations related to boiling

6.3.1 Introduction

Flow-induced vibrations in boiling systems may cause structural fatigue failure. Furthermore, such vibrations may compromise the ability to control system pressure, temperature, and flow rate. Other possible outcomes include temperature increase in pipe walls due to the decrease in burnout heat flux, and thermal fatigue failure due to temperature fluctuations on the pipe wall. In order to avoid vibrations in boiling systems, a thorough study to evaluate potential vibration mechanisms is required at the design stage.

A study by Ledinegg [105] on pulsations in a boiler header represents the first research work in the field of boiling two-phase flow. Thereafter, research on a fast-breeder reactor (FBR) steam generator, a liquefied natural gas (LNG) plant vaporizer, and a reboiler in a heat recovery system was conducted [106]. The review papers [107–110] present the historical background of this research.

6.3.2 Vibration mechanisms

The generation mechanisms of flow-induced vibrations related to boiling are varied. In many cases, vibration is caused by more than one mechanism. A classification based on the basic mechanisms is presented in Table 6.3 [111,112].

6.3.3 Analytical approach

Only a few estimation methods have been developed for the analysis of vibration in boiling systems. One of these methods, an analytical model for pressure-drop-type oscillations [113], is explained in this section as an example of a non-linear analysis. A single boiling piping system, which is fitted with a surge tank between the supply tank and the heating tube, is shown in Fig. 6.26. The relation between flow rate Q_2 and the pressure difference $\Delta P_2 = P_2 - P_e$ between the surge tank and the exit of the heating tube is shown in Fig. 6.27. The relation between the flow rate Q_1 at the valve upstream of the surge tank and the pressure drop $\Delta P_1 = P_1 - P_2$ corresponds to curve No. 1 in Fig. 6.28. On the other hand, the relation between the flow rate to the heating tube Q_2 and the pressure drop $\Delta P_2 = P_2 - P_e$ corresponds to line No. 2 in the same figure. To simplify the analysis, the pressure P_1 in the supply tank and the pressure P_e at the heating tube exit are assumed constant. Consider now the case where the condition of the system is somewhere between points B and D on curve No. 2 in Fig. 6.28. Perturbation of the pressure P_2 causes the flow rate Q_2 to decrease along the curve DB, thereby increasing the pressure P_2. Since the flow rate Q_1 (Fig. 6.26) becomes higher than the flow rate Q_2, the water level in the surge tank increases, resulting in an increase of the pressure P_2 to the point B. As the pressure on curve No. 1 corresponding to the flow rate at point B is different from that on curve

Table 6.3 Classification and physical mechanisms of flow-induced vibrations related to boiling

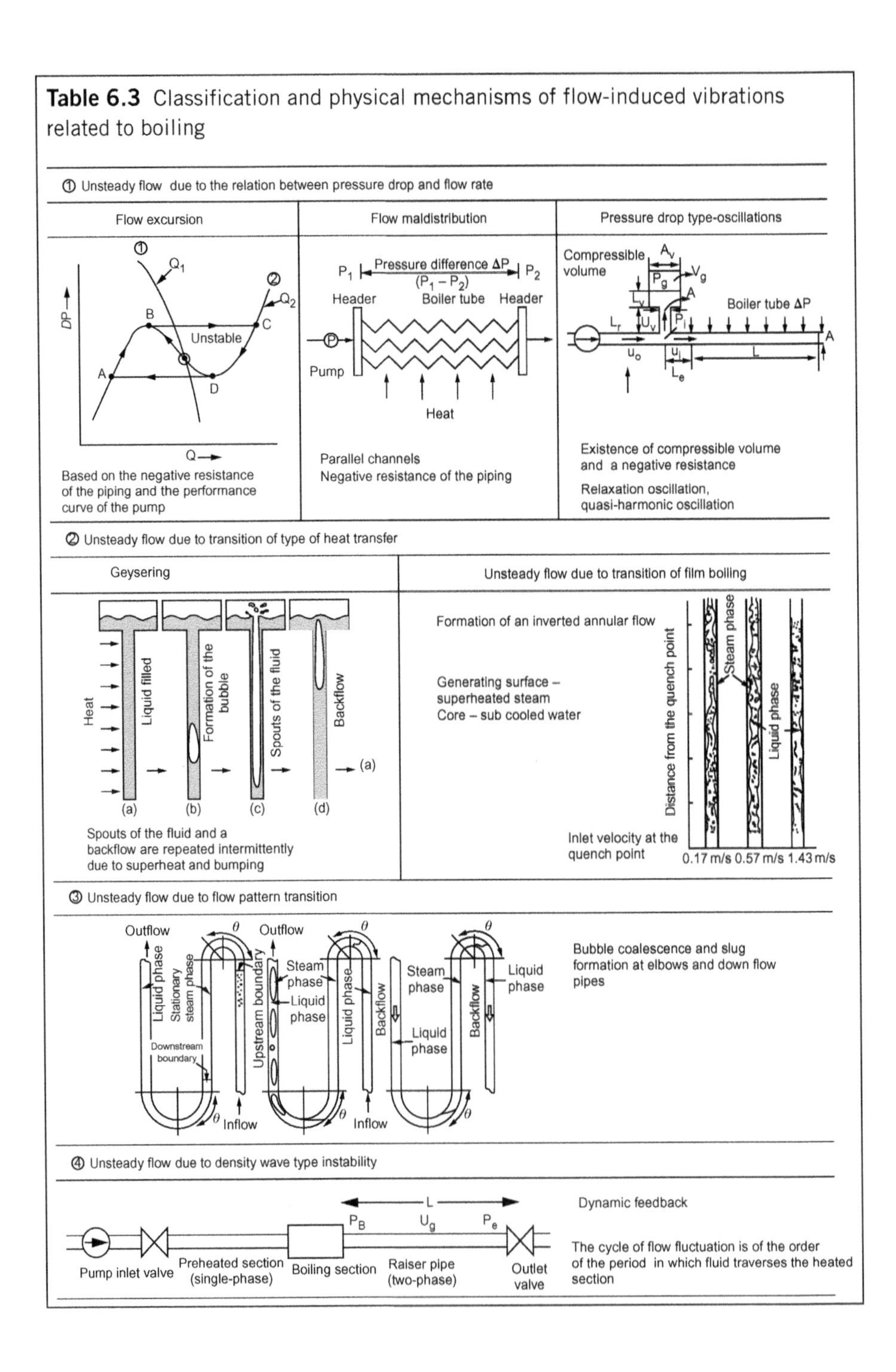

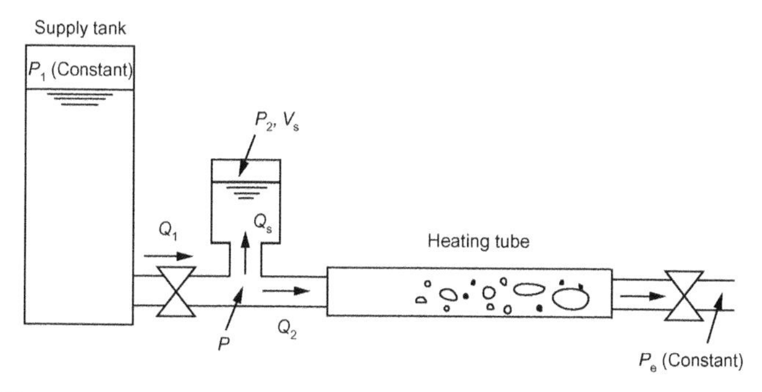

FIGURE 6.26

Single boiling piping system.

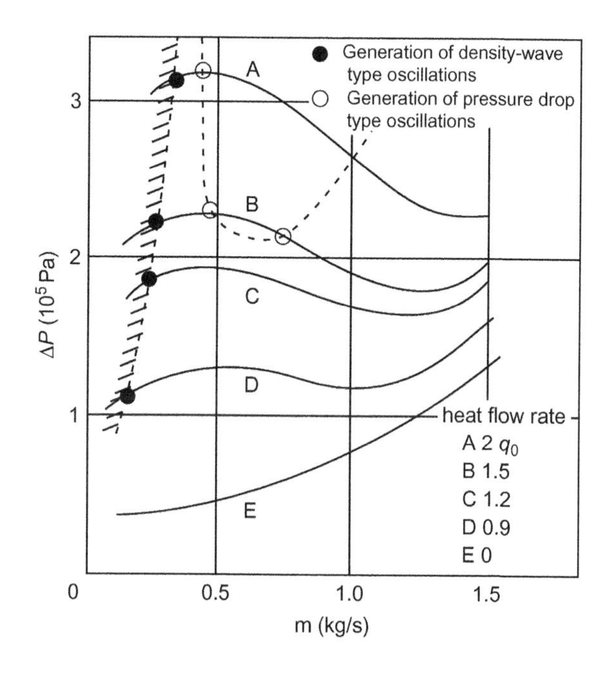

FIGURE 6.27

Characteristic curves of mass flow rate m in the heating tube versus the pressure difference ΔP between the inlet and outlet. The conditions for generation of density-wave-type oscillations (heat flow rate $q_0 = 2931$ W), and pressure drop type oscillations are also shown.

No. 2, point B cannot be stable. The operating point of the system consequently jumps from point B to point C, causing a sudden increase of the flow rate Q_2. At point C, since the flow rate Q_2 is higher than the flow rate Q_1, the water level in the surge tank decreases. This causes the pressure P_2 to decrease to point D.

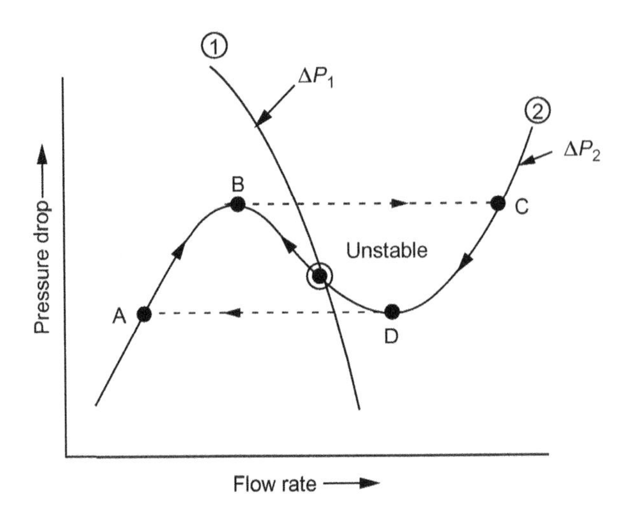

FIGURE 6.28

Relation between flow rate and pressure drop for pressure-drop-type oscillations. Limit cycle is formed.

Similarly to point B, the system is unstable at point D. The operating point therefore jumps from point D to point A with a sudden decrease of the flow rate Q_2. The flow rate Q_2 is, however, lower than the flow rate Q_1 at point A. The pressure in the surge tank P_2 increases to point B and the cycle is repeated, resulting in the generation of oscillations.

Hence, for pressure-drop-type oscillation, the existence of a compressible volume, such as a surge tank, and a negative resistance element in the system are dominant factors. Other than these factors, the following characteristics of pressure-drop-type oscillations are worth noting:

1. An increase of the negative gradient of the pressure versus flow rate characteristic, an increase of the heat duty, or an increase of the pressure drop at the exit all have a destabilizing effect.
2. A horizontal boiling piping system is more stable than a vertical one.

A numerical model for pressure drop oscillations taking into consideration the characteristics of the foregoing destabilization mechanism was proposed by Nakanishi et al. [113]. Particular attention was paid to the negative gradient of the pressure drop versus flow rate characteristics. The model was formulated using Van der Pol's equation (Eq. (6.8)), which is a well-known equation for modeling self-excited oscillations. The assumptions for this analysis are:

- The supply water flow rate Q_1 is constant.
- The pressure at the exit of the heating tube is constant.
- The acceleration of the fluid due to boiling is negligible.

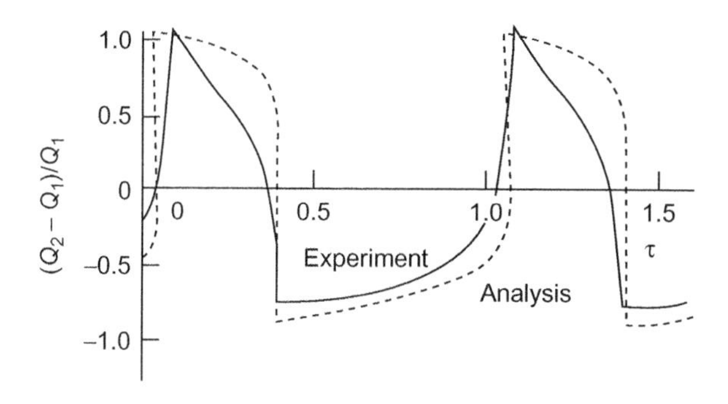

FIGURE 6.29

Waveform of flow rate Q_2 for pressure-drop-type oscillations. Experimental results and analytical results obtained by solution of Van der Pol's equation ($\varepsilon = 34.2$, $\beta = -0.263$) are compared [113]. τ is the period of oscillations.

- The inertial mass of the fluid in the boiling part is negligible.
- The inertial mass of the fluid under steady state conditions is used.

The non-linear model takes the form:

$$q - \varepsilon(1 - 2\beta q - q^2)\dot{q} + q = 0 \tag{6.8}$$

$$q = \frac{Q_2 - Q_1}{\sqrt{(Q_1 - Q_B)(Q_D - Q_1)}}, \quad \tau = \frac{t}{\sqrt{C_s(I_s + I)}}, \quad \varepsilon = \sqrt{\frac{C_s}{I + I_s}} \cdot \alpha(Q_1 - Q_B)(Q_D - Q_1)$$

$$\beta = \frac{2Q_1 - (Q_B + Q_D)}{2\sqrt{(Q_1 - Q_B)(Q_D - Q_1)}}, \quad C_s = \frac{\rho V_s}{kP_2}$$

where I is the inertial mass of the fluid in the heating tube, I_s the inertial mass of the fluid in the surge tank, ρ the density of water, k the isentropic exponent, and α a positive constant related to the slope of the $P - Q$ characteristic curve.

When $\beta = 0$, that is, $Q_B + Q_D = 2Q_1$ in Eq. (6.8), the oscillations of the flow rate Q_2 through the heating tube can be described by Van der Pol's equation; self-excited limit cycle oscillations are generated in the system. When ε is large, relaxation oscillations are observed. The positive agreement between the experimental results and analytical simulation by Nakanishi is shown in Fig. 6.29. In the analysis, the waveform of the flow rate Q_2 was obtained by solving Eq. (6.8), using values of ε and β corresponding to experimental conditions. A period of the pressure drop type oscillation corresponds to a complete cycle about the path ABCDA in Fig 6.28.

6.3.4 Vibration/oscillation problems and solutions

Boilers, nuclear steam generators, and BWRs are examples of facilities that may suffer flow-induced vibrations related to boiling [114]. In the 1960s, oscillations of the sodium exit temperature were observed in an FBR and, consequently, research on these systems has been performed in many countries [106]. Since numerous references for these problems may be found in [106] and [114], they will not be considered here. Instead, examples of problems experienced in other industries will be presented in what follows.

6.3.4.1 Thermosyphon reboiler

6.3.4.1.1 Geysering

Geysering is an unsteady two-phase flow phenomenon observed in systems such as those shown in Table 6.3. Spouts of gas and liquid, as well as reverse liquid flow, are repeated intermittently. The basic mechanism is the same as that of geysers observed in natural hot springs.

In oil refineries and gas and chemical plants, geysering can occur in the exit piping of thermosyphon-side reboilers if the elevation of the reboiler is sufficiently low [115]. Referring to Fig. 6.30, in the case of a low-load, bubbly flow or slug flow is formed in the exit piping. Consequently, only the gas flows up the column while the liquid level in the piping increases gradually. As the boiling point increases in the piping due to the increase of static pressure, heat is accumulated in the liquid. When liquid flow into the column commences, the static pressure and the boiling point decrease, resulting in evaporation in the spout. The liquid level decreases due to liquid outflow caused by the foregoing combination of a static pressure decrease and evaporation effects, and the system returns to the

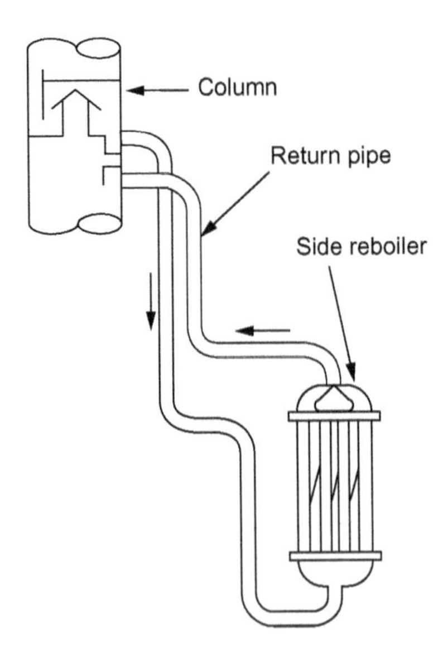

FIGURE 6.30

Geysering in thermosyphon side reboiler [115].

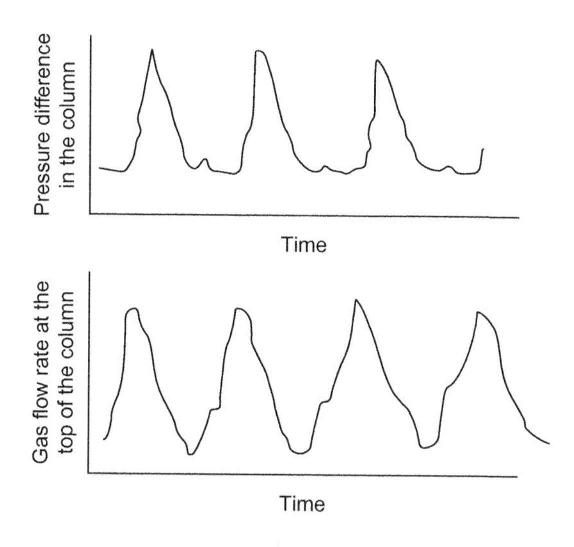

FIGURE 6.31

Flow rate fluctuation and pressure difference observed in the column due to geysering.

previous low liquid level. A cycle of flow fluctuation by geysering may last between 1 and 3 minutes (Fig. 6.31).

The effects of geysering on equipment include:

- Piping vibration due to sudden fluctuations in the internal flow momentum.
- Damage of internals in columns due to intermittent influx of two-phase flow.

As countermeasures, the following points should be considered in the design:

- Shortening length of return piping.
- Appropriate determination of pipe size to maintain high flow rates, thereby avoiding unstable flow conditions.

6.3.4.1.2 Density-wave type oscillations

Fatigue failure and leakage caused by thermal stress fluctuations induced by density-wave-type oscillations in a reboiler were experienced as reported in Ref. [116]. The pressure pulsation period at the exit of the reboiler was about 25 s; thus, it is a very low frequency phenomenon. Increasing the pressure drop in the inlet piping by installing a valve was effective in reducing the pressure pulsations.

The following measures may be effective against flow oscillations in thermosyphon reboilers.

- Shortening the piping.
- Increasing the pressure drop in the inlet piping, or decreasing the pressure drop in the exit piping.
- Increasing the static pressure in the system.

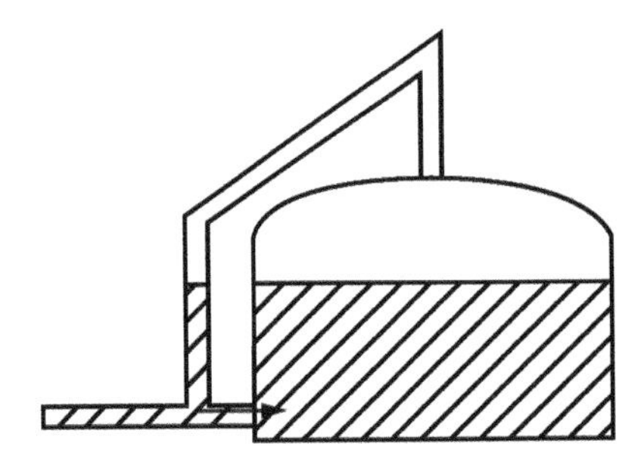

FIGURE 6.32

G-tank bottom-feed system.

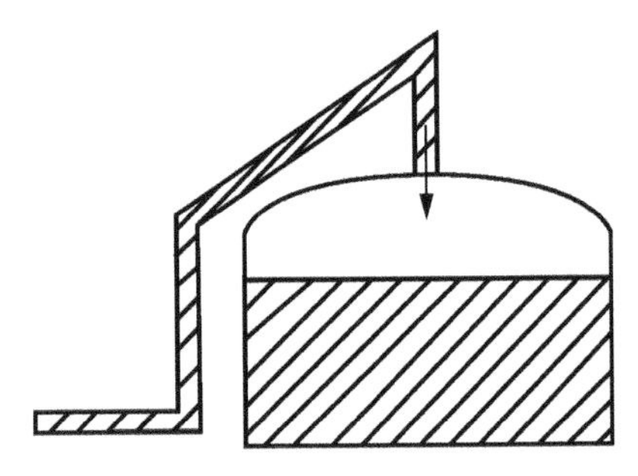

FIGURE 6.33

G-tank top-feed system.

6.3.4.2 LNG tank piping

Geysering is observed in LNG tank systems as well. Typical configurations of feed systems for a LNG tank are shown in Figs. 6.32 and 6.33. The height of the tank frequently exceeds 30 m. The liquid boiling point increases with increasing static pressure. Since external heat is accumulated in the liquid as latent heat, the system may become unstable. The possibility of geysering in both bottom-feed and top-feed systems has been confirmed by means of experiments with scale models [117].

6.3.4.3 Rocket engine feed system

Piping and supports are known to have been damaged by geysering in a liquid-oxygen feed system for a rocket and missile engine. As a geysering-suppression approach, cooling by helium injection was confirmed to be effective. The use of a concentric inner tube or a cross-feed recirculation system has been shown to effectively eliminate any geysering problems (see Figs. 6.34 [118] and 6.35 [119]).

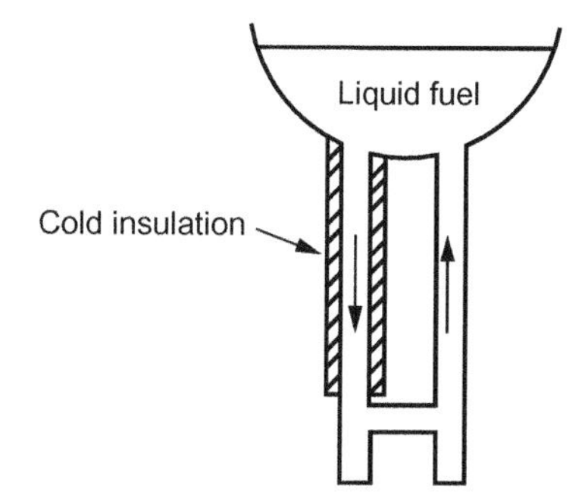

FIGURE 6.34

Geysering suppression by the use of cross-feed recirculation in the liquid fuel feed system.

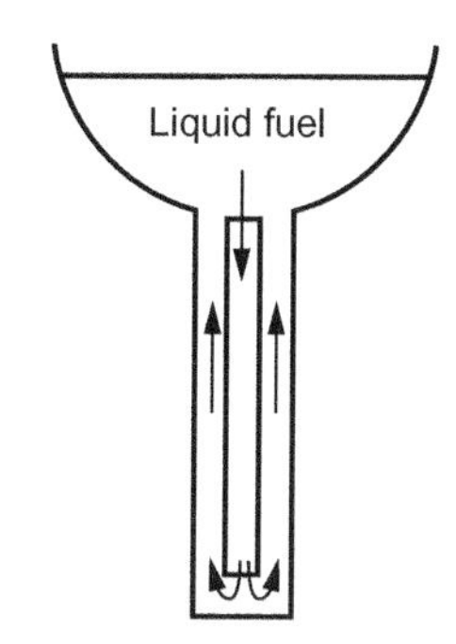

FIGURE 6.35

Geysering suppression by the use of concentric inner tube in the liquid fuel feed system.

References

[1] JSME, Report of the Research Committee RC-SC40 on Noise and Vibration in Combustion Facilities (in Japanese), 1978 (288).
[2] H.I. Ansoff, J. Appl. Mech. 71 (1949) 158.
[3] P.L. Rijke, Pogg. Ann. 107 (1859) 339.
[4] A.A. Putnum, W.R. Denis, J. Acoust. Soc. Am. 28 (2) (1956) 246.
[5] J.W.S. Rayleigh, 1877 The Theory of Sound, first ed., Dover Publications, 1945.
[6] W.H. Bragg, Library of world culture, No.30, (1963), Heibonsya (translated into Japanese by Kurihara).
[7] T. Saito, Trans. JSME 31 (221) (1965) 143.
[8] H. Madarame, Trans. JSME (in Japanese), C47-413, 10 (1st report, 1981), C47-416, 434 (2nd, 1981),C48-432, 1157 (3rd, 1982).
[9] T. Yamaguti, Dissertation, (in Japanese), Kansai University, 1996.
[10] H. Yamamoto, Report of Tanigawa Thermal Engineering Promotion Fund (in Japanese), 1995.
[11] Y. Katto, et al., *Trans. JSME* (in Japanese) 43 (365) (1977) 203.

[12] G.W. Swift, J. Acoust. Soc. Am. 84 (4) (1988) 1145.
[13] A.T. Jones, J. Acoust. Soc. Am. 16 (4) (1945) 254.
[14] Merk, H.J., Sixth Symposium on Combustion, The Combustion Institute, 1956, p. 500.
[15] A.A. Putnam, W.R. Dennis, J. Acoust. Soc. Am. 25 (5) (1954) 716.
[16] G.F. Carrier, Quart. Appl. Math. 12 (1955) 383.
[17] P.K. Baade, ASHRAE Trans. 78 (2) (1978) 449.
[18] T. Nagahiro, et al., JSME Paper (in Japanese) No. 760-4, 1976, p. 21.
[19] Y. Mizutani, et al., JSME Paper (in Japanese) No. 954-4, 1995, 84.
[20] T. Suga, et al., *J. Mach. Res.* (in Japanese) 47 (12) (1995) 44.
[21] S. Kadowaki, *Trans. JSME* (in Japanese) B56-527 (1990) 2104.
[22] H. Madarame, Trans. JSME (in Japanese), C48-432, p. 1157.
[23] T. Fujikawa, et al., JSME Paper (in Japanese), No. 981-2, 1998, 41, No. 98-7, 1998.
[24] F.L. Eisinger, J. Trans. ASME 121 (1999) 444.
[25] JSME, Report of Research Committee RC-SC40 (in Japanese), 1978, p. 10.
[26] A.A. Putnum, Combustion Driven Oscillations in Industry, American Elsevier, New York, 1971.
[27] J. Kameda, *Therm. Contr.* (in Japanese) 23 (5) (1970) 14.
[28] S. Kobayashi, JSME Kansai 173th Seminar Material (in Japanese), 1990, p. 79.
[29] Japan Burner Research Committee, J. JBRC (in Japanese), No. 33, 25, No. 35, p. 1.
[30] Aoki, A.K. Saima, Report of Nihon Daigaku Science and Engineering Institute (in Japanese), 1987, p. 496.
[31] I. Akiyama, Combustion Driven Oscillation on Multi-spad Burner, Hiroshima University Dissertation (in Japanese), 1995.
[32] S.A. Tabata, et al., *Tech. Rep. Tokyo Gas* (in Japanese) 24 (1949) 225.
[33] S. Kaneko, *JSME D&D00 Forum* (in Japanese) (2000) 25No. 00-6.
[34] T. Yamamoto, *Tech. Rep. Mitsubishi Heavy Indust.* (in Japanese) 5 (5) (1968).
[35] T. Hiruta, et al., *Trans. JSME* (in Japanese) B45 (398) (1979) 1557.
[36] H. Ikugoe, et al., *Trans. JSME* (in Japanese) B48 (426) (1982) 373.
[37] M. Ono, et al., JSME Paper (in Japanese), No. 940-261, 1994, p. 597.
[38] T. Ishii, et al., JSME Paper (in Japanese), No. 940-261, 1994, p. 93.
[39] X.T. Ishii, et al., *Trans. JSME* (in Japanese) B50 (449) (1984) 151.
[40] I. Akiyama, *Indust. Heat.* (in Japanese) 28 (4) (1991) 65.
[41] I. Akiyama, *Trans. JSME* (in Japanese) B61 (588) (1995) 3082.
[42] Y. Segawa, *Trans. JSME* (in Japanese) B53 (486) (1987) 642.
[43] A. Tabata, *Tech. Rep. Tokyo Gas* (in Japanese) 24 (1979) 225.
[44] K. Katayama, Acoustic Research Meeting in Hiroshima (in Japanese), 1978.
[45] S. Sato, *Trans. JSME* (in Japanese) 33 (252) (1967) 1260.
[46] M. Sato, The 29th Combustion Symposium (in Japanese), in Kyoto, 1991, p. 400.
[47] M. Sato, T. Sugimoto, 30th Symposium on Combustion (in Japanese), in Nagoya, 1992, p. 73.
[48] H. Madarame, Material of Non-linear Vibration Research Meeting (in Japanese), 1992, p. D1.
[49] M. Sato, 31st Symposium on Combustion (in Japanese), in Yokohama, 1993, p. 108.
[50] T. Saito, Preprint of the Second Japan Heat Transfer Symposium (in Japanese), 1970, p. 8.

[51] M. Nakamoto, et al., Energy and Resources Society, 13th Conference Paper, 1994, p. 153.

[52] T. Ochi, JSME Paper (in Japanese), No. 964-1, 1996, p. 611.

[53] M. Nakamoto, et al., JSME Paper (in Japanese), No. 964, 1996, p. 613.

[54] R. Becker, R. Gunter, 13th International Symposium, the Combustion Institute, 1971, p. 517.

[55] G. Wolfbrandt, et al., ASHRAE Trans. 78 (2) (1978) 466.

[56] Y. Matsui, Combust. Flame 43 (1981) 199.

[57] T. Sugimoto, Y. Matsui, 19th International Symposium on Combustion, the Combustion Institute, 1982, p. 245.

[58] Y. Segawa, JSME Int. J. 30 (267) (1987) 1443.

[59] J.A. Carvalo, et al., Combust. Flame 76 (17-27) (1989) 17.

[60] N. Friker, A. Roberts, Gas Warme 28 (1979) 13.

[61] J.G. Seebold, ASME Paper, 72-PET-19, 1972, p. 1.

[62] A. Thomas, G. Wiliam, Proc. Roy. Soc. (London) A294 (1966) 449.

[63] T.J. Smith, J.K. Kilham, J. Acoust. Soc. Am. 35 (5) (1963) 715.

[64] I.R. Hurle, Proc. Roy. Soc. (London) A303 (1968) 409.

[65] W.C. Strahe, J. Sound Vib. 23 (1) (1972) I13.

[66] B.N. Shivashankara, et al., AIAA Paper, No. 73-1025, 1973.

[67] K. Suzuki, *J. JSME* (in Japanese) 80 (708) (1977) 1188.

[68] S. Kotake, *J. Fuel. Soc.* (in Japanese) 50 (535) (1971) 832.

[69] T. Urie, K. Sato, *Rep. Osaka Prefecture Eng. Inst.* (in Japanese) (1974) 63.

[70] M. Kazuki, et al., *J. High Temp. Soc.* (in Japanese) 15 (3) (1985) 117.

[71] K. Kishimoto, *Combust. Res.* (in Japanese) 100 (1955)35–44 and 101 (1955), 35.

[72] A.A. Putnum, ASME Paper, 81-WA/Fu-8, 1982, p. 1.

[73] T. Kanabe, et al., JSME Paper (in Japanese), 770-14, 1977, p. 22.

[74] G. Bitterlich, Noise Cont. Eng. 14 (1) (1980) 1.

[75] T. Okuzono, *Piping Technol.* (in Japanese) (1775) I15.

[76] S. Watanabe, et al., 27th Symposium on Combustion (in Japanese), 1989, p. 101.

[77] T. Saito, Indust. Heat. 17 (3) (1980) 11 (in Japanese).

[78] Y. Ogisu, JSME Paper (in Japanese), No. 760-4, 1976, p. 17.

[79] T. Azuma, et al., *Rep. Osaka Prefecture Eng. Inst.* (in Japanese) 75 (1979) l.

[80] T. Azuma, *Pollution* (in Japanese) 15 (2) (1980) 15.

[81] M. Nishimura, *Trans. JSME* (in Japanese) 29 (207) (1963) 1844.

[82] Y. Mizutani, J. Mach. Res. 40 (7) (1988) 777.

[83] Y. Oiwa, et al., *Trans. JSME* (in Japanese) B55 (517) (1989) 2824.

[84] M. Nakamoto, 31st Symposium on Combustion (in Japanese), in Yokohama, 1993, p. I17.

[85] J. Sugimoto, JSME Paper (in Japanese), No. 964. 1, 1996, p. 609.

[86] M. Kazuki, et al., JSME Paper (in Japanese), No. 964-1, 1966, p. 585.

[87] S. Kagiya, K. Hase, JSME Paper (in Japanese), No. 964-1, 1996, p. 615.

[88] S. Kotake, K. Takamoto, *J. Sound Vib.* (in Japanese) 112 (2) (1987).

[89] H.H. Chiu, M. Summerfield, Acta Astonica 1 (1973) 967.

[90] B.N. Shivashankara, Acta Astonica 1 (1974) 985.

[91] B.N. Shivashankara, AIAA Paper, 76-586, 1976, p. 1.

[92] H.A. Hassan, J. Fluid Mech. 66 (3) (1974) 445.

[93] D.C. Methews, N.F. Rekos, AIAA Paper, 76-579, 1976, p. 1.

[94] K. Okada, *Sound Environment and Control Engineering* (in Japanese), Fujitechnosystem, 2000.
[95] MXA-1-301, Aktiebolaget Atomenergi, Sweden, 1974.
[96] I. Aya, H. Nariai, *JSME Int. J.* Ser. II 31 (3) (1988) 461–468.
[97] S. Kamei, M. Hirata, *Trans. JSME* (in Japanese) 49 (438) (1983) 483.
[98] S. Fukuda, S. Saitoh, *J. AESJ* (in Japanese) 24 (5) (1982) 372.
[99] E. Taira, H. Kasashima, et al., *Rep. Indust. Res. Inst. Miyazaki* (in Japanese) (39) (1994) 65–68.
[100] M. Narasaki, et al., Proceedings of ASME-JSME, Thermal Engineering Joint Conference, 1987, 381–388.
[101] Handbook of Gas/Liquid Two-Phase Flow (in Japanese), CORONA Publishing, 1989.
[102] T. Fujii, K. Akagawa, Y. Itoh, *Dynamic Piping Design of Gas/Liquid Two-Phase Flow* (in Japanese), Nikkan Kogyo Shimbun Ltd., 1999.
[103] S. Katajara, Nurkkala, et al., ICONE 6th-6485, 1998.
[104] G.B. Wallis, P.H. Rothe, M.G. Izenson, NRC Report NUREG-5220, vol. 1–2, 1988.
[105] M. Ledinegg, Die Warme 61 (8) (1938) 891–898.
[106] The JSME research subgroup report on the dynamics of the two-phase flow (in Japanese), 2P-SC26, 1977.
[107] J.A. Boure, et al., Nucl. Eng. Des. 25 (1973) 165.
[108] K. Akagawa, *Turbomachinery* (in Japanese) 3 (6) (1975) 13.
[109] M. Arinobu, et al., *J. JSME* (in Japanese) 82 (728) (1979) 391.
[110] R.T. Lahey Jr., Nucl. Eng. Des. 95 (1986) 5–34.
[111] T. Fujii, et al., Piping Plan for Gas-Liquid Two-Phase Line (in Japanese), Nikkan Kogyo Shimbun Ltd., 1999.
[112] Handbook of Gas-Liquid Two-Phase Flow Technology (in Japanese), edited by the USME, Corona Publishing, 1989.
[113] S. Nakanishi, et al., *Trans. JSME* (in Japanese) 44 (388) (1978) 4252.
[114] State of the Art Report in Boiling Water Reactor Stability, OECD-NEA, 1996.
[115] S. Kaneko, H. Matsuda, Diagnosis and Countermeasures for the Plant Equipment and Piping Systems (in Japanese), JSME, No. 99–20, 1999, pp. 57–77.
[116] The JSME 69th Plenary Meeting, V_BASE forum (in Japanese), No. 24, 1991.
[117] M. Morioka, et al., LNG8 Los Angeles, USA, Session III, Paper 13, 1986.
[118] S.K. Morgan, H.F. Brady, Adv. Cryo. Eng. 7 (1962) 206–213.
[119] H.S. Howard, Adv. Cryo. Eng. 18 (1972) 162–169.

CHAPTER

Vibrations in Rotary Machines

7

Chapter 7 addresses flow-induced vibrations occurring in fluid machinery with the focus being on vibrations of blades and cascades, vibrations of rotors partially filled with fluid, and vibrations induced by confined flows. Rotary machines often encounter serious vibration problems. For example, blades, which are critical components, may fail due to self-excited vibrations caused by the wake from preceding blades or inlet flow distortion. Destabilizing forces from seals may also cause large-amplitude vibrations of the rotary shaft. In special cases, production downtime may be caused by unstable vibration of rotors partially filled with fluid that are included in automatic washing machines or centrifuges.

This chapter includes sections on the various vibration phenomena encountered in rotating machines. The mechanisms behind different vibration problems are discussed and evaluation methods and preventive measures are presented.

7.1 Vibration of blades and cascades

7.1.1 Overview and categorization of phenomena

Fluid machinery such as turbine compressors are typically designed by first deciding on the specifications of the blades according to the required performance, and then considering the strength issues. Conventionally, the consideration of blade strength was limited to investigation of static centrifugal and steady bending stresses and avoiding dynamic resonances. Flow in turbomachinery is generally considered to be steady with the exception of phenomena such as rotating stall and surging. However, rotating cascades are subjected to a relatively unsteady flow field due to the wake from preceding stators and struts or the inlet distortion from inlet conditions. As a result, fluctuating fluid forces act on the cascades and induce vibrations. These external forces are periodic and may cause resonance. Even a small fluctuating fluid force may result in large vibration stresses on the blades, and long-term operation may cause fatigue failure.

Longer and lighter (thinner) blades are desirable for improving the performance of turbomachinery. However, this means a reduction in the rigidity of the blades, and as a consequence, vibrations due to fluctuating fluid forces can occur more readily. It is therefore essential to investigate vibration properties in order to improve safety and reliability, especially at the development stage of turbomachinery.

The following is a typical example. The twelfth-stage rotor blade in the low pressure stage of Reactor 5 of the Hamaoka Nuclear Power Plant (operated by Chubu Electric Power Company) failed in an accident in 2006. Investigations concluded that the cause was random vibrations caused by fluid excitation forces generated by unsteady flow during low load operation and due to flashback steam from the extraction steam pipe during load cut-off testing. The fluid excitation forces caused excessive repeated stresses that initiated and propagated cracks in the fork-shaped attachment (pinhole) in the twelfth-stage rotor blade [1].

As in the above example, blade failure is mainly caused by repeated stresses from fluid excitation forces. Resonance, where the forcing frequency matches a blade natural frequency, is the most dangerous condition. Vibration phenomena resulting in failure often arise from inadequate consideration during the design stage coupled with a lack of knowledge of blade vibration. This chapter provides design engineers with the minimum knowledge required.

Blade vibration can be categorized by vibration pattern as depicted in Fig. 7.1. The frequencies of the excitation forces during forced vibrations are clear, as seen in the figure. The absolute rule is therefore to avoid matching between the blade natural frequencies and the excitation force frequencies (i.e. to avoid resonance). However, the precise natural frequencies of the blades are difficult to calculate,

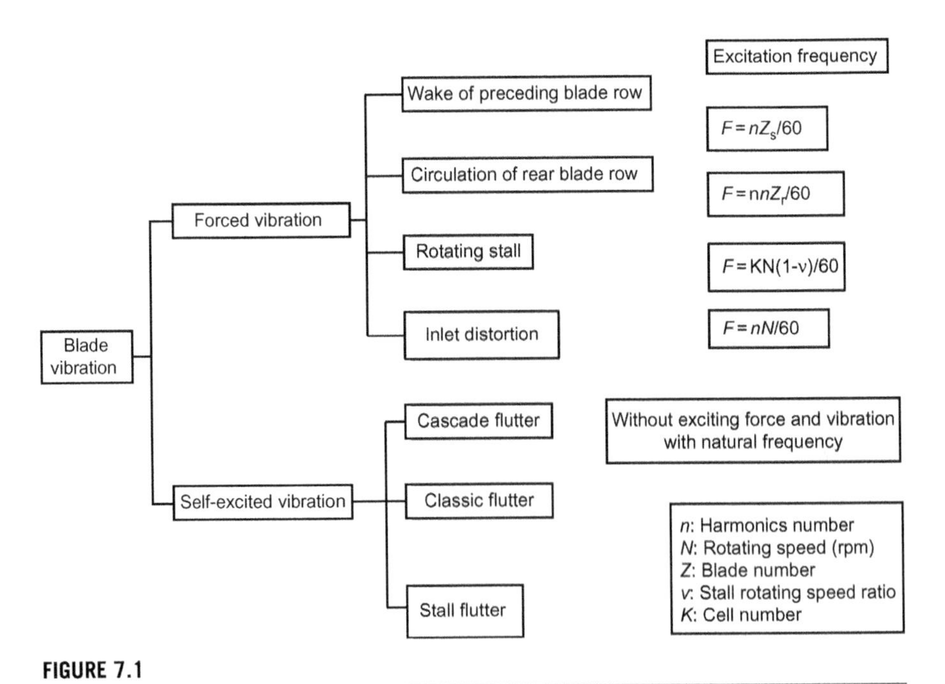

FIGURE 7.1

Categories of blade vibration.

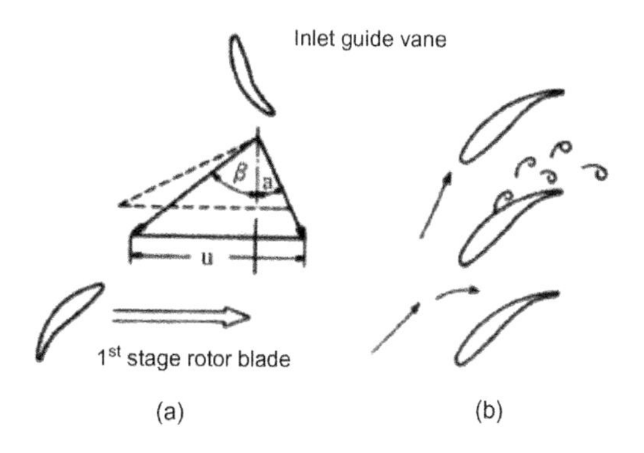

FIGURE 7.2

Velocity triangle of the bucket entrance and state of stall [2].

especially because the boundary conditions are not well-defined. Therefore, an error margin of 5–10% must be taken into account.

Rotating stall can occur in compressors with decelerating cascades; however, predicting the occurrence limit and the number and width of stall cells can be difficult. Confirming safety is also difficult. One approach involves detecting rotating stall during test runs, thus plotting the rotating stall zone in a pressure-versus-flow diagram and avoiding operation in this zone. Rotating stall arises through the mechanism described next. Figure 7.2(a) shows the velocity triangle of the bucket entrance prior to a stall. A decrease in the flow rate will result in a decrease of axial flow velocity, but the circumferential velocity does not change; therefore, the velocity triangle will change to that shown by the dotted line. The result is an increase in the angle of attack (β) relative to the rotor blade. When this angle reaches a certain critical value, the air current can no longer flow through the back of the rotor, resulting in a separated flow. This is the condition of stall. An axial compressor cut at a cylindrical surface of a given radius reveals blade cross-sections positioned at even intervals, as shown in Fig. 7.2(b) [2]. Assume that a blade has stalled for some reason. The flow separation decreases the effective width of the low pressure side flow path 'behind' the blade. Some of the fluid that flowed into this path is diverted further toward the rear blade while some flows through the high pressure side flow path on the 'front' side of the blade. The blade behind the stalled blade therefore has an increased angle of attack because of the flow from below, and, if already close to stalling, stalls because of this higher angle. For the blade in front, the diverted flow decreases the effective angle of attack thus reducing the likelihood of stall in this case. As a consequence, the stall zone propagates along the cascade to the back of the blades. Figure 7.3 shows the pressure waveform and flow rate at the casing wall on the high and the low pressure sides of the blades [4].

Flutter can occur in blades with a small mass ratio $m/\pi\rho b^2$ (m, mass of blade per unit length; ρ, fluid density; b, half-chord length) such as those used in

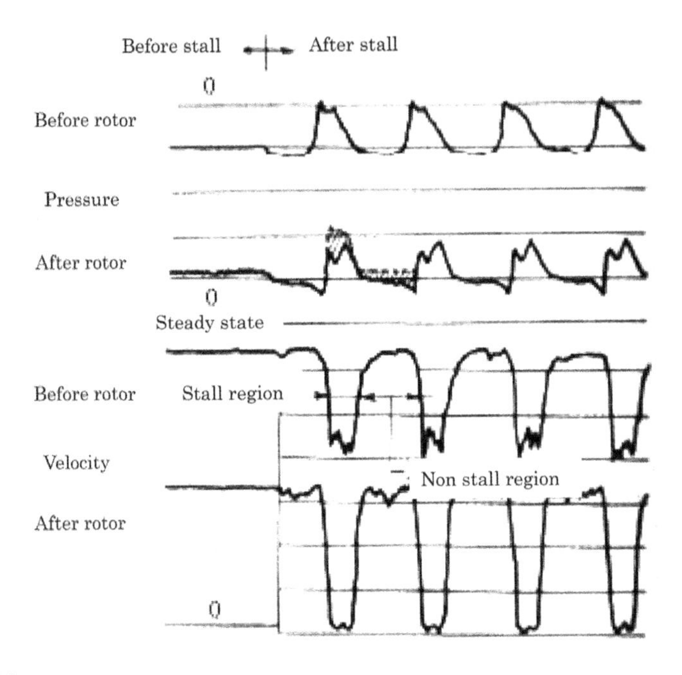

FIGURE 7.3

Change in pressure at the casing wall on the high and the low pressure sides [3].

aircraft gas turbines, but rarely becomes an issue for industrial blades because of their large mass ratio. Flutter results in large-amplitude vibrations and thus, the stability limit must be estimated in advance.

Figure 7.4 shows a Campbell diagram [5] used in the design phase to investigate the possibility of blade resonance. The radii of the circles in the figure represent the amplitude of vibration. This magnitude is not known during the design phase, and is obtained through actual measurements.

7.1.2 Vibration of blades under gust loading (forced vibration)

7.1.2.1 History of research and evaluation

Gusts are velocity fluctuations, for instance those acting on blades passing through the wake from preceding stators or non-uniform inlet flow. There are two main areas of research on fluctuating fluid forces: one analyzing the fluctuating fluid forces generated when blades vibrate in a uniform flow, and the other where gusts act on a static blade. The first research work in the latter field was on the analysis of unsteady flow around a blade by Karman and Sears [6] in 1938. Sears [7] later derived the fluctuating fluid forces on a single blade subjected to sinusoidal gusts normal to the blade. Later, Kemp and Sears [8,9] used these results to

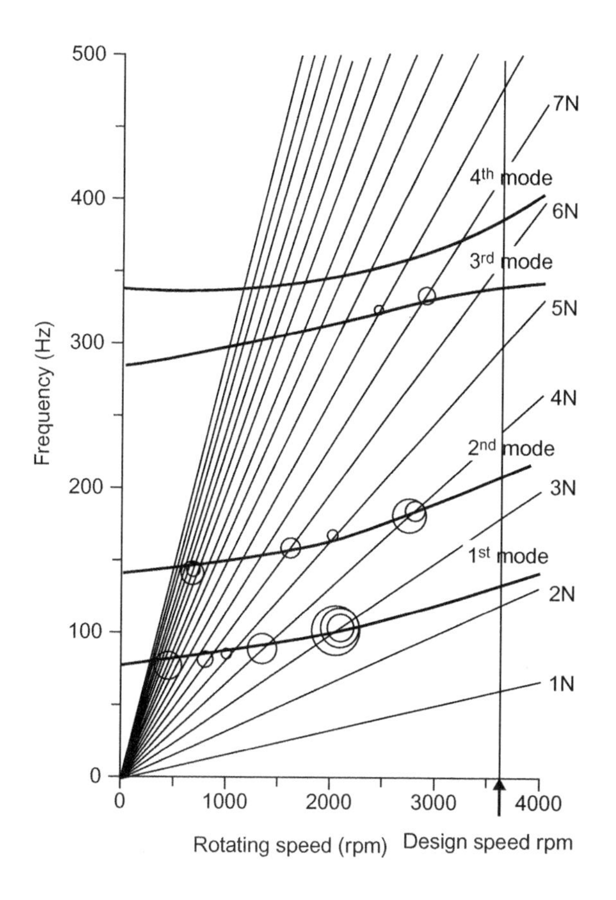

FIGURE 7.4

Campbell diagram.

analyze the interference of the potential flow between cascades and the fluctuating lift on blades as a result of the viscous wake from preceding stators.

The above analysis covered gusts normal to the blades. Horlock [10] showed that the component of gusts parallel to the chord had a strong influence on the fluctuating fluid forces when there was an angle of attack. Nauman and Yeh [11] obtained an almost exact analytical solution of the fluctuating fluid forces on a single blade under periodic gusts by taking the angle of attack and camber into account, and applied the solution to a cascade wake. Ishihara and Funakawa [12] further analyzed the fluctuating fluid forces when the gust amplitude decayed along the chord in an actual blade.

Research on cascades includes work by Whitehead [13] on the analysis of fluctuating fluid forces on blades in static and vibrating cascades under sinusoidal gusts, and by Murata and Tsujimoto [14], who analyzed the unsteady flow in parallel cascades using conformal mapping to investigate and clarify the influence of the cascade parameters on the fluctuating fluid forces. Nishiyama and Kobayashi [15] further performed analysis that considered compressibility.

Theoretical research on rotating stall mainly focused on modeling the dynamic properties of cascades that cover stall cells. For instance, Emmons et al. [16] developed a model describing a stall as the closure of flow paths in a cascade. Marble [17] carried out analysis using a characteristic curve that is discontinuous at the stall point, and showed that the time delay at the boundary layer and the fluid inertia in cascades are two of the factors that affect rotating stall. Sears [18] found, through analysis based on actuator theory, that the phase between changes in angle of attack and lift is important. However, as these analyses cannot predict the number of stall cells, and reach different conclusions on rotating speed, they are not sufficient for describing rotating stall.

In an experimental study by Iura and Rannie [19], a systematic investigation using a three-stage axial compressor showed that rotating stalls can be categorized into small and large stalls, and that there is hysteresis between these two types. Sovran [20] experimented with different distances between the inlet guide vane and rotor, and found that interference between the cascades is an important factor affecting the number of stall cells. Takata [21] published the paper *Rotating stall of multistage axial compressors* in 1961. The linear theory presented in the paper considers cascade interference, and shows that the inertia of the fluid in and between the cascades and the time delay at the boundary layer are more important factors in governing the rotating speed compared to the inertia of fluid outside the cascades. Furthermore, the theory showed that the rotating speed becomes one-half of the rotation speed in multistage systems and less than one-half with a reduced number of stages. Interference between cascades is thought to be important in determining the number of stall cells; however, this is an assumption as stalling is a nonlinear phenomenon and is also dependent on cascade geometry and dimensions. Takata and Nagano [22] stepped away from linear theory and developed a theory that considers constant amplitude or nonlinear cascade properties that determine the amplitude and nonlinear equations of motion with finite fluctuation. As a result, various properties of rotating stall including number of stall cells; magnitude and waveform of fluctuation; rotating speed; and the mechanism behind these became clear. This was impossible to achieve with the previous linear theories.

7.1.2.2 Evaluation methods

The three most fundamental evaluation methods for unsteady aerodynamic forces are: (1) evaluation of fluctuating fluid forces on a single blade based on the unsteady blade theory, (2) evaluation of fluctuating fluid forces on a cascade of flat blades and, (3) evaluation of excitation fluid forces from a rotating stall.

(1) Evaluation of fluctuating fluid forces on a single blade based on the unsteady blade theory

Figure 7.5 shows a cross-section of an actual gas turbine, while in Fig. 7.6 the arrangement of blades in a cascade is shown. The procedure for obtaining the

FIGURE 7.5

Internal structure of a gas turbine.

fluctuating fluid forces on a rotor when the rotor passes through the wake from preceding stators is described below.

A thin blade with camber and angle of attack as shown in Fig. 7.7 is considered. The unsteady lift is given by the following equation [12]:

$$F = \pi\rho c V w_p \sin\phi[\cos\phi\{fF_f(k) + \alpha F_\alpha(k)\} + S(k)]e^{i\omega t} \tag{7.1}$$

where ρ is the fluid density, V the relative inlet velocity of the rotor, w_p the gust amplitude, and ϕ the angle between rotor and stator. Other parameters are, f the camber (blade height/half-chord length), a the angle of attack, and c the chord length. The variable ω is the angular frequency of excitation ($\omega = 2\pi NZ/60$), where N is the speed of revolution and Z the number of blades. The reduced frequency $k = \omega c/2V$. $F_f(k)$ and $F_a(k)$ are unsteady lift coefficients from camber and angle of attack, respectively, and are given by the following equations [23]:

$$F_f(k) = \frac{4}{k}J_1(k)\left\{2C(k) - 1\right\}\left[J_0(k) - \frac{J_1(k)}{k} - jJ_1(k)\right] - \left[J_0(k) - \frac{J_1(k)}{k} + jJ_1(k)\right] \tag{7.2}$$

$$F_\alpha(k) = J_0(k) + jJ_1(k) \tag{7.3}$$

$S(k)$ is the Sears function given in Eq. (7.4):

$$S(k) = [J_0(k) - jJ_1(k)]C(k) + jJ_1(k) \tag{7.4}$$

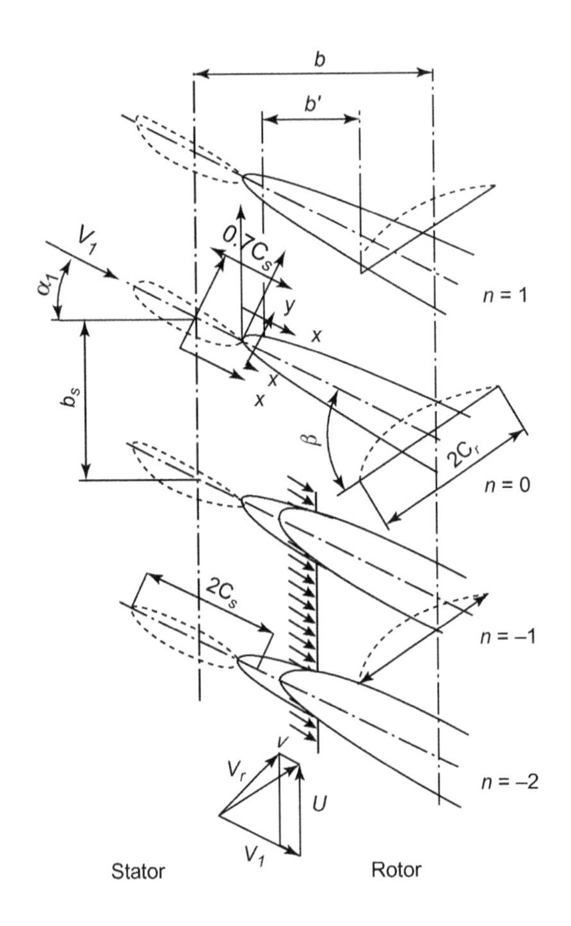

FIGURE 7.6

Blade configuration in a cascade.

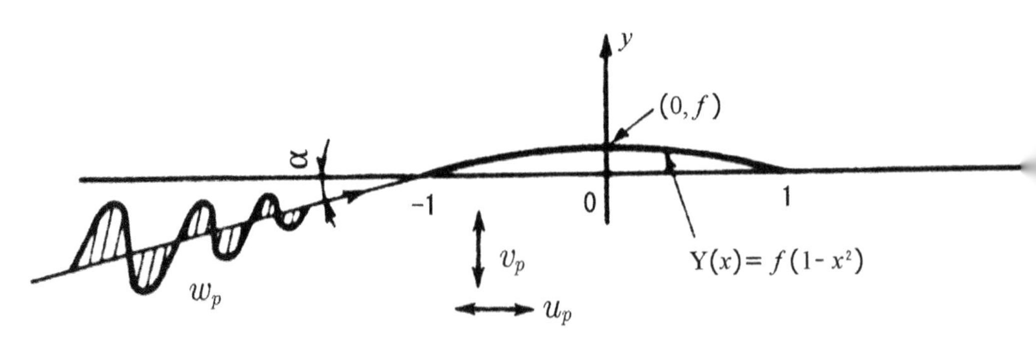

FIGURE 7.7

Blade under gust loading.

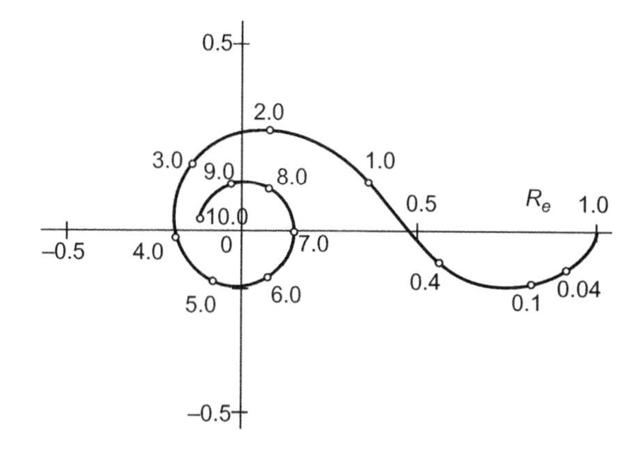

FIGURE 7.8

Sears function.

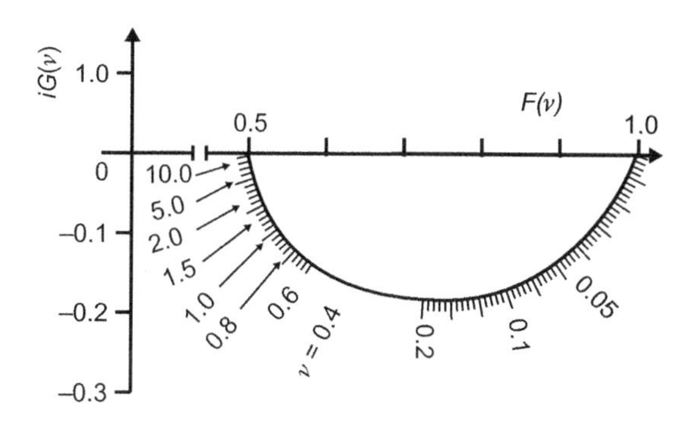

FIGURE 7.9

The Theodorsen function.

$$C(k) = H_1^{(2)}(k)/\{H_1^{(2)}(k) + jH_0^{(2)}(k)\} \tag{7.5}$$

In the equations above, $C(k)$ is the Theodorsen function. $J_0(k)$ and $J_1(k)$ are Bessel functions of the first kind, and $H_0(k)$ and $H_1(k)$ are Hankel functions. The Sears and Theodorsen functions are shown in Figs 7.8 and 7.9, respectively. The functions F_f and F_a that give the effects of camber and angle of attack are as shown in Figs 7.10 and 7.11, respectively.

The following equations by Silverstein et al. [24] are used for the gust amplitude w_p:

$$\omega_p = -u_c \sin\phi \frac{2\sqrt{\pi}}{K} e^{\frac{\pi^2 m^2}{K^2}} \tag{7.6}$$

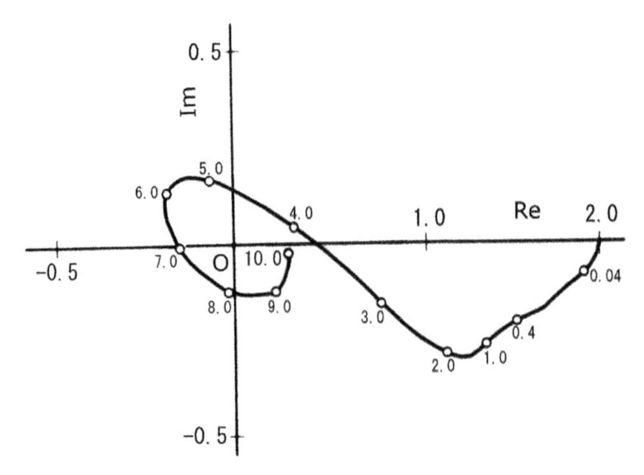

FIGURE 7.10

The function F_f.

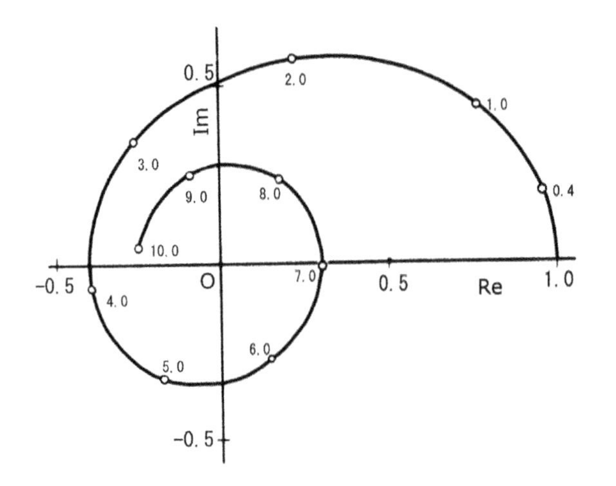

FIGURE 7.11

The function F_a.

$$\frac{u_c}{V} = -\left(2.42\sqrt{C_D}\right)/\left(\frac{x}{c} + 0.6\right) \tag{7.7}$$

$$\frac{Y}{c} = 0.68\sqrt{2C_D\left(\frac{x}{c} + 0.3\right)} \tag{7.8}$$

$$K^2 = \pi\cos^2\alpha \cdot \left(\frac{d}{Y}\right)^2 \tag{7.9}$$

where C_D is the drag coefficient, Y the half-width of the wake, and m the order of the Fourier component. The fluid excitation force obtained by Eq. (7.1) is used to

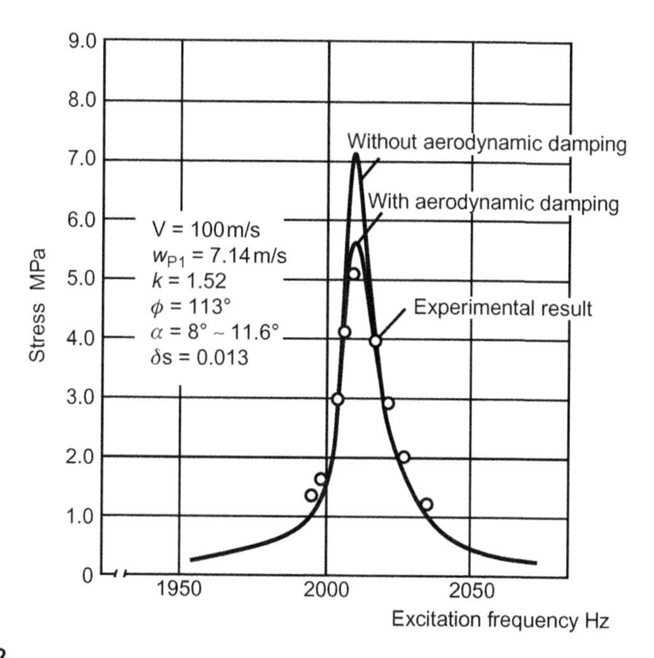

FIGURE 7.12

Vibration stress level in a blade (comparison of theory and expreiment).

analyze the vibration stress. A comparison with experimentally measured values is shown in Fig. 7.12 [12]. The good agreement confirms the effectiveness of the theory.

(2) Evaluation of fluctuating fluid forces on a cascade of flat blades [4]

Blades are arranged in a cascade as shown in Fig. 7.13. Unlike the single blade case, the fluctuating fluid force is affected by cascade parameters such as chord-pitch ratio (solidity) c/s, stagger angle ξ, and difference in phase of gusts flowing into adjacent blades. Research by Murata et al. [14] suggests the following dependencies:

1. Higher solidity results in lower fluctuating fluid forces; however, the phase difference between blade motion and gust loading does not change significantly.
2. The fluctuating fluid forces increase with increasing stagger angle if the solidity is unchanged; however, the gust-blade motion phase difference does not change much.
3. The blade motion phase difference with gusts flowing into adjacent blades results in larger fluctuating fluid forces.

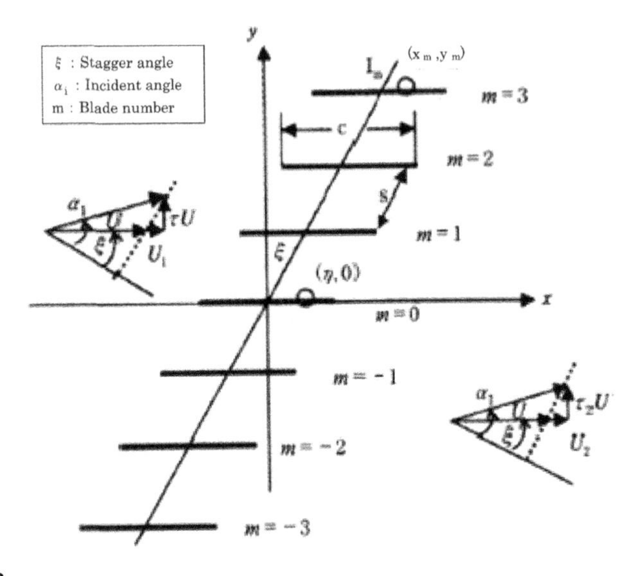

FIGURE 7.13

Cascade of flat blades.

4. The dependencies above become more significant when the dimensionless frequency is small.

A typical axial turbomachine has a solidity of about 1 and a stagger angle of 30–60°. The number of stators and rotors also differ. Therefore, there is usually a phase difference between gusts flowing into adjacent blades. The fluctuating fluid force coefficient is similar to that of a single blade, as shown in Fig. 7.14; the assessment method for single blades is therefore also effective for cascades.

(3) Evaluation of excitation fluid force from rotating stall

Turbomachinery that makes use of decelerating cascades, such as blowers and compressors, may have a region where pressure increases with decreasing flow as shown in Fig. 7.15 [25]. Surging and rotating stall occur at these operating points. Rotating stall is described here because it has a strong relation to blade vibration. Rotating stall is an issue for blade vibration because stalled and unstalled cells form in an annular flow path. These rotate at a slower speed than the rotating speed, in the direction of rotation; the blade is therefore subjected to periodic fluctuating fluid forces. Actual blade failures and previous stress measurements showed that the fluctuating forces result in extremely large stresses at the resonance point. It would be ideal if non-stalling machines could be designed, but this is not possible due to the nature of rotating machines. The region of rotating stall occurrence must therefore be determined during test runs and avoided during

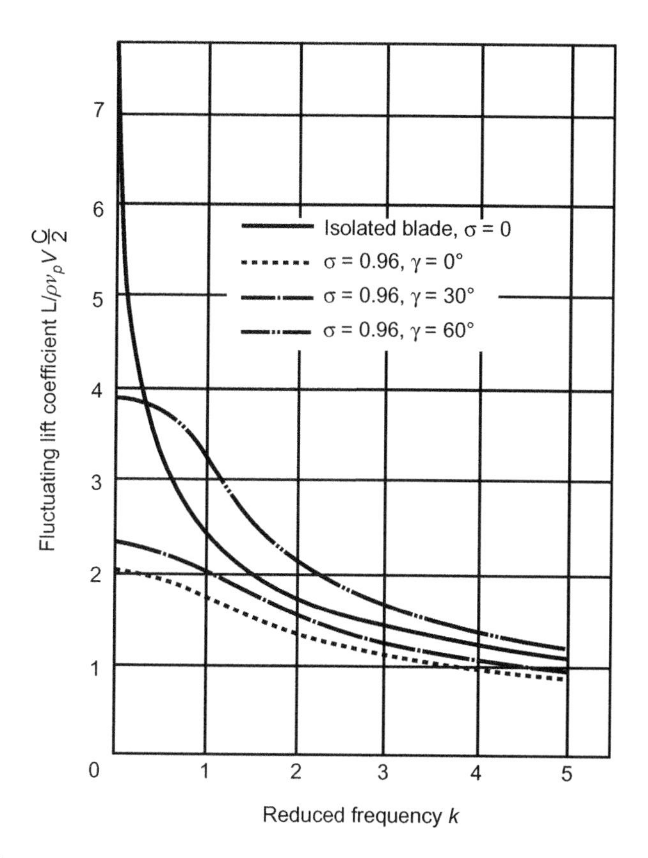

FIGURE 7.14

Cascade unsteady lift coefficient [4].

normal operation. However, the region cannot be avoided in transitional states such as starting and stopping.

Rotating stall is a form of forced vibration; therefore, avoidance of resonance and increasing blade damping are important measures. It is, however, difficult to predict the excitation frequency at the design stage because the rotating speed ratio and number of stall cells (described below) cannot be theoretically determined.

Propagation of stall cells along the rotation direction after a rotating stall can be observed by inserting two hot-wire probes behind a cascade; the rotating speed ratio and number of stall cells can then be obtained. Stall cells propagate at a slower speed (angular frequency ω_s) compared to the rotor speed (ω_r). If the number of stall cells is given by K (with multiple stall cells spaced at equal intervals), then the frequency of the resultant periodic fluctuating fluid force on the rotor is $f_c = nK(1-\nu)N/60$. Here, $\nu = \omega_s/\omega_r$ is the rotating speed ratio, which is typically in the range 0.3 to 0.8 [26]. The fluctuating fluid force from a rotating stall

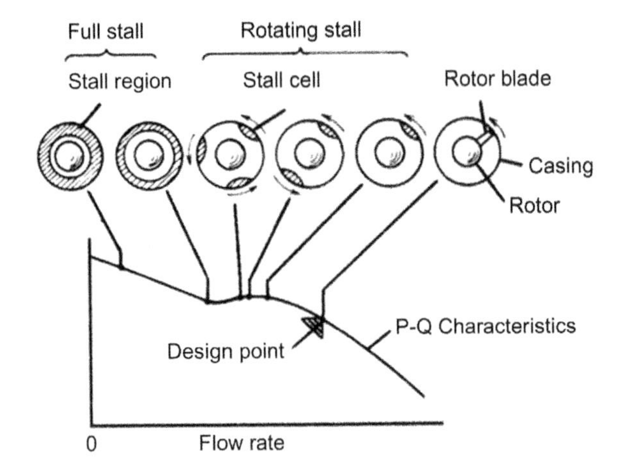

FIGURE 7.15

Change in stall cells around a rotor.

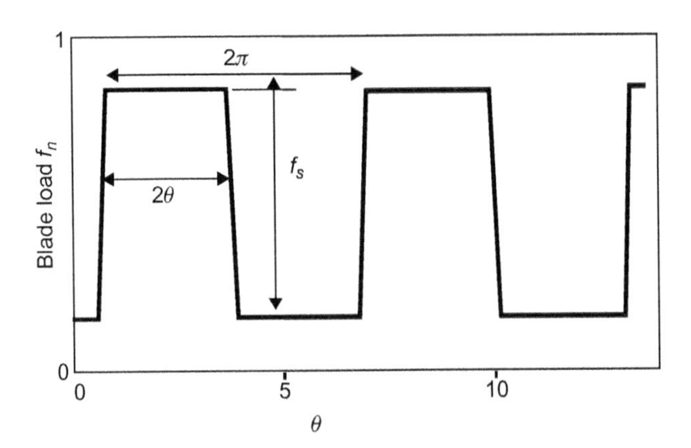

FIGURE 7.16

Resultant fluctuating fluid force from rotating stall [26,27].

may be considered as a square wave, as shown in Fig. 7.16, with a difference of the resultant fluctuating fluid force during stall and non-stall f_s. Fourier components can be roughly evaluated by Eq. (7.10), where $\nu' = \omega_r - \omega_s$:

$$F(t) = \sum_N F_n \cos(n\nu' t - \phi_n) \tag{7.10}$$

$$F_n = \frac{2f_s}{\pi} \frac{\sin(n\theta)}{n} \tag{7.11}$$

7.1.2.3 Example of blade failure and possible countermeasures

Figure 7.17 shows the fatigue failure of blades in a five-stage axial compressor caused by rotating stall. The left figure is a cross-section of the failed blades. A beach-mark pattern, which is unique to fatigue failure, can be observed. Figure 7.18 shows the vibration stress during test runs on a blast furnace compressor blade. The stress level is high during rotating stall when the blade resonates.

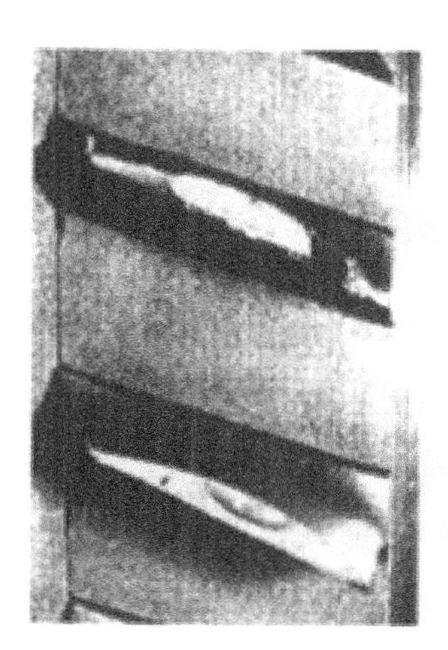

FIGURE 7.17

Blade failure due to rotating stall in an axial compressor [2].

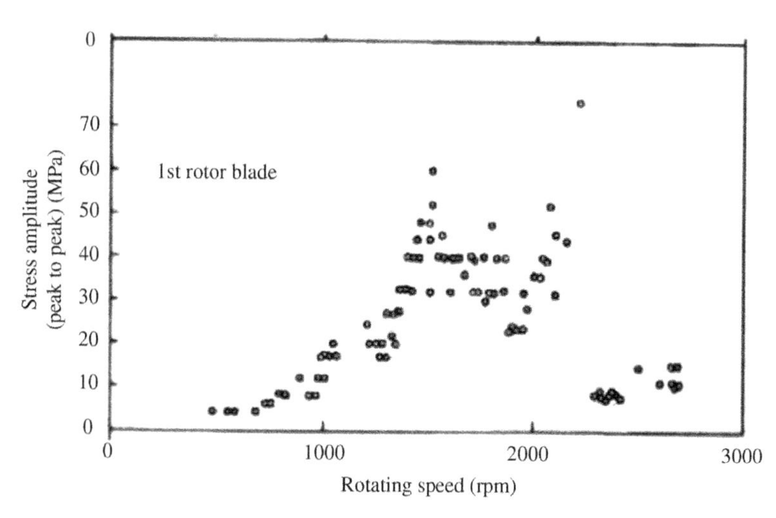

FIGURE 7.18

Actual measured vibration stress in an axial compressor blade [2].

A resonance stress level of about 50 MPa was observed in these measurements, suggesting that failure may have been caused by resonance.

The ultimate countermeasure is complete prevention of resonance. However, it is practically impossible to avoid resonance in all stages under gust-forced vibration even for a fixed rotation speed. In practice, resonance with periodic forcing from a rotating stall cannot be avoided as long as the number of stall cells and the rotating speed ratio are unknown. The conclusion is therefore that high-strength blades that do not suffer fatigue failure in resonance should be designed.

7.1.3 Flutter of blades and cascades (self-excited vibration) [3,28]

7.1.3.1 History of research and evaluation

In the present context there are two types of flutter: coupled-mode (potential) flutter and stall flutter. Fluid forces from the phase differences between degrees-of-freedom of blade motion (bending and torsion modes) under conditions of no stall act as excitation forces in coupled-mode flutter. This type of flutter becomes an issue when the mass ratio ($m/\pi\rho b^2$) is small, and must be considered in applications where weight reduction is important, such as for aircraft wings and aircraft gas turbines. Lane's [29] finite pitch cascade theory is based on an analysis of coupled mode flutter that has a degree-of-freedom representing the difference in vibration phase between adjacent blades; it considers also the stagger angle β. The theory matches experimental measurements when $\beta = 0$; however, the blade turning angle is not considered. Hanamura et al. [30] suspected that the deflection angle would have a large effect on the fluid forces, and analyzed coupled mode flutter when the turning angle is not zero (i.e. steady circulation exists). Their numerical analysis assumes two-dimensionality and incompressible potential flow. The analysis is, however, quite accurate even with these assumptions. The above-mentioned classic flutter involves coupling between the bending and torsion modes, and can be observed in single blades. The finite pitch cascade theory shows that pure bending flutter, which cannot occur in single blades, may take place in cascades. In the theory, each blade is represented by a concentrated vortex.

The frequency equation, which considers the phase differences between blade deflections and fluid forces, forms a circulatory matrix; the problem can therefore be cast mathematically as a complex eigenvalue problem.

Generally, flutter occurs when the energy delivered by the fluid flow surpasses what can be dissipated by damping in and about the blades.

7.1.3.2 Evaluation method

Pure bending flutter of a cascade of flat blades, such as shown in Fig. 7.13, is described here. For single blades, the reader is referred to Section 3.2.1.

The unsteady aerodynamic forces acting on a flat blade in a cascade with an angle of attack is calculated when the blades are vibrating with a given amplitude and phase. The blades are flat; therefore, the blades and wakes can be represented by vortex sheets. The strength of the vortex on a blade has two components: a small, steady component ζ and a finite, unsteady component γ. The component ζ induces a steady velocity on the blade whereas γ induces an unsteady velocity. Vibrations of the blade change ζ, resulting in an unsteady velocity (of first order in magnitude). On the other hand, the unsteady velocity induced by γ is a small second-order term and is disregarded. Integral equations for ζ and γ can be obtained by considering the boundary conditions at the blade surface where the induced upwash velocity is zero. The values of ζ and γ can be obtained by solving these integral equations such that the Kutta-Joukowski condition is satisfied at the trailing edge of the blade. The forces acting on the surface of a blade can then be evaluated using ζ and γ. These forces are proportional to the vibration velocity of the blade. The sign of the real part of the unstable lift coefficient (a complex number) indicates whether there is aerodynamic damping, when negative, or excitation (flutter), when positive. The reader is referred to references [3,29] for details.

7.1.3.3 Actual failure examples and possible countermeasures

Figure 7.19 shows the calculation results for five cascades with properties given in Table 7.1. The dimensionless frequency λ ($=\omega c/U$) is used as a parameter,

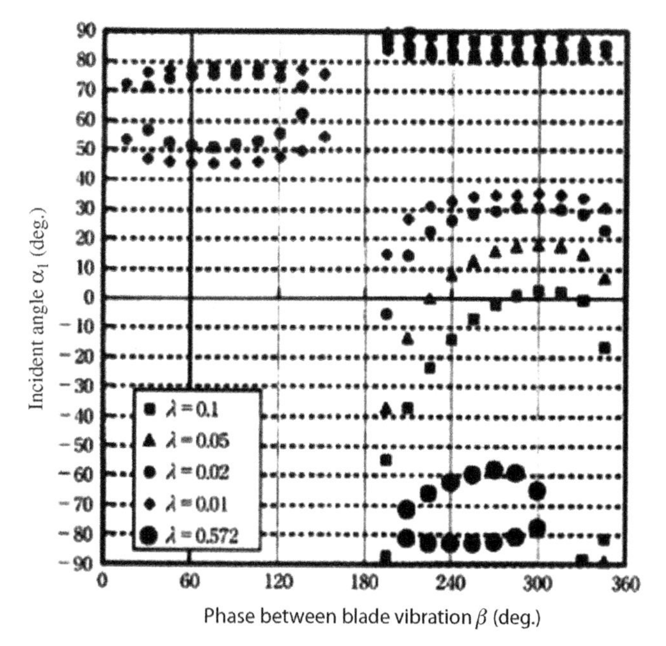

FIGURE 7.19

Flutter region for turbofan blades.

Table 7.1 Calculated properties of cascades

Symbol	Unit	Case 1	Case 2	Case 3	Case 4	Case 5
C	mm	27.74	27.74	27.74	27.74	27.74
S	mm	35.52	35.52	35.52	35.52	35.52
α_1	deg.	49.31	49.31	49.31	49.31	49.31
ξ	deg.	39.87	39.87	39.87	39.87	39.87
λ		0.572	0.01	0.02	0.05	0.10
U	m/s	190.0	190.0	190.0	190.0	190.0

and the regions surrounded by symbols (planes formed by the difference in oscillation phase between blades and angle of attack) are flutter regions. A flutter region becomes larger when the dimensionless frequency is reduced. The solid circles represent an actual cascade. This flutter region is small and therefore does not pose a problem in practice.

The most important countermeasure is to increase the dimensionless frequency. The next important countermeasure is the addition of damping. This is effective in both forced and self-excited vibration, and is therefore discussed in a separate (section 7.1.4) below. Another approach is blade mistuning, or using blades with different mass, rigidity, and aerodynamic properties in the cascade to increase the flutter limit. Intentional mistuning is called detuning, and has the following properties:

- Blades with different eigenfrequencies and mass are positioned alternately. Making blades with higher natural frequency heavier and lower frequency blades lighter increases the effectiveness of the countermeasure.
- Change in eigenfrequencies is utilized, which is more practical than changing the material of the blades.

7.1.4 Blade damping

Blade vibrations, whether forced or self-excited, can be prevented by increasing the damping. Blade damping in general consists of (1) material damping, (2) structural damping, and (3) fluid damping. Structural damping cannot be expected at high rotation speeds because the centrifugal forces increase and the blade root becomes nearly rigid. Although material damping is dependent on the material and the vibrational stresses, even an overestimate typically gives a small logarithmic decrement of about 0.02, and the resonance response magnification factor can reach 150. Some type of damping is therefore actively added to prevent resonance. Efforts have also been made to increase the fluid damping. Ishihara's research on damping of blades is discussed in detail in Ref. [31]; the characteristics of structural damping are also described therein.

Figure 7.20 shows an experimental apparatus used in the development of gas turbines. Examples of damping ratios obtained using the apparatus are also

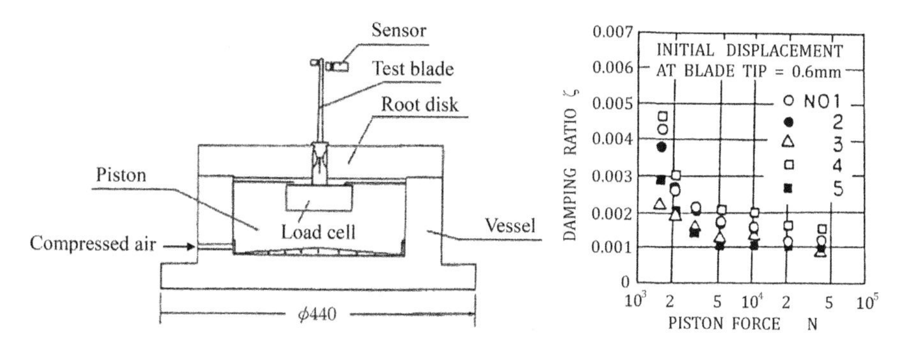

FIGURE 7.20

Damping measurement apparatus and typical measurement results.

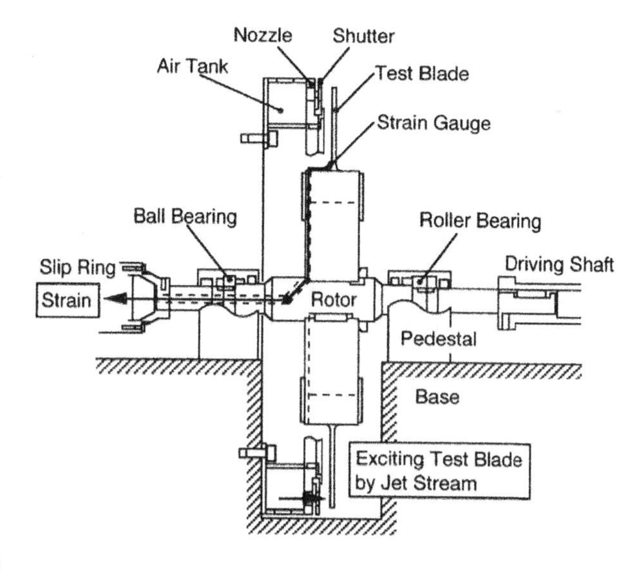

FIGURE 7.21

Rotary test machine.

shown [32]. A blade is impulsively excited (using an impact hammer) while the blade root is secured by uplift using air pressure. The logarithmic decrement of damping is obtained from the waveform of the resulting free damped vibrations. In the graph on the right the horizontal axis represents the uplift. Increased uplift results in lower saturation damping. This method takes into account the effect of the centrifugal force acting at the blade root, but not the centrifugal force in the axial direction of the blade.

Figure 7.21 shows a method that can take into account the effects of the centrifugal forces in the axial direction of the blade. Here, a blade is attached to a

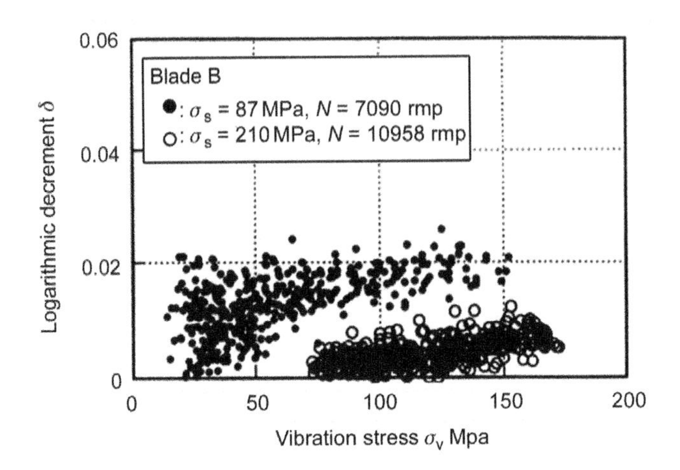

FIGURE 7.22

Logarithmic decrement of damping versus vibration stress.

rotary test machine and is hit while rotating. The damping ratio is obtained from the waveform of free damped vibrations [33]. Figure 7.22 shows some measurement results. Damping is lower when static stresses (from centrifugal force) are large, while it increases with increasing vibration stresses. The scatter or variation in the measured damping ratios is relatively large, as can be seen in the figure.

7.1.5 Numerical approach to evaluation of blade vibration

7.1.5.1 Unsteady fluid forces

Numerical modeling of unsteady flows and unsteady fluid forces has become a general approach following recent progress in computational fluid dynamics (CFD). Although large-scale computing facilities are still needed to carry out precise three-dimensional analyses including small- to large-scale turbulence and acoustics effects, it is possible to use desktop workstations or personal computers to carry out numerical simulations of unsteady flows in rotating blades due to periodic or large-scale disturbances such as rotor-stator interactions and inlet distortions.

Typical examples of numerical methods are shown in Table 7.2. Numerous methods have been developed and applied to various problems. Among them, the finite volume method is widely used for the investigation of unsteady flows in rotating machinery because of its convenience when modeling flows in complicated geometries using unstructured grids.

For flows in rotating machinery such as compressors, pumps, etc., the Reynolds number is large, hence the influence of turbulence is considerable. However, turbulence characteristics usually include very small scales in space and time (small-scale turbulence structures). It is still expensive to simulate such

Table 7.2 Numerical flow solution methods [34]

Finite difference method (FDM)	Spatial discretization is carried out using a structured grid in which the grid nodes are ordered. Many kinds of high accuracy schemes are available
Finite volume method (FVM)	Computational domains are discretized into small control volumes and the governing equations are described in integral form. Both structured and unstructured grids are available. Use of unstructured grids is convenient for the discretization of complicated geometries, such as those present in rotating machinery. However, the spatial discretization accuracy is usually restricted to third order using the unstructured grid
Finite element method (FEM)	Computational domains are discretized into finite elements using unstructured grids. FEMs are often used for the analyses of the flowstructure interactions due to the high compatibility with the structural simulation solver

Table 7.3 Turbulence models [2]

Reynolds-averaged Navier–Stokes simulation (RANS)	RANS models are widely used for practical problems and numerous models have been developed. Turbulence effects are modeled in terms of time-averaging theory. The dissipation due to small turbulent eddies is described as a local viscosity, termed the eddy viscosity. RANS models are classified according to the number of transport equations that are solved to obtain the eddy viscosity. For example, the Baldwin–Lomax model is an algebraic model in which transport equations are not used. The Spalart–Allmaras model and the k-ε models are representative examples of one-equation and two-equation models, respectively. Although RANS models are not suited to solving highly unsteady problems, they can nonetheless be used with care for unsteady analysis of rotating machinery.
Large eddy simulation (LES)	In LES, the turbulent eddies of scales smaller than the grid size are filtered out and instead represented as sub-grid scale (SGS) eddy viscosity. A typical SGS model is the Smagorinsky model. Various improvements have been made, including dynamic models. Generally, the eddy viscosity provided by the SGS model in LES is much smaller than that provided by RANS. For this reason, highly accurate schemes and highly resolved computational grids are required to carry out LES.

flows in detail. In practice, turbulence models are used to simulate high Reynolds number flows efficiently.

Turbulence models are classified into two types: the first is the Reynolds-averaged Navier–Stokes Simulation (RANS) based models; the second is large eddy simulation (LES) modeling, as shown in Table 7.3. Additionally, hybrid

LES/RANS models and detached eddy simulation (DES) have also been developed recently, taking advantage of both RANS and LES modeling. Although turbulence models have significant advantages in cutting the computational costs, every model has, at the same time, some disadvantages. Turbulence models should therefore be selected with care and with sufficient understanding of the detailed characteristics of each modeling approach.

7.1.5.2 Blade vibration

The finite element method is widely used to simulate blade vibration and is implemented in various commercial software.

One of the approaches to simulation of blade vibration in rotating machinery is to use a coupled solver, as shown in Fig. 7.23(a). In this method, solutions of the flow solver and the structural solver are combined through mesh deformations, boundary conditions, and forces, at each time step of the unsteady calculation. The coupled solver can solve complicated flow–structure interaction

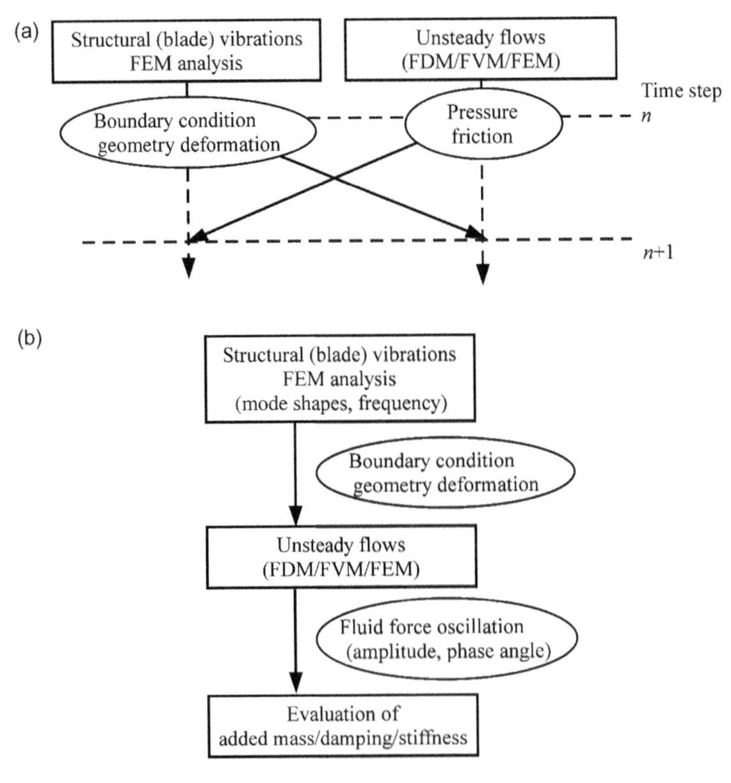

FIGURE 7.23

Flow-chart of flowstructure interaction solver. (a) coupled solver, (b) decoupled solver.

problems, including non-linear phenomena, and the results typically agree well with experiments. However, the coupled solver also has some disadvantages, in the form of expensive computational costs, because two kinds of solvers run simultaneously. For this reason, a decoupled solver is often more useful in practice. Figure 7.23(b) shows one example of a decoupled solver. The blade deformations at lower-order vibration modes are computed using the structural solver as the first step. In the flow solver, the unsteady flow response is calculated by implementing the blade deformation through the mesh deformation, boundary conditions, or source terms. The fluid dynamic damping forces can be estimated using relationships between unsteady fluid forces and blade motions, such as the phase differences and energy balances in one cycle of the blade vibrations [35]. Although the simplification of the decoupled solver restricts the types of problems which can be treated, the simplification often becomes advantageous in understanding the physical mechanisms of the underlying phenomena because the fluid forces can be divided into the added damping and stiffness components.

7.1.5.3 Evaluation of blade vibration

The flow-blade coupled oscillation problem can be solved in the time domain [36,37] or in the frequency domain [38,39]. The choice of domain depends mainly on the type of flow solver, which is computationally more expensive than the structural solver.

Time domain flow solvers typically adopt the non-linear equations of motion for the flow and can solve for non-linear phenomena, such as vortex-induced vibrations, as well as periodic oscillations such as those related to rotor–stator interaction. By using a frequency domain solver for the structural calculations, unsteady flow-induced vibrations with non-linear flow phenomena can be computed with reduced computational costs. The flow and structure interactions can be implemented using any of the methods shown in Fig. 7.23.

Frequency domain flow solvers usually adopt the linearized Euler equations or the linearized Navier–Stokes equations. Variables are divided into time-mean components and small perturbation components. The time-mean components are calculated using non-linear equations. The complex amplitudes of the small perturbation components are computed using linearized equations with the assumption that the solutions have the form of harmonic functions. Both components are usually obtained using the steady-state calculation method, thus saving computation time.

It is, however, difficult to apply a linearized solver to problems with complicated flow oscillations or to non-linear flows. To overcome this difficulty, the development of advanced frequency domain solvers, which model non-linear effects, is the subject of much recent research. The fluid–structure interactions are usually calculated using a decoupled method, as shown in Fig. 7.23(b), in order to reduce the computational costs.

7.2 Vibrations of rotating bodies partially filled with liquid

7.2.1 Summary of phenomena

Automatic washing machines [40–44], centrifuges [45], continuous annealing furnace cooling towers [46], and fluid couplings [46] are examples of rotating bodies partially filled with liquid. In some cases, water or oil may also infiltrate the rotating body by condensation or other causes [47]. The displacement of the liquid in these bodies causes significant unbalance forces [43].

Figure 7.24 shows a cutaway drawing of an automatic washing machine [42]. A fluid balancer (a ring-like container partially filled with water) is attached to the top of the basket. During the centrifuging process, when the basket rotates at super-critical speed, the water moves to the position diametrically opposite to the unbalance caused by the laundry by a self-centering effect [48]. As a result, the unbalance forces of the laundry – and thus the vibration level of the basket – can be reduced.

For a rotating body partially filled with liquid, self-excited vibrations may occur for rotation speeds slightly higher than the critical speed. This happens when the natural frequency of the liquid oscillations and that of the rotating body are nearly equal [49–51]. For the liquid oscillations, jump phenomena [52] and generation of isolated waves [40–42], which are peculiar to nonlinear oscillating systems, may occur. To prevent self-excited vibrations, baffle plates, which introduce resistance to the circumferential flow are, for instance, inserted in the fluid balancers of automatic washing machines.

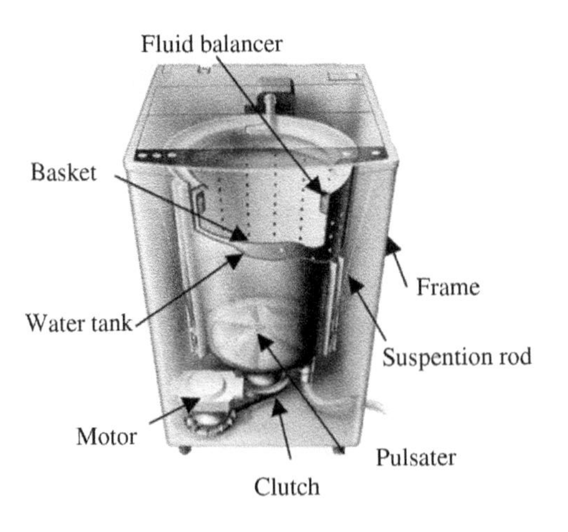

FIGURE 7.24

A cutaway drawing of an automatic washing machine.

7.2.2 Research history

An early case of self-excited vibrations of a rotating body partially filled with liquid is that presented by Ehrich [47]. The author introduced an example of self-excited vibrations which occurred when a small amount of oil infiltrated the rotor of a jet engine [47]. Thereafter, rotor stability assessment became the subject of much research. Wolf presented analysis based on inviscid flow [53], while Saito et al. presented a finite difference method-based analysis considering viscosity [54–60]. Kaneko et al. presented a theoretical analysis also taking viscosity into consideration [49–51]. Subsequently, Yasuo et al. investigated the stabilization effect caused by baffle plates [61,62].

Nonlinear oscillations of liquids in rotating bodies have also been the subject of significant research. Jinnouchi et al. investigated the hydraulic jump [52], while Kasahara et al. investigated the isolated (solitary) wave [40–42]. Yoshizumi considered jump phenomena and the free surface waveform of the liquid using an eddy viscosity model [63,64]. Visualization tests of multiphase systems (i.e. two liquids; liquid and gas; solid in liquid, or granular matter in liquid) in a rotating cavity were also conducted by Kozlov et al. [65–67]. In addition, experimental work on stabilization control utilizing magnetic dampers or magnetic bearings was presented by Matsushita et al. [45]. Experimental and theoretical research was also conducted by Zhang et al. [68,69].

Research has also been done on unbalance force-induced vibrations of rotating bodies partially filled with liquid. Katayama et al. [46] experimentally studied the dynamics of a cooling roll, found in a continuous annealing furnace, and a fluid coupling, while Takahashi et al. [70] developed a calculation method using a finite element method. Ida et al. [43] presented an analytical model for an automatic washing machine. The liquid flow inside one- and two-race fluid balancers with baffles was computed using CFD by Jung et al. [44].

7.2.3 Self-excited vibration of a cylindrical rotating body partially filled with liquid

A cylindrical rotating body partially filled with liquid may suffer self-excited vibration as shown in Fig. 7.25 [51]. Here, $\tilde{\Omega} = \Omega/\Omega_0$ is a reduced rotation speed, where Ω is the angular speed, and Ω_0 the undamped natural angular frequency of the rotating body without liquid. In Fig.7.25, 'A' is the vibration amplitude of the rotating body and $\tilde{\omega} = \omega/\Omega_0$ is the reduced frequency, where ω is the angular frequency. The ratio $R = b/a$ is the reduced liquid depth, a being the inner radius of the rotating body and b the radius of the free surface of the liquid. The dynamic viscosity of the liquid is given by ν.

Self-excited vibrations occur when the rotation speed is higher than natural frequency, i.e. $\tilde{\Omega} > 1$ (the reduced rotating speed, $\tilde{\Omega}$, is approximately 1.4 in Fig. 7.25). Following the onset of unstable vibrations, jump and hysteresis phenomena may appear. These result from the nonlinear characteristics of liquid oscillations.

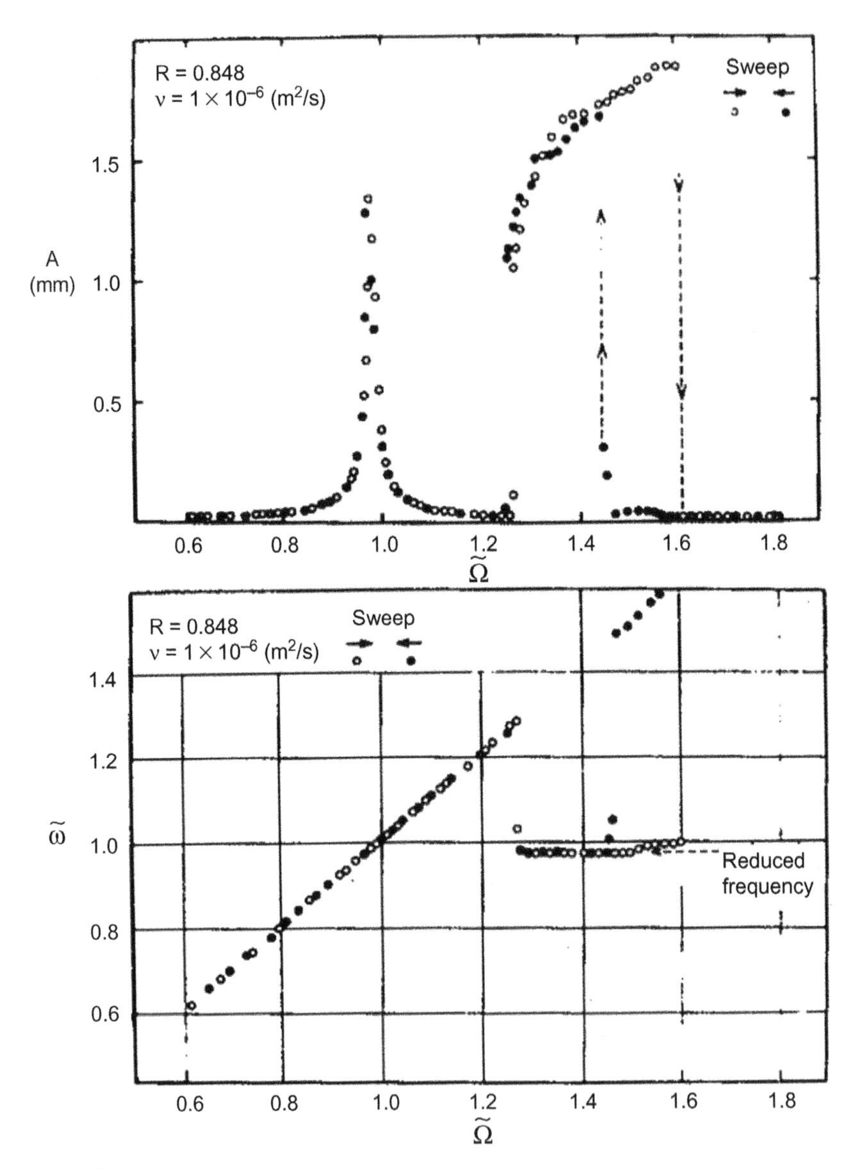

FIGURE 7.25

Amplitude and frequency of vibration of rotating body (water) [51].

7.2.3.1 Basic analysis model [49]

Figure 7.26 shows a basic analysis model for a rotating body partially filled with liquid. Here, $2h$ is the length of the cylindrical rotating body, ρ is the fluid density, p the fluid pressure, u, v, w the velocity components of the liquid, in a

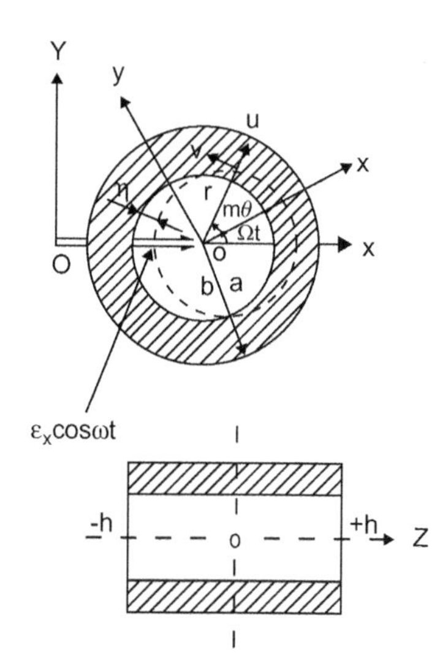

FIGURE 7.26

Basic analysis model of rotating body partially filled with liquid [49].

cylindrical coordinate system $o\text{-}r\theta z$ fixed to the rotating body, and t the time. The continuity equation and the inviscid form of the Navier–Stokes equations are written as follows:

$$\frac{\partial u}{\partial r} + \frac{u}{r} + \frac{\partial v}{r\partial \theta} + \frac{\partial w}{\partial z} = 0 \tag{7.12}$$

$$\frac{\partial u}{\partial t} - \frac{v^2}{r} + u\frac{\partial u}{\partial r} + \frac{v}{r}\frac{\partial u}{\partial \theta} + w\frac{\partial u}{\partial z} - r\Omega^2 - 2\Omega v = -\frac{\partial p}{\rho \partial r} \tag{7.13}$$

$$\frac{\partial v}{\partial t} + \frac{uv}{r} + u\frac{\partial v}{\partial r} + \frac{v}{r}\frac{\partial v}{\partial \theta} + w\frac{\partial v}{\partial z} + 2\Omega u = -\frac{\partial p}{\rho r \partial \theta} \tag{7.14}$$

$$\frac{\partial w}{\partial t} + u\frac{\partial w}{\partial r} + \frac{v}{r}\frac{\partial w}{\partial \theta} + w\frac{\partial w}{\partial z} = -\frac{\partial p}{\rho \partial z} \tag{7.15}$$

Let η be the surface displacement of the liquid. The boundary conditions can then be written as follows:

$$u = 0 \quad (r = a) \tag{7.16}$$

$$w = 0 \quad (z = \pm h) \tag{7.17}$$

$$u - \frac{\partial \eta}{\partial t} - \frac{v}{r}\frac{\partial \eta}{\partial \theta} - w\frac{\partial \eta}{\partial z} = 0 \quad (r = b + \eta(\theta, z, t)) \tag{7.18}$$

$$p = 0 \quad (r = b + \eta(\theta, z, t)) \tag{7.19}$$

7.2.3.2 Natural frequencies of liquid oscillations

A change in the depth of the liquid alters its natural frequency of oscillation σ, as shown in Fig. 7.27 [10]. Here, m is the circumferential wave number. In a coordinate system fixed to the rotating body, the solution consists of a backward-travelling wave ($\sigma > 0$), which propagates in the direction opposite of the rotation, and a forward-travelling wave ($\sigma < 0$), which propagates in the direction of rotation. If the liquid depth is increased (R decreases), the difference in the natural frequencies of the forward- and backward-travelling waves increases.

By linearizing and solving (equations 7.12) to (7.19) using a perturbation method [49], the natural angular frequency of the liquid oscillations can be expressed as:

$$\sigma = \Omega(-1 \pm \sqrt{1 + m\gamma})/\gamma \tag{7.20}$$

where $\gamma = (1 + R^{2m})/(1 - R^{2m})$.

7.2.3.3 Forced vibration response of the liquid

Figure 7.28 shows an example of the displacement amplitude of the surface of the liquid when the rotating body is excited by a sinusoidal force in one direction [49]. Here, $\overline{\omega} = \omega/\Omega$ is the reduced excitation frequency corresponding to the angular excitation frequency, ω. The resonance peaks appear at the natural frequencies of the backward-travelling wave ($\overline{\omega} < 1$) and the forward-travelling wave ($\overline{\omega} > 1$). Moreover, near the resonance frequencies, jump phenomena accompanied by hysteresis occur. This is, again, due to the nonlinear characteristics of the fluid.

By adding the excitation force terms $\omega^2\varepsilon_x\{\cos(\sigma_{(+)}t + m\theta) + \cos(\sigma_{(-)}t + m\theta)\}/2$ and $-\omega^2\varepsilon_x\{\sin(\sigma_{(+)}t + m\theta) + \sin(\sigma_{(-)}t + m\theta)\}/2$ to the right-hand sides of

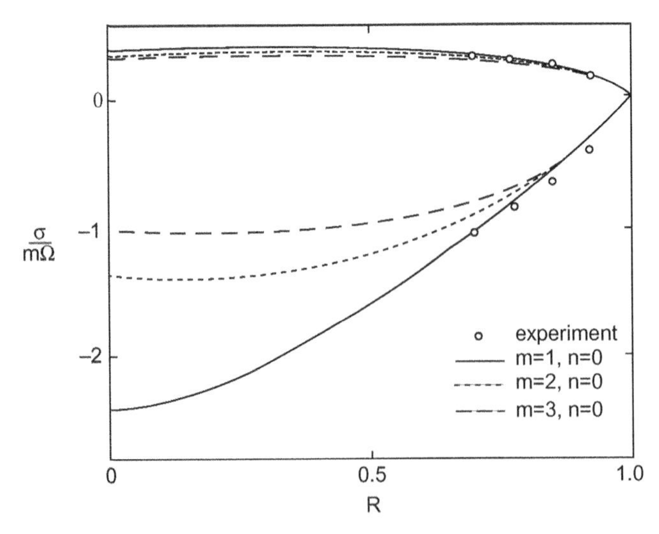

FIGURE 7.27

Relationship between the depth and the natural frequency of a liquid [49].

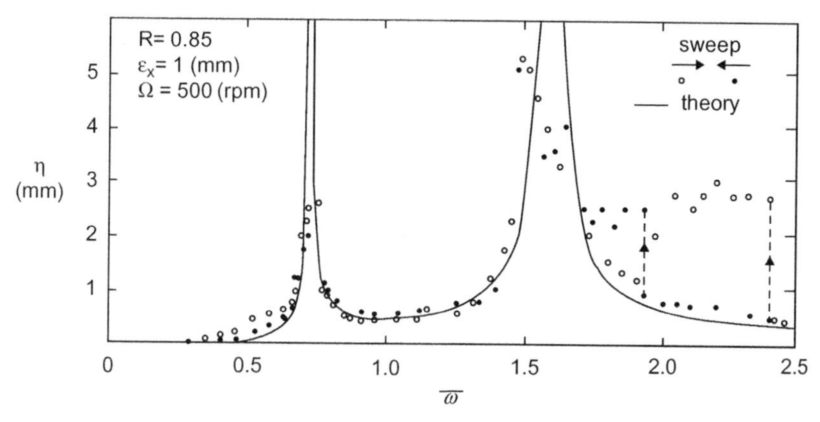

FIGURE 7.28

Resonance response of liquid surface displacement (water) [49].

(equations 7.13) and (7.14), respectively; and by linearizing and solving these equations using the perturbation method, the approximate response of the liquid can be found. When the rotating body is excited at the frequency ω, the liquid surface displacement in a coordinate system at rest (for the circumferential wave number $m = 1$) takes the form [49]:

$$\overline{\eta}(R, \theta, t) = -\frac{2\omega^2}{\sigma_{(\pm)}(K_{(\pm)}^2 - 4K_{(\pm)} - 4\gamma)}\cos(\sigma_{(\pm)}t + \theta) \tag{7.21}$$

where $\overline{\eta} = \eta/\varepsilon_x$, $\sigma_{(\pm)} = \Omega \pm \omega$, $K_{(\pm)} = 2\Omega/\sigma_{(\pm)}$, and ε_x is the amplitude of the displacement excitation.

7.2.3.4 Resultant fluid force acting on the rotating body

Figure 7.29 shows the resultant fluid force

$$F_X = \rho a^2 h\pi\varepsilon_x\omega^2(f_1\cos\omega t + f_2\sin\omega t) \tag{7.22}$$

which acts in the excitation direction of the rotating body when it is excited by a sinusoidal force [50]. Here, f_1 and f_2 are the cosine component and the sine component, respectively.

The magnitude of the fluid force changes suddenly at the backward wave resonance frequency ($\overline{\omega} < 1$) and at the forward wave resonance frequency ($\overline{\omega} > 1$). The change is steep when the dynamic viscosity ν is small. Moreover, the sign of f_2 changes at the border of the synchronous point ($\overline{\omega} = 1$). The fluid force will initiate self-excited vibrations at the backward wave resonance frequency, as the phase of fluid force progresses to match that of the displacement.

The fluid force components f_1 and f_2 can be computed by adding a viscous term and an excitation force term to (equations 7.12)–(7.19), followed by

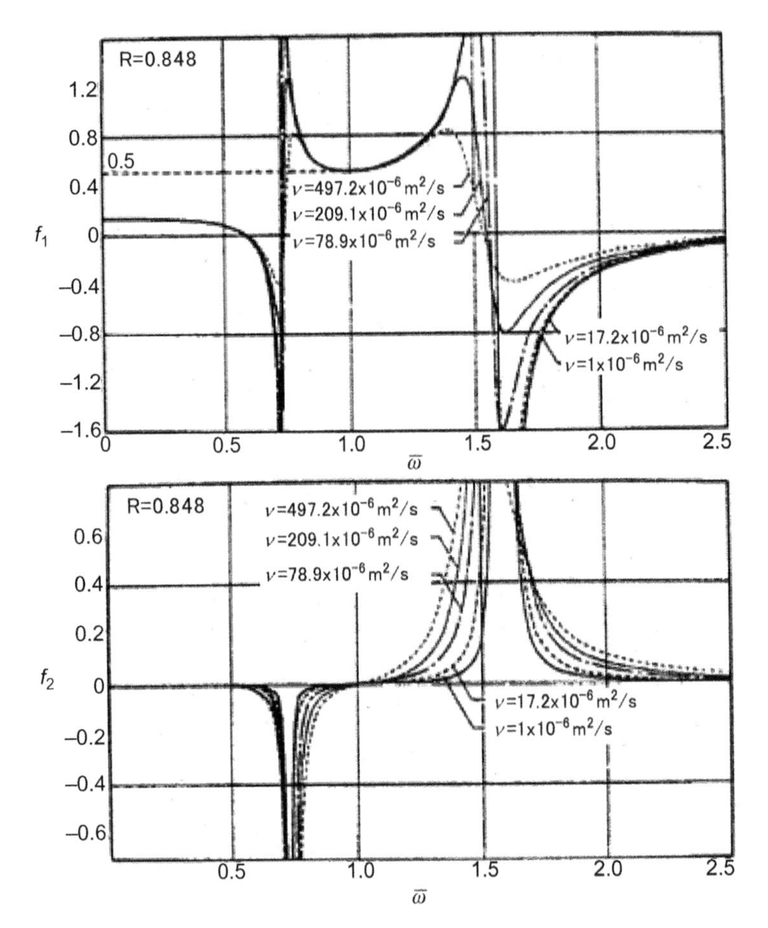

FIGURE 7.29

Example of calculated fluid force [50].

application of the perturbation method [50]. If the viscosity outside the boundary layer is omitted, the force components take the following forms:

$$f_1 = \frac{2Rg_1/K}{g_1^2 + g_2^2} + \frac{1}{2} - \text{sgn}(K)\frac{\lambda ER(g_1 - g_2)}{g_1^2 + g_2^2} \tag{7.23}$$

$$f_2 = \frac{-Rg_2/K}{g_1^2 + g_2^2} - \frac{\lambda ER(g_1 + g_2)}{g_1^2 + g_2^2} \tag{7.24}$$

Here, $K = 2/(1 - \overline{\omega})$, $E = \nu/(a^2\Omega)$, $\lambda = 1/\sqrt{|K|E}$, $g_1 = f + g(-1 + 1/\lambda)$, $g_2 = g(-1 + 1/\lambda)/\lambda$, $f = R(2 - 2/K - K/2)$, and $g = (2 + 2/K - K/2)/R$. E is a non-dimensional number, the so-called Ekman Number, which expresses the ratio between viscous forces and Coriolis forces.

7.2.3.5 Vibration stability of the rotating body

Figure 7.30 shows an example of a stability limit diagram, displaying the critical rotation speed as a function of the external damping parameter, ζ_e [51]. Within the hatched domain, small disturbances generate unstable vibrations. The unstable region becomes smaller if the external damping is increased. However, the influence of external damping is relatively small when the viscosity of the liquid is low, e.g. in the case of water [51].

Related to the above-mentioned fluid force, the undamped natural angular frequency Ω_R and the reduced damping ratio ζ are given by the following [51]:

$$\Omega_R = \sqrt{\frac{k}{m + \frac{\rho a^2 h\pi}{2}}} \tag{7.25}$$

$$\zeta = \frac{(c + \rho a^2 h\pi f_2 \Omega_R)\Omega_R}{2k} \tag{7.26}$$

Here, m is mass of the rotating body, and, c and k are viscous damping coefficient and spring constant of the rotor support, respectively. The second term in the denominator of equation (7.25) represents the added mass of the liquid. It is seen that the natural frequency falls with increasing added mass but the added mass is not related to the liquid depth. Therefore, even if just a small amount of liquid enters the rotating body, the natural frequency of the rotor system is the same, as in the case where the rotating body is completely filled with liquid. Finally, note that equation (7.26) can be used as a stability criterion since self-excited vibrations occur whenever ζ becomes negative.

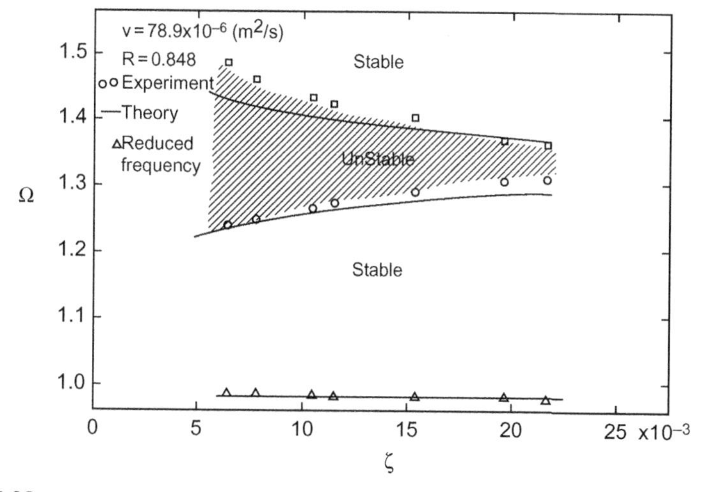

FIGURE 7.30

Stability diagram when external damping is changed (compressor oil) [51].

7.2.3.6 *Stability criterion using the characteristic equation*

Unstable vibrations occur when a dynamic system has a complex eigenvalue λ with a positive real part. Assuming that the displacements can be written in vector form as $\{x\} = \{X\}\exp(\lambda t)$, the equations of motion can be written in matrix form as follows:

$$[D]\{X\} = \{0\} \tag{7.27}$$

A non-trivial solution exists when:

$$|D| = 0 \tag{7.28}$$

where $|D|$ denotes the determinant of the matrix $[D]$. The complex eigenvalues λ can be computed using, for example, the Newton–Raphson method.

Figure 7.31 illustrates the influence of the reduced liquid viscosity coefficient $\overline{\nu}$ on the leading complex eigenvalue. In this example, the rotor is supported by magnetic bearings [68]. The horizontal axis in the figure represents the reduced rotating speed, while the vertical axis represents the imaginary part of the reduced complex eigenvalue $\overline{\lambda}$ divided by the imaginary unit. Unstable oscillations exist if $\mathrm{Im}(\overline{\lambda}/j) < 0$. When the viscosity coefficient is increased the unstable region narrows, while the stable region in the high-speed domain expands.

7.2.3.7 *Nonlinear vibration behavior accompanied by a hydraulic jump*

When the rotation speed is close to the natural frequency of the system, a large wave is formed, accompanied by a hydraulic jump, as shown in Fig. 7.32 [52]. Here, H is the liquid depth and θ the circumferential position.

Figure 7.33 shows an example of the relationship between the wave height and the rotating speed [52]. In this case, the rotating body is excited at a

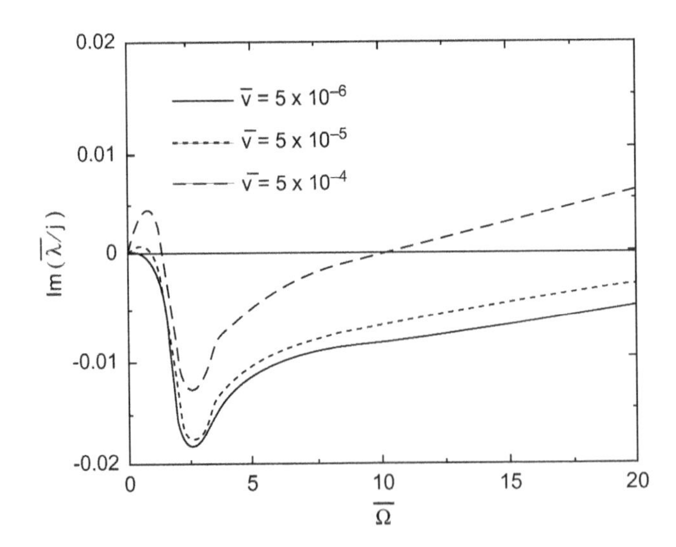

FIGURE 7.31

Influence of the reduced viscosity coefficient on the reduced complex eigenvalue [68].

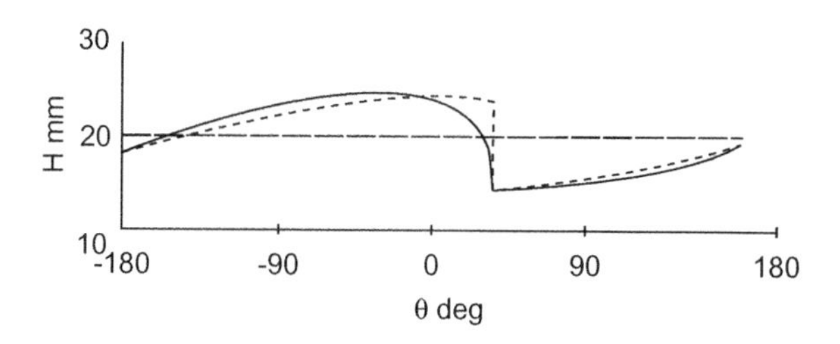

FIGURE 7.32

Free surface form of the liquid near a resonant point (water) [52].

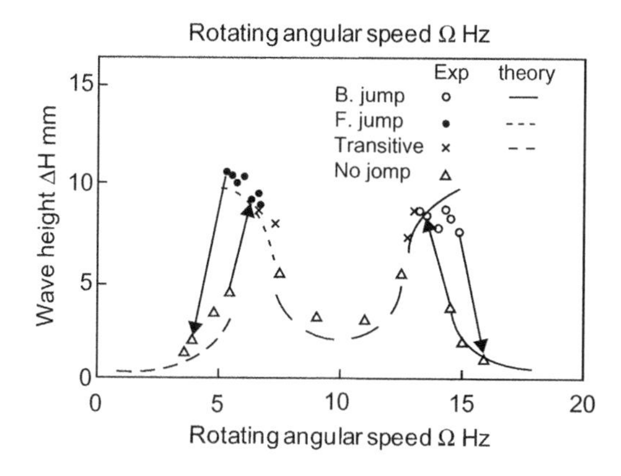

FIGURE 7.33

Wave height and position of a hydraulic jump [52].

frequency of 10 Hz, and the vibration response of the liquid becomes almost symmetrical about the synchronous rotation speed ($\Omega = 10$ Hz). As the rotation speed moves away from 10 Hz, large wave motions appear, accompanied by a hydraulic jump, which initially starts as a gently-sloping wave.

If the rotation speed moves even further away from 10 Hz, the hydraulic jump disappears. A jump and a hysteresis phenomenon occur as shown in Fig. 7.33 near 5 Hz and 15 Hz.

The nonlinear equation governing the stationary solution is obtained by applying the shallow water approximation and hydraulic jump theory to the Navier–Stokes equations. A numerical solution is then obtained by integrating the nonlinear equation using the Runge–Kutta method [52].

7.2.3.8 The isolated wave near a resonance frequency

Figure 7.34 shows the surface wave on the liquid layer, in the circumferential direction. Here the rotating body is excited with a frequency near the resonance frequency of a primary forward travelling wave of 15.33 Hz [42]. The fluid moves as a whole, and the isolated wave (resulting from higher-order components peculiar to nonlinear vibrations) comes into existence near the resonance frequency.

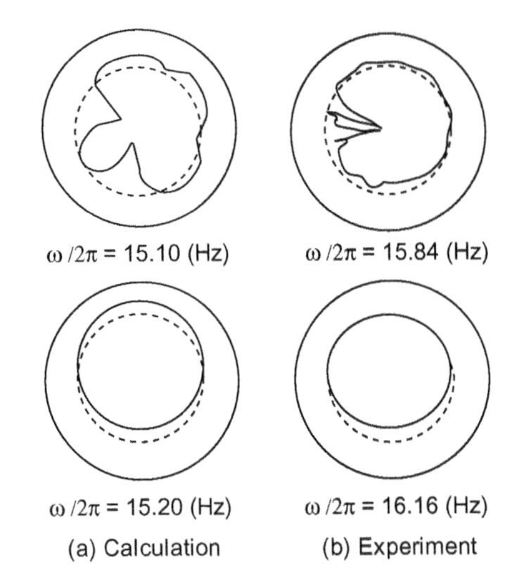

FIGURE 7.34

Surface deformation in circumferential direction near resonance frequency of the primary forward wave mode [42].

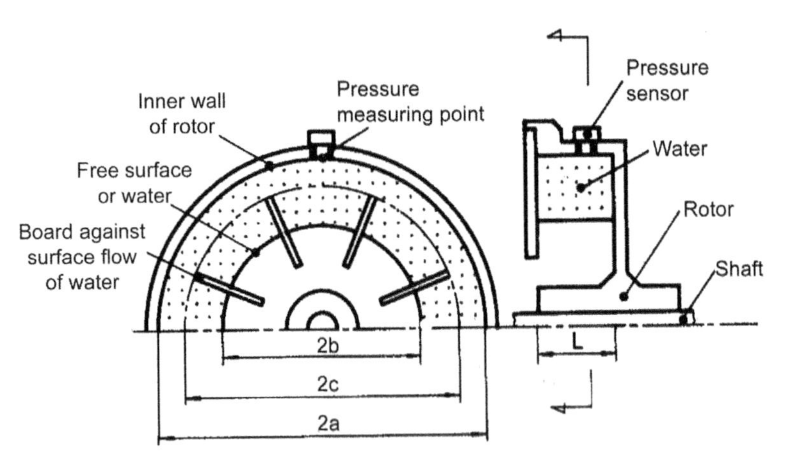

FIGURE 7.35

Partially liquid-filled rotating body fitted with baffle plates [61].

In Refs. [40]–[42] the vibration response was calculated by applying the shallow water assumptions to the Navier–Stokes equations. A time history analysis was carried out using the Crank–Nicholson method for the time domain and a staggered grid approach for the space domain.

7.2.3.9 Influence of baffle plates

Vibration stability can be improved by installing baffle plates, as shown in Fig. 7.35, which prevent flow of the surface layer [61]. If such plates are installed the unstable region becomes smaller, as shown in Fig. 7.36 [62], because the fluid

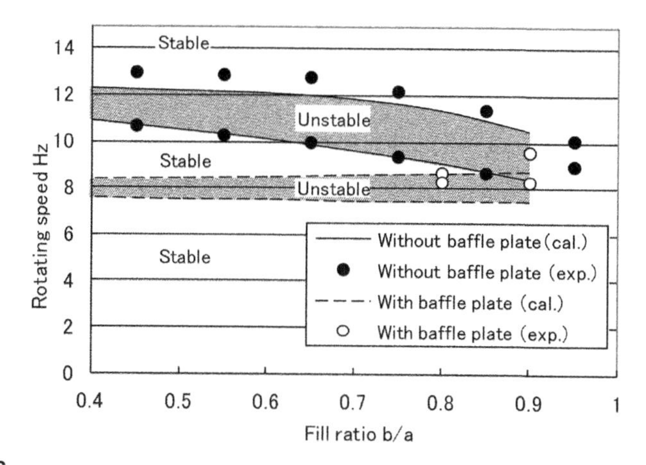

FIGURE 7.36

Influence of baffle plates on unstable region (water) [62].

Table 7.4 Hints for countermeasures

Item	Countermeasure	Memo
Rotating speed	Prevent operation in the unstable rotating speed range ($\overline{\Omega} = 1 \sim 2$)	Refer to Fig. 7.25
External damping	Increase the external damping	Refer to Fig. 7.30 However, if the viscosity of the liquid is low, the effect of external damping is small [12]
Liquid depth	Adjust the range of unstable rotating speeds by changing the depth of the liquid. If the liquid depth is large, the unstable rotating speed increases	Refer to Fig. 7.36
Baffle plates	Install baffle plates	Refer to Fig. 7.36
Prevention of liquid infiltration	Keep unnecessary liquid from infiltrating the rotating body	An example where self-excited vibration occurred when oil infiltrated a jet engine has been reported in [47]

flow is prevented. The effect of the baffle plates becomes more pronounced as the amount of contained fluid is increased. Therefore, if the ratio b/a is made small (see Fig. 7.35) the unstable vibration domain may disappear altogether. This has been verified experimentally.

7.2.4 Hints for countermeasures

Table 7.4 summarizes possible countermeasures against self-excited vibrations of rotating bodies partially filled with liquid.

7.3 Vibration induced by annular flow in seals

7.3.1 Seal-related self-exciting vibration of rotors

Turbo-machines may suffer large abnormal vibrations similar to oil whip, which occur suddenly when the load or the output is increased. Since the vibration is caused by fluid flow, it is often referred to as steam whirl (whip) for steam turbines. Analogously, we propose the terms hydro-whirl (whip) in the case of centrifugal pumps, and gas whirl (whip) in the case of centrifugal compressors. The instability is usually caused by both the working components such as blades or impeller, and/or by seals depending on the performance conditions. In this section, this type of vibration is explained, concentrating mainly on seal-related instabilities.

In rotor dynamics analysis, seals are modeled based on eight parameters, representing fluid stiffness and fluid damping. The related coefficient matrices generally contain cross terms in the orthogonal, x and y directions. The analysis method is similar to the approach taken for oil film bearings. Stability analysis is done taking into consideration the seal dynamic characteristics represented by the eight model parameters. The key difference between the seal and the oil film bearing, the latter having circumferential flow, is the existence also of axial flow in the seal, induced by the difference in pressure between seal inlet and outlet. The inlet axial flow contains, in addition, a swirl component generated by the blades or the impeller. This inlet swirl increases the effective rigidity. However, it is also responsible for the destabilizing cross-stiffness terms. To increase the seal pressure drop, several seal designs exist besides the standard straight annular seal. These include the stepped seal, the screw seal and the labyrinth seal types.

7.3.2 Historical background

After the Second World War, plant production capacity steadily increased. This improvement in production capacity was accompanied by increasingly frequent cases of occurrence of abnormal vibrations in turbo-machines, limiting plants from reaching full production capacity. To solve this problem, Thomas [71] and Black et al. [72,73] undertook research work to determine the dynamic characteristics of the blades, the impellers and annular seals in turbomachinery. Thomas et al. worked primarily on turbines while Black et al. concentrated on pumps. In the late 1970s, during the development of the space shuttle, vibration problems in the main engine hydrogen pump were encountered. This is said to have caused a 1-year delay in the shuttle development program. Following the occurrence of this problem NASA conferences on Rotordynamics Instability Problems in High-Performance Turbomachinery were initiated. It was at one of these conferences that Child and Dressman [74] presented research results on the dynamic characteristics of several kinds of seal. One of the first models was by Kostyuk [75].

Furthermore, Iwatsubo et al. [76] developed a solution method to determine the eight force coefficients in the Kostyuk labyrinth seal model. Brennen et al. [77], Ohashi et al. [78], and Tsujimoto et al. [79] performed in-depth research on pump impeller dynamic characteristics. As a result of these research efforts, it was possible to design turbo-machines which were dynamically stable against rotor dynamics whirl instabilities.

7.3.3 Stability analysis methods

7.3.3.1 Rotor dynamics model of turbo-machine

To analyze the rotor dynamics, bearings have conventionally been considered to be the only elements between the rotor and casing. Recently, to solve the rotor dynamics stability problem, seals, impellers etc. are also modeled (similarly to bearings) using eight flow coefficients, as shown in Fig. 7.37. In the case of liquid flow, seals are modeled using 12 parameters including fluid inertia effects as shown in equation (7.29)

$$\begin{bmatrix} F_X \\ F_Y \end{bmatrix} = -\begin{bmatrix} k_{XX} & k_{XY} \\ k_{YX} & k_{YY} \end{bmatrix} \cdot \begin{bmatrix} X \\ Y \end{bmatrix} - \begin{bmatrix} c_{XX} & c_{XY} \\ c_{YX} & c_{YY} \end{bmatrix} \cdot \begin{bmatrix} \dot{X} \\ \dot{Y} \end{bmatrix} - \begin{bmatrix} m_{XX} & m_{XY} \\ m_{YX} & m_{YY} \end{bmatrix} \cdot \begin{bmatrix} \ddot{X} \\ \ddot{Y} \end{bmatrix} \tag{7.29}$$

where X-Y is the fixed coordinate system and F the total reaction force. The matrices on the right-hand side of (7.29) are the fluid stiffness, the fluid damping and the added mass matrix, respectively.

7.3.3.2 Dynamics of annular seals

Black et al. [72,73] derived the analytical equations giving the dynamics coefficients for annular seals, such as shown in Fig. 7.38. The equations are based on lubrication theory. The Lomakin effect, which produces positive stiffness, was introduced by incorporating inlet swirl in the model. The analytical model and the

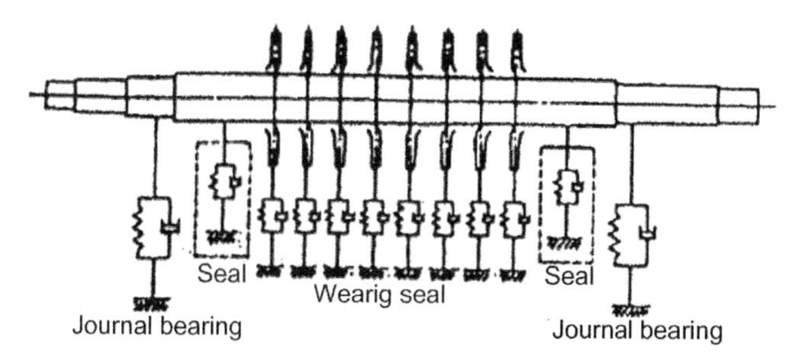

FIGURE 7.37

Model of centrifugal pump.

mechanism of the Lomakin effect are explained in what follows. The fluid force characteristic matrices are:

$$\begin{bmatrix} k_{XX} & k_{XY} \\ k_{YX} & k_{YY} \end{bmatrix} = \begin{bmatrix} \frac{\pi R \Delta p}{\lambda}\left[\mu_0 - \frac{1}{4}\mu_2 \Omega^2 T^2\right] & \frac{\pi R \Delta p}{\lambda}\left[\frac{1}{2}(\mu_1 - \mu_s)\Omega T + \mu_2 \Omega_0 T\right] \\ -\frac{\pi R \Delta p}{\lambda}\left[\frac{1}{2}(\mu_1 - \mu_s)\Omega T + \mu_2 \Omega_0 T\right] & \frac{\pi R \Delta p}{\lambda}\left[\mu_0 - \frac{1}{4}\mu_2 \Omega^2 T^2\right] \end{bmatrix}$$

$$\begin{bmatrix} c_{XX} & c_{XY} \\ c_{YX} & c_{YY} \end{bmatrix} = \begin{bmatrix} \frac{\pi R \Delta p}{\lambda}\mu_1 T & \frac{\pi R \Delta p}{\lambda}\mu_2 \Omega T^2 \\ -\frac{\pi R \Delta p}{\lambda}\mu_2 \Omega T^2 & \frac{\pi R \Delta p}{\lambda}\mu_1 T \end{bmatrix}$$

$$\begin{bmatrix} m_{XX} & m_{XY} \\ m_{YX} & m_{YY} \end{bmatrix} = \begin{bmatrix} \frac{\pi R \Delta p}{\lambda}\mu_2 T^2 & 0 \\ 0 & \frac{\pi R \Delta p}{\lambda}\mu_2 T^2 \end{bmatrix} \tag{7.30}$$

where R is the radius of the rotor shaft, Ω the rotational angular velocity, Δp the pressure drop in the seal, Ω_0 the inlet swirl angular velocity of the fluid, λ the friction coefficient of the annular gap and T the average fluid passage time interval. The dimensionless quantities μ_0, μ_1, μ_2 in equation (7.30) are obtained using narrow gap fluid tribology theory in the non-rotational state. μ_s is the non-dimensional cross-stiffness due to inlet swirl, also obtained using the same theory. The non-dimensional quantities are presented in Refs. [72] and [73]. For example, the value of μ_0 is found to be in the range 0–0.35, $\mu_1 \sim 0-1.6$, $\mu_2 \sim 0-0.08$ and $\mu_s \sim 0-0.5$.

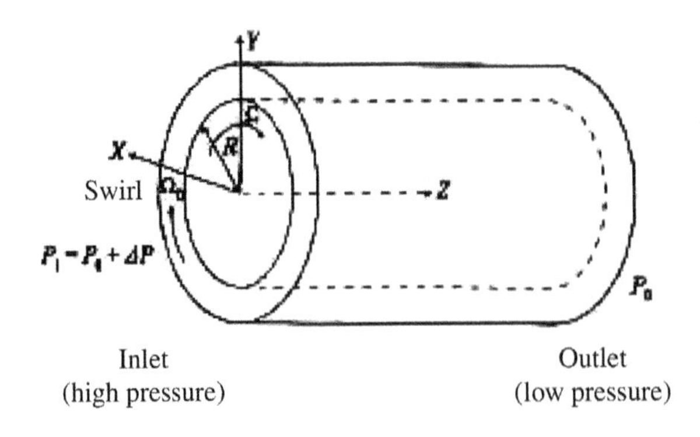

FIGURE 7.38

Annular seal.

The component $\pi R \cdot \Delta p \cdot \mu_0 / \lambda$ in the stiffness matrix diagonal terms is the stiffness due to the Lomakin effect. The component $\pi R \cdot \Delta p \cdot \mu_s \cdot \Omega \cdot T / \lambda$ in the stiffness matrix cross terms represents the inlet swirl effect while the term $\pi R \cdot \Delta p \cdot (\mu_1 - \mu_s) \cdot \Omega \cdot T / (2\lambda)$ is the effect of the rotation itself.

Figure 7.39 shows the mechanism of the Lomakin effect. The pressure difference between the inlet and outlet in the annular seal equals the summation of the pressure drop at the seal entrance and the pressure drop due to leakage flow resistance in the narrow gap of the seal. When the rotor shaft is eccentric, the gap in the same direction is narrower, which increases the pressure drop due to leakage flow resistance. This causes the inlet pressure drop on the eccentric side to decrease. As a result, the pressure on the eccentric side is then higher than on the opposite side as indicated by the shaded area at the bottom of Fig. 7.39. A restoring force therefore acts as a positive stiffness, to center the rotor shaft.

When there is no pressure drop along the seal axis, only a cross-stiffness effect exists. This effect generates a destabilizing force. The presence of a pressure variation, on the other hand, induces a positive stiffness effect. As a result, the natural frequency of the rotor shaft increases. Note that since the damping coefficient of the oil film bearing is independent of frequency, the effective damping performance increases with cross-stiffness.

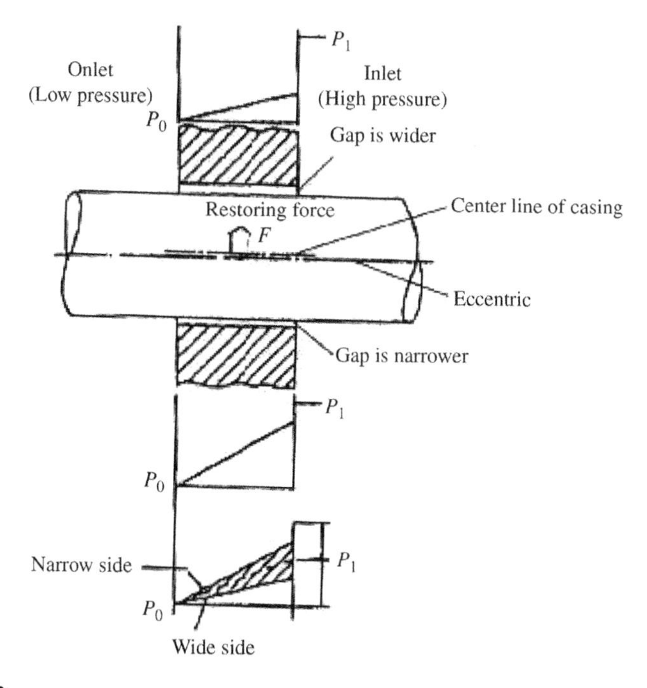

FIGURE 7.39

Lomakin effect mechanism.

7.3.3.3 Dynamic characteristics of labyrinth seals

The labyrinth seal shown in Fig. 7.40 is usually used for compressible fluids. It has been confirmed experimentally that this seal has cross-stiffness forces induced by leakage flow with inlet swirl. Kostyuk [75] presented a labyrinth seal model in which the seal fin is represented by an orifice and the seal chamber by an annular volume. Iwatsubo et al. [76] determined the model's eight parameters using the perturbation method. Based on the work by Kostyuk and Iwatsubo et al. it became possible to quantitatively assess the system stability. It was also shown, for instance, that the sign of the cross-stiffness changes from + to − depending on the direction of the inlet swirl. Employing the inlet swirl efficiency, presented by Jenny [80], in the Iwatsubo solution method, the dynamic characteristics of several types of labyrinth seals can be calculated.

The relation between inlet swirl and flow in labyrinth seals is also shown in Fig. 7.40. Gas enters the seal with a given swirl (circumferential) velocity. The circumferential flow velocity in the seal chamber is about half the rotor circumferential velocity. The flow velocity distribution when inlet swirl velocity is higher or

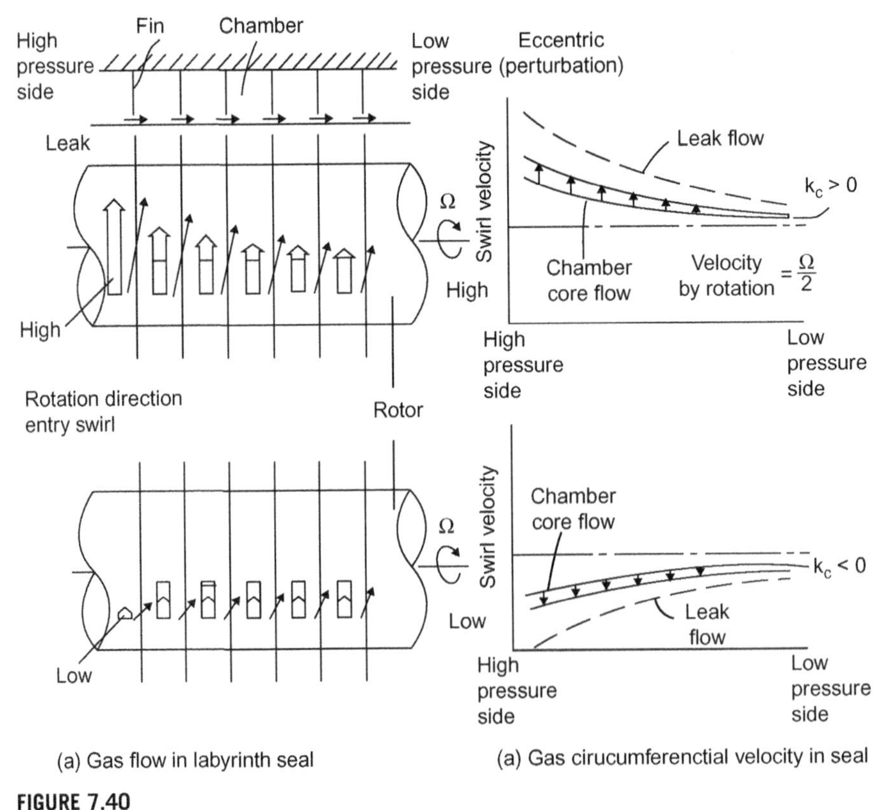

FIGURE 7.40

Labyrinth seal and cross-stiffness generation mechanism.

lower than the flow velocity in the chamber changes as shown in Fig. 7.40. If leakage gas with faster swirl than the chamber mean velocity by rotation in the forward direction enters the chamber, the upper figure of Fig. 7.40(a), the cross-stiffness increases unstable in the forward direction, the upper figure of Fig. 7.40(b). On the other hand, if the swirl velocity is lower, the lower figure of Fig. 7.40(a), the circumferential flow velocity in the chamber decreases, the lower of Fig. 7.40(b); the sign of the cross-stiffness is then opposite and the force acts in the backward direction. The total cross-stiffness therefore decreases for the forward direction.

For a seal in which the inlet swirl is larger than half the rotational speed of the rotor, a swirl canceller should be used to reduce the inlet swirl.

7.3.4 Examples of practical problems

7.3.4.1 Space shuttle high pressure hydrogen pump (hydro-whirl [whip])

Growing vibrations, as shown in Fig. 7.41, occurred in the liquid hydrogen fuel pump, leading to damage of seals and other components. As a result, it is said that the development of the space shuttle was delayed by more than a year. As a countermeasure, the type of seal was changed from the stepped labyrinth seal, Fig. 7.42(a), to the straight seal having the geometry shown in Fig. 7.42(b). As a result of using the straight seal, the natural frequency of the rotor system

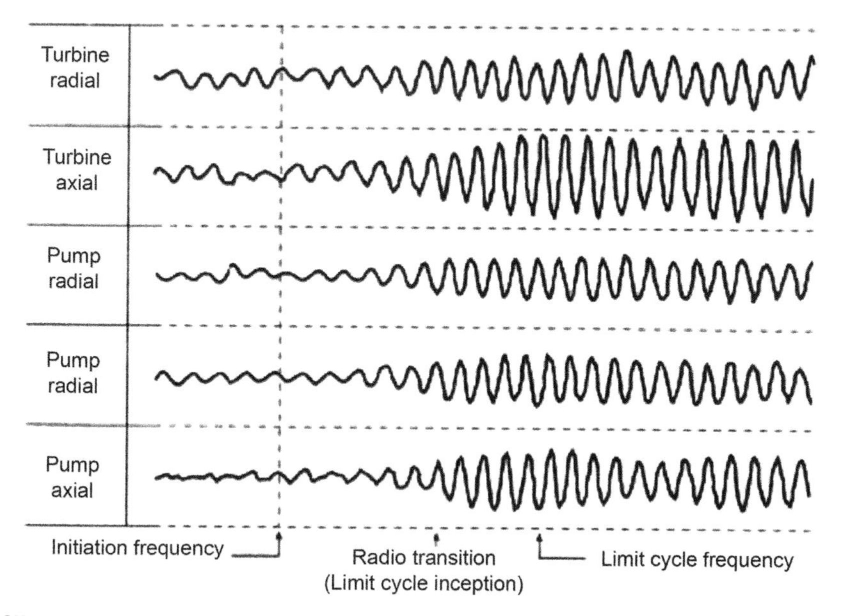

FIGURE 7.41

Abnormal vibration signature.

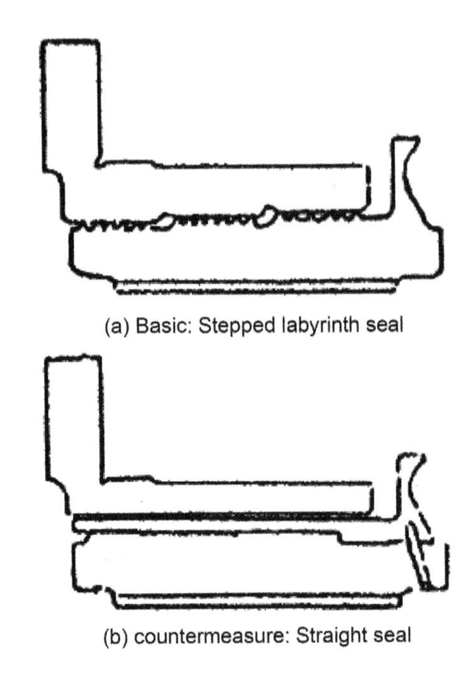

FIGURE 7.42

Seal, (a) before and, (b) after countermeasure.

increased. This stabilized the rotor and thus making it possible to operate at the rated rotation speed.

7.3.4.2 Steam generator turbine (steam whirl [whip])

The 15 MW steam generator turbine which started operation in the early 1970s suffered abnormal vibrations when the output surpassed 12 MW following modifications to improve performance, about 20 years later. Figure 7.43 shows the waterfall diagram of rotor shaft vibration when the abnormal vibrations occurred. The frequency of vibration was 36 Hz, which corresponded to the natural frequency of the rotor-bearing system. The stability evaluation result for this eigenfrequency is shown in Fig. 7.44 – where the ' × ' indicates the output power at which abnormal vibrations occurred. Since the rotor was supported by annular oil film bearings, the usual countermeasure involved replacing these by tilting pad bearings. However, due to the slender rotor shaft, the damping factor of the rotor bearing system did not improve. Stability analysis indicated that the most suitable solution was to reduce the cross-stiffness of the gland-seal. As a countermeasure then, a swirl canceller was installed at the seal inlet. This stabilized the rotor system as confirmed in Fig. 7.44.

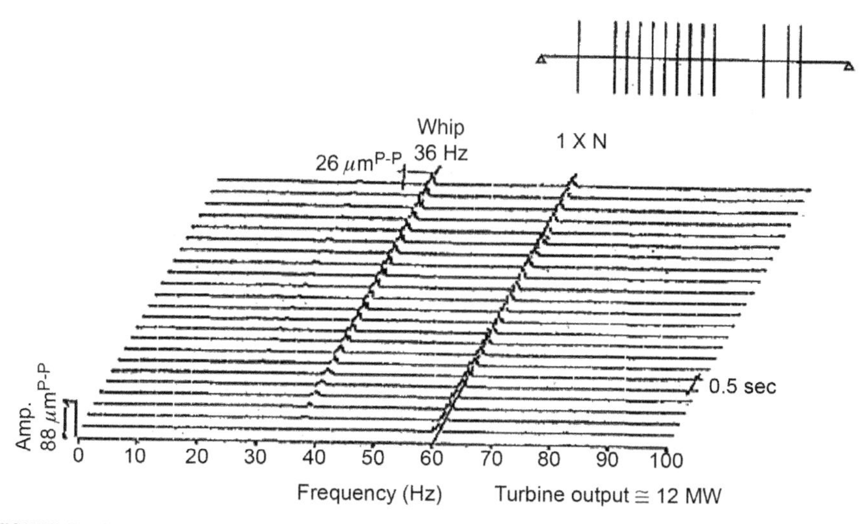

FIGURE 7.43

Abnormal vibration waterfall diagram.

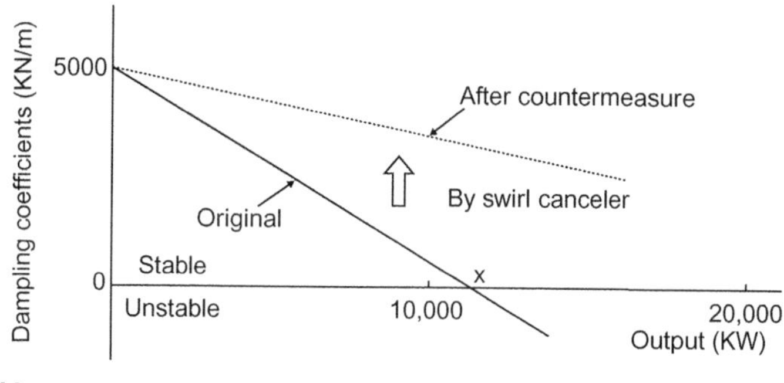

FIGURE 7.44

Stability discriminant result.

7.3.4.3 Synthetic gas compressor (gas whirl [whip])

Whirl instability occurred in a centrifugal compressor train which compressed hydrogen and nitrogen, the component gases for ammonia production, from 20 bar to 300 bar. The compressor train is shown in Fig. 7.45. At 95% load capacity and a rotation speed of 10,750 rpm growing vibrations suddenly occurred, mainly in the LP compressor. The vibration signatures are shown in Fig. 7.45. As a result of the vibrations the plant production had to be limited to below 95% of full capacity. The frequency of the vibration was initially 74 Hz but later increased to 80 Hz. The direction of orbital whirl was forward, hence in the rotational direction.

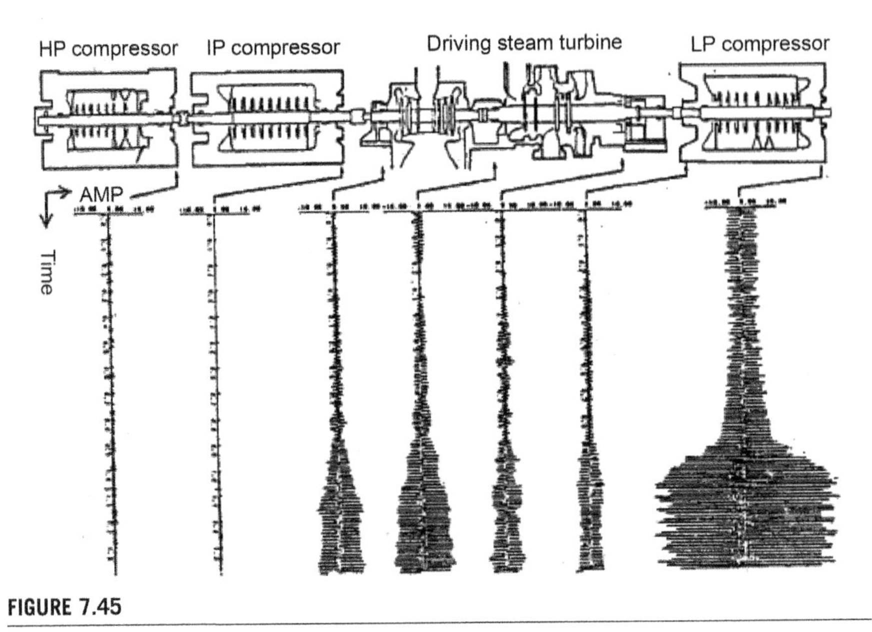

FIGURE 7.45

Compressor train and abnormal vibration signature.

In order to determine the type and source of instability causing the abnormal vibration, the block diagram shown in Fig. 7.46 was used. Firstly, an eigenvalue analysis of the rotor system considering only high-stiffness bearings and oil film seals was done. The stability was next evaluated based on stiffness magnitudes obtained by modal analysis. The dynamic characteristics of the oil film seal were incorporated based on bearing theory. For the remaining seals, which were all labyrinth type, the dynamic characteristics discussed above were applied. For the impeller, the Watchel equation was used but modified by the following Thomas equation:

$$k_{XY} = a \cdot \frac{P \cdot \beta}{\Omega \cdot D \cdot H} \tag{7.31}$$

where P is the blade power, Ω the angular velocity, D the pitch circle diameter of the blades, H the blade height, β the non-dimensional seal coefficient and a a non-dimensional constant.

Modal analysis was applied in the stability analysis. The analysis results are presented in Fig. 7.47. For each compressor, the contribution ratio of the various seals to system instability (excitation) is represented by the height of the corresponding right-hand bar while the contribution to system stability (damping) is represented by the height of the left-hand bar. From the stability analysis results, only the LP compressor rotor is found to be unstable. This conclusion agreed with the observed vibration behavior. The location where the countermeasure

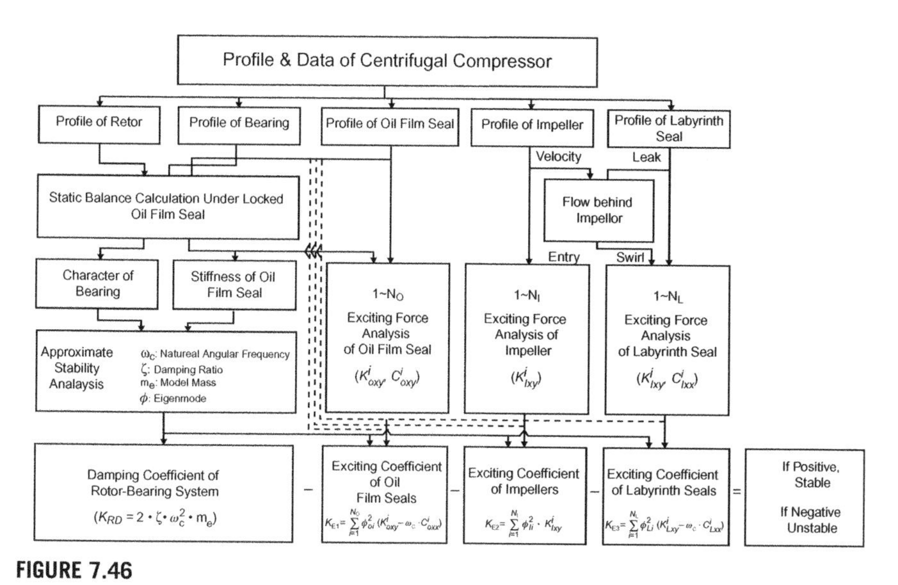

FIGURE 7.46

Compressor system stability analysis block diagram.

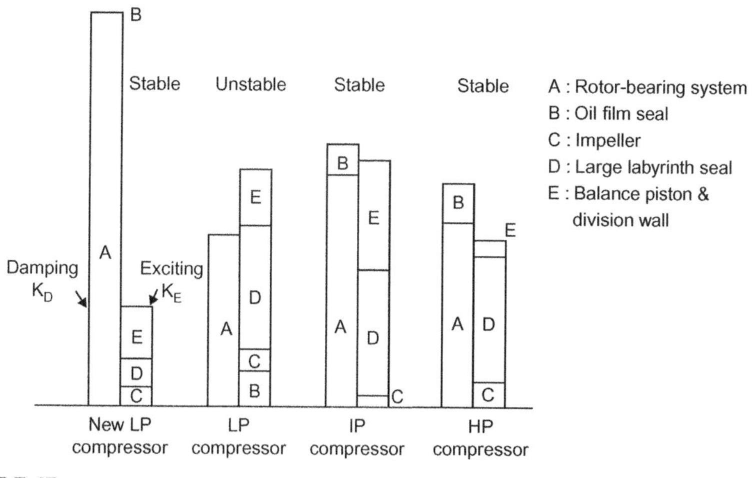

FIGURE 7.47

Stability discriminant result based on modal analysis.

should be installed was therefore clearly determined. Since the damping performance of the LP compressor rotor-bearing system proved to be inadequate, the LP compressor was replaced by a high-stability compressor supported by damper bearings. As a result, the plant could operate at over 100% capacity.

References

[1] Ministry of Economy, Trade and Industry, News Release, 2006, pp. 1–11.
[2] R. Umakoshi, et al., Kawasaki Tech. Rep. 62 (1976) 247–251.
[3] K. Ishihara, Kawasaki Tech. Rep. 149 (2002) 58–65.
[4] K. Ishihara, Gakuironnbun, Osaka University, 1980. p. 78.
[5] N. Sakai, Turbomachinery 30 (8) (2002) 448–490.
[6] T.H. Von Karman, W.R. Sears, J. Aeronaut. Sci. 5–10 (1938) 379–390.
[7] W.R. Sears, J. Aeronaut. Sci. 8-3 (1941) 104–108.
[8] N.H. Kemp, W.R. Sears, J. Aeronaut. Sci. 20-7 (1953) 585–597.
[9] N.H. Kemp, W.R. Sears, J. Aeronaut. Sci. 220-7 (1955) 477–483.
[10] J.H. Horlock, *Trans. ASME*, Ser. D 90-4 (1968) 494–500.
[11] H. Naumann, H. Yeh, *Trans. ASME*, Ser. D 90-4 (1973) 1–10.
[12] K. Ishihara, M. Funakawa, Trans. JSME 45 (397) (1979) 1213–1223.
[13] D.S. Whitehead, Report and Memoranda, A.R.C., No. 3254, 1960, pp. 1–37.
[14] S. Murata, et al., Trans. JSME 42 (353) (1976) 161–170.
[15] T. Nishiyama, et al., Trans. JSME 41 (345) (1975) 1457–1469.
[16] H.W. Emmons, C.E. Rearson, H.P. Grant, Trans. JSME 77-4 (1955) 455–469.
[17] F.E. Marble, J. Aeronaut. Sci. 22-8 (1955) 541–554.
[18] W.R. Sears, J. Appl. Mech. 20-1 (1953) 57–62.
[19] T. Iura, W.D. Rannie, Trans. ASME 76 (1954) 463–471.
[20] G. Sovran, *Trans. ASME*, Ser. A 81 (1959) 24–34.
[21] H. Takata, Japan Aerosp. Explor. Agency 2 (6) (1961).
[22] H. Takata, S. Nagano, Trans. JSME 37 (296) (1971) 687–695.
[23] K. Ishihara, M. Funakawa, Trans. JSME 44 (384) (1978) 2717–2725.
[24] A. Silverstein, S. Katzoff, W.K. Bullivant, NACA Rep. 651 (1939).
[25] Turbomachinery, Introduction of Turbomachinery, Nihon Kougyou Syuppan, 2005, p. 98.
[26] Y. Tanida, JSME P-SC10, 1980, pp. 235–254.
[27] K. Ishihara, M. Funakawa, Trans. JSME 45 (395) (1979) 933–941.
[28] D.S. Whitehead, Reports and Memoranda, No. 3386, 1962.
[29] F. Lane, WADC Tech. Rep. (1954) 54–449.
[30] Y. Hanamura, et al., *Trans. JSME* 32 (244) (1966) 1823–1841.
[31] K. Ishihara, JSME SC10, pp. 57–64, 1980.
[32] M. Tanaka, et al., Proceedings of International Conference on Rotor Dynamics, JSME, IFToMM, 1986, pp. 307–312.
[33] K. Ishihara, et al., Trans. JSME 64 (624, C) (1998) 2908–2914.
[34] J.H. Ferziger, M. Peric, Computational Methods for Fluid Dynamics, third ed., Springer-Verlag, Berlin, 2002.
[35] F.O. Carta, J. Eng. Power 89-3 (1967) 419–426.
[36] R. Srivastava, T.G. Keith Jr., J. Propul. Power 21-1 (2005) 167–174.
[37] A.I. Sayma, M. Vahdati, M. Imregun, J. Fluids Struct. 14-1 (2000) 87–101.
[38] W.S. Clark, K.C. Hall, J. Turbomachinery 122-3 (2000) 467–476.
[39] T. Chen, P. Vasanthakumar, L. He, J. Propul. Power 17-3 (2001) 651–658.
[40] M. Kasahara, et al., *Trans. JSME* (in Japanese) 66 (646, C) (2000) 1762–1768.
[41] M. Kasahara, et al., CDROM Proceedings of Dynamics and Design Conference 2000 (in Japanese), No. 338, JSME, 2000 (on CD ROM).

[42] M. Kasahara, et al., CDROM Proceedings of Dynamics and Design Conference 2000 (in Japanese), No. 339, JSME, 2000 (on CD ROM).
[43] M. Ida, et al., The collection of the draft papers presented in the 72nd National Conference of the JSME (in Japanese), No. 73-4, 1996, pp. 538–539.
[44] C.H. Jung, et al., J. Mech. Sci. Tech. 25 (6) (2011) 1465–1474.
[45] O. Matsushita, et al., *Trans. JSME* (in Japanese) 53 (496, C) (1987) 2453–2458.
[46] K. Katayama, et al., *Mitsubishi Heavy Ind. Tech. Rev.* (in Japanese) 24 (6) (1987) 605–610.
[47] F.F. Ehrich, *J. Eng. Ind.* ASME 89 (B-4) (1967) 806–812.
[48] JSME, *Mechanical Engineering Handbook* (in Japanese), Maruzen, Japan, 2004. p. α2–150.
[49] S. Kaneko, S. Hayama, *Trans. JSME* (in Japanese) 49 (439, C) (1983) 370–380.
[50] S. Kaneko, S. Hayama, *Trans. JSME* (in Japanese) 49 (439, C) (1983) 381–391.
[51] S. Kaneko, S. Hayama, *Trans. JSME* (in Japanese) 51 (464, C) (1985) 765–772.
[52] Y. Jinnouchi, et al., *Trans. JSME* (in Japanese) 51 (467, C) (1985) 1463–1471.
[53] J.A. Wolf, Trans. ASME 35 (4, E) (1968) 676–682.
[54] S. Saito, T. Someya, *Trans. JSME* (in Japanese) 44 (388, C) (1978) 4115–4122.
[55] S. Saito, T. Someya, *Trans. JSME* (in Japanese) 44 (388, C) (1978) 4123–4129.
[56] S. Saito, T. Someya, Pap. ASME (1979) 1–8 No. 79-DET-62.
[57] S. Saito, T. Someya, *Trans. JSME* (in Japanese) 45 (400, C) (1979) 1325–1331.
[58] S. Saito, et al., *Trans. JSME* (in Japanese) 48 (427, C) (1982) 321–327.
[59] S. Saito, *Trans. JSME* (in Japanese) 48 (429, C) (1982) 656–661.
[60] S. Saito, *Trans. JSME* (in Japanese) 48 (435, C) (1982) 1722–1728.
[61] A. Yasuo, et al., *Trans. JSME* (in Japanese) 51 (462, C) (1985) 265–271.
[62] A. Yasuo, et al., *Trans. JSME* (in Japanese) 55 (551, C) (1989) 602–610.
[63] F. Yoshizumi, *Trans. JSME* (in Japanese) 73 (729, C) (2007) 1338–1345.
[64] F. Yoshizumi, *Trans. JSME* (in Japanese) 73 (735, C) (2007) 2900–2908.
[65] V.G. Kozlov, N.V. Kozlov, Fluid Dyn. 43 (1) (2008) 9–19.
[66] A. Salnikova, et al., Microgravity Sci. Technol. 21 (1-2) (2009) 83–87.
[67] V.G. Kozlov, A.A. Ivanova, Microgravity Sci. Technol. 21 (4) (2009) 339–348.
[68] H. Zhang, et al., *Trans. JSME* (in Japanese) 58 (548, C) (1992) 1012–1017.
[69] H. Zhang, et al., *Trans. JSME* (in Japanese) 58 (556, C) (1992) 3456–3460.
[70] R. Takahashi, et al., *Turbomachinery* (in Japanese) 24 (3) (1996) 136–142.
[71] H.J. Thomas, Bulletin de l' A.I.M/71, pp. 1039–1063.
[72] H.F. Black, D.N. Jessen, ASME Paper, 71-WA/FE-38, 1971, pp. 1–5.
[73] H.F. Black, P.E. Allaire, L.E. Barrett, Inlet Flow Swirl in Short Turbulent Annular Seal Dynamics, Nineth International Conference on fluid Sealing, Paper D4 1981, pp. 141–152.
[74] D. Child, J. Dressman, Testing of Turbulent Seals for Rotordynamic Coefficients, NASA Conference Publication 2250, 1982, pp. 157–171.
[75] A. Kostyuk, Theoretical Analysis of Aerodynamic Forces in Labyrinth Grands of Turbomachines, Teplonergetika, 1972, pp. 39–44.
[76] T. Iwatsubo, et al., Flow Induced Force of Labyrinth Seal, NASA Conference Publication 2250 1982, pp. 205–222.
[77] C.E., Brennen, A.J. Acosta, T.K. Caughey, NASA Conference Publication 2436, 1986, pp. 270–295.

[78] H. Ohashi, A. Sakurai, J. Nishihara, NASA Conference Publication 3026, 1988, pp. 285–306.

[79] T. Tsujimoto, A.J. Acosta, T.K. Caughey, NASA Conference Publication 3026, 1988, pp. 307–322.

[80] R. Jenny, Labyrinths as a Cause of Self-Excited Rotor Oscillations in Centrifugal Compressor, 4, Sulzer Technical Review, 1980, pp. 149–146.

CHAPTER

Vibrations in Fluid–Structure Interaction Systems

When a body vibrates in a fluid, or a body containing fluid vibrates, the fluid in contact with the body is forced to move. The motion of the fluid induced by the body vibration in turn generates a fluid force acting on the body itself. Such fluid-dynamic feedback effects should be taken into consideration in the analysis of fluid-structure systems.

8.1 Summary

When a submerged body moves in a fluid, forces depending on its acceleration and velocity act on the body itself. It is necessary to solve the equations of motion of the structure appropriately by taking these fluid forces into account when analyzing various phenomena of fluid–structure interaction, as described in this book.

To derive closed-form expressions for the fluid forces, it is necessary to solve the governing equations of the fluid flow, considering the structural motion as boundary conditions, and to integrate shear stresses and pressure on the structural surface. However, analytical closed-form expressions for the fluid forces are generally hard to obtain. For the many cases where closed-form expressions for the fluid forces cannot be derived, solutions need to be obtained using CFD (computational fluid dynamics) techniques or by experimental measurements. These forces can then be fed into the structural equations of motion.

When solving fluid–structure interaction problems using CFD techniques, time-integration is usually applied to the discretized governing equations of the fluid and the structure, and time-traces of the response are obtained. In the analysis described above, structural motions are taken into consideration as boundary conditions for the governing equation of the fluid, and the equations of motion of the structure include the fluid forces derived from the integration of the surface stresses and pressure. In general, the method of solving the governing equations of the fluid and structure simultaneously is called the strong coupling (or direct coupling) approach. On the other hand, solving them alternately, exchanging the data on the boundary, is called the weak coupling (or the sequential coupling) algorithm. In the former method a more exact solution can be obtained since all the variables in the equations are simultaneously solved for. However, the approach is limited to potential flow problems when using commercial simulation codes. The latter method can usually be used with commercial simulation codes for the fluid and structure, although the method is less exact than the direct coupling approach. At present, coupled analysis using CFD is rarely applied in the practical design process since the calculation load is too heavy, and a general understanding of the phenomena involved is hard to obtain.

In the FIV analyses mainly used for practical design processes, empirical or analytical closed-form expressions of the fluid forces, which are obtained by experimental measurement, simplification of the structural geometry, or using potential flow theory, are usually used. These forces are then introduced in the equations of motion of the structure. Taking the fluid–structure interaction system shown in Fig. 8.1 (consisting of a cylindrical structure elastically supported in a cross flow) as an example, the governing equation can be expressed as

$$m_s\ddot{x} + c_s\dot{x} + k_s x = F(\ddot{x}, \dot{x}, x) + F_f(t), \tag{8.1}$$

where x is the cylinder displacement, the dots denote derivatives with respect to time t, and m_s, c_s, k_s are the cylinder mass, structural damping coefficient and stiffness, respectively. The right-hand side of the equation expresses the fluid forces acting on the cylinder. The first term is the unsteady fluid force induced by the cylinder acceleration, velocity and displacement, while the second term is the extraneous fluid force independent of the cylinder motion. The first-order Taylor series expansion of the first term is expressed as follows:

$$\begin{aligned} F &= \frac{\partial F}{\partial \ddot{x}}\ddot{x} + \frac{\partial F}{\partial \dot{x}}\dot{x} + \frac{\partial F}{\partial x}x + \cdots \\ &\approx -m_a\ddot{x} - c_a\dot{x} - k_a x. \end{aligned} \tag{8.2}$$

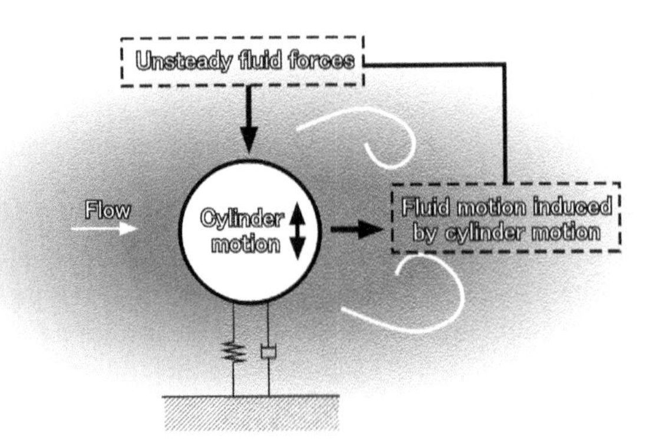

FIGURE 8.1

Typical fluid-structure system.

Thus, the unsteady fluid force is generally expressed in terms of components in phase with the cylinder acceleration $\ddot{x}$, the velocity $\dot{x}$ and the displacement x. Eq. (8.1) can then be rearranged as

$$(m_s + m_a)\ddot{x} + (c_s + c_a)\dot{x} + (k_s + k_a)x = F_f(t). \tag{8.3}$$

The cylinder oscillates as if the cylinder mass, the damping coefficient and the stiffness were $m_s + m_a$, $c_s + c_a$ and $k_s + k_a$, respectively. In general, m_a, c_a and k_a are known as the added mass, the added damping and the added stiffness, respectively. These quantities need to be determined by using appropriate analysis or experiments when evaluating flow-induced vibrations.

This chapter describes evaluation methods for the added mass and damping, for cylindrical structures dealt with in Chapters 2 and 3. The chapter also describes evaluation methods for the case of free surface oscillations, also called "sloshing," and the vibration of vessels coupled with the contained fluid as typical examples for application of potential flow theory.

8.2 Added mass and damping

8.2.1 Structures under evaluation

When a body immersed in a fluid is accelerated the fluid surrounding the body also needs to be accelerated. Consequently, the body is subjected to a fluid force depending on the acceleration as a reaction force. Letting the additional force on the body be F_M and the body acceleration be $\ddot{x}$, the effect of the force is then equivalent to increasing the body mass by $-F_M/\ddot{x}$. The nominal increment of the mass, $-F_M/\ddot{x}$, is generally called the added (or virtual) mass. This section describes methods for evaluating the added mass of cylindrical structures and three-dimensional bodies of basic geometry.

Similarly, the fluid force proportional to the body velocity introduces an added damping, which governs the growth and attenuation of the flow-induced vibrations. The damping force originates from the fluid viscosity for a body vibrating in a quiescent fluid. For a body vibrating in a steady flow the damping force is associated with the change in relative velocity between the body and the steady flow. This section also describes estimation methods of the fluid damping, used for evaluation of the forced vibration response of cylindrical structures.

8.2.2 Evaluation methods

8.2.2.1 Added mass

The added mass should be evaluated appropriately when analyzing flow-induced vibrations for the following reasons:

1. The most important effect of the added mass is lowering the natural frequency. If the added mass is not taken into account the calculated natural

frequency is overestimated, which generally leads to a decrease in safety margin.

2. Another important component of the added mass is that due to cross-coupling. Cross-coupling is often observed for multi-degree-of-freedom non-symmetric bodies where the translational and rotational motions interact with each other through the inertial forces. The cross-coupling effect is also seen in a tube bundle. In the latter case vibration of an arbitrary tube in the tube bundle induces acceleration of the surrounding fluid and consequently the other tubes are excited by the induced fluid motion. The analysis results may markedly differ depending on whether the fluid inertia coupling effect is taken into consideration or not.

The added mass for structures of typical shapes can be obtained analytically by dealing with the surrounding fluid motion as a potential flow. The added mass for a polygonal cylinder can also be obtained by using mathematical mapping techniques. For example, the added mass of a two-dimensional circular cylinder in an infinite and quiescent fluid and the resulting inertial fluid force are expressed as follows [1]:

$$m_a = \rho \frac{\pi}{4} d^2, \tag{8.4}$$

$$F = -m_a \ddot{x}, \tag{8.5}$$

where m_a expresses the added mass per unit cylinder length, ρ is the fluid density, and d the cylinder diameter. The added mass coefficient C_M is the ratio of the added mass to the displaced fluid mass m_f. For the case above, the displaced fluid mass is $\rho\pi d^2/4$ and therefore the added mass coefficient is 1.0.

The added mass coefficient is not always unity but rather depends on the body shapes, the fluid viscosity, and the fluid compressibility. The effective added mass coefficient of a body in a steady flow is different from the value in still fluid. Furthermore, it depends on the vibration amplitude since flow separation and vortex shedding occur when the displacement amplitude is large relative to the cylinder diameter, even for the case of still fluid. The added mass coefficients of a single body with typical geometry vibrating in a non-viscous, incompressible, infinite and still fluid are given in paragraphs (1) and (2) below. The effect of neighboring structures and the case of a tube bundle are described in paragraphs (3) and (4), respectively. The effects of fluid viscosity and compressibility are discussed in paragraph (5).

(1) Added mass of a body vibrating in a quiescent fluid

Table 8.1 presents the added mass of cylindrical structures of various cross-sections [2–5]. The tabulated values are per unit length for a cylinder of infinite span length vibrating translationally in a non-viscous, incompressible, infinite and still fluid.

Table 8.2 shows the added mass for three-dimensional bodies [2,3,6–8] . The added mass is for bodies vibrating translationally in one direction in an infinite

Table 8.1 Added mass of cylindrical structures per unit length [2–5].

<table>
<tr><th>Configuration of cross section</th><th>Added mass</th></tr>
<tr><td>Circular [2]</td><td>$\rho\pi a^2$</td></tr>
<tr><td rowspan="2">Elliptic [2]</td><td>$\rho\pi a^2$</td></tr>
<tr><td>$\rho\pi b^2$</td></tr>
<tr><td>Rectangular [3]</td><td>$\rho\pi a^2 K$
<table>
<tr><td>b/a</td><td>0</td><td>0.1</td><td>0.2</td><td>0.5</td><td>1</td><td>2</td><td>5</td><td>10</td></tr>
<tr><td>K</td><td>1.00</td><td>1.14</td><td>1.21</td><td>1.36</td><td>1.51</td><td>1.70</td><td>1.98</td><td>2.23</td></tr>
</table></td></tr>
<tr><td>Diamond [3]</td><td>$\rho\pi a^2 K$
<table>
<tr><td>b/a</td><td>0.5</td><td>1</td><td>2</td><td>5</td></tr>
<tr><td>K</td><td>0.85</td><td>0.76</td><td>0.67</td><td>0.61</td></tr>
</table></td></tr>
<tr><td>Flat plate [2]</td><td>$\pi\rho a^2$</td></tr>
<tr><td>Cruciform [4]</td><td>$\pi\rho a^2$, where $b \ll a$</td></tr>
<tr><td rowspan="2">Arc [5]</td><td>$\frac{\pi\rho a^2}{2}\left(1+\frac{1}{\cos^2\alpha}\right)$</td></tr>
<tr><td>$\frac{\pi\rho a^2}{2}\tan^2\alpha$</td></tr>
<tr><td rowspan="2">Joukowski symmetrical aerofoil [5]</td><td>$\frac{\pi\rho a^2}{4}\{4+(k-2)(k+1)\}$</td></tr>
<tr><td>$\frac{\pi\rho a^2}{2}(k-2)(k+1)$</td></tr>
<tr><td colspan="2">ρ: fluid density, arrow: vibration direction</td></tr>
</table>

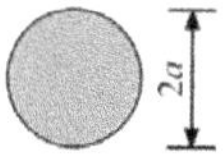

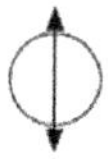

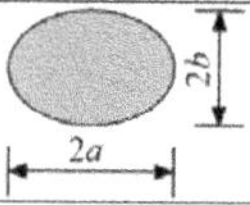

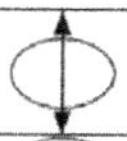

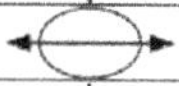

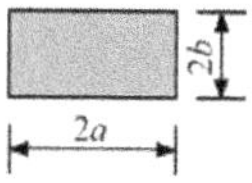

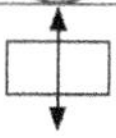

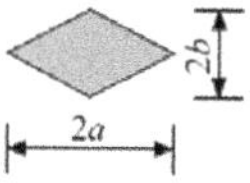

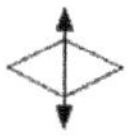

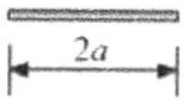

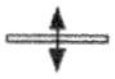

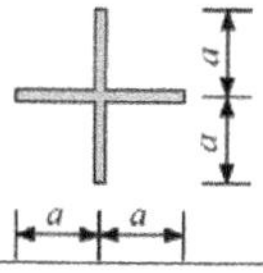

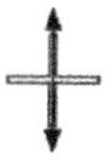

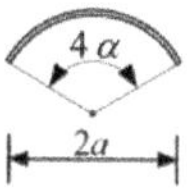

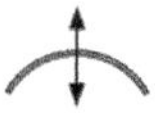

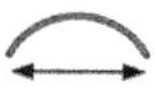

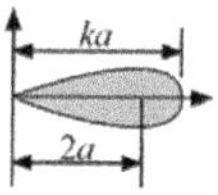

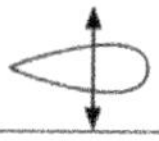

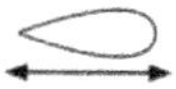

Table 8.2 Added mass of three-dimensional body in an infinite fluid [2,3,6–8].

Configuration of cross section		Added mass
Circular disk [2]	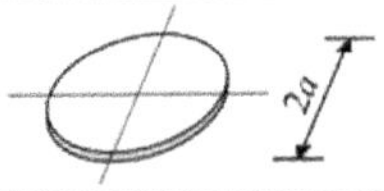	 $\frac{8}{3}\rho a^3$
Ellipse plate [6]	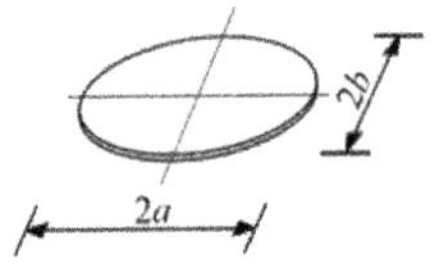	 $\rho ab^2 K_1$
Rectangular plate [7]	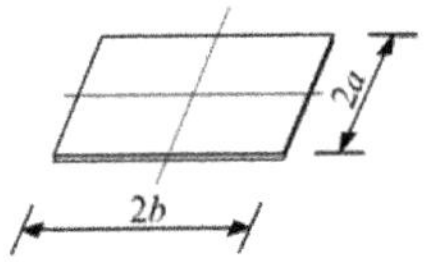	 $\rho\pi ab^2 K$

b/a	1	2	3	∞
K	0.47	0.84	1.00	1.00

Configuration of cross section		Added mass	
Sphere [2]	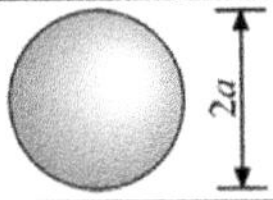	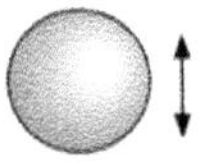 $\frac{2}{3}\rho\pi a^3$	
Ellipsoid [8]	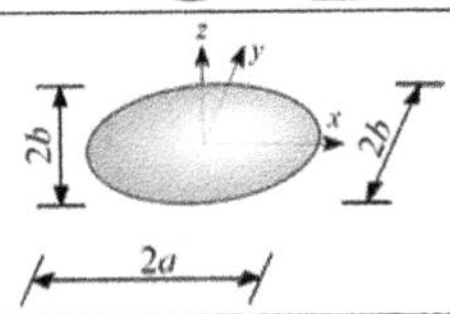	x direction	$\rho ab^2 K_2$
		y direction z direction	$\rho ab^2 K_3$
Circular cylinder of finite span		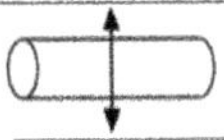 $\rho\pi a^2 LK$	

$L/2a$	1.2	2.5	5.0	9.0	∞
K	0.62	0.78	0.90	0.96	1.0

b/a	K_1	K_2	K_3	b/a	K_1	K_2	K_3
0	4.1846	0	1.3333	0.6	3.2819	0.3543	0.8706
0.1	4.1228	0.02761	1.2803	0.7	3.1130	0.4306	0.8101
0.2	3.9874	0.07883	1.1923	0.8	2.9538	0.5083	0.7565
0.3	3.8202	0.1406	1.1012	0.9	2.8051	0.5870	0.7090
0.4	3.6404	0.2084	1.0158	1.0	2.6667	0.6667	0.6667
0.5	3.4588	0.2800	0.9389				

ρ: fluid density, arrow: vibration direction

and still fluid. Note that the added mass of a sphere is one-half of the displaced fluid mass.

The two tables show the added masses of vibrating rigid bodies. The evaluation method of the added mass for vibrating elastic structures, e.g. a flat plate under bending vibrations, is presented in the handbook by Blevins [2]. The handbook also presents the added mass for semi-submerged bodies such as ships and marine structures.

(2) Added mass moment of inertia

When a body vibrates rotationally in a fluid, an effect equivalent to an increase in the mass moment of inertia comes into play. The nominal increase in the moment of inertia is called the added mass moment of inertia. Table 8.3 shows the added moment of inertia for cylindrical structures of typical cross-sectional shapes [2–4], while in Table 8.4 the added moment of inertia for three-dimensional bodies is presented [2,6]. The added moment of inertia is zero when a circular cylinder rotationally vibrates about its central axis and when a sphere vibrates about its center point. For other shapes, the added moment of inertia is generally non-zero.

Table 8.3 Added mass moment of inertia of cylindrical structures [2–4].

Configuration of cross section	Added mass moment of inertia
Circle [2] (diameter $2a$)	0
Ellipse [2] (axes $2a$, $2b$)	$\frac{\pi}{8}\rho\left(a^2-b^2\right)^2$
Flat plate [2] (width $2a$)	$\frac{\pi}{8}\rho a^4$
Rectangle [3] (sides a, a, b, b)	$\rho\pi a^4 K$
Cruciform [4] (arms a)	$\frac{2}{\pi}\rho a^4$

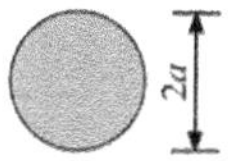

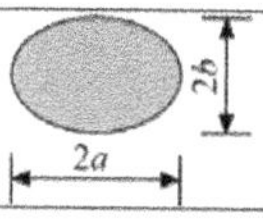

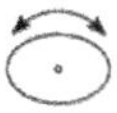

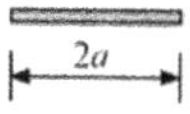

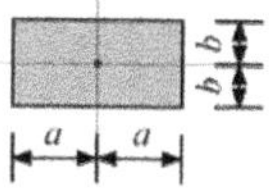

Rectangle [3]:

b/a	0	0.1	0.2	0.5	1
K	0.125	0.147	0.15	0.15	0.234

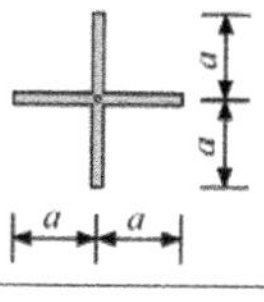

ρ: fluid density, arrow: vibration direction

Table 8.4 Added mass moment of inertia of three-dimensional body [2,6].

Configuration of cross section	Added mass moment of inertia						
Circular disk [2]	$\frac{16}{45}\rho a^5$						
Elliptic plate [6]	$\rho a^3 b^2 K$						
	b/a	0.1	0.2	0.3	0.4	0.5	1
	K	0.8033	0.7398	0.6713	0.6067	0.5489	0.3556

ρ: fluid density, arrow: vibration direction

(3) Effect of neighboring wall on the added mass [1]

It should be noted that the added mass increases due to blockage effects in the case of a body close to a wall and for closely packed tube bundles. Ignoring the effect of blockage by other structures results in underestimation of the added mass and consequently overestimation of the natural frequency. This often leads to a decrease in the safety margin when evaluating flow-induced vibrations. For example, the added mass coefficient of a circular cylinder close to a rigid wall is given by

$$C_M = 1 + 4\sinh^2\alpha\left(\sum_{n=1}^{\infty} n\frac{e^{-3n\alpha}}{\sinh(n\alpha)}\right), \tag{8.6}$$

$$\alpha = \ln\left(1 + \frac{G}{r} + \sqrt{\left(1+\frac{G}{r}\right)^2 - 1}\right), \tag{8.7}$$

where G is the gap width between the cylinder and rigid wall, and r the radius of the cylinder [9]. Figure 8.2 shows the added mass coefficient expressed by Eq. (8.6) [10]. The added mass coefficient is seen to increase as the gap width decreases.

As another example, Fig. 8.3 shows a narrow passage with parallel walls where the upper wall vibrates translationally. The motion corresponds to that in the simplest case of leakage-flow-induced vibration described in Section 3.3. For the case where the steady flow component is zero, the unsteady fluid force proportional to the wall acceleration is given by

$$F = -\frac{\rho L^3}{12H_0}\ddot{h}, \tag{8.8}$$

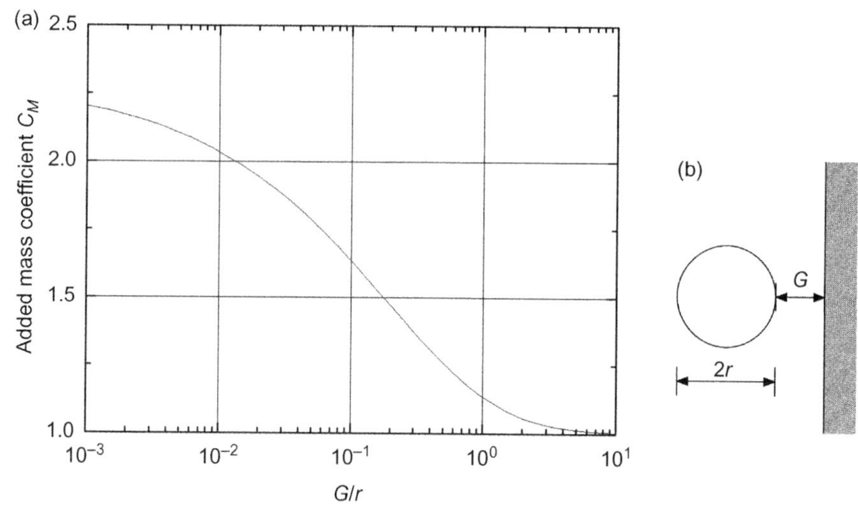

FIGURE 8.2

The added mass coefficient expressed by Eq. (8.6).

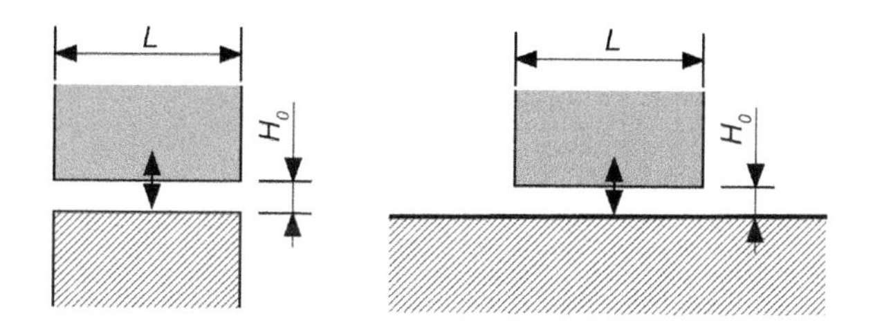

FIGURE 8.3

A vibrating wall close to a stationary one.

where ρ, L, H_0 and $\ddot{h}$ are the fluid density, the length of the passage, the gap width and the acceleration of the upper wall, respectively. Equation (8.8) is valid for $H_0 \ll L$. The added mass of the vibrating wall, which equals $\rho L^3/12H_0$, is inversely proportional to the gap width.

Table 8.5 presents the added mass for structures such as a plate close to a rigid wall, a circular cylinder vibrating in a concentric rigid tube and so on [11–13]. Generally, the added mass tends to increase as the gap width decreases.

(4) Cross-coupling effect of fluid inertia

In the paragraphs (1)–(3), the added mass in the translational direction and the added mass moment of inertia in the rotational direction are described. For asymmetric structures, however, the vibrations in the translational and rotational directions interact with each other due to fluid inertia coupling effects [2,14].

Table 8.5 Added mass of a structure close to a rigid wall or other structure [11–13].

Configuration of cross section	Added mass
Circular cylinder in a concentric tubular space [11]	$\rho \pi a^2 \dfrac{b^2 + a^2}{b^2 - a^2}$
Sphere in a concentric spherical space [12]	$\dfrac{2}{3} \rho \pi a^3 \left(\dfrac{b^3 + 2a^3}{b^3 - a^3} \right)$
Flat plate close to a rigid wall	$\rho \pi a^2 K$
Circular cylinder close to another rigid cylinder [13]	$\rho \pi a^2 K$

Flat plate close to a rigid wall:

$h/2a$	∞	2.5	5
K	1.00	1.03	1.165

Circular cylinder close to another rigid cylinder:

b/a	∞	1.2	0.8	0.4	0.2	0.1
K	1.0	1.024	1.044	1.096	1.160	1.224

ρ: fluid density, arrow: vibration direction

The fluid inertia forces due to the added mass and the added mass moment of inertia can be expressed for asymmetric bodies in the following matrix form:

$$\begin{bmatrix} F_x \\ F_y \\ F_\theta \end{bmatrix} = - \begin{bmatrix} m_{xx} & m_{xy} & m_{x\theta} \\ m_{yx} & m_{yy} & m_{y\theta} \\ m_{\theta x} & m_{\theta y} & m_{\theta\theta} \end{bmatrix} \begin{bmatrix} \ddot{x} \\ \ddot{y} \\ \ddot{\theta} \end{bmatrix} \tag{8.9}$$

Equation (8.9) shows that the acceleration of a body in the x direction induces fluid inertia forces not only in the x direction but also in the y and θ directions due to the cross-coupling effect. The off-diagonal terms, representing the coupling between motions in the translational and rotational directions or in the two translational directions x and y, sometimes play an important role in the vibration mechanism.

In the same manner, cross-coupling effects in the added mass should be considered for multiple cylinders since a cylinder is excited by the motion of other cylinders through the fluid. Figure 8.4 shows two circular cylinders close to each other as a typical example. The cylinders are elastically supported in the fluid and can oscillate only in the y direction. When they have the same spring constant k_s, cylinder mass m_s and diameter d, the equation of motion of the two-cylinder system can be expressed in the following matrix form [1]:

$$\begin{bmatrix} m_s & 0 \\ 0 & m_s \end{bmatrix}\begin{bmatrix} \ddot{y}_1 \\ \ddot{y}_2 \end{bmatrix} + \begin{bmatrix} k_s & 0 \\ 0 & k_s \end{bmatrix}\begin{bmatrix} y_1 \\ y_2 \end{bmatrix} = -\begin{bmatrix} m_{a0}\alpha_{11} & m_{a0}\alpha_{12} \\ m_{a0}\alpha_{21} & m_{a0}\alpha_{22} \end{bmatrix}\begin{bmatrix} \ddot{y}_1 \\ \ddot{y}_2 \end{bmatrix} \tag{8.10}$$

The right-hand side expresses the added mass of the coupled system. Taking into account the symmetry of the system, $\alpha_{11} = \alpha_{22}$ and $\alpha_{12} = \alpha_{21}$. The natural frequencies of the system are easily shown to be

$$f_1 = \frac{1}{2\pi}\sqrt{\frac{k_s}{m_s + m_{a0}(\alpha_{11} + \alpha_{12})}}, \quad f_2 = \frac{1}{2\pi}\sqrt{\frac{k_s}{m_s + m_{a0}(\alpha_{11} - \alpha_{12})}} \tag{8.11}$$

The natural frequencies f_1 and f_2 correspond to the in-phase and out-of-phase natural modes of vibration, respectively. The terms m_{a0} $(\alpha_{11} + \alpha_{12})$ and m_{a0} $(\alpha_{11} - \alpha_{12})$ are the effective modal added masses of the coupled system. It should be noted that there exist two natural frequencies and one of them is lower than that of the single isolated cylinder.

Figure 8.5 shows the upper and lower bounds of the effective added mass coefficients as functions of pitch-to-diameter ratio s/d for a tube bundle having a hexagonal arrangement [15]. The abscissa shows the values of the effective added mass normalized by that of a single cylinder. The upper bound increases and the lower bound decreases due to the coupling effect as s/d becomes smaller. The upper and lower bounds for the 7-cylinder case do not differ from those of the 19- and 37-cylinder cases although the values of the upper bound are larger and those of the lower bound are smaller as the number of tubes increases. This indicates that the coupling effect of neighboring cylinders is dominant. The added

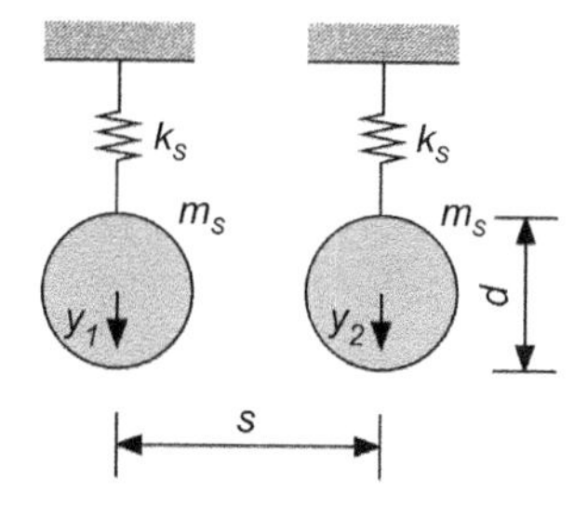

FIGURE 8.4

Two circular cylinders elastically supported in a fluid.

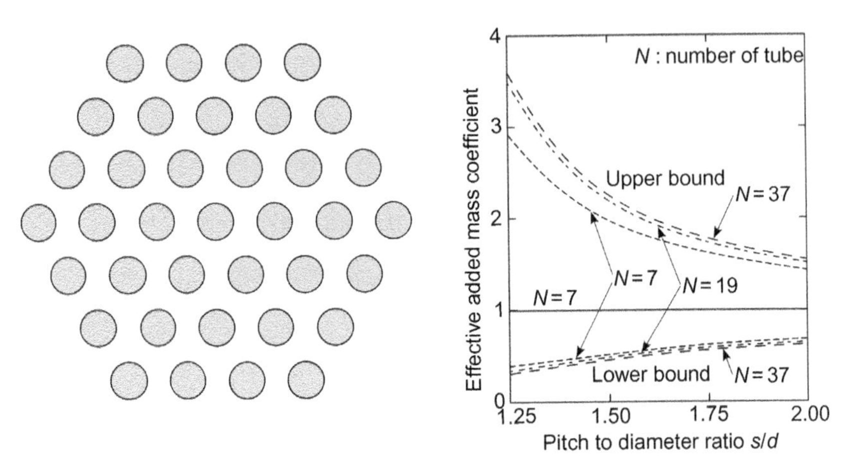

FIGURE 8.5

Tube bundle in a hexagonal arrangement and the effective added mass coefficient [15].

mass and the natural vibration modes of tube arrays including more than two cylinders are described in detail by Chen [1].

(5) Effect of viscosity and compressibility

In paragraphs (1)–(4), the added mass of structures vibrating in an inviscid and incompressible fluid is presented. As the fluid viscosity increases, however, the added mass tends to be larger because of momentum diffusion over a wider area. In the case of a circular cylinder in a concentric tubular space, vibrating in the direction normal to the longitudinal axis with an angular frequency ω, for example, the added mass coefficient can be expressed as [11]

$$C_M = \frac{b^2 + a^2}{b^2 - a^2} + \frac{2}{a}\sqrt{\frac{2\nu}{\omega}}, \tag{8.12}$$

The second term on the right-hand side expresses the effect of the fluid viscosity, ν. The added mass coefficient increases with the fluid viscosity. In general, the fluid viscosity effect is represented by the kinematic Reynolds number Re_d $(=\omega a^2/\nu)$. Using Re_d, Eq. (8.12) can be rearranged as

$$C_M = \frac{b^2 + a^2}{b^2 - a^2} + 2\sqrt{\frac{2}{Re_d}}. \tag{8.13}$$

The equation above can be applied to cases where $\mathrm{Re}_d \gg 1$. Figure 8.6 shows the added mass calculated using Eq. (8.13).

For the cases where the kinematic Mach number M_d $(=\omega d/c$; ω, angular frequency; d, representative diameter; c, acoustic velocity) is large, the added mass also depends on fluid compressibility. The effect of compressibility on the added

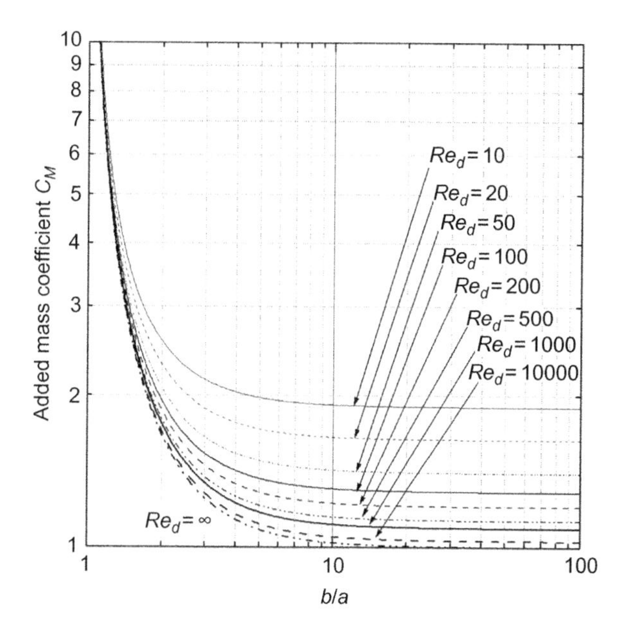

FIGURE 8.6

Added mass of a circular cylinder in a concentric tubular passage as a function of diameter ratio b/a and the kinematic Reynolds number Re_d [11].

mass can be ignored for $M_d < 0.1$. For practical plant structures, the kinematic Mach number is usually very small. The effects of the fluid viscosity and compressibility are also described in the book by Chen [1].

8.2.2.2 Fluid damping

The added damping force plays an important role of governing damping and growth of flow-induced vibrations. As mentioned in Chapter 1, self-excited vibrations occur when the sum of the structural damping and the added damping is negative. It is also important to evaluate the damping of the system appropriately when calculating the forced vibration response of systems with positive total damping, where self-excited vibrations do not arise. How to deal with the added damping depends on the variety of mechanisms of flow-induced vibration. The reader may refer to the mechanisms and phenomena discussed in Chapters 2–7. This section describes estimation methods for the fluid damping which are used for evaluation of the forced vibration responses of cylindrical structures. The fluid damping of structures vibrating in a quiescent fluid, which originates from the fluid viscosity, is given in paragraph (1) below, while that in a steady flow is described in paragraphs (2) and (3). The fluid damping in two-phase flows is presented in paragraph (4).

(1) Fluid damping of a cylindrical structure vibrating in still fluid [16]

When a cylindrical structure vibrates in still fluid with a small amplitude, a shear stress force generated by the fluid viscosity acts on the structural surface. In this case, the force on the structure is called the viscous drag. On the other hand, a pressure drag force acts on a structure vibrating with large amplitude, resulting from flow separation and vortex shedding. For a cylindrical structure vibrating in the direction normal to the longitudinal axis, the unsteady fluid force is expressed as follows:

$$F_D = -\frac{1}{2}\rho d\dot{x}|\dot{x}|C_D - m_a\ddot{x} \tag{8.14}$$

where F_D is the drag force per unit length, C_D the drag coefficient, ρ the fluid density, d the diameter of the cylindrical structure, and, $\dot{x}$ and $\ddot{x}$ the velocity and the acceleration of the structure, respectively.

As mentioned in Section 1.3, an approximate expression for the term $\dot{x}|\dot{x}|$ can be derived for a structure undergoing sinusoidal vibration of the form $x = x_0\sin\omega t$. The result is

$$\dot{x}|\dot{x}| = x_0{}^2\omega^2|\cos\,\omega t|\cos\,\omega t \approx \frac{8}{3\pi}x_0{}^2\omega^2\cos\,\omega t = \frac{8}{3\pi}x_0\omega\dot{x}. \tag{8.15}$$

Hence, Eq. (8.14) can be expressed as

$$F_D = -\,c_a\dot{x} - m_a\ddot{x}, \tag{8.16}$$

$$c_a = \frac{4}{3\pi}\rho d C_D x_0\omega. \tag{8.17}$$

Using Eqs. (8.16) and (8.17), with $m = m_s + m_a$, and $k_a = 0$, the equation of motion (8.1) can be re-written as

$$m\ddot{x} + \left(c_s + \frac{4}{3\pi}\rho d C_D x_0\omega\right)\dot{x} + k_s x = 0,$$

or more simply

$$\ddot{x} + 2(\zeta_s + \zeta_a)\omega_n\dot{x} + \omega_n{}^2x = 0, \tag{8.18}$$

where the substitutions $\zeta_s = c_s/(2m\omega_n)$ and $\omega_n{}^2 = k_s/m$ have been used. ζ_a in Eq. (8.18) is the fluid damping ratio. The added damping coefficient c_a in Eq. (8.17) is reduced to the fluid damping ratio to give:

$$\zeta_a = \frac{2}{3\pi}\frac{\rho d^2}{m}\frac{x_0}{d}C_D \tag{8.19}$$

To calculate the damping ratio ζ_a, the drag coefficient C_D therefore needs to be known. The drag coefficient depends on the cross-section shape of the cylinder, fluid viscosity, vibration amplitude and frequency. With the exception of the circular cylinder, there are few data available for evaluating C_D for general cross-sectional shapes.

Under the assumption that the vibration amplitude is small enough for flow separation not to occur, the drag coefficient C_D for a circular cylinder is theoretically derived by Wang [17] as

$$C_D = \frac{3\pi^3}{2K}\left\{(\pi\beta)^{-\frac{1}{2}} + (\pi\beta)^{-1} - \frac{1}{4}(\pi\beta)^{-\frac{3}{2}} + \cdots\right\}. \tag{8.20}$$

K and β in the above equation are, respectively, the Keulegan–Carpenter number and the Stokes parameter defined as follows:

$$K = 2\pi x_0/d, \ \beta = fd^2/\nu, \tag{8.21}$$

where f and ν are the frequency and the fluid viscosity, respectively. For $\beta \gg 1$ the first term in the brackets in Eq. (8.20) is dominant, and the equation can therefore be simplified to

$$C_D \approx \frac{3\pi^{\frac{5}{2}}}{2K\sqrt{\beta}}. \tag{8.22}$$

Substituting Eq. (8.22) into Eq. (8.19), the following expression can be derived for the case $\omega = \omega_n$.

$$\zeta_a = \frac{\pi}{2}\frac{\rho d^2}{m}\left(\frac{\nu}{\pi f d^2}\right)^{\frac{1}{2}} \tag{8.23}$$

Several measurement results for the drag coefficient C_D, expressed as a function of K and β, have also been reported. They are typically derived from damped free vibration tests [18,19]. C_D can also be obtained from measurements of the drag force on a stationary cylinder in a pulsating flow [20,21], which is equivalent to a free vibration test measurement. According to experimental measurements, C_D is inversely proportional to K as predicted by Eq. (8.22) but the absolute value of C_D is about one to three times the theoretical one, on the condition that K is lower than a certain critical value. Where K is beyond the critical value, C_D rapidly increases and is in the range 1–2 for $K > 20$. According to Hall [22], the critical Keulegan–Carpenter number, K_{crit}, is given as a function of β by the following equation:

$$K_{crit} = \frac{5.78}{\beta^{1/4}}\left(1 + \frac{0.21}{\beta^{1/4}} + \cdots\right), \tag{8.24}$$

which agrees well with actual measurement results. For $K < 5$ and large β, Bearman and Russell [23] proposed the following empirical expression:

$$C_D = \frac{3\pi^{\frac{5}{2}}}{K\sqrt{\beta}} + 0.08K. \tag{8.25}$$

The second term on the right-hand side expresses the increase in C_D due to vortex shedding. The drag force coefficient also depends on the surface roughness. Measurement results are also reported by Sarpkaya [20].

For shapes other than the circular cylinder, several measurement data for C_D are available. Examples include a flat plate and a rectangular cylinder in a pulsating flow. For a square cylinder, C_D is in the range 2–3 for $K > 1$. Test results in a pulsating flow have also been reported for the drag force coefficient of a square cylinder as a function of attack angle and that of a rectangular cylinder as a function of aspect ratio [24–28]. Using these measurement values and Eq. (8.19), the fluid damping for cylindrical structures vibrating in a quiescent fluid can be estimated.

(2) Fluid damping in a cross flow [16]

The fluid damping originating from the relative flow velocity change acts on a cylindrical structure vibrating in a cross flow. Taking a circular cylinder oscillating in the streamwise direction in a cross flow as an example, the relative flow velocity is written as $U_{rel} = U - \dot{x}$, using the incident flow velocity U and the cylinder vibration velocity $\dot{x}$. Based on the quasi-steady assumption, the drag force on the cylinder can, for $U \gg \dot{x}$, be expressed as follows:

$$F_x = \frac{1}{2}\rho {U_{rel}}^2 dC_D \approx \frac{1}{2}\rho U^2 dC_D\left(1 - 2\frac{\dot{x}}{U}\right). \tag{8.26}$$

This equation includes an unsteady component proportional to the cylinder vibration velocity $-\rho U d C_d \cdot \dot{x}$, in addition to the steady component. Converting the unsteady component into the damping ratio in the same manner as in Eqs. (8.16)–(8.18), the following expression is obtained:

$$\zeta_x = \frac{1}{4\pi} V_r \frac{\rho d^2}{m} C_D, \tag{8.27}$$

where $V_r = U/f_n d$.

For transverse vibration with the cylinder velocity $\dot{y}$, the relative flow velocity and attack angle, as shown in Fig. 8.7 are given by

$${U_{rel}}^2 = U^2 + \dot{y}^2, \tag{8.28}$$

$$\sin\alpha = -\dot{y}/U_{rel}. \tag{8.29}$$

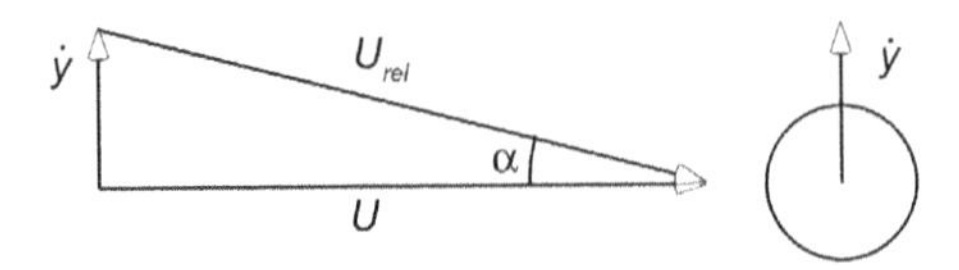

FIGURE 8.7

Relative flow velocity and angle of attack.

For $U \gg \dot{y}$, these equations can be expressed as $U_{rel}{}^2 \approx U^2$ and $\sin\alpha \approx -\dot{y}/U$, respectively. The unsteady fluid force on the cylinder in the transverse direction is hence written as follows:

$$F_y = \frac{1}{2}\rho U_{rel}{}^2 d C_D \sin\alpha \approx -\frac{1}{2}\rho U^2 d C_D \frac{\dot{y}}{U} \tag{8.30}$$

Thus, the unsteady fluid force proportional to cylinder velocity is obtained, and can be expressed in terms of the damping ratio

$$\zeta_y = \frac{1}{8\pi} V_r \frac{\rho d^2}{m} C_D. \tag{8.31}$$

It should be noted that the Eqs. (8.27) and (8.31) are derived via linearization for cases where the absolute value of the cylinder velocity is much smaller than the incident flow velocity. The application of these expressions should therefore be limited. They can be used on the condition that the reduced velocity is much higher than the value where vortex-induced vibrations occur. In the reduced velocity range lower than that of significant vortex-induced vibrations, the fluid damping is nearly proportional to the reduced velocity; however, these expressions overestimate the fluid damping, according to Chaplin [19]. In the JSME guideline S-012, for evaluation of flow-induced vibrations of a cylindrical structure [29], it is recommended to evaluate random vibration responses conservatively with zero fluid damping.

(3) Fluid damping in a parallel flow [16]

For bending vibrations of a cylindrical structure in a parallel flow, the fluid damping ratio is given by

$$\zeta = \frac{1}{8\pi} c_N C_M V_r \frac{\rho D^2}{m}, \tag{8.32}$$

$$V_r = \frac{U}{f_n D}, \tag{8.33}$$

where C_M is the added damping coefficient, f_n is the natural frequency of the structure, and c_N is the friction factor, defined as follows:

$$c_N = \begin{cases} 0.04 & \text{Paidoussis [30]}, UD/\nu = 9 \times 10^4 \\ 0.02 \sim 0.1 & \text{Chen [31]} \\ 1.3\,(UD/\nu)^{-0.22} & \text{Connors et al. [32]} \end{cases} \tag{8.34}$$

(4) Fluid damping in two-phase flow [33]

The fluid damping in two-phase flow is generally higher than in the case of single-phase flow discussed in paragraphs (1)–(3). Pettigrew proposes that the fluid damping in two-phase flow be expressed as the sum of the damping due to the fluid viscosity that depends on a steady flow component, and that originating from two-phase flow shown in Figure 2.22 in Section 2.3. A detailed description of two-phase flow damping is given in the review paper by Pettigrew and Taylor [33].

8.3 Sloshing and bulging

8.3.1 Overview

Liquid storage containers are found, for example, in chemical process plants, thermal power stations, rockets etc. When designing such vessels, dynamic loads caused by earthquakes or other external excitations have to be taken into account, in addition to static liquid pressure, gas pressure and deadweight. When vessels containing liquid are subjected to seismic excitation free surface oscillations of the contained liquid called "sloshing" can be generated. These oscillations generally have low frequency and large amplitude. Vessel wall vibrations with relatively high frequency, called "bulging," can also occur. In this section, sloshing and bulging occurring in liquid storage vessels are discussed. This section presents practical examples of damage to actual tanks due to sloshing, and also discusses sloshing in cases with steady flow in vessels. Methods of suppression of sloshing, and damping devices utilizing sloshing are also presented.

8.3.2 Description and historical review of vibration phenomena

When a liquid storage tank is subjected to seismic excitation, free surface oscillations called sloshing are generated. This phenomenon can be easily modeled using linear theory when the wave height is small. For the case of large wave height, however, nonlinear effects have to be taken into account to describe the response, since the system behaves as a nonlinear oscillator of soft or hard spring type. When the liquid depth is very low, the standing wave assumption cannot be used since a travelling wave is generated. In this case, a formulation using shallow-water theory should be used. For the case where vessel walls are flexible, considerable vibration of the walls, called "bulging," occurs. In this case the system should be dealt with as a coupled liquid-wall (fluid-structure) system.

There are two main methods for evaluating sloshing. One is Housner's theory developed in the 1950s, where sloshing is dealt with as a phenomenon in a lumped parameter system. The other is potential flow theory, where sloshing is dealt with as a continuous system phenomenon. Overturning moment and base shear stress, which are necessary for vessel and support structure design, can be calculated by using these theories. In the theories, the deformation of the vessel itself is not taken into account. There were, however, improvements in the 1970s. The Flügge shell theory or the finite element method (FEM) is applied to model the structure in the improved methods, where the vessel is considered as an axisymmetric shell, and the behavior of the contained liquid is analyzed using potential flow theory. After the 1980s, a method taking the nonlinearity of liquid motion into consideration and a second for evaluating the case accompanied by a steady flow were developed.

In the next subsection, Housner's theory [34], where sloshing is modeled using a lumped parameter approach, is first presented. Following this, the potential flow theory approach, where the liquid is modeled as a continuous system, is described.

8.3.3 Evaluation methods

8.3.3.1 Housner's theory

In Housner's theory, it is assumed that an incompressible liquid is contained in a rectangular vessel composed of rigid walls, and that the wave height is sufficiently small. First, the (resultant) fluid force acting on the vessel subjected to an external acceleration in the horizontal direction and its point of action are obtained. Next, a spring mass system is determined where the force and moment equivalent to those in the case above are applied under the same external acceleration. The key feature of the theory is that the liquid pressure in the vessel can be decomposed into two components, one caused by the impulsive force and the other due to the liquid oscillations. It should be noted that the force and its point of action are easily determined but the sloshing wave height cannot be accurately derived from the theory.

(1) Pressure due to impact

Here we consider the case where an impulsive acceleration $\ddot{u}_0$ in the x direction acts on the rectangular vessel at rest as shown in Fig. 8.8 where the velocities in the x and y directions are u and v, respectively.

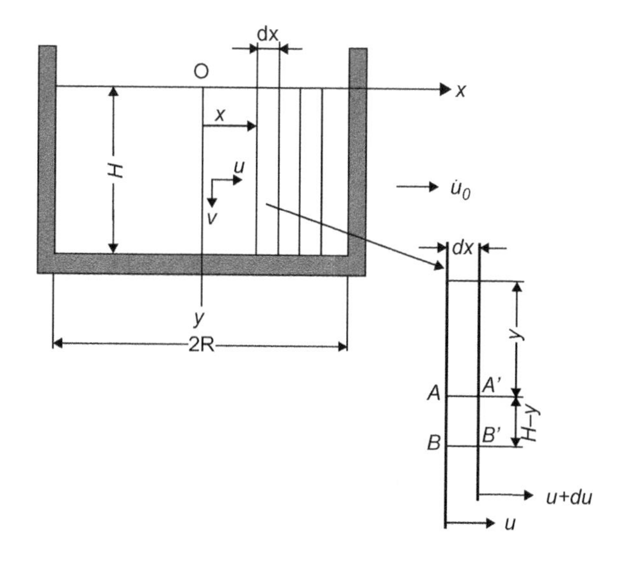

FIGURE 8.8

Rectangular vessel and fluid element.

The liquid in the vessel is discretized into small elements of width dx. From the balance of flow into and out from the fluid element $AA'BB'$, the continuity equation yields the relation, $v = (H - y)\frac{du}{dx}$. The relation between the acceleration $\dot{v}$ in the y direction and the pressure p can be expressed as $\frac{\partial p}{\partial y} = -\rho\dot{v}$, where ρ is the fluid density. The force P, due to the pressure p which acts on the fluid element, is related to the acceleration $\dot{u}$ in the x direction as $\rho H\dot{u} = -\dfrac{dP}{dx}$. Using the relations above, a differential equation for $\dot{u}$ can be written. The pressure distribution is obtained by solving this equation using the boundary conditions $\dot{u} = \dot{u}_0$ at $x = \pm R$. Integrating the pressure distribution across the wall surface gives the following impact force P_i acting on the vessel wall:

$$P_i = \frac{\rho\dot{u}_0 H^2}{\sqrt{3}}\tanh\frac{\sqrt{3}R}{H} \tag{8.35}$$

The action point (height) H_0 of the force is obtained from the moment balance relation below.

$$\int_0^H p(H - H_0 - y)dy = 0; \quad \therefore H_0 = \frac{3}{8}H \tag{8.36}$$

(2) Pressure due to oscillations

Here we consider only oscillations in the first mode, and assume the motion to be approximately rotational about the z axis, Fig. 8.9. In this case, the velocity in the y direction is expressed as $v = x\dot{\theta}$, where θ is the rotational angle around the z axis and is a function of the elevation y and time t. Expressing the angular natural frequency of oscillation as ω, the rotational angle is given by $\theta = \theta_0(y)\sin\omega t$. Calculating the kinetic and potential energies of the liquid contained in the vessel

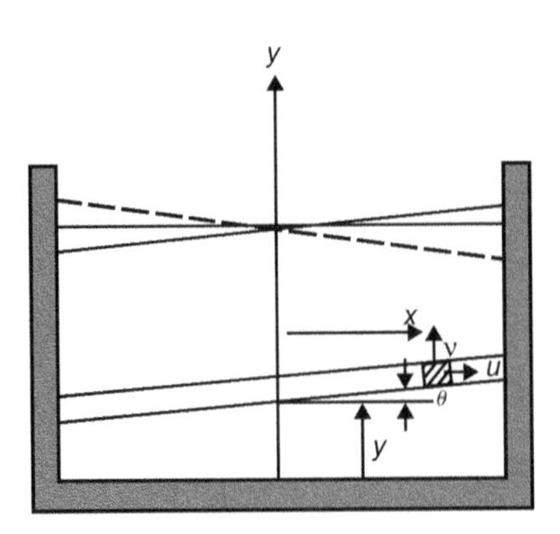

FIGURE 8.9

Vibration mode.

under the boundary conditions $\theta_0 = \overline{\theta}_0$ at $y = H$ and $\theta_0 = 0$ at $y = 0$, the following equation for the angular natural frequency is derived via energy conservation:

$$\omega^2 = \sqrt{\frac{5}{2}}\frac{g}{R}\tanh\sqrt{\frac{5}{2}}\frac{H}{R} \tag{8.37}$$

Integrating the oscillation-induced pressure acting on the total wall surface, the oscillation-induced fluid force P_v can be obtained as follows:

$$P_V = \int_0^H p_{(x=R)}dy = \frac{1}{3}\rho\omega^2 R^3 \overline{\theta}_0 \sin \omega t \tag{8.38}$$

The action point, H_1, of the force is also derived from the moment balance and is given by

$$H_1 = H\left(1 - \frac{\cosh\left(\sqrt{5/2}H/R\right) - 1}{\sqrt{5/2}(H/R)\sinh\left(\sqrt{5/2}H/R\right)}\right) \tag{8.39}$$

(3) Spring-mass system model

Having obtained the fluid forces, the liquid in the vessel is next modeled as a spring-mass system. It is supposed that the fluid force can be decomposed into two components: one caused by the impact the other due to liquid oscillations. The former is modeled as a mass m_0 rigidly fixed to the vessel. The latter is modeled as a mass m_1 elastically connected to the vessel walls by springs.

The mass m_0 is obtained by dividing the impulsive force on the vessel by its acceleration thus

$$m_0 = \frac{2P_i}{\ddot{u}_0} = \frac{2\rho H^2}{\sqrt{3}}\tanh\frac{\sqrt{3}R}{H} \tag{8.40}$$

On the other hand, the mass m_1 is derived by imposing the following two conditions to be satisfied. One is that the force acting on the wall through the springs, caused by the oscillation of the mass m_1, matches the fluid force due to sloshing. The second is that the maximum kinetic energy of the spring mass system equals the maximum potential energy of the fluid in the vessel during oscillation. The following result for the mass m_1 is then obtained:

$$m_1 = \frac{2}{3}\sqrt{\frac{5}{2}}\rho R^2 \tanh\sqrt{\frac{5}{2}}\frac{H}{R} \tag{8.41}$$

This completes the outline of Housner's theory.

When the vessel shape and water depth are given, the liquid in the vessel can be modeled as the system shown in Fig. 8.10 by using Housner's theory. The force in the horizontal direction and the moment acting on the vessel can be calculated for an arbitrary external acceleration.

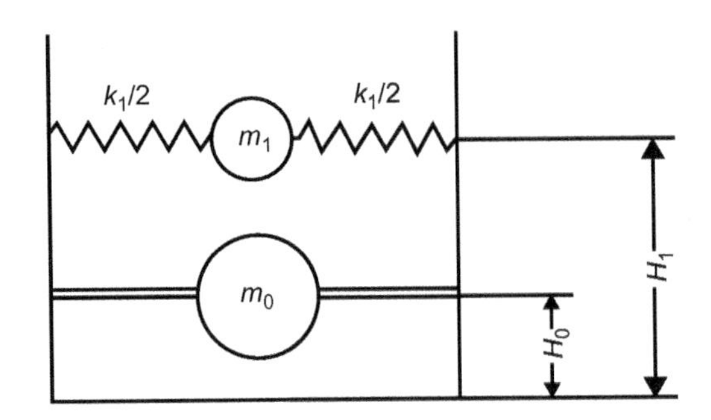

FIGURE 8.10

Spring-mass system model.

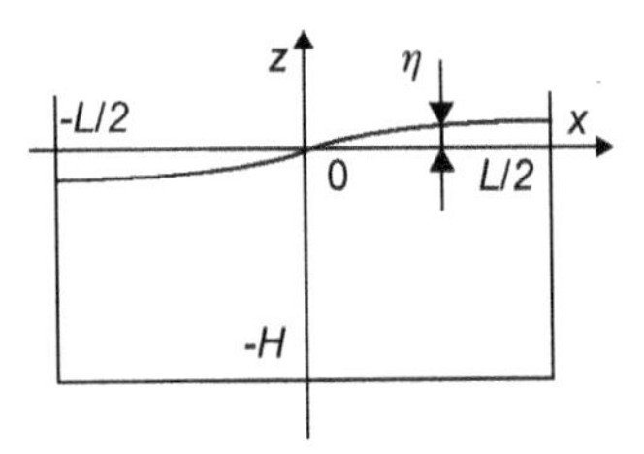

FIGURE 8.11

Sloshing in a two-dimensional rectangular vessel.

8.3.3.2 Potential flow theory [35,36]

In this subsection, the analysis of sloshing as a continuous system phenomenon is described. Here the liquid is assumed to be inviscid and incompressible. The motion of the liquid is supposed to be irrotational, and the surface tension negligible.

(1) Sloshing in a rectangular tank

Consider a rectangular vessel of depth H and width L, as shown in Fig. 8.11. Assuming the wave amplitude of sloshing is sufficiently small, the velocity potential ϕ satisfies the equations and boundary conditions below:

The continuity equation

$$\frac{\partial^2 \phi}{\partial x^2} + \frac{\partial^2 \phi}{\partial z^2} = 0 \tag{8.42}$$

The non-penetration condition at the vessel floor

$$\frac{\partial \phi}{\partial z} = 0 \text{ at } z = -H \tag{8.43}$$

Condition of zero velocity normal to side walls

$$\frac{\partial \phi}{\partial x} = 0 \text{ at } x = \pm L/2 \tag{8.44}$$

The kinematic boundary condition at the liquid free surface

$$\frac{\partial \varphi}{\partial z} = \frac{\partial \eta}{\partial t} + \frac{\partial \varphi}{\partial x}\frac{\partial \eta}{\partial x} \text{ at } z = \eta \tag{8.45}$$

The dynamic boundary condition at the liquid free surface:

$$\frac{\partial \phi}{\partial t} + \frac{1}{2}\left|\nabla \phi\right|^2 + g\eta = 0 \text{ at } z = \eta \tag{8.46}$$

The kinematic boundary condition at the liquid free surface, Eq. (8.45), is derived based on the following assumptions: the liquid molecules existing on the free surface do not break away from nor slip into the liquid itself, and the velocity vector of their motion coincides with that of the velocity of the free surface. Equation (8.46) is the generalized Bernoulli's equation, expressing the balance of forces at the free surface.

The following equation can be derived from Eqs. (8.45) and (8.46) by expanding the derivatives of the velocity potential ϕ with respect to the variables z and t in Taylor series about the equilibrium $z = 0$, and assuming the higher-order terms to be negligible:

$$\frac{\partial^2 \phi}{\partial t^2} + g\frac{\partial \phi}{\partial z} = 0 \text{ at } z = 0 \tag{8.47}$$

Solving the continuity Equation (8.42) with the boundary conditions of Eqs. (8.43), (8.44) and (8.45) being satisfied, the general solution for the velocity potential can be obtained as

$$\phi = \frac{gA\cosh k(z + H)}{\omega \cosh kH} \sin kx \cos \omega t \tag{8.48}$$

Disregarding the second term on the right-hand side of Eq. (8.45), the generalized solution of the displacement η of the free surface is given by

$$\eta = A\sin kx \sin \omega t, \tag{8.49}$$

where A is a constant and $k = n\pi/L$ is the wave number. The natural frequency is expressed as follows:

$$f = \frac{1}{2\pi}\sqrt{gk\tanh kH} \tag{8.50}$$

In the case where a steady flow exists in the vessel, the response of the free surface is obtained as outlined below [36–38]. Supposing that a steady surface flow of velocity U is present and the flow velocity due to sloshing is significantly lower than the steady flow velocity, the velocity potential is assumed to have the form

$$\phi = Ux + \frac{gA\cosh k(z+H)}{\omega\cosh kH}\sin kx\cos\omega t \tag{8.51}$$

where the wave number is expressed as $k = n\pi/L$. Substituting Eq. (8.51), with $z = 0$, into the boundary conditions at the free surface and ignoring higher-order terms gives the following angular natural frequency

$$\omega = \sqrt{gk\tanh kH - k^2U^2} \tag{8.52}$$

The reason for the decrease in the natural frequency of sloshing due to the surface flow of velocity U is that the flow along the deformed shape of the free surface causes a centrifugal force which acts on the fluid in the direction opposite to the restoring force of gravity. The damping ratio tends to be higher when a steady flow exists along the free surface.

(2) Sloshing in a cylindrical tank

Next, sloshing in a cylindrical tank is considered. Here we evaluate the sloshing response in a cylindrical tank subjected to a displacement excitation u_g in the x direction as shown in Fig. 8.12. In the same manner as in the case of a rectangular tank, the equation of motion of the system is obtained by using the velocity potential ϕ. The boundary conditions at the free surface, side walls and bottom are similar to Eqs. (8.43), (8.44) and (8.47). Solving the continuity equation with these boundary conditions gives the following result for the fundamental natural frequency [36].

$$f = \frac{1}{2\pi}\sqrt{\frac{1.84g}{R}\tanh\left(\frac{1.84H}{R}\right)} \tag{8.53}$$

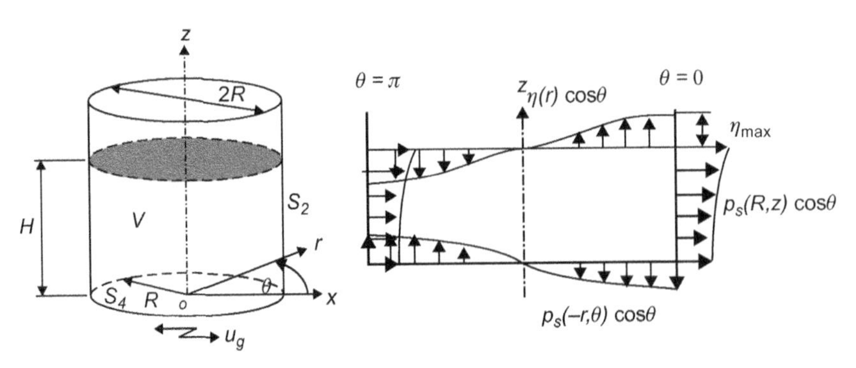

FIGURE 8.12

Sloshing in a cylindrical tank and its pressure distribution [39].

The value 1.84 in the above equation is the smallest root of the derivative $J_1'(r)$ of the first order Bessel function of the first kind. Supposing that the response of the first mode is predominant, thus making the contributions of the second and higher modes negligibly small, the wave height can be expressed as follows, in terms of the acceleration response spectrum S_A[39]:

$$\eta(r) = 0.837\frac{R}{g}\frac{J_1(\varepsilon_1 r/R)}{J_1(\varepsilon_1)}S_A \tag{8.54}$$

The liquid dynamic pressure is expressed in terms of the maximum wave height η_{max} at $r = R$ as

$$p_s(r,z) = \rho_L g \eta_{max}\frac{R}{g}\frac{J_1(\varepsilon_1 r/R)}{J_1(\varepsilon_1)}\frac{\cosh(\varepsilon_1 z/R)}{\cosh(\varepsilon_1 H/R)} \tag{8.55}$$

The overturning moment and base shear can be obtained by integrating the liquid dynamic pressure in the above equation.

8.3.3.3 Evaluation method for bulging [39]

Here we evaluate the dynamic response of a cylindrical tank subjected to a displacement excitation u_g in the x direction as shown in Fig. 8.12. Supposing that the velocity potential ϕ exists, the dynamic liquid pressure p can be related to the velocity potential according to Bernoulli's theorem as

$$p = -\rho\frac{\partial\theta}{\partial t} \tag{8.56}$$

The liquid motion is governed by the following four equations:

The equation of continuity

$$\frac{\partial^2 p}{\partial r^2} + \frac{1}{r}\frac{\partial p}{\partial r} + \frac{1}{r^2}\frac{\partial^2 p}{\partial \theta^2} + \frac{\partial^2 p}{\partial z^2} = 0 \tag{8.57}$$

The boundary condition at the liquid free surface

$$\frac{\partial p}{\partial z} = 0 \quad \text{at } z = H \tag{8.58}$$

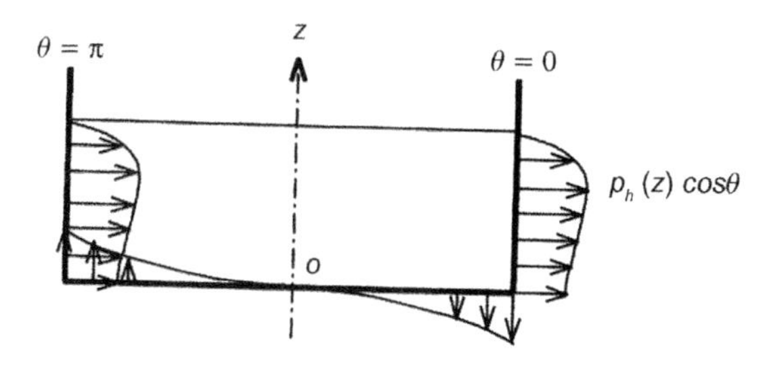

FIGURE 8.13

Distribution of dynamic liquid pressure due to bulging [39].

The boundary condition at the side wall

$$\frac{\partial p}{\partial r} = \rho(\ddot{u}_r + \ddot{u}_g)\cos\theta \text{ at } r = R \tag{8.59}$$

The boundary condition at the bottom

$$\frac{\partial p}{\partial z} = 0 \text{ at } z = 0 \tag{8.60}$$

Equation (8.59) expresses the condition that the liquid velocity normal to the side wall coincides with the velocity of the tank. Here u_r is the elastic deformation of the side wall, which is related to the coupling between tank vibration and liquid motion. An eigenvalue analysis of the equations above yields the natural frequencies and mode shapes. The first natural period T_b of bulging is given approximately by the following equation, where the wall thickness variation is taken into account based on shell theory [40]:

$$T_b = \frac{2}{\lambda}\sqrt{\frac{W_L}{\pi E t_h}} \quad \text{for} \quad 0.15 \leqq \frac{H}{D} \leqq 2.0, \tag{8.61}$$

$$\lambda = 0.067\left(\frac{H}{2R}\right)^2 - 0.30\left(\frac{H}{2R}\right) + 0.46, \; W_L = \rho_L \pi R^2 H \tag{8.62}$$

where W_L and E are the liquid mass and Young's modulus of elasticity of the side wall, respectively. When the wall thickness of the cylindrical tank is variable, the thickness at the height $H/3$ from the bottom is used for calculation of the natural period. After calculating the natural period, the maximum acceleration response of the tank can be obtained by using a seismic input acceleration and the spectrum of the acceleration response.

For a cylindrical liquid storage tank, the inertia due to the contained liquid mass is overwhelmingly larger than that due to the mass of the tank itself. The pressure on the wall caused by the inertia due to the liquid mass is called the dynamic liquid pressure. The dynamic liquid pressure is expressed as a sum of the pressure due to rigid motion and the pressure due to elastic deformations by the following equation [39]:

$$p_h(z) = \rho_L H\left\{\sum_{i=0}^{5} C_{0i}\left(\frac{z}{H}\right)^i\right\}\ddot{u}_g + \rho_L H\left\{\sum_{i=0}^{5} C_{1i}\left(\frac{z}{H}\right)^i\right\}(\ddot{u}_r - \ddot{u}_g) \tag{8.63}$$

where C_{0i} and C_{1i} are coefficients expressing the dynamic liquid pressure due to the rigid motion and that due to the elastic deformations, respectively. The overturning moment and base shear can be obtained by integrating the dynamic liquid pressure.

One of the typical types of damage to tanks is failure of the tank ceiling due to sloshing. It is therefore important to estimate the impact load on the tank ceiling due to sloshing. The impact load can be shown to be lower when the deformation of the ceiling is taken into consideration than when the ceiling is assumed to be rigid [41].

8.3.3.4 Natural frequency of sloshing

When the sloshing amplitude is sufficiently small, for a certain mode of free oscillations in a vessel of constant depth, the velocity potential is given by

$$\phi = \frac{gA\cosh\omega(z+H)/c}{\omega\cosh\omega H/c}\tilde{\phi}\sin(\omega t+\varphi) \tag{8.64}$$

where the wave velocity c is a function of the wave frequency, the liquid depth, and the gravitational acceleration, and is obtained by solving the following characteristic equation:

$$c = \frac{g}{\omega}\tanh\frac{\omega H}{c} \tag{8.65}$$

The equation above can be considerably simplified when the liquid depth in the vessel is much larger or much smaller than the wave length ($2\pi c/\omega$), i.e., when the wave velocity can be approximately expressed as $c=(gH)^{1/2}$ for the case of shallow liquid ($\omega H/c<\pi/10$) and as $c=g/\omega$ for the case of a deep tank ($\omega H/c>\pi$).

For the cases of shallow liquid, the wave velocity increases as the liquid depth increases until it reaches the critical liquid depth $H_{crit}=\pi c/10\omega$. The natural frequency of sloshing in shallow liquid cases where $H<\pi c/10\omega$ tends to increase with increasing liquid depth since the natural frequency of sloshing is proportional to the wave velocity.

Table 8.6 shows the natural frequencies and modes for vessels of various shapes. This table is a summary for moderate liquid depth (cases not shallow enough or deep enough). Table 8.7 presents the natural frequency for cases where the vessel cross-section varies in the depth direction. As shown in the table, the variation of cross-section in the depth direction has little effect on the natural frequency. Hence the natural frequencies for vessels having variable cross-sections in the depth direction can be roughly estimated by using the values for constant cross-section vessels of identical depth.

8.3.3.5 Sloshing amplitude

(1) Rectangular vessel

In this section we turn to the determination of the sloshing amplitude when a rectangular vessel is excited. In the case of large-amplitude waves, nonlinearities should be considered in order to estimate sloshing amplitude precisely. This section outlines an evaluation method for obtaining the sloshing amplitude of the first mode with wave number $k=\pi/L$ when the vessel (shown in Fig. 8.8) is subjected to a sinusoidal excitation $X_0\sin\omega t$ in the x direction [43]. The perturbation method is used since the nonlinearities are weak. Introducing a small parameter ε the variables ω, ϕ, and η are expanded in powers of this parameter and the basic equations then solved for the respective orders of ε, namely ε^0, ε^1, and ε^2.

Table 8.6 Natural frequencies and modes for vessels of various shapes [42].

Shape	Natural frequency	Natural mode
	$f_{ij} = \frac{g^{1/2}}{2\pi^{1/2}}\left[\left(\frac{i^2}{a^2}+\frac{j^2}{b^2}\right)^{1/2}\tanh\pi h\left(\frac{i^2}{a^2}+\frac{j^2}{b^2}\right)^{1/2}\right]^{1/2}$ $\frac{1}{10} < \left(\frac{i^2}{a^2}+\frac{j^2}{b^2}\right)^{\frac{1}{2}} < 1$	$\eta_{ij} = \cos\frac{i\pi x}{a}\cos\frac{j\pi y}{b}$
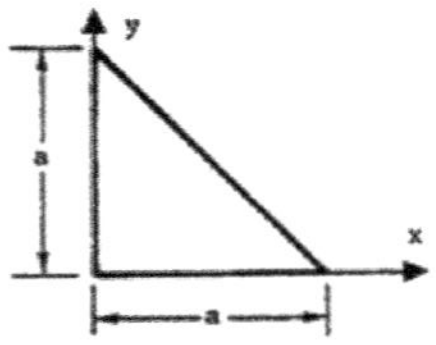	$f_{ij} = \frac{g^{1/2}}{2\pi^{1/2}a}\left[\left(i^2+j^2\right)^{1/2}\tanh\frac{\pi h}{a}\left(i^2+j^2\right)^{1/2}\right]^{1/2}$ $\frac{1}{10} < \left(i^2+j^2\right)^{\frac{1}{2}} < 1 i \neq j$	$\eta_{ij} = \cos\frac{i\pi x}{a}\cos\frac{j\pi y}{a}$ $\pm\cos\frac{j\pi x}{a}\cos\frac{i\pi y}{a}$ $-\ldots i+j = even$ $+\ldots i+j = odd$
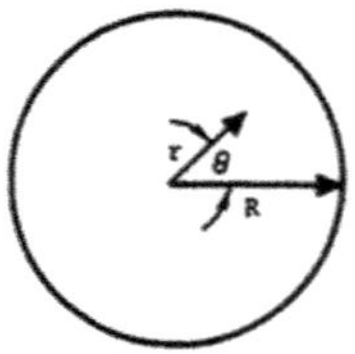	$f_{ij} = \frac{1}{2\pi}\left(\frac{\lambda_{ij} g}{R_1}\tanh\frac{\lambda_{ij} h}{R_1}\right)^{1/2}$ $\frac{\pi}{10} < \frac{h}{R} < \pi$, where $J_1{}'(\lambda_{ij}) = 0$ (see λ_{ij} table below)	$\eta_{ij} = J_i\left(\lambda_{ij}\frac{r}{R}\right)\begin{Bmatrix}\sin i\theta \\ or \\ \cos i\theta\end{Bmatrix}$

λ_{ij}

j	i = 0	1	2	3
0	0	1.8412	3.0542	4.2012
1	3.8317	5.3314	6.7061	8.0152
2	7.0156	8.5363	9.9695	11.3459
3	10.173	11.7060	13.1704	14.5859

(Continued)

Table 8.6 (Continued)

Shape	Natural frequency	Natural mode
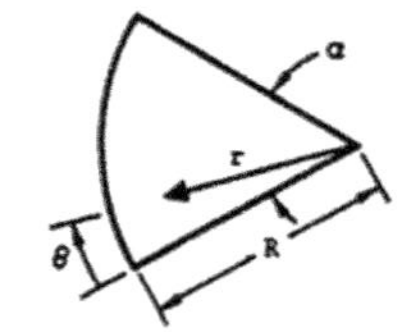	The same equation as that in frame 3 with the index i: $i = \frac{180n}{\alpha} \qquad 0 \leqq \alpha \leqq 360$ j is given in frame 3. λ_{ij} for non-integer i can be interpolated from the table in frame 3.	$\eta_{ij} = J_i\left(\lambda_{ij}\frac{r}{R}\right)\cos i\,\theta$
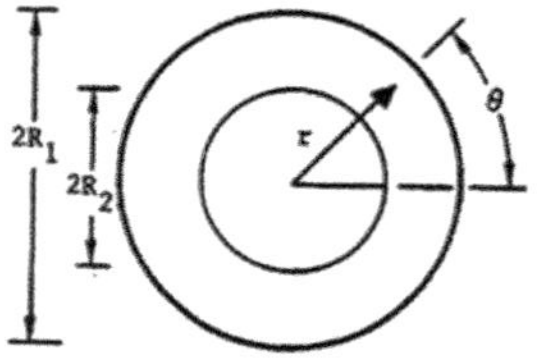	$f_{ij} = \frac{1}{2\pi}\left(\frac{\lambda_{ij}g}{R_1}\tanh\frac{\lambda_{ij}h}{R_1}\right)^{1/2} \quad \frac{\pi}{10} < \frac{h}{R_1} < \pi$ (see λ_{ij} table below)	$\eta_{ij} = G_{ij}(r)\begin{Bmatrix}\sin i\theta \\ or \\ \cos i\theta\end{Bmatrix}$ $G_{ij}(r) = Y_i'(\lambda_{ij})J_i\left(\lambda_{ij}\frac{r}{R_1}\right) - J_i'(\lambda_{ij})Y_i\left(\lambda_{ij}\frac{r}{R_1}\right)$
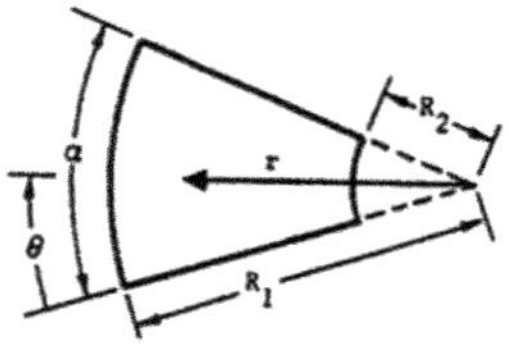	The same equation as that in frame 5 with the index i: $i = \frac{180n}{\alpha} \qquad 0 \leq \alpha \leq 360$ j is given in frame 5. λ_{ij} for non-integer i can be interpolated from the table in frame 5.	$\eta_{ij} = G_{ij}(r)\cos i\,\theta$

$\frac{R_2}{R_1}$	j	λ_{ij}, i = 0	1	2	3
0.3	0	0	1.5821	2.9685	4.1801
	1	4.7058	5.1374	6.2738	7.7213
0.5	0	0	1.3547	2.6812	3.9577
	1	6.3932	6.5649	7.0626	7.8401

(Continued)

Table 8.6 (Continued)

Shape	Natural frequency	Natural mode
$\frac{x^2}{a^2}+\frac{y^2}{b^2}=1$; 2a; 2b	$f=\frac{(gh)^{1/2}}{2\pi a}\left[\frac{18+6(b/a)^2}{5+2(b/a)^2}\right]^{1/2}$ $\frac{h}{a}<0.1$ Shallow liquid	$\eta = Ax$ $A = constant$
L	$f=\begin{cases}\frac{(gh)^{1/2}}{2L} & \frac{h}{L}<\frac{1}{10}\\ \frac{1}{2}\left(\frac{g}{\pi L}\right)^{1/2} & \frac{h}{L}>1\end{cases}$	L = typical maximum lateral dimension

Table 8.7 The effect of vessel shape (in the depth direction) on the natural frequency of sloshing [42].

Shape	Natural frequency
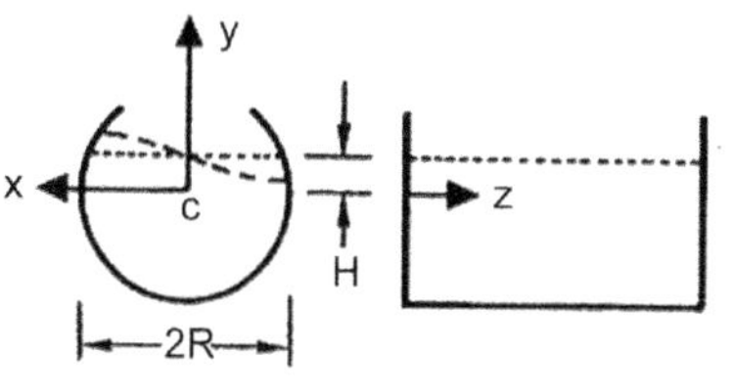	$f_i = \frac{\lambda_i^{1/2}}{2\pi}\left(\frac{g}{R}\right)^{1/2}$ (see values below)
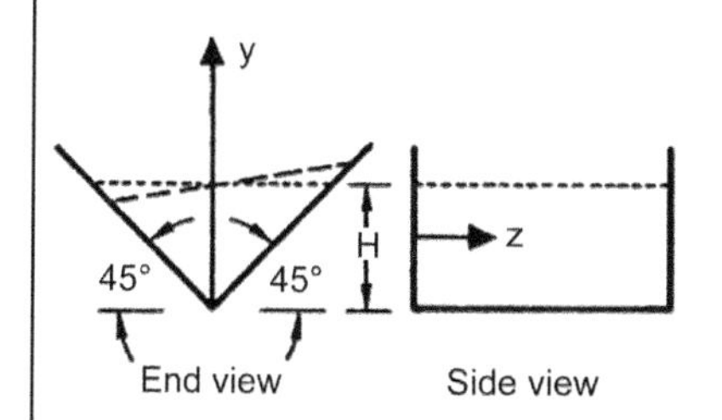	$f_i = \frac{\lambda_i^{1/2}}{2\pi}\left(\frac{g}{H}\right)^{1/2}$; $\lambda_1 = 1.0$, $\lambda_2 = 2.324$, $\lambda_3 = 3.9266$, $\lambda_i = \alpha \tanh \alpha,\ i > 1$, where $\cos 2\alpha \cosh 2\alpha = 1$.
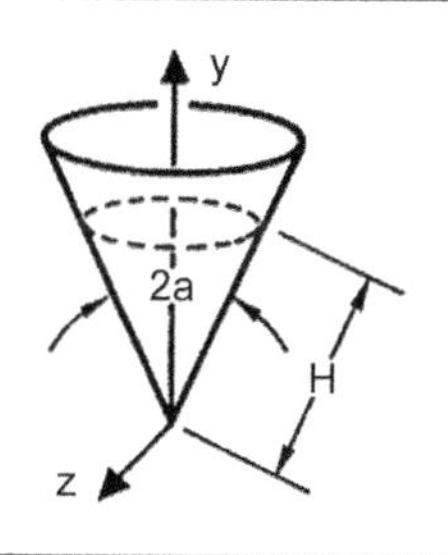	$f_i = \frac{\lambda_i^{1/2}}{2\pi}\left(\frac{g}{H}\right)^{1/2}$ 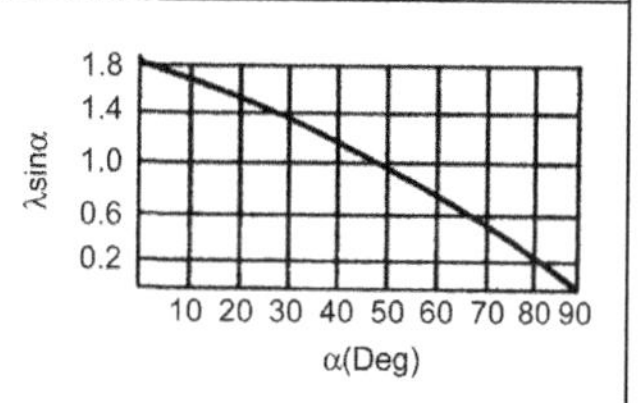
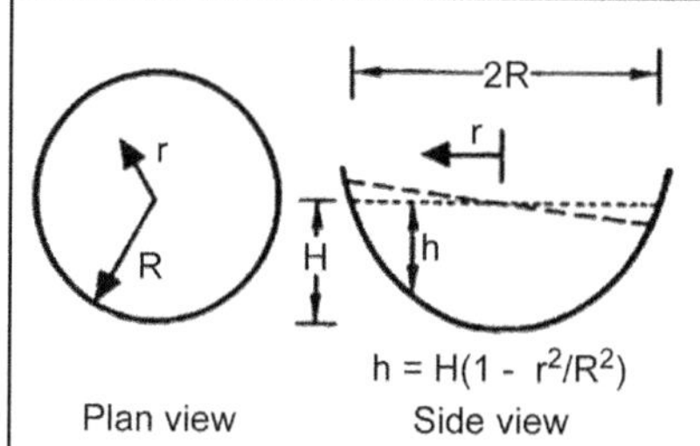	$f_i = \lambda_{ij}^{1/2}\frac{(gH)^{1/2}}{R}$

H/R	λ_1	λ_2	λ_3
−1.0	1.0	6.0	15.0
−0.8	1.045	5.38	10.85
−0.6	1.099	4.97	9.13
−0.4	1.165	4.74	8.33
−0.2	1.249	4.65	7.99
0.0	1.360	4.70	7.96
0.2	1.513	4.91	8.23
0.4	1.742	5.34	8.89
0.6	2.13	6.22	10.28
0.8	3.04	8.42	13.84
1.0	∞	∞	∞

(Continued)

Table 8.7 (Continued)

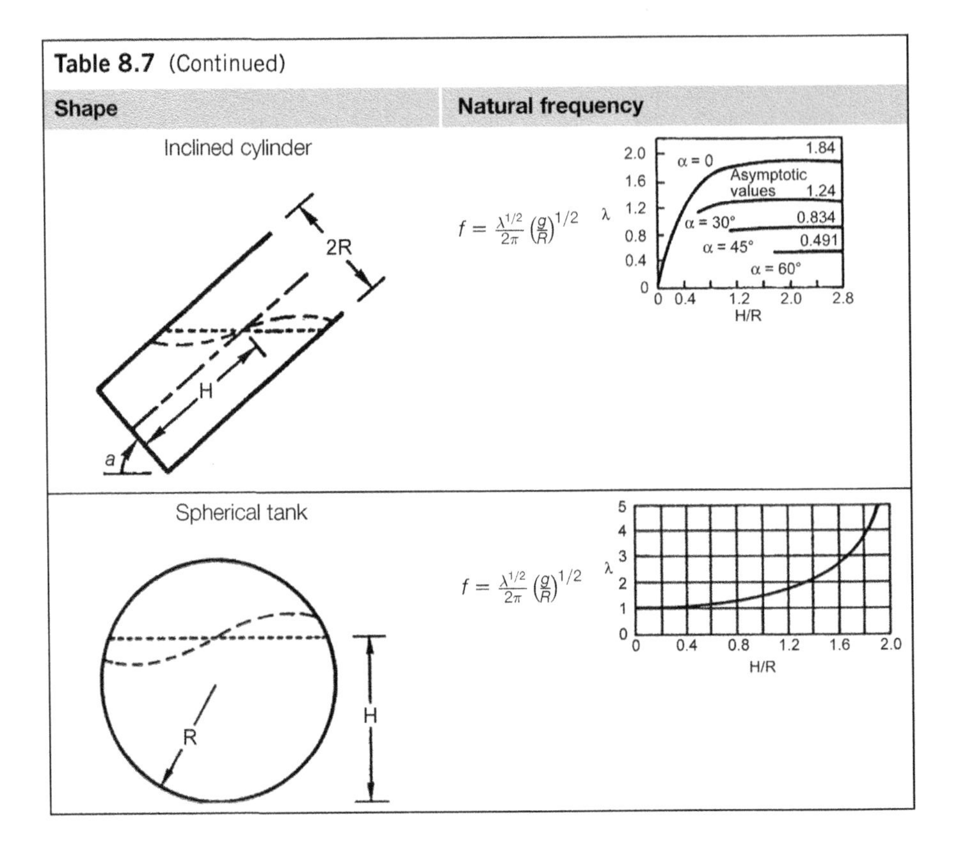

The resulting relationship between frequency (ω) and amplitude (A) is represented in the following equation:

$$1 - \frac{\omega}{\omega_1} + \left(\frac{A}{L}\right)^2 K = \frac{2T_H X_0}{\pi A} \tag{8.66}$$

where $T_H = \tanh kH$, and $\omega_1 = \sqrt{gkT_H}$. K is given by the following equation:

$$K = \frac{\pi^2}{64}\left(9T_H^{-4} - 12T_H^{-2} - 2T_H^2 - 3\right) \tag{8.67}$$

The oscillatory response of the liquid surface is then given by

$$\eta = A\sin kx \sin \omega t + (kA^2/8)\cos 2kx\{(T_H - T_H^{-1}) - (3T_H^{-3} - T_H^{-1})\cos 2\omega t\} \tag{8.68}$$

Figure 8.14 shows the resonance vibration response of liquid surface. K, in Eq. (8.67), is positive when $H/L < 0.337$ (and negative when $H/L > 0.337$). For a shallow tank, as shown in Fig. 8.14(a), a hard-spring type of resonance curve is obtained, where the resonance frequency increases with the excitation amplitude.

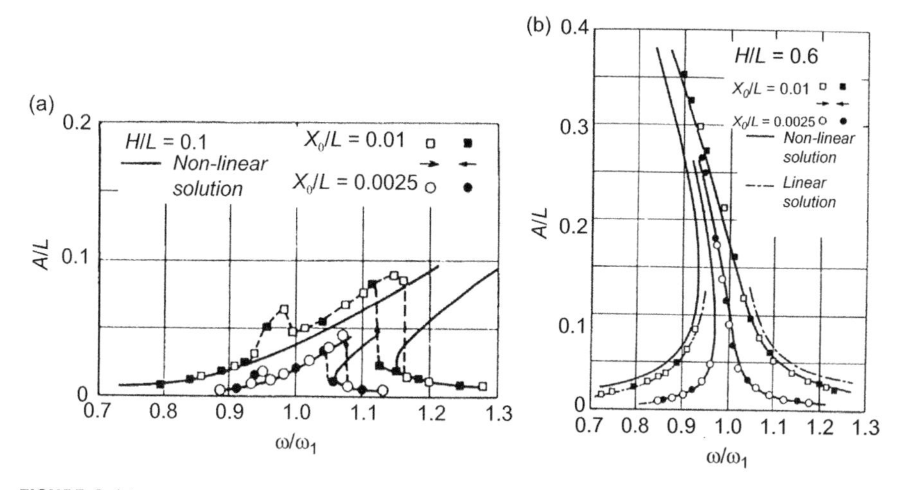

FIGURE 8.14

Resonance vibration response of liquid surface [43].

In the case of a deep tank, as shown in Fig. 8.14(b), a soft-spring type of resonance curve is obtained. In the case of a deep tank, a standing wave is assumed for the derivation of the sloshing response equation. On the other hand, for very shallow cases, wave theory is needed to formulate the sloshing response since traveling waves may then exist.

(2) Cylindrical vessel [36,44,45]

Identical natural frequencies and mode shapes in two orthogonal directions exist for sloshing in a cylindrical vessel. For a linear system, the vibration response can be obtained by a simple summation in the respective directions. However, the sloshing responses in the two directions can interfere with each other due to the nonlinearity of sloshing. When the cylindrical vessel is excited in the x direction, the liquid surface also oscillates in the y direction due to a parametric excitation. The result is a rotating (or swirling) motion of the liquid surface. Figure 8.15 shows an example of the liquid surface response spectrum for a cylindrical vessel.

The sloshing response occurs in one direction for low excitation frequency as seen on the lower branch in Fig. 8.15. When the excitation frequency exceeds a certain limiting value, the sloshing amplitude increases rapidly and unidirectional and rotating sloshing can simultaneously exist. On further increasing the excitation frequency, a second transition point is reached past which the liquid motion reverts to low-amplitude unidirectional sloshing. The response curve of unidirectional sloshing has the characteristics of a soft spring when $H/D > 0.3$. On the other hand, the rotating type of sloshing has the characteristics of a soft spring regardless of the magnitude of H/D.

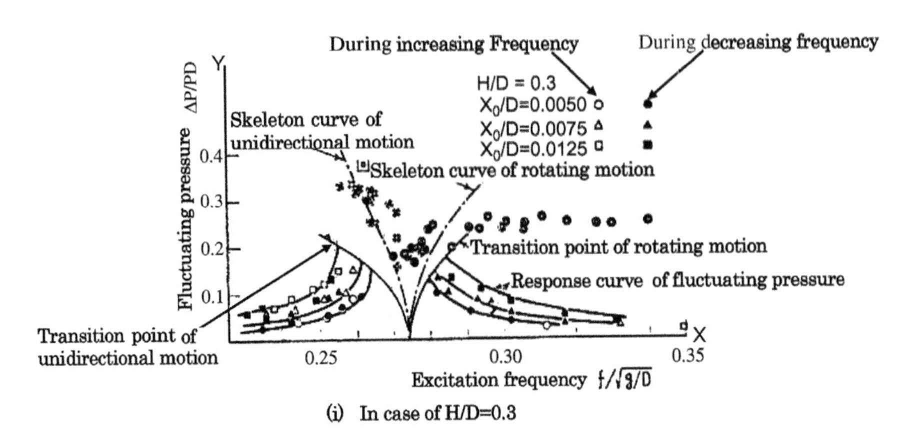

FIGURE 8.15

Resonance response of the liquid surface in a cylindrical vessel [45].

(3) Sloshing due to vertical excitation [46,47]

Liquid vessels in liquid container ships and rockets are typically excited in the vertical direction. Sloshing can therefore occur in these internal vessels. The natural circular frequencies of sloshing in the vertical direction are expressed by the following equation [46]:

$$\omega_{mn}^2 = \left\{ \left[\frac{2\xi_{mn}}{D} \right]^3 \frac{\sigma}{\rho} + \frac{2g\xi_{mn}}{D} \right\} \tanh \frac{2H\xi_{mn}}{D} \tag{8.69}$$

where ξ_{mn} is an eigenvalue which satisfies the equation $J'_m(\xi_{mn}) = 0$. Here J'_m is the derivative of the m-th order Bessel function of the first kind. Concentrically symmetric surface waves correspond to $m = 0$, and ξ_{0n} takes the values 3.83, 7.02 and 10.17, respectively, for $n = 1$, 2 and 3. When the excitation amplitude exceeds a certain threshold value the liquid free surface transitions to a 1/2-subharmonic wave. Figure 8.16 shows the stability boundary chart for a 1/2-subharmonic surface mode $(1,1)_{\mathrm{m}}$ for water. As the acceleration of the vertical oscillations increases, the free surface disintegrates and spray particles are ejected from the surface. The liquid free surface then presents a complicated liquid flow mechanism with interference between the liquid free surface and spray droplets.

(4) Sloshing subjected to pitching excitation [48]

Liquid tanks and fuel tanks in tankers (tankships) are often subjected to pitching and roll motion about the horizontal axes. It is important to investigate the characteristics of sloshing in a vessel subjected to pitching (or rolling) excitation from the viewpoint of ship safety. Even in the case where a vessel is subjected to excitation about one horizontal axis, sloshing modes in directions other than the excitation direction can be parametrically excited via nonlinear coupling.

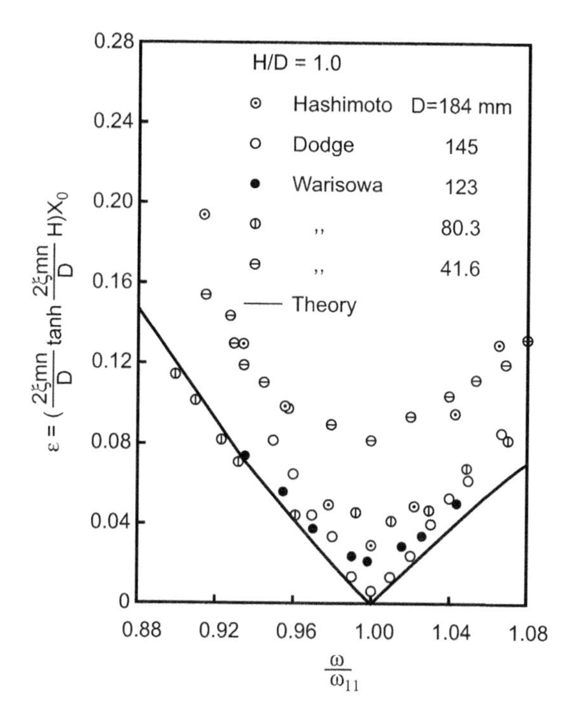

FIGURE 8.16

Stability boundary of 1/2 subharmonic resonance (1,1)m [46].

8.3.4 Examples of sloshing problems and countermeasures

8.3.4.1 Damage of oil storage tanks with floating roofs in the 2003 Tokachi Oki earthquake [49]

The severe Tokachi Oki earthquake occurred in the area of Hokkaido, Japan (after which it is named) on 26 September in 2003. Several large tanks with floating roofs (Fig. 8.17) were damaged, with the resulting damage causing fires. The floating roof has little suppression effect on liquid sloshing. Being a floating structure itself the roof moves together with the liquid surface during sloshing under earthquake excitation. The floating roof often causes damage due to impact and friction between the roof and the tank itself or peripheral equipment. In this earthquake, a ground motion with periods of oscillation ranging from 5 to 10 seconds caused sloshing. The sloshing in turn caused large compression stresses at the pontoon of the floating roof in the circumferential direction and buckling of the upper and lower plates of the pontoon.

Measures against the damage included: (i) structural strengthening to prevent buckling of the pontoon, (ii) increasing buoyancy of the pontoon, and (iii) attachment of reinforcements to the pontoon.

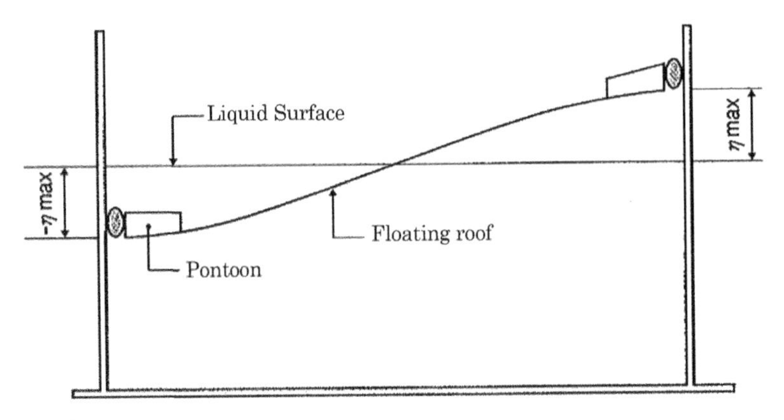

FIGURE 8.17

Sloshing of tanks with floating roofs [49].

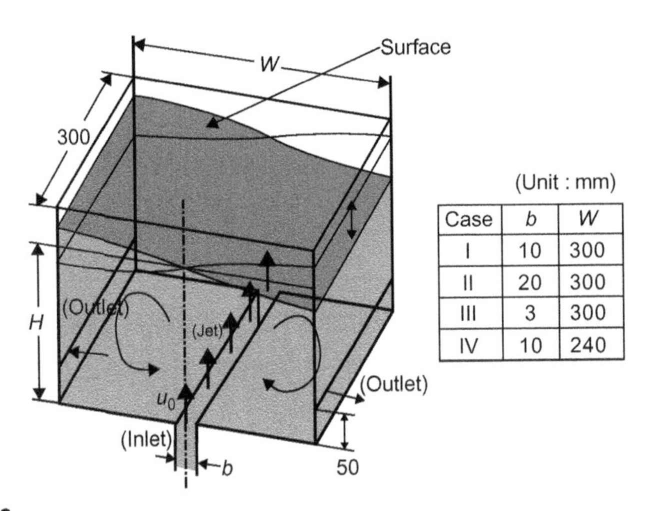

Case	b	W
I	10	300
II	20	300
III	3	300
IV	10	240

FIGURE 8.18

Self-induced sloshing caused by a jet in a tank [50].

8.3.4.2 Self-induced sloshing caused by ingress of a jet in a tank [36,50,51]

Self-induced sloshing can be caused by ingress of a jet in a "two-dimensional" rectangular tank as shown in Fig. 8.18. Small initial pressure fluctuations due to sloshing motion cause fluctuations of the jet in horizontal direction. The jet fluctuations transport momentum in the direction orthogonal to the jet axis. In regions that gain momentum, the resulting jet force acts toward the direction of the momentum gradient to increase the sloshing velocity. In regions that lose momentum the opposite effect occurs (i.e. a decrease in the sloshing velocity). When the jet force acts towards the direction where the sloshing is induced or amplified, self-induced sloshing occurs.

Self-induced sloshing can also be caused by a jet ingress in a cylindrical tank. A swell due to the jet occurs in the liquid as shown in Fig. 8.19. As the jet moves (away) in the horizontal direction, the swell loses the momentum supply from the jet, resulting in a drop in liquid surface elevation. This is repeated periodically, thus leading to sloshing.

8.3.4.3 Weir vibrations coupled to plenum sloshing [52,53]

Self-excited sloshing due to fluid discharge over a flexible cylindrical weir occurred in the cooling circuit of Super-Phenix-1, a fast breeder reactor in France (Fig. 8.20). In this system, the fluid supplied from the bottom of the upstream plenum flows upward along the cylindrical weir, whose bottom is fixed. The fluid then overflows over the top of the weir, and flows downward along the weir into the downstream plenum. Self-excited oscillations involving a coupled mode

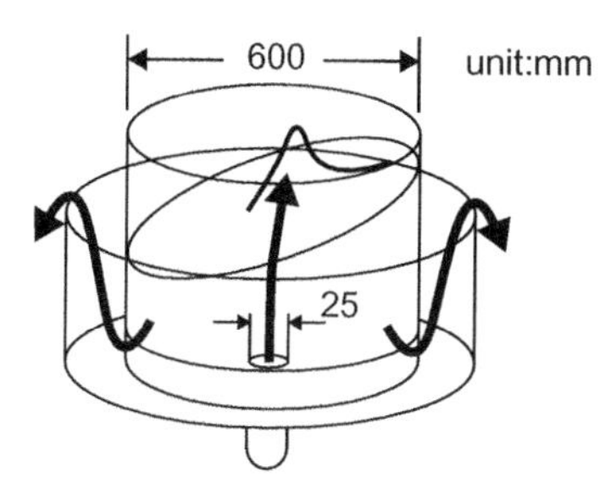

FIGURE 8.19

Self-induced sloshing in a cylindrical vessel [51].

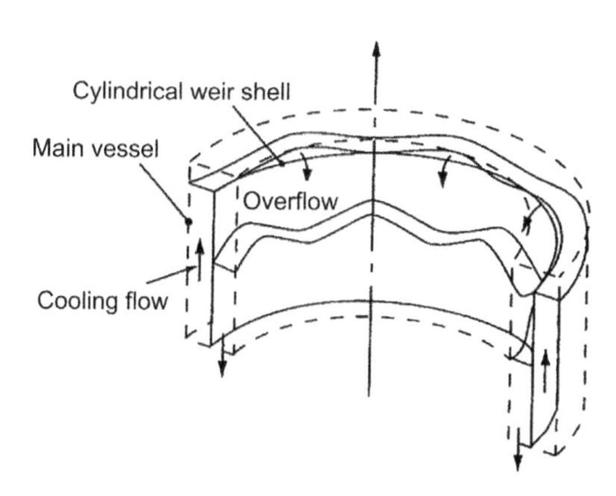

FIGURE 8.20

Weir vibration coupled to sloshing [53].

between sloshing in the upper and lower plenums and ovalling vibrations of the cylindrical weir can occur. The stability of the system depends on the phase difference between liquid surface oscillation in the downstream plenum and fluctuation of the flow rate into the plenum. The phase difference is governed by the first-order delay of the fluctuating overflow rate relative to the weir vibration. The phase difference also depends on the dead-time required for fluid at the top of the weir to reach the surface of the downstream plenum.

8.3.4.4 Vibration suppression liquid sloshing damper [54]

In order to suppress vibrations of structures, such as very tall buildings and towers subjected to strong winds, some vibration suppression dampers utilizing the fluid forces generated by liquid sloshing are installed in these structures which have very low natural frequencies (Fig. 8.21).

The sloshing damper utilizes the same mechanism as a classic dynamic damper. In the dynamic damper, an additional system with a mass, spring and damping is installed at the top of the building. The system parameters are chosen to match the natural frequency of the additional system with the natural frequency of the building.

The sloshing damper utilizes the fluctuating pressure due to liquid sloshing. The natural frequency of sloshing is adjusted by changing the water depth. Meshes or nets in the direction orthogonal to the flow due to sloshing are installed in the vessel to introduce additional damping. The sloshing damper has the

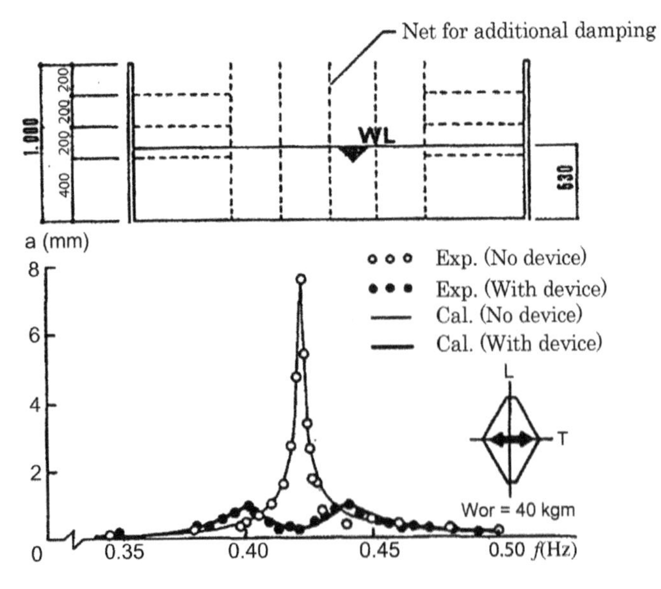

FIGURE 8.21

Vibration suppression liquid sloshing damper [54].

following merits: (i) the damper can suppress even small-amplitude building vibrations, (ii) the damper system is simple because of having only a few mechanical parts, (iii) the natural frequency of sloshing can also be adjusted.

8.3.4.5 Sloshing suppression control in a moving cylindrical container [55]

In pipe-less plants used for production of chemicals and foods, raw materials are transported from storage tanks to the mixing vessel on transportation carts using liquid containers instead of piping. The methodology for designing an optimal driving pattern of the moving liquid container, to reduce residual free surface oscillations after a rapid access operation, and to stop the container at an exact target location, is required for pipe-less plants. Some research into sloshing suppression control for the moving liquid container has been performed along the following lines. A driving pattern – based on optimal control theory – is proposed, in order to reduce residual free surface oscillations. The magnitude of control input is selected as a cost function (Fig. 8.22). A method which does not use any sensors to measure the height of the liquid surface has also been proposed, based on feed-forward control.

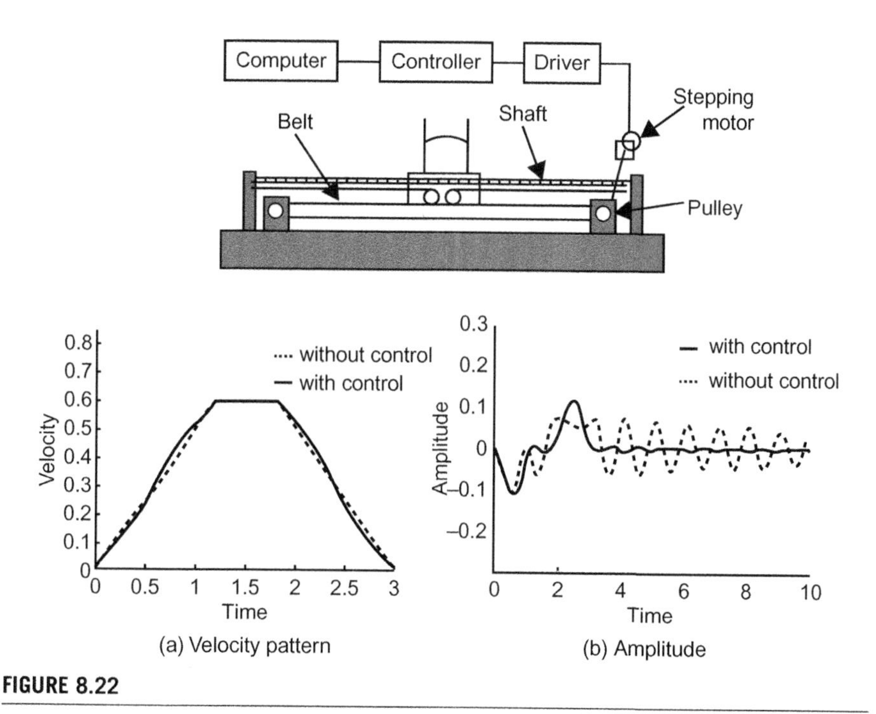

FIGURE 8.22

Sloshing suppression control of liquid in a moving cylindrical container [55].

Table 8.8 Proposed methods for sloshing suppression.

No.	Schematic of suppression method	Explanation of sloshing suppression mechanism
1. Installation of a reversed U-tube in a liquid container as a liquid dynamic damper [57,58].	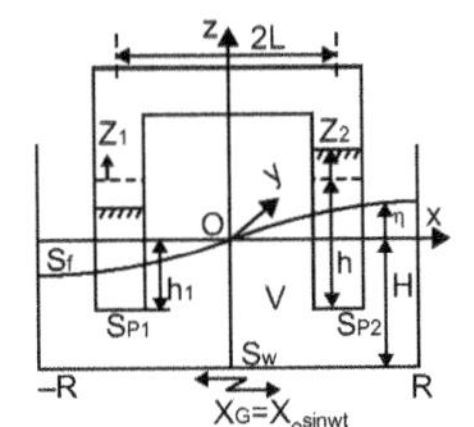	The reversed U-tube is installed, as a sub-system, into the liquid container, which is the main system. The reversed U-tube plays the role of a dynamic damper when its natural frequency is almost the same as that of sloshing in the liquid container. It is possible to reduce the sloshing amplitude of the main system to 1/6 ~ 1/10 when the ratio of the spring coefficients is in the range of 3.5 ~ 4.5%.
2. Utilization of fluid resistance caused by a bulkhead.	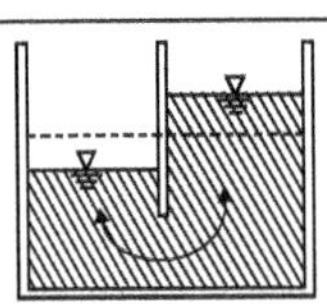 (a) U tube mode 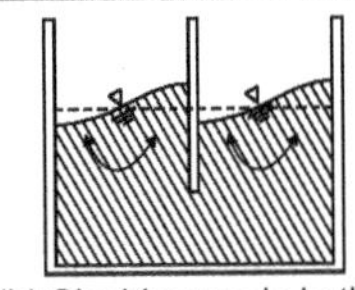 (b) Sloshing mode in the separated vessel	The sloshing mode is separated into two modes by the bulkhead. One is the sloshing mode of the water column in a U-tube. The other is the sloshing mode in the sector tank. The vertical bulkheads can suppress sloshing by changing the natural frequencies of sloshing and decreasing the participation factor of the U-tube shaped sloshing mode. The damping characteristics of the U-tube shaped sloshing mode are determined by vortices generated below the bulkhead [59,60].

(Continued)

Table 8.8 (Continued)

No.	Schematic of suppression method	Explanation of sloshing suppression mechanism
3. Sloshing suppression method employing tube bundles in a cylindrical liquid container.	y y x x x	Wave height and fluid forces that act on tube bundles due to sloshing decrease and fluid damping increases as the number of tubes increases. Fluid forces acting on the tubes are largest near the free surface and decreases rapidly in the depth direction [56].
4. Active sloshing suppression method using controlled gas bubble injection [61].		Sloshing in liquid container can be suppressed by controlling the timing and duration of periodic intermediate gas bubbles injection on the side? of the liquid container.

8.3.4.6 Sloshing suppression methods

The suppression of sloshing in liquid containers is important in order to secure the seismic reliability of these containers. However, there are still many problems to be solved, for the reasons that follow. Viscous damping is not effective for the suppression of the liquid surface oscillations because the energy dissipation at the container wall surfaces is very small. It is also almost impossible to adopt, for the liquid container, the same viscous damper as that used for mechanical vibration damping.

Some suppression methods for sloshing in liquid containers are proposed in Table 8.8. One suppression method adopts a reversed U-tube in the liquid container as a dynamic damper. Another method utilizes fluid resistance caused by tube bundles and/or a bulkhead which divide the liquid container vertically into several sectors. Active sloshing control by periodic gas bubble injection has also been proposed.

References

[1] S.S. Chen, Flow-Induced Vibration of Circular Cylindrical Structures, Hemisphere Publishing, 1987.
[2] R.D. Blevins, *Formulas for Natural Frequency and Mode Shape*, Robert E. Krieger Publishing, 1979.
[3] K. Wendel, Jahrbuch Sciffbautechnisches Gesellschaft vol. 44 (1950) 207.
[4] A.E. Bryson, J. Aeron. Sci. 21 (1954) 424–427.
[5] L.I. Sedov, Two-Dimensional Problems in Hydrodynamics and Aerodynamics, Interscience, 1950. p. 29.
[6] M.M. Munk, Aerodynamic Theory, Vol.1, Julius Springer, Berlin, 1934. p. 302.
[7] W.K. Meyrhoff, J. Ship Res. 14 (1970) 100–111.
[8] H. Lamb, Hydrodynamics, sixth ed., Dover Publications, New York, 1945, pp. 152–156, 700-701.
[9] V.Y. Mazur, Izv. Akad. Nauk SSSR, Mekhan. Zhidk. i Gaza 3 (1966) 75–79.
[10] S.S. Chen, H. Chung, ANL-CT-76-45, Argonne National Laboratory, Argonne, IL, 1976.
[11] S.S. Chen, M.W. Wambsganss, J.A. Jendrzejczyk, Trans. ASME J. Appl. Mech. 43 (1976) 325–329.
[12] N.L. Ackermann, A. Arbhabhirama, J. Eng. Mech. Div. Am. Soc. Civil Engrs. 90 (1964) 123–130. EM4.
[13] S.S. Chen, J. Press. Vessel Tech. 97 (1975) 78–83.
[14] E. Naudascher, R.D. Rockwell, Flow-Induced Vibrations – An Engineering Guide, A.A. Balkema, Rotterdam, 1994.
[15] S.S. Chen, J. Fluids Eng. 99 (1977) 462–469.
[16] R.D. Blevins, Flow-Induced Vibration, second ed., Van Nostrand Reinhold, 1990.
[17] C.-Y. Wang, *J. Fluid Mec*h. 32 (1968) 55–68.
[18] L. Johanning, et al., J. Fluids Structs. 15 (2001) 891–908.
[19] J.R. Chaplin, J. Fluids Structs. 14 (2000) 1101–1117.
[20] T. Sarpkaya, J. Fluids Structs. 15 (2001) 909–928.
[21] T. Sarpkaya, J. Fluid Mech. 165 (1986) 61–71.
[22] P. Hall, J. Fluid Mech. 146 (1984) 337–367.
[23] Bearman, P.W. and Russell, M.P., *Proc. 21st Symp. Naval Hydrodynamics*, 1996, pp. 622–634.
[24] J.M.R. Graham, J. Fluid Mech. 97 (1980) 331–346.
[25] P.W. Bearman, et al., J. Fluid Mech. 154 (1985) 337–356.
[26] A. Okajima, et al., Trans. JSME Ser. B 63 (1997) 3548–3556. (in Japanese)
[27] A. Okajima, et al., Trans. JSME Ser. B 65 (1999) 2243–2250. (in Japanese)
[28] A. Okajima, et al., Trans. JSME Ser. B 65 (1999), pp. 3941–2949. (in Japanese)
[29] JSME, JSME Standard, Guideline for Evaluation of Flow-Induced Vibration of a Cylindrical Structure in a Pipe, JSME S 012, 1998.
[30] M.P. Paidousiss, J. Fluid Mech. 26 (1966) 737–751.
[31] S.S. Chen, Nucl. Eng. Des. 63 (1981) 81–100.
[32] Connors, et al., ASME PVP conf. 63 (1982) 109–124.
[33] M.J. Pettigrew, C.E. Taylor, J. Press. Vessel Technol. 116 (1994) 233–253.
[34] G.W. Housner, Bull. Seism. Soc. Amer. 47 (1957) 15–35.
[35] I. Imai, *Fluid Dynamics* (in Japanese), Iwanami Shoten, 1993.

[36] H. Madarame, Text of course, Jpn. Soc. Mech. Eng. (No. 97-28) (1997) 21–26.

[37] F. Hara, *Trans. JSME* (in Japanese) 53 (491, C) (1987) 1358–1362.

[38] F. Hara, *Trans. JSME* (in Japanese) 54 (504, C) (1988) 1637–1645.

[39] JSME, *Handbook of Vibration and Buckling of Shell* (in Japanese), Gihodoshuppan, 2003.

[40] F. Sakai, *J. High Press. Inst. Jpn.* (in Japanese) 18 (4) (1980) 184–192.

[41] C. Minowa, et al., *Trans. JSME* (in Japanese) 65 (631, C) (1999) 923–931.

[42] R.D. Blevins, Formulas for Natural Frequency and Mode Shape, Van Nostrand Reinhold, 1979, pp. 364–375

[43] S. Hayama, K. Aruga, T. Watanabe, Nonlinear responses of sloshing in rectangular tanks (1st Report, Nonlinear Responses of Surface Elevation) Bull. JSME 26 (219) (September 1983) 1641–1648.

[44] N. Kimura, H. Ohasi, *Trans. JSME* (in Japanese) 44 (385) (1978) 3024–3033.

[45] N. Kimura, H. Ohasi, *Trans. JSME* (in Japanese) 44 (386) (1978) 3446–3454.

[46] H. Hashimoto, S. Sudo, Dynamic behavior of liquid free surface in a cylindrical container subject to vertical vibration, Bull. JSME 27 (227) (May 1984) 923–930.

[47] K. Kimura, et al., *Trans. JSME* (in Japanese) 60 (578, C) (1994) 3259–3267.

[48] K. Kimura, et al., *Trans. JSME* (in Japanese) 62 (596, C) (1996) 1285–1294.

[49] F. Sakai, *Jpn. Soc. Steel Constr.* (in Japanese) (No. 52) (2004) 20–25.

[50] M. Fukaya, et al., *Trans. JSME* (in Japanese) 62 (594, B) (1996) 541–548.

[51] M. Iida, et al., *Trans. JSME* (in Japanese) 61 (585, B) (1995) 1669–1676.

[52] H. Nagakura, S. Kaneko, Trans. ASME, J. Pressure Vessel Tech. 122 (2000) 33–39.

[53] K. Hirota, et al., *Proc. Dyn. Des. Conf.* (in Japanese) 1 (1995) 123–126.

[54] T. Noji, et al., J. Struct. Constr. Eng. AIJ No. 419 (1991) 145–152.

[55] H. Yamagata, S. Kaneko, *Trans. JSME* (in Japanese) 64 (621, C) (1998) 1676–1684.

[56] S. Shintaku, et al., *Trans. JSME* (in Japanese) 56 (521, C) (1990) 8–14.

[57] Y. Inoue, S. Hayama, A Study on a anti-sloshing device in a liquid tank, JSME Int. J. Ser. III 31 (3) (1988) 545–553.

[58] S. Hayama, M. Iwabuchi, A study on the suppression of sloshing in a liquid tank (1st Report,Suppression of Sloshing by Means of Reversed U-tube) Bull. JSME 29 (252) (1986) 1834–1841.

[59] N. Kobayashi, et al., *Trans. JSME* (in Japanese) 62 (594, C) (1996) 482–487.

[60] M. Watanabe, et al., *Trans. JSME* (in Japanese) 67 (657, C) (2001) 1422–1429.

[61] F. Hara, H. Shibata, Experimental study on active suppression by gas bubble injection for earthquake induced sloshing in tanks, JSME Int. J. 30 (260) (1987) 318–323.

Index

Note: Page numbers followed by "*f*", and "*t*" refers to figures and tables respectively.

G

H

I

J

K

L

M

Q

R

S

T

Printed and bound by CPI Group (UK) Ltd, Croydon, CR0 4YY
08/06/2025
01896872-0018